Waste as a Resource

Series Editor

Cristina Trois, Engineering, University of KwaZulu-Natal, Durban, South Africa

The Waste as a Resource series explores state-of-the-art knowledge and innovation in waste management strategies focused on approaches and technologies that ensure maximum value recovery from waste through materials recycling and energy production. The series aims to navigate the transition toward a circular economy by presenting techniques and strategies for waste minimization, effective waste diversion from landfills, and, most importantly, exploring the possible pathways for treating waste as a resource. The series has a global scope, focusing on developing countries and emerging economies. This will assist in filling the gap in current publications that need to adequately explore valorization technologies that are appropriate for and adaptable in low-income countries or informal contexts.

Jaya Arora • Abhishek Joshi • Ramesh C. Ray
Editors

Transforming Agriculture Residues for Sustainable Development

From Waste to Wealth

Springer

Editors
Jaya Arora
Laboratory of Biomolecular Technology
Department of Botany
University College of Science
Mohanlal Sukhadia University
Udaipur, Rajasthan, India

Abhishek Joshi
Laboratory of Biomolecular Technology
Department of Botany
University College of Science
Mohanlal Sukhadia University
Udaipur, Rajasthan, India

Ramesh C. Ray
Center for Industrial Biotechnology Research
Siksha 'O' Anusandhan (SOA)
deemed to be University
Khandagiri, Bhubaneswar, India

Formerly, ICAR-Central Tuber Crops
Research Institute (Regional Center)
Bhubaneswar, Odisha, India

ISSN 2731-8222 ISSN 2731-8230 (electronic)
Waste as a Resource
ISBN 978-3-031-61132-2 ISBN 978-3-031-61133-9 (eBook)
https://doi.org/10.1007/978-3-031-61133-9

This Springer imprint is published by the registered company Springer Nature Switzerland AG
The registered company address is: Gewerbestrasse 11, 6330 Cham, Switzerland

If disposing of this product, please recycle the paper.

Preface

Agricultural residues are a significant waste product of various agriculture practices. These residues mainly include crop residues, industrial processing wastes, livestock wastes, and fruit and vegetable wastes. Global trends in agricultural waste production reveal a complex landscape influenced by various factors. Rapid population growth, urbanization, and changing dietary preferences contribute to increased demand for food, amplifying agricultural activities. This in turn leads to higher waste production, encompassing crop residues, post-harvest losses, and livestock waste. Farmers in the countryside burn agricultural waste on their fields after each harvest season. As a result, it has polluted the country's rural areas, water, and air. Efforts to address food security may inadvertently escalate waste volumes. Moreover, as countries modernize, the shift toward intensive farming practices and agro-industrial activities intensifies waste generation. These residues are usually left to decompose, or dumped in landfills, leading to environmental degradation and health hazards. However, with the growing demand for sustainable agriculture practices, there is a need to find innovative ways to utilize these residues. This book aims to provide a comprehensive exploration of the potential of agriculture waste valorization, showcasing innovative technologies and applications for transforming waste materials into valuable resources such as value-added metabolites, bioenergy, etc. Addressing various aspects of the agricultural waste-to-wealth paradigm, it will serve as a valuable resource for researchers, policymakers, and industry professionals seeking sustainable solutions for agricultural residue management and the transition to a more circular economy.

There are 17 chapters in this book, divided into 4 parts. Part I is the *Introduction* consisting of two chapters. Chapter 1 provides an overview of the impacts of agricultural waste on the environment, and Chap. 2 provides a glimpse of various valorization strategies in agricultural residue management. Part II (*Routes for Valorization to Agricultural Wastes*) consists of four chapters describing the biochemical and thermochemical conversion strategies, emerging technologies for the extraction of value-added compounds from agriculture wastes along with the integration of cross-industry technology to increase precision agriculture output and efficiency, including the Internet of Things, artificial intelligence, blockchain, machine learning, unmanned aerial vehicles, and artificial neural networks for the transformation of agricultural activities on digital platforms. One of the chapters in this section has explored the possibility of biodegradable trash as a feasible substitute for developing a maintainable economy that aids society by providing safer and less hazardous greenways for the manufacture of nanoparticles. Part III (*Recent Trends in the Valorization of Agricultural Wastes*) consists of nine chapters covering a wide range of aspects on agro-waste valorization, such as the production of single-cell protein, biofertilizers, and application in soil amendment, high-value animal feed, sustainable packaging materials, i.e., bioplastics, biopolymers, and biocomposites, sustainable building materials, and production of bioenergy. A chapter is dedicated to dealing with agricultural waste applications in water and wastewater treatment. Part IV (*Public Policy and Circular Economy*) has two chapters discussing the challenges and perspectives in agri-waste valorization, taking sugarcane trash as a case study, and the last chapter on economic viability and policy implications of agriculture waste valorization: social, economic impacts, policy frameworks, and regulations, with a case study in Veracruz, Mexico.

The book provides up-to-date information on research and development in the field of agricultural residue transformation into the production of high-value products and biocommodities, from waste to wealth. The book will be useful for all those involved in the environment, disposal of agricultural waste, rural economy, production of useful metabolites, and energy. Editors wish to thank all the international contributors who have put their time at our disposal and made serious efforts to put all the latest information in one place for this book. We are thankful to Prof. Cristina Trois, Director of the Centre for Renewable and Sustainable Energy Studies, South African Research Chair in Waste and Climate Change (SARCHI), School of Engineering, University of KwaZulu-Natal, for accepting the book proposal in the "Waste as a Resource" Book Series. We are also thankful to the staff at Springer editorial and production house for their professional support.

Udaipur, Rajasthan, India — Jaya Arora
Udaipur, Rajasthan, India — Abhishek Joshi
Bhubaneswar, Odisha, India — Ramesh C. Ray
March 2024

Contents

About the Editors

Jaya Arora, Assistant Professor, Department of Botany, Mohanlal Sukhadia University, Udaipur, India has been teaching Botany since 2012. She did her MSc and PhD in Botany with a specialization in plant tissue culture and secondary metabolite production from Mohanlal Sukhadia University. She was awarded the Maharana Fateh Singh Award in 2000 for meritorious performance during her matriculation and the Gargi Award and Scholarship for matriculation by Govt. of Rajasthan, India, 2000. She has 14 years of research experience with one UGC start-up completed project (2013–2015) and mentored one Women Scientists Scheme-A project (2016–2019) funded by the Ministry of Science and Technology, Dept. Science and Technology, Govt. of India. Currently, she is Co-PI in two different projects funded by MHRD RUSA 2.0, India. The research group of Dr. Arora has published research papers in various international and national repute journals with high impact factors, such as *Acta Physiologiae Plantarum*, *In Vitro Cell Dev Biol – Plant*, *Bioengineered*, *Frontiers in Plant Science*, *Nanomaterials*, *Biofuels*, *Biologia Futura*, *Horticulturae*, *Plants*, dealing with plant biotechnology, proteomics (bioactive peptides), natural products, secondary metabolite production, seed biology studies, and conservation of medicinal plants. Dr. Arora has edited and authored five books, including publications with prestigious publishing houses mainly Springer, Cham. This work was funded by various funding agencies like UGC, CSIR, and DST. Currently, she is supervising two PhD students and two students have been awarded PhD degree .

Abhishek Joshi, is currently associated with research and teaching at the Department of Botany, Mohanlal Sukhadia University, Udaipur, India. He holds a Master's degree in Botany with a specialization in Plant tissue culture and halophyte biology. His doctoral research was focused on unraveling the biochemical and molecular mechanisms that enable halophytes to thrive in saline conditions. He had 8 years of research experience of halophytes occurring in Southern Rajasthan of Indian subcontinent with 1 e-book and 26 publications in the international repute journals and books.

Ramesh C. Ray FNAAS, FNABS is currently working as Adjunct Professor, Center for Industrial Biotechnology Research, Siksha 'O' Anusandhan (SOA) deemed to be University, Khandagiri, Bhubaneswar, India, former principal scientist (Microbiology) and head of the ICAR-Central Tuber Crops Research Institute (Regional Centre), Bhubaneswar, India. He has 40 years of research experience in agriculture, food, and industrial microbiology; published 145 research and review papers in national and international journals; 75 books chapters, edited/co-edited 24 books published by Springer, CRC, and Elsevier; and co-authored three books and received more than 8250 citations. He was a guest faculty to Department of Botany and Microbiology, Utkal University, Orissa (India), and visiting professor/scientist at several universities and research institutes in India and abroad and consultant to International Potato Centre, Lima, Peru, on sweet potato. He is a distinguished fellow of the prestigious National Academy of Agricultural Sciences, New Delhi, and the National Academy of Biological Sciences, Chennai, India.

Part I
Introduction

Chapter 1
Agricultural Waste and Its Impact on the Environment

Chitra Yadav, Pooja Yadav, Abhishek Joshi, Mukesh Meena, Harish, and Jaya Arora

Abbreviations

BT	Billion tons
CH4	Methane
CO	Carbon monoxide
CO_2	Carbon dioxide
EEA	European Environment Agency
Eq	equivalence
FAOSTAT	Food and Agriculture Organization of The United Nations Statistics
GDP	Gross domestic product
GHGs	Greenhouse gases
KT	Kilotons
MT	Million tons
N	Nitrogen
NO2	Nitrous oxides
P	Phosphorus
PAHs	Polycyclic aromatic hydrocarbons
PM	Particulate matter
SDGs	Sustainable development goals

C. Yadav · P. Yadav · A. Joshi · J. Arora (✉)
Laboratory of Biomolecular Technology, Department of Botany, University College of Science, Mohanlal Sukhadia University, Udaipur, India
e-mail: jayaarora@mlsu.ac.in

M. Meena
Laboratory of Phytopathology and Microbial Biotechnology, Department of Botany, Mohanlal Sukhadia University, Udaipur, Rajasthan, India

Harish
Plant Biotechnology Laboratory, Department of Botany, Mohanlal Sukhadia University, Udaipur, Rajasthan, India

J. Arora et al. (eds.), *Transforming Agriculture Residues for Sustainable Development*, Waste as a Resource, https://doi.org/10.1007/978-3-031-61133-9_1

SO2	Sulfur dioxides
UNEP	United Nations Environment Programme
VOCs	Volatile organic compounds

1.1 Introduction

The world's population marked 8 billion people on November 15, 2022, and with this milestone of growing population, it is expected to increase to 9.7 billion by 2050 and 11.2 billion by 2100, respectively (United Nations Department of Economic and Social Affairs, Population Division, 2022). The world population is increasing daily, and fulfilling the food requirements of this growing human population is a major challenge worldwide (Arora et al., 2023; Koul et al., 2022; Riseh et al., 2024). To combat this food insecurity, global food production has increased significantly as a result of the use of modern agricultural techniques. Particularly in developing nations like India and Mexico, the development of dwarf varieties of rice and wheat has a major positive impact on food grain production (Duque-Acevedo et al., 2020). For farm scientists, achieving SDG 2 (Sustainable Development Goal 2) with this growing population is a difficult task. Another important factor is to achieve a balance between the production of nutritious food production and farmer income. To feed this population, agroecological and organic agricultural concepts are integrated (Giller et al., 2021). It is imperative to control population growth while managing agro-industry waste efficiently. Therefore, it is necessary to use all resources sustainably (Ramawat et al., 2023). China, India, and Africa have seen incredible population and economic growth over the past century, along with an increase in the production of agricultural waste, with China holding the top position and India being the second-largest country in agricultural waste generation (Kaul et al., 2022). The agricultural wastes which are the by-products of producing and processing agricultural products mainly staple food crops, fruits, vegetables, and livestock manure produce 998 million tons every year at the global level (Joshi & Arora, 2023; Raut et al., 2023; Tripathi et al., 2019, Fig. 1.1). These waste products came from arable land and horticulture practices. Stems, branches, and leaves make up the majority of crop residue (Mohite et al., 2022; Ray, 2022). Unfortunately, only a small proportion of this waste is utilized in the form of fodder for cattle, used as manure, shed building, low-quality paper, and matchsticks making, while most of it is being disposed of through random burning or landfilling activities (Maji et al., 2020; Patel et al., 2022; Phiri et al., 2023). Burning agricultural waste has traditionally been used to prepare land for the next crop cycle (Arora et al., 2023; Choochuay et al., 2022). They may contain materials useful beings, but their economic value is less than the cost of collection (Obi et al., 2016).

The disposal of agricultural waste through open burning, landfilling, and direct dumping significantly contributes to the emission of various gaseous and solid pollutants including carbon dioxide (CO_2), carbon monoxide (CO), methane

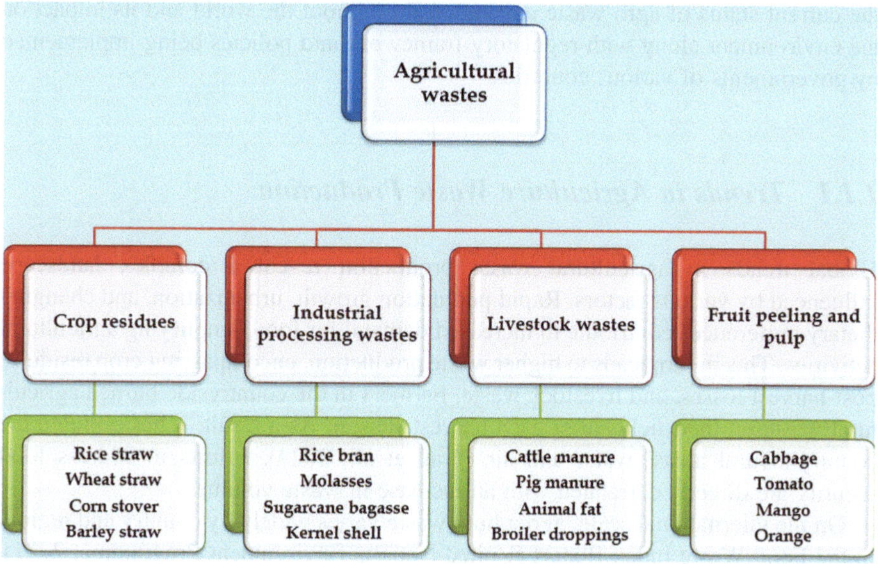

Fig. 1.1 Classification and major examples of agricultural wastes

(CH_4), nitrogen oxides (NOx), sulfur oxides (SOx), particulate matter (PM_{10} and $PM_{2.5}$), and polycyclic aromatic hydrocarbons (PAHs). These pollutants can cause serious environmental damage and pose a risk to human health and safety (Kaab et al., 2019; Koul et al., 2022). Besides, disposal of agrochemicals (fertilizer, pesticides, herbicides veterinary antibiotics) rich residue could result in the deterioration of natural resources, particularly soil and water at a regional scale, and may threaten people's health (He et al., 2019).

Another important issue related to agricultural waste is related to the economy of a country. Despite agricultural food loss, the food waste generated in the food chain comprised 42% from domestic activities, 38% from food dispensation, and 20% from other procedures (Capanoglu et al., 2022; Bancal & Ray, 2022). Such food waste has been estimated at around 57 million tons costing EUR 130 billion in the European Union (EAA, 2024). Thus, food waste reduction can be used to feed the starving deprived people of the growing population. It is a critical issue in terms of the economy, environment, and society. To combat the situation of increasing agro-waste, many countries started working on 5R strategies (Refuse, Reduce, Reuse, Repurpose, and Recycle). The agriculture waste is now being reused as the substrate for producing various value-added compounds (Arora et al., 2023; Bala et al., 2023, see Chap. 2 for details). Agricultural waste can be valorized through various strategies to produce the desired compound and achieve the goals of sustainable development (see Chap. 2). Besides this reduction in food loss is also a big measure being taken up by many countries to fill the food gap, generated by 2050 for feeding approximately 9 billion population (Capanoglu et al., 2022). This chapter deals with the current status of agro-waste generation throughout the world and its impact on the environment along with regulatory framework and policies being implemented by governments of various countries.

1.1.1 Trends in Agriculture Waste Production

Global trends in agricultural waste production reveal a complex landscape influenced by various factors. Rapid population growth, urbanization, and changing dietary preferences contribute to increased demand for food, amplifying agricultural activities. This in turn leads to higher waste production, encompassing crop residues, post-harvest losses, and livestock waste. Farmers in the countryside burned agricultural waste on their fields after each harvest season. As a result, it has polluted the country's rural areas' water and air (Tran et al., 2024). Efforts to address food security are directly correlated with an increase in waste volumes.

On the international scale, agriculture waste varies widely by country and region. In the Food Waste Index Report (United Nations Environment Programme, 2021), studies were carried out with 152 food waste data points identified in 54 countries representing 75% of the world population. The three main sectors were identified for food waste calculation, comprising household, food service, and retail. The contribution of high-income, upper-middle income, and lower-middle-income countries in household waste generation is 79, 76, and 91 kg/capita/year, respectively, while for low-income group, data is insufficient. The high-income countries' contribution to food service and retail is 26 kg/capita/year and 13 kg/capita/year, while for the rest groups, the data is insufficient. Overall, the per capita per year food waste generation is similar in high, upper, and lower-middle-income countries (United Nations Environment Programme, 2021). The estimated waste, arising from food supply chains mainly produced in special geographical conditions, are vegetable oil in the UK approx. 50,000–100,000 tons per year, tomato pomaces in Europe approx. 4,000,000 tons per year, wheat straw in the USA approx. 57,000 tons per year, cereal waste in Europe approx. 40,000–45,000 tons per year, orange peel in the USA approx. 700 tons per year, grape pomace in France approx. 700 tons per year, olive waste worldwide 2,881,500 tons per year, 3,000,000–4,200,000 tons per year apple pomace worldwide, and 70–140 tons per year potato peel worldwide (Capanoglu et al., 2022).

China being the most populous country is occupying the top position in agriculture waste generation. Recently China has generated 870 million tons of straw comprising 230 million tons (MT) of rice straw and 1.06 billion tons (BT) of poultry manure, livestock along 160 MT of household waste as agriculture waste. China is making policies, generating funds to reuse these agricultural wastes, and diverting towards bioenergy generation to fulfill the ever-increasing demand for energy (Liu et al., 2022; Cong et al., 2023). India is the second-largest country in terms of agricultural output, with 70% of rural households still depending on agriculture for their livelihood.

India produces 350–990 MT/year of agricultural waste comprising 130 MT/year of paddy straw (Koul et al., 2022). In a study done by Jain et al. (2014), the major agricultural waste residue comprised of cereals including millet husk, rice husk and oilseeds including groundnut cake and rapeseed mustard cake, fibers with jute, cotton stalk, coconut husk, and sugarcane bagasse. The edible part of crop, the residue-to-crop ratio, and the dry matter fraction in the crop biomass were used to estimate the amount of crop residue generated. For cereal crops, the residue-to-crop ratio ranged from 1.5 to 1.7, for fiber crops from 2.15 to 3.0, for oilseed crops from 2.0 to 3.0, and for sugarcane, it is 0.4. Nine major crops produced 620.4 MT of dry crop residue in total. Uttar Pradesh made the largest contribution followed by Punjab, West Bengal, Haryana, Maharashtra, and Andhra Pradesh in crop residue generation and further burning (Jain et al., 2014).

Malaysia, having good agricultural practices, alone produces almost 1.2 MT of agricultural waste with pineapple leaves as one of the major agricultural waste (Sarangi et al., 2023). Nowadays to achieve zero agriculture waste target, Malaysia has taken the initiative to derive commercial products from pineapple leaves including high-quality fibers, bioethanol, biofertilizer, and hydrogels (Sarangi et al., 2023). Iran produced approximately 3.1 MT of rice (paddy) in 2019, according to FAO statistics. Thus, the production of rice husk and straw will be approximately 620 thousand tons and 4.65 MT, respectively, in the coming years (Pashaki et al., 2024).

In EU countries, agricultural waste, including food waste and related unused products, was estimated at around 2.6 BT/year, from the year 2010 to 2016, which was greater than the waste generated in all other sectors (i.e., industrial solid waste and municipal solid wastes) (Bedoić et al., 2019). Eurostat estimated that 0.088 BT of agriculture waste was generated in the year 2018. EU is contributing approximately 1.3 BT/year to global food waste generation according to the United Nations' Food and Agriculture Organization (FAO) (Duquennoi & Martinez, 2022). Europe including the former Russian Federation contributed 14% to global food wastage and hold the third position worldwide in this sector comparable to North America and Oceania (also 14%). In the same year, Industrialized Asia and South and Southeast Asia contributed to 28% and 23% of food waste generation (Lipinski et al., 2013).

Agriculture waste generation is an indispensable part of countries where either agriculture plays a major role in its gross domestic product (GDP) or has the burden to feed the ever-increasing population. Worldwide various organizations are keeping an eye on that sector to make it more sustainable in terms of the generation of valuable by-products and safe disposal of such agricultural waste. This is not the point in debating which country is producing which waste more as the data points are random and not comparable; the time is to take action to achieve SDG 12.3, SDG 12.4, and SDG 12.5 of Agenda 2023 by ensuring sustainable consumption and production of food.

1.2 Impact of Agriculture Waste on the Environment

Various agricultural activities are correlated with an increase in waste biomass generation at different levels. The improper disposal results in various hazardous environmental impacts (Fig. 1.2). The following sections deal with some of them in detail.

1.2.1 Emissions of Greenhouse Gases (GHGs)

Open burning is the primary cause of air pollution, climate change, and the melting of ice and snow in regions such as the cryosphere (Estrellan & Iino, 2010). Worldwide, farmers frequently burn cultivated fields to remove stalks, weeds, and leftovers before planting new crops. Stubble burning is a substantial contributor to air pollution, particularly in South Asia (Abdurrahman et al., 2020). Several recent studies have reported that open burning of crop residues releases substantial amounts of PM, black carbon, PAHs, volatile organic compounds (VOCs), and other toxic pollutants into the atmosphere, which adversely affect air quality, produce photochemical smog, and reduce visibility (Mehmood et al., 2020; Jaffe et al., 2020; Bahşi et al., 2023). In a recent study, Phuong et al. (2021) investigated the impact of open rice straw burning on the air quality in the Mekong Delta of Vietnam. This study reported a substantial increase in the concentrations of PM, PAHs, VOCs, and

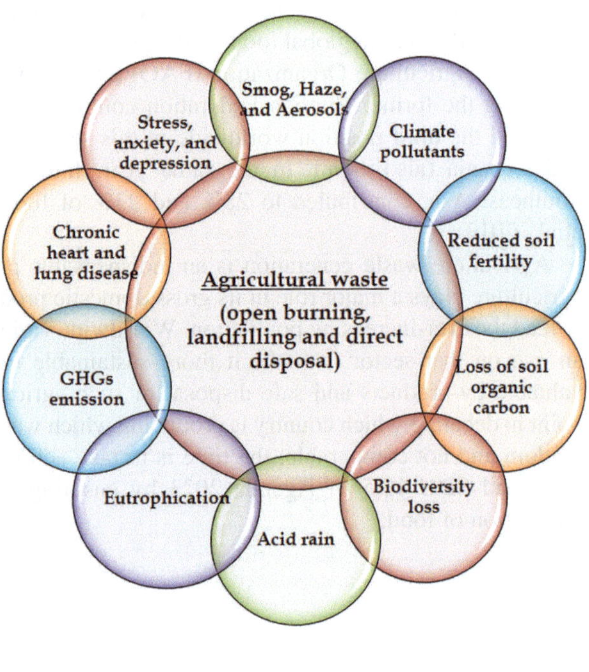

Fig. 1.2 Impacts of agricultural waste on Ecosystems and human health

Table 1.1 Estimated region-wise GHGs emissions by the crop residue-burning practices in the year 2021

	Estimated emission (KT)		
Region	CH_4	N_2O	CO_2 eq (AR5)
Africa	151.3226	3.923	5276.6278
America	275.3083	7.1374	9600.0434
Asia	522.663	13.5506	18,225.473
Europe	121.9359	3.1608	4251.8172
Oceania	14.6803	0.3807	511.9339
World (total)	1085.9101	28.1525	37,865.8953

Data generated from agri-food systems disseminated in the FAOSTAT Climate Change Emissions domains

primary gaseous pollutants. Furthermore, a 1000-fold increase in PAH concentration was reported during this study. The latest gridded emission inventory of $PM_{2.5}$ emissions from residue burning in India reported 990.68 kilotons (KT or Gg) of $PM_{2.5}$ emissions (Sahu et al., 2021), with rice contributing 41% followed by wheat (27%) sugarcane (14%), maize (8%), and coarse cereal (7%).

Also, open burning practices lead to the release of gaseous pollutants, including CO_2, CH_4, and N_2O, which are potent GHGs. These GHGs have detrimental impacts on global warming and climate change (Jogdand, 2020; Mar et al., 2022). Both CO_2 and CH_4 emissions from burning can be attributed to combustion processes, whereas N_2O emissions can be attributed to the decomposition of nitrogen-containing compounds in soil after a burning event (Jiménez et al., 2015). As reported in a recent study by Abdurrahman et al. (2020), burning 63 MT of crop stubble releases 3.4 metric tons of CO, 0.1 metric tons of NOx, 91 metric tons of CO_2, and 0.6 metric tons of CH_4 into the atmosphere.

The most recent data from FAOSTAT indicate that the global emissions of CH_4, N_2O, and CO_2 (eq) resulting from crop residue burning in 2021 are estimated to be 1085.91, 445.17, and 396,983 KT, respectively (Anonymous, 2024). Furthermore, Asia is the largest emitter, accounting for over 48% of total CH_4, N_2O, and CO_2(eq) emissions, followed by Africa, America, Europe, and Oceania (Table 1.1). A recent study that used a novel satellite-based approach to track crop residue burning reported that GHG emissions from agricultural residue burning across India have increased by 75% (1.3 ton CO_2e per hectare) since 2011 (Deshpande et al., 2023).

Other major agri-food system practices such as agri-food waste disposal systems, food processing, and manure management have also produced organic waste, which is commonly disposed of in landfills and composted. A variety of pollutants are released during these processes, including GHGs, phenols, heavy metals, and acid-sensitive compounds (Gaur et al., 2020). When GHGs react with water, they create acidulous compounds such as nitric acid, sulfuric acid, and ammonium nitrate, which later disperse as acid rain that harms every component of both terrestrial and aquatic ecosystems (Debnath et al., 2021; Bhardwaj & Saxena, 2023).

Based on the most recent data from FAOSTAT, it is estimated that the global emissions of CH_4 in 2021 accounted for 42,283.84 KT from agri-food waste disposal, 159.30 KT from food processing, and 9964.68 KT from manure

Table 1.2 Estimated region-wise GHGs emissions by the different practices of agri-food systems in the year 2021

Agri-food system practices	Region	Estimated emission (KT)		
		CH_4	N_2O	CO_2 eq (AR5)
Agri-food waste disposal	Africa	5824.572	29.4668	170,896.7215
	Americas	9395.6797	37.9916	276,465.4041
	Asia	22,565.3707	137.1134	689,840.3073
	Europe	4268.7354	38.7371	137,033.9414
	Oceania	229.4889	1.6641	7023.7549
	World (total)	42,283.8467	244.973	1281,260.1293
Food processing	Africa	1.5101	0.1409	10,164.5488
	Americas	15.8731	0.8762	145,437.1035
	Asia	121.0629	4.0253	295,689.5149
	Europe	19.3422	1.2858	198,386.6935
	Oceania	1.5093	0.0848	9326.5353
	World (total)	159.2977	6.413	659,004.396
Manure management	Africa	769.3245	28.8953	10,164.5488
	Americas	2356.2583	88.1227	145,437.1035
	Asia	4061.8643	224.2352	295,689.5149
	Europe	2463.3302	96.601	198,386.6935
	Oceania	313.9044	7.3231	9326.5353
	World (total)	9964.6817	445.1773	659,004.396

Data generated from agri-food systems disseminated in the FAOSTAT Climate Change Emissions domains

management practices (Anonymous, 2024). Similarly, global N_2O emissions were estimated to be 244.97 KT from agri-food waste disposal, 6.413 KT from food processing, and 445.18 KT from manure management practices. Meanwhile, global CO_2(eq) emissions were estimated to be 1,281,260.13 KT from agri-food waste disposal, 659,004.39 KT from food processing, and 659,004.39 KT from manure management practices (Table 1.2). In the current report published by the European Environment Agency (EEA), in 2024, agriculture in the European Union contributed to 11% of GHGs, with a significant contribution of harmful air pollutants such as ammonia.

1.2.2 Pollution of Land and Water Resources

Agricultural activities generate a significant amount of organic waste, which can have significant impacts on terrestrial as well as aquatic ecosystems if improper landfilling and dumping are not managed (Siddiqua et al., 2022).

Undoubtedly, the agriculture sector consumes 70% of the global surface water supply and is the primary cause of the deprivation of both surface and groundwater resources (Jury & Vaux, 2007). Over the past decades, there has been a significant increase in the use of fertilizers, pesticides, herbicides, and other synthetic agri-chemicals to meet the growing food demand in moving dietary preferences and population growth (Verma et al., 2023). When these chemicals are applied beyond the capabilities of the soil to retain, agrochemical runoff during the production of crops, vegetables, and fruits can lead to water pollution (Pericherla et al., 2020; Verma et al., 2022; Zahoor & Mushtaq, 2023).

Although, nitrogen (N) and phosphorus (P) are crucial macronutrients for plant growth, an excess amount of these nutrients in water can have severe ecological consequences. The presence of excessive amounts of nitrate (NO_3^-) and phosphate (PO_4^{3-}) in surface waters can result in eutrophication, which is caused by excess nutrients stimulating the growth of plants and algae (Liu et al., 2021; Akinnawo, 2023).

It has been shown that the groundwater can be contaminated by synthetic pesticides and herbicides through a process called leaching, which implicates the downward crusade of chemicals from the soil surface by rainfall, infiltration, and irrigation (Sadegh-Zadeh et al., 2017). As a result of the intensive use of pesticides in agriculture for high yields, water bodies have been polluted by runoff from terrestrial environments, as well as the direct disposal of chemical waste. Besides deteriorating water quality, these chemicals also pose a threat to non-target organisms, including fish, shrimp, phytoplankton, and zooplankton (Kumar et al., 2023).

Similarly, veterinary antibiotics are widely used in livestock farming, and their utilization is anticipated to increase by over 60% shortly (Spielmeyer, 2018). A substantial portion of such antibiotics added to animal feed is excreted in the urine or feces. Studies have indicated that many antibiotics are strongly adsorbed in soils and are not easily degraded. Upon excretion, antibiotics can pollute the surface or groundwater via nonpoint source pollution from manure-applied land (Peña et al., 2020; Kokotović et al., 2024). Despite the low concentrations, manure-applied antibiotics could lead to a decline in microbial population by shrinking food sources and affecting decomposition and mineralization processes (Zaller et al., 2021; Kaur et al., 2021).

The pollutants generated from agri-food and livestock production practices have detrimental effects on aquatic ecosystems, resulting in a decline in fish populations and alterations to the composition of aquatic plants and animals (Akinbile et al., 2016; Zhao et al., 2020). Furthermore, these pollutants can disrupt biodiversity by eliminating weeds and insects, which can have cascading effects on the ecological food chain (Mahmood et al., 2016; Naz et al., 2023).

Food processing industries, such as oil mills and paper and pulp industries, have detrimental effects on both land and water resources. The waste produced by these industries contains a mixture of organic and inorganic contaminants, primarily arising from utensils and floor cleaning rather than food waste (Singh & Chandra, 2019). This wastewater, commonly referred to as effluent, ultimately flows into rivers and lakes, leading to pollution of aquatic ecosystems (Asgharnejad et al., 2021). Furthermore, the waste generated from the lye peeling of vegetables and fruits, which utilizes high salt concentrations for pickling and food processing, has been shown to adversely affect soil fertility. Several recent studies have demonstrated that residual waste and effluents produced by food processing industries adversely impact aquatic and terrestrial ecosystems, leading to changes in water and soil quality as well as a decline in aquatic and terrestrial creature populations (Nguyen et al., 2020; Read et al., 2020; Shabir et al., 2023).

1.2.3 Impact on Human Health and Well-Being

The burning of crop residues releases a mixture of hazardous pollutants that can exacerbate respiratory diseases, including asthma, chronic obstructive pulmonary disease, and bronchitis (Parikh et al., 2021). Recent studies have also linked exposure to chronic pollutants to adverse cardiovascular outcomes, such as an increased risk of heart attack, stroke, and cardiovascular mortality (Adetona et al., 2020; Chukwuemeka et al., 2021).

The abundant release of pollutants like black carbon, $PM_{2.5}$, and some gaseous pollutants (SO_2, CO, N_2O, etc.,) during stubble burning can induce systemic inflammation, oxidative stress, and endothelial dysfunction, contributing to the development and progression of cardiovascular and neurodegenerative disorders (Jadhav et al., 2022; Kusumawardani et al., 2023). Long-term exposure to these pollutants can also increase stress, anxiety, and depression as well as decrease cognitive function and productivity (Lai et al., 2022; Ramadan et al., 2023). It has been shown that psychologically vulnerable populations, such as children, the elderly, and individuals with preexisting health conditions, are disproportionately affected by these pollutants, further contributing to social disparities in health outcomes (Ajay et al., 2022; Khurana et al., 2023).

Landfilling and disposal practices for agro-industrial waste pose significant risks to human populations. Residents living near landfills may experience respiratory issues and infections owing to chronic exposure to landfill emissions, particularly VOCs and PM (Ogbuehi et al., 2022). Agro-industrial waste contains organic pollutants, pathogens, and heavy metals, which can be leached into water bodies, causing waterborne diseases such as diarrhea and drowsiness, especially in populations that rely on untreated or inadequately treated water sources (Kumari et al., 2017; Asomaku, 2023).

During the cultivation practices, agricultural workers are frequently exposed to synthetic agrochemicals, which can negatively impact both workers and surrounding communities. Through skin contact, exposure to these chemicals may cause serious health problems, such as diabetes, reproductive disorders, neurological dysfunction, cancer, and respiratory illness (Rani et al., 2021).

1.3 Regulatory Framework and Policies

Regulatory frameworks and policies are crucial for managing the environmental impact of agricultural waste and promoting sustainable waste management practices. International agreements, conventions, and policies provide a global cooperative framework for environmental issues, while national legislation and regulations establish standards and requirements for waste management (Joshi et al., 2019; Zorpas, 2020).

The Basel Convention, established in 1989, aims to control the transboundary movement of hazardous wastes and ensure environmentally sound management. It provides guidelines for waste classification, minimization, and disposal, emphasizing the prevention of pollution and the protection of human health and the environment. Regulating the international trade of hazardous waste helps to prevent dumping and improper disposal practices (Krueger, 2001).

The Stockholm Convention, adopted in 2001, targets persistent organic pollutants (POPs) related to agriculture, such as pesticides and agrochemicals. Essentially, this convention aims to eliminate or restrict the production, use, and release of POPs (Rayfuse, 2016). It contributes to protecting biodiversity and public health by minimizing the release of POPs into the environment (Wang et al., 2022).

International cooperation and coordination regarding environmental issues, including agricultural waste management, is greatly facilitated by the UN Environment Programme (UNEP). The UNEP promotes integrated approaches to addressing agricultural runoff and waste disposal pollution (Ijaiya, 2017). By raising awareness, providing technical assistance, and supporting capacity-building efforts, UNEP helps countries develop and implement policies and strategies for sustainable agricultural waste management, thus reducing environmental degradation and promoting ecosystem resilience (Citaristi, 2022). Further, see Chap. 17 of this book for details on policy frameworks and regulations related to agricultural waste management.

1.4 Conclusion

The impact of agricultural waste on soil, water, air, biodiversity, and climate poses significant challenges to both environmental sustainability and human well-being, highlighting the importance of waste management in agriculture. Policy instruments and incentives encouraging farmers to adopt environmentally friendly practices coupled with RandD initiatives supporting innovation and technological advancement in waste management can help mitigate these challenges. For instance, the implementation of "no burn" and 5R (Refuse, Reduce, Reuse, Repurpose, and Recycle) strategies has the potential to reduce GHGs and black carbon emissions by half, while also providing economic and social benefits not only to farmers but to the entire population. By implementing comprehensive regulatory frameworks and policies, policymakers can protect the environment, promote public health, and ensure the sustainability of agricultural production systems.

References

Abdurrahman, M. I., Chaki, S., & Saini, G. (2020). Stubble burning: Effects on health and environment, regulations and management practices. *Environmental Advances, 2*, 100011.

Adetona, O., Ozoh, O. B., Oluseyi, T., Uzoegwu, Q., Odei, J., & Lucas, M. (2020). An exploratory evaluation of the potential pulmonary, neurological and other health effects of chronic exposure to emissions from municipal solid waste fires at a large dumpsite in Olusosun, Lagos, Nigeria. *Environmental Science and Pollution Research, 27*, 30885–30892.

Ajay, S. V., Kirankumar, P. S., Varghese, A., & Prathish, K. P. (2022). Assessment of dioxin-like POP's emissions and human exposure risk from open burning of municipal solid wastes in streets and dump yard fire breakouts. *Exposure and Health, 14*, 763–778.

Akinbile, C. O., Erazua, A. E., Babalola, T. E., & Ajibade, F. O. (2016). Environmental implications of animal waste pollution on agricultural soil and water quality. *Soil and Water Research, 11*(3), 172–180. https://doi.org/10.17221/29/2015-SWR

Akinnawo, S. O. (2023). Eutrophication: Causes, consequences, physical, chemical and biological techniques for mitigation strategies. *Environmental Challenges, 12*, 100733.

Anonymous. (2024). *Food and agriculture organization of the United Nations, FAOSTAT*. Emissions totals. https://www.fao.org/faostat/en/#data/GT

Arora, J., Ramawat, K. G., & Mérillon, J. M. (2023). Disposal of agricultural waste and its effects on the environment, production of useful metabolites and energy: Potential and challenges. In K. G. Ramawat, J. M. Mérillon, & J. Arora (Eds.), *Agricultural waste: Environmental impact, useful metabolites and energy production* (pp. 3–20). Springer Nature Singapore.

Asgharnejad, H., Khorshidi Nazloo, E., Madani Larijani, M., Hajinajaf, N., & Rashidi, H. (2021). Comprehensive review of water management and wastewater treatment in food processing industries in the framework of water-food-environment nexus. *Comprehensive Reviews in Food Science and Food Safety, 20*(5), 4779–4815.

Asomaku, S. O. (2023). Quality assessment of groundwater sourced from nearby abandoned landfills from Industrial City in Nigeria: Water pollution indices approach. *HydroResearch, 6*, 130–137.

Bahşi, K., Ustaoğlu, B., Aksoy, S., & Sertel, E. (2023). Estimation of emissions from crop residue burning in Türkiye using remotely sensed data and the Google Earth Engine platform. *Geocarto International, 38*(1), 2157052.

Bala, S., Garg, D., Sridhar, K., Inbaraj, B. S., Singh, R., Kamma, S., Tripathi, M., & Sharma, M. (2023). Transformation of agro-waste into value-added bioproducts and bioactive compounds: Micro/Nano formulations and application in the agri-food-pharma sector. *Bioengineering (Basel, Switzerland), 10*(2), 152. https://doi.org/10.3390/bioengineering10020152

Bancal, V., & Ray, R. C. (2022). Overview of food loss and wastes in fruits and vegetables: From issues to resources. In R. C. Ray (Ed.), *Fruits and vegetable wastes: Valorization to bioproducts and platform chemicals (.)* (pp. 3–32). Springer.

Bedoić, R., Ćosić, B., & Duić, N. (2019). Technical potential and geographic distribution of agricultural residues, co-products and by-products in the European Union. *Science of the Total Environment, 686*, 568–579.

Bhardwaj, A., & Saxena, A. (2023, September). A review: Acid rain; its emission, deposition and effect on ecosystem. In *AIP Conference Proceeding* (Vol. 2735, p. 030008). AIP Publishing. https://doi.org/10.1063/5.0141305

Capanoglu, E., Nemli, E., & Tomas-Barberan, F. (2022). Novel approaches in the valorization of agricultural wastes and their applications. *Journal of Agricultural and Food Chemistry, 70*(23), 6787–6804.

ChooChuay, C., Pongpiachan, S., Tipmanee, D., Deelaman, W., Iadtem, N., Suttinun, O., & Cao, J. (2022). Effects of agricultural waste burning on PM2. 5-bound polycyclic aromatic hydrocarbons, carbonaceous compositions, and water-soluble ionic species in the ambient air of Chiang-Mai, Thailand. *Polycyclic Aromatic Compounds, 42*(3), 749–770.

Chukwuemeka, I., Njoku, V., Arinze, C., Chizoruo, I., & Blessing, E. A. (2021). A review: Effects of air, water and land dumpsite on human health and analytical methods for determination of pollutants. *AMEC J, 4*, 80–106.

Citaristi, I. (2022). United Nations Environment Programme—UNEP. In *The Europa directory of international organizations 2022* (pp. 193–199). Routledge Publishing.

Cong, H., Meng, H., Chen, M., Song, W., & Xing, H. (2023). Co-processing paths of agricultural and rural solid wastes for a circular economy based on the construction concept of "zero-waste city" in China. *Circular Economy, 100065.*

Debnath, B., Sikdar, A., Islam, S., Hasan, K., Li, M., & Qiu, D. (2021). Physiological and molecular responses to acid rain stress in plants and the impact of melatonin, glutathione and silicon in the amendment of plant acid rain stress. *Molecules, 26*(4), 862.

Deshpande, M. V., Kumar, N., Pillai, D., Krishna, V. V., & Jain, M. (2023). Greenhouse gas emissions from agricultural residue burning have increased by 75% since 2011 across India. *Science of the Total Environment, 904*, 166944.

Duque-Acevedo, M., Belmonte-Ureña, L. J., Cortés-García, F. J., & Camacho-Ferre, F. (2020). Agricultural waste: Review of the evolution, approaches and perspectives on alternative uses. *Global Ecology and Conservation, 22*, e00902.

Duquennoi, C., & Martinez, J. (2022). European Union's policymaking on sustainable waste management and circularity in agroecosystems: The potential for innovative interactions between science and decision-making. *Frontiers in Sustainable Food Systems, 6*, 937802.

EEA. (2024). *Agriculture and food systems*. https://www.eea.europa.eu/en/topics/in-depth/agriculture-and-food

Estrellan, C. R., & Iino, F. (2010). Toxic emissions from open burning. *Chemosphere, 80*(3), 193–207.

Gaur, V. K., Sharma, P., Sirohi, R., Awasthi, M. K., Dussap, C. G., & Pandey, A. (2020). Assessing the impact of industrial waste on environment and mitigation strategies: A comprehensive review. *Journal of Hazardous Materials, 398*, 123019.

Giller, K. E., Delaune, T., Silva, J. V., Descheemaeker, K., van de Ven, G., Schut, A. G., & van Ittersum, M. K. (2021). The future of farming: Who will produce our food? *Food Security, 13*(5), 1073–1099.

He, K., Zhang, J., & Zeng, Y. (2019). Knowledge domain and emerging trends of agricultural waste management in the field of social science: A scientometric review. *Science of the Total Environment, 670*, 236–244.

Ijaiya, H. (2017). The role of the United Nations Environmental Programme (UNEP) to the development of international environmental law. *KIU Journal of Social Sciences, 3*(2), 185–192.

Jadhav, K., Ray, E., Vaghasiya, K., & Verma, R. K. (2022). Neurodegenerative disorders due to inhalation of various small particles. In *Nanomedical drug delivery for neurodegenerative diseases* (pp. 41–54). Academic Press.

Jaffe, D. A., O'Neill, S. M., Larkin, N. K., Holder, A. L., Peterson, D. L., Halofsky, J. E., & Rappold, A. G. (2020). Wildfire and prescribed burning impacts on air quality in the United States. *Journal of the Air and Waste Management Association, 70*, 583–615.

Jain, N., Bhatia, A., & Pathak, H. (2014). Emission of air pollutants from crop residue burning in India. *Aerosol and Air Quality Research, 14*(1), 422–430. https://doi.org/10.4209/aaqr.2013.01.0031

Jiménez, E., Tapiador, F. J., & Sáez-Martínez, F. J. (2015). Atmospheric pollutants in a changing environment: Key issues in reactivity and monitoring, global warming, and health. *Environmental Science and Pollution Research, 22*, 4789–4792.

Jogdand, O. K. (2020). Study on the effect of global warming and greenhouse gases on environmental system. In A. Satish, D. Ravindra, S. Shinde, S. C. Ameta, & A. K. Haghi (Eds.), *Green chemistry and sustainable technology* (pp. 275–306). Apple Academic Press.

Joshi, A., & Arora, J. (2023). Recent trends and future prospects of nanotechnology for agro-waste valorization into biofuels. In K. G. Ramawat, J. M. Mérillon, & J. Arora (Eds.), *Agricultural waste: Environmental impact, useful metabolites and energy production* (pp. 559–582). Springer Nature.

Joshi, A., Kanthaliya, B., & Arora, J. (2019). Current scenario of potential renewable energy sources for sustainable development in India. *The Journal of Plant Science Research, 35*(2), 205–214.

Jury, W. A., & Vaux, H. J., Jr. (2007). The emerging global water crisis: Managing scarcity and conflict between water users. *Advances in Agronomy, 95*, 1–76.

Kaab, A., Sharifi, M., Mobli, H., Nabavi-Pelesaraei, A., & Chau, K. W. (2019). Combined life cycle assessment and artificial intelligence for prediction of output energy and environmental impacts of sugarcane production. *Science of the Total Environment, 664*, 1005–1019.

Kaur, R., Singh, D., Kumari, A., Sharma, G., Rajput, S., Arora, S., & Kaur, R. (2021). Pesticide residues degradation strategies in soil and water: A review. *International Journal of Environmental Science and Technology*, 1–24.

Khurana, R., Kumar, N., & Gupta, L. (2023). Air pollution as a risk for mental health problems: An exploratory study of direct and indirect pathways. *European Economic Letters (EEL), 13*(5), 131–141.

Kokotović, I., Veseli, M., Ložek, F., Karačić, Z., Rožman, M., & Previšić, A. (2024). Pharmaceuticals and endocrine disrupting compounds modulate adverse effects of climate change on resource quality in freshwater food webs. *Science of the Total Environment, 912*, 168751.

Koul, B., Yakoob, M., & Shah, M. P. (2022). Agricultural waste management strategies for environmental sustainability. *Environmental Research, 206*, 112285.

Krueger, J. (2001). The Basel convention and the international trade in hazardous wastes. In *Yearbook of international co-operation on environment and development* (pp. 43–51). Routledge.

Kumar, H., Prasad, K., Kumar, M., Sawarkar, A. D., Kumar, M., & Singh, L. (2023). Pesticide pollution in freshwater: Occurrence, distribution, impact, and remediation. In J. Singh, A. Pandey, S. Singh, V. K. Garg, & P. Ramamurthy (Eds.), *Current developments in biotechnology and bioengineering* (pp. 53–98). Elsevier INC.

Kumari, P., Gupta, N. C., & Kaur, A. (2017). A review of groundwater pollution potential threats from municipal solid waste landfill sites: Assessing the impact on human health. *Avicenna Journal of Environmental Health Engineering, 4*(1), 11525.

Kusumawardani, I. A. J., Indraswari, P. G., & Komalasari, N. L. G. Y. (2023). Air pollution and lung cancer. *Journal Respirasi, 9*, 150–158.

Lai, W., Li, S., Li, Y., & Tian, X. (2022). Air pollution and cognitive functions: Evidence from straw burning in China. *American Journal of Agricultural Economics, 104*(1), 190–208.

Lipinski, B., Hanson, C., Waite, R., Searchinger, T., & Lomax, J. (2013). *Reducing food loss and waste*. https://www.wri.org/research/reducing-food-loss-and-waste?page=1

Liu, L., Zheng, X., Wei, X., Kai, Z., & Xu, Y. (2021). Excessive application of chemical fertilizer and organophosphorus pesticides induced total phosphorus loss from planting causing surface water eutrophication. *Scientific Reports, 11*(1), 23015.

Liu, Q., Pan, S., Zhou, L., Feng, L., Li, Z., Du, L., & Wei, Y. (2022). Improving the biogas potential of rice straw through microwave-assisted ammoniation pretreatment during anaerobic digestion. *Bioenergy Research, 15*(2), 1240–1250.

Mahmood, I., Imadi, S. R., Shazadi, K., Gul, A., & Hakeem, K. R. (2016). Effects of pesticides on the environment. In K. Hakeem, M. Akhtar, & S. Abdullah (Eds.), *Plant, soil and microbes* (pp. 253–269). Springer.

Maji, S., Dwivedi, D. H., Singh, N., Kishor, S., & Gond, M. (2020). Agricultural waste: Its impact on environment and management approaches. In R. N. Bharagava (Ed.), *Emerging eco-friendly green technologies for wastewater treatment* (pp. 329–351). Springer.

Mar, K. A., Unger, C., Walderdorff, L., & Butler, T. (2022). Beyond CO_2 equivalence: The impacts of methane on climate, ecosystems, and health. *Environmental Science and Policy, 134*, 127–136.

Mehmood, K., Wu, Y., Wang, L., Yu, S., Li, P., Chen, X., & Seinfeld, J. H. (2020). Relative effects of open biomass burning and open crop straw burning on haze formation over central and eastern China: Modeling study driven by constrained emissions. *Atmospheric Chemistry and Physics, 20*(4), 2419–2443.

Mohite, A. S., Jagtap, A. R., Avhad, M. S., & More, A. P. (2022). Recycling of major agriculture crop residues and its application in polymer industry: A review in the context of waste to energy nexus. *Energy Nexus, 100134*.

Naz, S., Iqbal, S. S., Manan, A., Chatha, M., & Zia, M. (2023). The web of life: Role of pesticides in the biodiversity decline. *International Journal of Forest Sciences, 3*(2), 72–94.

Nguyen, T. T., Rosello, C., Bélanger, R., & Ratti, C. (2020). Fate of residual pesticides in fruit and vegetable waste (FVW) processing. *Food, 9*(10), 1468.

Obi, F. O., Ugwuishiwu, B. O., & Nwakaire, J. N. (2016). Agricultural waste concept, generation, utilization and management. *Nigerian Journal of Technology, 35*(4), 957–964.

Ogbuehi, N. C., Orji, M. C., & Afolabi, O. O. (2022). Health exposure and environmental challenges of households living nearby an open landfill system in Nigerian Urban Centre. *International Archives Public Health and Community Medicine, 6*(3), 1–7.

Parikh, R., Rao, S. R., Kukde, R., O'Connor, G. T., Patel, A., & Hibberd, P. L. (2021). Assessing the respiratory effects of air pollution from biomass Cookstoves on pregnant women in rural India. *International Journal of Environmental Research and Public Health, 18*(1), 183.

Pashaki, S. G. A., Khojastehpour, M., Ebrahimi-Nik, M., & Tedesco, S. (2024). Potential of ash from agricultural waste as substitute of commercial $FeCl_3$ in primary treatment of landfill leachate. *Journal of Environmental Management, 351*, 119932.

Patel, N., Feofilovs, M., & Blumberga, D. (2022). Agro biopolymer: A sustainable future of agriculture–state of art review. *Environmental and Climate Technologies, 26*(1), 499–511.

Peña, A., Delgado-Moreno, L., & Rodríguez-Liébana, J. A. (2020). A review of the impact of wastewater on the fate of pesticides in soils: Effect of some soil and solution properties. *Science of the Total Environment, 718*, 134468.

Pericherla, S., Karnena, M. K., & Vara, S. (2020). A review on impacts of agricultural runoff on freshwater resources. *International Journal on Emerging Technologies, 11*, 829–833.

Phiri, R., Rangappa, S. M., Siengchin, S., Oladijo, O. P., & Dhakal, H. N. (2023). Development of sustainable biopolymer-based composites for lightweight applications from agricultural waste biomass: A review. *Advanced Industrial and Engineering Polymer Research, 6*, 436–450.

Phuong, P. T. H., Nghiem, T. D., Thao, P. T. M., Pham, C. T., Thi, T. T., & Dien, N. T. (2021). Impact of rice straw open burning on local air quality in the Mekong Delta of Vietnam. *Atmospheric Pollution Research, 12*(11), 101225.

Ramadan, B. S., Rosmalina, R. T., Khair, H., Rachman, I., & Matsumoto, T. (2023). Potential risks of open waste burning at the household level: A case study of Semarang, Indonesia. *Aerosol and Air Quality Research, 23*(5), 220412.

Ramawat, K. G., Mérillon, J. M., & Arora, J. (Eds.). (2023). *Agricultural waste: Environmental impact, useful metabolites and energy production*. Springer.

Rani, L., Thapa, K., Kanojia, N., Sharma, N., Singh, S., Grewal, A. S., & Kaushal, J. (2021). An extensive review on the consequences of chemical pesticides on human health and environment. *Journal of Cleaner Production, 283*, 124657.

Raut, N. A., Kokare, D. M., Randive, K. R., Bhanvase, B. A., & Dhoble, S. J. (2023). Introduction: Fundamentals of waste removal technologies. In N. A. Raut (Ed.), *360-degree waste Management* (pp. 1–16). Elsevier B.V.

Ray, R. C. (2022). *Fruits and vegetable wastes: Valorization into bioproducts and platform chemicals* (p. 448). Springer.

Rayfuse, R. (2016). Principles of international environmental law applicable to waste management. In *Waste management and the green economy* (pp. 11–32). Edward Elgar Publishing.

Read, Q. D., Brown, S., Cuéllar, A. D., Finn, S. M., Gephart, J. A., Marston, L. T., & Muth, M. K. (2020). Assessing the environmental impacts of halving food loss and waste along the food supply chain. *Science of the Total Environment, 712*, 136255.

Riseh, R. S., Vazvani, M. G., Hassanisaadi, M., & Thakur, V. K. (2024). Agricultural wastes: A practical and potential source for the isolation and preparation of cellulose and application in agriculture and different industries. *Industrial Crops and Products, 208*, 117904.

Sadegh-Zadeh, F., Abd Wahid, S., & Jalili, B. (2017). Sorption, degradation and leaching of pesticides in soils amended with organic matter: A review. *Advances in Environmental Technology, 3*(2), 119–132.

Sahu, S. K., Mangaraj, P., Beig, G., Samal, A., Pradhan, C., Dash, S., & Tyagi, B. (2021). Quantifying the high resolution seasonal emission of air pollutants from crop residue burning in India. *Environmental Pollution, 286*, 117165.

Sarangi, P. K., Singh, A. K., Srivastava, R. K., & Gupta, V. K. (2023). Recent progress and future perspectives for zero agriculture waste technologies: Pineapple waste as a case study. *Sustainability, 15*(4), 3575.

Shabir, I., Dash, K. K., Dar, A. H., Pandey, V. K., Fayaz, U., Srivastava, S., & Nisha, R. (2023). Carbon footprints evaluation for sustainable food processing system development: A comprehensive review. *Future Foods, 7*, 100215.

Siddiqua, A., Hahladakis, J. N., & Al-Attiya, W. A. K. (2022). An overview of the environmental pollution and health effects associated with waste landfilling and open dumping. *Environmental Science and Pollution Research, 29*(39), 58514–58536.

Singh, A. K., & Chandra, R. (2019). Pollutants released from the pulp paper industry: Aquatic toxicity and their health hazards. *Aquatic Toxicology, 211*, 202–216.

Spielmeyer, A. (2018). Occurrence and fate of antibiotics in manure during manure treatments: A short review. *Sustainable Chemistry and Pharmacy, 9*, 76–86.

Tran, T. H., Nguyen, T. B. T., Le, H. S. T., & Phung, D. C. (2024). Formulation and solution technique for agricultural waste collection and transport network design. *European Journal of Operational Research, 313*(3), 1152–1169.

Tripathi, N., Hills, C. D., Singh, R. S., & Atkinson, C. J. (2019). Biomass waste utilisation in low-carbon products: Harnessing a major potential resource. *npj Climate and Atmospheric Science, 2*(1), 35.

United Nations Department of Economic and Social Affairs, Population Division. (2022). *World population prospects 2022: Summary of results*. UN DESA/POP/2022/TR/NO. 3.

United Nations Environment Programme. (2021). *Food waste index report 2021*. Nairobi.

Verma, K. K., Song, X. P., Joshi, A., Rajput, V. D., Singh, M., Sharma, A., Singh, R. K., Li, D. M., Arora, J., Minkina, T., & Li, Y. R. (2022). Nanofertilizer possibilities for healthy soil, water, and food in future: An overview. *Frontiers in Plant Science, 13*, 865048.

Verma, K. K., Song, X. P., Degu, H. D., Guo, D. J., Joshi, A., Huang, H. R., & Li, Y. R. (2023). Recent advances in nitrogen and nano-nitrogen fertilizers for sustainable crop production: A mini-review. *Chemical and Biological Technologies in Agriculture, 10*(1), 111.

Wang, Z., Adu-Kumi, S., Diamond, M. L., Guardans, R., Harner, T., Harte, A., & Yuan, B. (2022). Enhancing scientific support for the Stockholm Convention's implementation: An analysis of policy needs for scientific evidence. *Environmental Science and Technology, 56*(5), 2936–2949.

Zahoor, I., & Mushtaq, A. (2023). Water pollution from agricultural activities: A critical global review. *International Journal Of Chemical And Biochemical Sciences, 23*(1), 164–176.

Zaller, J. G., Weber, M., Maderthaner, M., Gruber, E., Takács, E., Mörtl, M., & Székács, A. (2021). Effects of glyphosate-based herbicides and their active ingredients on earthworms, water infiltration and glyphosate leaching are influenced by soil properties. *Environmental Sciences Europe, 33*, 1–16.

Zhao, Q., De Laender, F., & Van den Brink, P. J. (2020). Community composition modifies direct and indirect effects of pesticides in freshwater food webs. *Science of the Total Environment, 739*, 139531.

Zorpas, A. A. (2020). Strategy development in the framework of waste management. *Science of the Total Environment, 716*, 137088.

Chapter 2
Valorization Strategies for Agriculture Residue: An Overview

Abhishek Joshi, Pooja Yadav, Chitra Yadav, Bhanupriya Kanthaliya, Krishan K. Verma, and Jaya Arora

Abbreviations

AI	Artificial intelligence
ANFs	Antinutritional factors
CBP	Consolidated bioprocessing
CNDs	Carbon nanodots
DPPH	2, 2-diphenyl-1-picrylhydrazyl
H_2	Hydrogen
IoT	Internet of things
K	Potassium
ML	Machine learning (ML)
MNPs	Magnetic nanoparticles
N	Nitrogen
NCPs	Nanocomposites
NPs	Nanoparticles
P	Phosphorus
SSF	Simultaneous saccharification and fermentation
USD	United States Dollar

A. Joshi · P. Yadav · C. Yadav · B. Kanthaliya · J. Arora (✉)
Laboratory of Biomolecular Technology, Department of Botany, Mohanlal Sukhadia University, Udaipur, Rajasthan, India
e-mail: jayaarora@mlsu.ac.in

K. K. Verma
Key Laboratory of Sugarcane Biotechnology and Genetic Improvement (Guangxi), Ministry of Agriculture and Rural Affairs/Guangxi Key Laboratory of Sugarcane Genetic Improvement/Sugarcane Research Institute, Guangxi Academy of Agricultural Sciences, Nanning, China

J. Arora et al. (eds.), *Transforming Agriculture Residues for Sustainable Development*, Waste as a Resource, https://doi.org/10.1007/978-3-031-61133-9_2

2.1 Introduction

According to the most recent Global Waste Management Outlook published by the United Nations Environment Programme and the International Solid Waste Association in 2024, global municipal waste generation is anticipated to rise from 2.3 billion tons in 2023 to 3.8 billion tons by 2050. This report also estimated the global annual cost of waste management, which was 252 billion United States dollars (USD) in 2020 and is projected to reach a stunning USD 640.3 billion by 2050 if immediate action is not taken (Anonymous, 2024).

Agricultural residue, commonly referred to as agro-waste, constitutes a substantial portion of the world's solid waste output and is expected to almost double by 2050 as a consequence of existing economic and population growth (Agapkin et al., 2022). These wastes are defined as vestige from the cultivation and processing of raw agricultural products, including fruits, vegetables, meat, poultry, dairy products, and crops. They are the discretionary outputs of agro-industrial processes, and their impact on the triple planetary crisis of climate change, biodiversity loss, and environmental pollution is well-documented (Arora et al., 2023).

In addition to this, these waste may contain various inherent bio-based molecules and phytochemicals, such as carbohydrates, fibers, proteins, lipids, polyphenols, peptides, minerals, and other compounds that can be harnessed through various biochemical, thermochemical, biotechnological, and innovative valorization strategies (Meena et al., 2020; Capanoglu et al., 2022; Khan et al., 2023). Biochemical processes involve the use of biological agents, such as enzymes or microorganisms, to transform biomass into valuable biofuels, biochemicals, and other products. These methods include enzymatic hydrolysis, fermentation, and anaerobic digestion (da Silva et al., 2020; Chaudhary et al., 2024). Thermochemical processes involve the use of heat and/or catalysts to convert biomass into valuable chemicals such as biofuels, biochar, and syngas. These processes include pyrolysis, gasification, and hydrothermal liquefaction (Oliveira et al., 2021).

Biotechnology offers unique opportunities to produce novel molecules with high specificity and efficiency via microbial, enzymatic, and metabolic engineering strategies. Microorganisms, such as bacteria, yeast, and fungi, can be engineered to produce target compounds, whereas enzymes derived from diverse biological sources can catalyze selective transformations, enabling the synthesis of specialty compounds with minimal waste and energy consumption (Wan et al., 2024). Metabolic engineering techniques can also be used to manipulate cellular pathways and enhance the production of target compounds (Joshi et al., 2022).

Advances in technology, including nanotechnology, artificial intelligence (AI), machine learning (ML), and the Internet of Things (IoT), have transformed the way we discover and optimize valorization processes, leading to improved bioavailability, stability, and applicability of functional products (Magnus et al., 2021; Snoussi et al., 2023). These advancements have been made possible by computational tools and high-throughput screening methods that allow rapid evaluation of enzyme activity, substrate specificity, and reaction conditions, ultimately accelerating the development of biocatalytic processes for the synthesis of bioproducts and biomaterials (Hong et al., 2023).

The economic viability and environmental impact of agricultural waste valorization strategies depend on several factors, such as feedstock availability, technology maturity, market demand, and policy support. Recent studies have assessed the techno-economic feasibility and environmental sustainability of different valorization pathways using life cycle assessment, economic modeling, and sensitivity analysis (Dastjerdi et al., 2021; Aldama et al., 2023). Furthermore, economic modeling and sensitivity analysis have demonstrated the potential for cost savings and revenue generation through the valorization of agricultural waste, particularly in regions with abundant biomass resources and supportive policy frameworks (Roy et al., 2023).

The goal of this chapter is to provide readers with a comprehensive understanding of the current state of agricultural residue valorization and the potential opportunities it presents for a more sustainable future.

2.2 Approaches for Agriculture Waste Transformation

Agricultural biomass primarily consists of complex carbohydrates and aromatic polymers such as cellulose, hemicellulose, lignin, and pectin. The intrinsic packaging of these polymers, the crystallinity and degree of polymerization of cellulose, the presence of acetyl groups, hydroxyl groups, and the syringyl-guaiacyl(S/G) ratio in lignin contribute to their structural complexity and recalcitrance, which makes it difficult to convert them into valuable products (Joshi et al., 2021). Several methods, including biochemical, thermochemical, biotechnological, and innovative conversion techniques, have been employed to extract value from agricultural residues. This section highlights a few recent accomplishments (discussed in Sects. 2.2.1, 2.2.2, and 2.2.3) to provide a detailed overview of the strategies used to transform agricultural waste.

2.2.1 Biochemical and Thermochemical Approaches

Pretreatment is a crucial step for breaking down the structural complexity of the agriculture residue, improving access to fermentable sugars, and enhancing the overall conversion efficiency. Various physical (i.e., mechanical comminution, high energy radiation), chemical (i.e., ionic liquids, dilute alkali, organosolv), and biological (i.e., rot fungi and bacteria) pretreatment methods have been employed to increase the efficiency of cellulose separation from lignin and other components (Joshi et al., 2021). Moreover, consolidation of pretreatment by combining more than one method from the same or different categories is also commonly used. In recent years, green and novel low-cost pretreatment strategies, such as deep eutectic solvents, hydrothermal pretreatment, microwaves, and ultrasonication, have gained importance due to their remarkable biomass deconstruction properties and potential to increase the overall productivity and economy of the bioprocess (Ab Rasid et al., 2021; Chaudhary et al., 2024). However, there is no single best pretreatment method, as each treatment has its detriments, and rigorous development is necessary to make downstream processes economically feasible.

Enzymatic hydrolysis, also known as saccharification, is a crucial biochemical process that converts complex carbohydrates found in agricultural waste into simple sugars using hydrolytic enzymes such as cellulases, hemicellulases, ligninases, and pectinases. The efficiency of these processes has been improved through advancements in enzyme activities and bioprocess optimization, leading to lower processing costs (da Silva et al., 2020). Techniques, such as immobilization, and enzyme cocktails have been explored to enhance enzyme stability, substrate specificity, and reaction kinetics. Furthermore, consolidated bioprocessing (CBP) strategies have emerged to streamline the enzymatic hydrolysis process and reduce the overall production costs (Woo et al., 2024).

Fermentation of hydrolyzed waste sugars offers a pathway for the production of biofuels, biochemicals, and biopolymers. Recent advancements in fermentation technology have focused on optimizing microbial strains, fermentation conditions, and downstream processing techniques to enhance product yield and purity (Rahmati et al., 2020). Metabolic engineering approaches have been used to develop microbial platforms that can efficiently convert diverse feedstock into target products with high specificities and yields (Joshi et al., 2022, 2023). Additionally, innovative fermentation processes such as simultaneous saccharification and fermentation (SSF) and CBP have been developed to simplify process flows, reduce capital costs, and increase overall process efficiency (Joshi et al., 2021).

The thermochemical conversion techniques primarily involve pyrolysis and gasification. Pyrolysis refers to the thermal decomposition of organic materials in the absence of oxygen, which leads to the production of biochar, bio-oil, and syngas. Researchers have made advancements in pyrolysis technology to enhance product quality, process efficiency, and reactor design (Alcazar-Ruiz et al., 2022). For instance, catalytic pyrolysis with zeolites, metal catalysts, and biochar-derived catalysts has been explored to improve the yield and quality of bio-oil and syngas products (Joshi & Arora, 2023a). Additionally, integrated pyrolysis systems that incorporate heat recovery and gas cleaning technologies have been developed to optimize energy utilization and minimize environmental emissions (Ge et al., 2023).

The gasification process involves partial oxidation of biomass, resulting in the production of syngas, a combustible gas mixture. Advances in gasification technology have focused on improving process efficiency, fuel flexibility, and syngas quality. Integrated gasification combined cycle systems have been developed to efficiently utilize syngas for power generation while capturing and sequestering carbon dioxide emissions (Oliveira et al., 2021). Furthermore, the combination of biomass gasification with carbon capture and utilization technologies presents a promising pathway for achieving negative carbon emissions and carbon-neutral energy production (Fernandez et al., 2020; Wang & Cheng, 2023).

An extensive literature of biochemical and thermochemical approaches for the transformation of agriculture waste has been discussed in Chap. 3 of this book.

2.2.2 Biotechnological Approaches for Agriculture Waste Transformation

The incorporation of biotechnology in waste transformation has led to a shift from conventional methods to innovative solutions, including genome editing, pathway engineering, and high-throughput screening. A range of synthetic biology tools, such as clustered regularly interspaced short palindromic repeats-associated proteins, transcription activator-like effector nucleases, and others have been utilized to design microbial strains with enhanced substrate utilization, product specificity, and tolerance to inhibitors (Hamdan et al., 2022). Several vital phenotypes from the bacterial, fungal, and yeast domains have been developed with improved substrate spectrum, and resistance to hydrolyzate toxins (Wan et al., 2024). In addition, these tools have been employed to modify the compositional characteristics of crop plants, resulting in amended conversion into valuable products. For instance, barley (*Hordeum vulgare*) was genetically engineered by downregulating the ferulate 5-hydroxylase gene, which affects the lignin S/G ratio, preserves plant traits, and improves straw properties for bioconversion (Shafiei et al., 2023). A recent engineering effort produced a new sugarcane variety (named KK3) by targeting the limonene synthase gene through *Agrobacterium*-mediated transformation. This process resulted in a significant reduction in lignin content (9.74–51.46%), making the sugarcane an ideal substrate for bioethanol production (Laksana et al., 2024). Process intensification strategies such as in situ product removal and cell immobilization have also been explored to enhance fermentation productivity and product purity (Santos et al., 2021).

Chapter 4 in this book extensively deals with the biotechnological approaches for agriculture waste transformation.

2.2.3 Emerging Approaches for Agriculture Waste Transformation

Nanotechnology, with its unique properties and tunable functionalities, has emerged as a powerful tool for enhancing the efficiency and selectivity of waste transformation processes. In the catalytic conversion process, magnetic nanoparticles (MNPs) have been commonly used as nanocatalysts due to their ability to immobilize enzymes and to be recycled and reused. Recent studies have demonstrated that MNPs can be coated or modified with mesoporous materials (e.g., zeolites and metal-organic frameworks), hierarchical structures (e.g., nanozeolites and core-shell nanoparticles), and functionalized surfaces (e.g., metal oxides and carbonaceous materials) to improve their catalytic activity for biomass pyrolysis, hydrodeoxygenation, polymerization, and reforming reactions (Thanigaivel et al., 2022; Snoussi et al., 2023). Advances in nanostructured catalysts have provided tailored pore size distributions, surface chemistry, and mechanical properties, enabling precise control over the reaction kinetics, diffusion limitations, and catalyst poisoning effects (Dey et al., 2022). In addition, the integration of nanocatalysts with advanced configurations such as microreactors, membrane reactors, and nanosensors has enabled continuous and scalable production processes with improved energy savings, reduced environmental impact, and enhanced process scalability (Preethi et al., 2024).

In recent years, AI has become increasingly popular at various stages of the process involved in waste transformation. This includes the selection of feedstock and optimization of processes as well as the analysis and quality control of the resulting products. ML algorithms, such as supervised learning algorithms, unsupervised learning algorithms, and deep learning algorithms, are the core of AI-driven waste valorization (Khanal et al., 2023). Spectroscopic techniques, such as near-infrared spectroscopy and hyperspectral imaging, when combined with ML algorithms, allow for the non-destructive analysis of biomass constituents, providing valuable insights into the variability, quality, and suitability of feedstock for specific valorization processes (Magnus et al., 2021; Tao et al., 2023). ML models trained using experimental data can predict optimal formulation and processing conditions for the synthesis of biopolymers, biofuels, biochar, and specialty chemicals. These models utilize computational tools such as kinetic models, thermodynamic models, and process simulators to simulate and optimize bioprocessing workflows (Kanthasamy et al., 2023; Alloun & Calvio, 2024).

Similarly, IoT has emerged as a transformative technology with significant implications for the conversion of agricultural waste into bioproducts. IoT-enabled sensors, actuators, and communication networks have been employed to monitor, control, and optimize bioprocessing operations in real time, leading to improved resource utilization, product quality, and sustainability (Hussain et al., 2023). These sensing and monitoring systems facilitate data-driven decision-making and adaptive control strategies, leading to enhanced process efficiency, reliability, and safety (Hong et al., 2023). Additionally, IoT-enabled global positioning systems and radio frequency identification technologies enable traceability and logistics management of biomass feedstock from the field to biorefinery, thereby reducing transportation costs and environmental impact (Fontaine et al., 2024).

Chapter 5 in this book extensively deals with the emerging approaches for agriculture waste transformation

2.3 Recent Trends in Valorization of Agricultural Waste into Bioproducts and Biomaterials

This section provides an in-depth analysis of recent trends in the valorization of agricultural waste into bioproducts and biomaterials. Successful examples of agricultural waste transformation into different value-added bioproducts and biomaterials are summarized in Table 2.1.

2.3.1 Nanoparticles (NPs) and Nanocomposites (NCPs)

The use of agricultural waste for the synthesis of NPs and NCPs has become a widely accepted approach for sustainable waste management and the development of eco-friendly materials with a variety of applications. Currently, agriculture residues are widely used as cheap, abundant precursors for developing both metallic and non-metallic NPs as well as hybrid NCPs (Nath et al., 2023; Joshi & Arora, 2023b).

A reducing agent is typically required to produce metallic NPs. Extensive research has been conducted on the use of agro-waste as a stabilizing or reducing agent for the synthesis of various metallic NPs, including copper (CU), gold (Au), silver (Ag), silicon (Si), and bimetallic alloy NPs, for applications in agriculture, food, and pharmaceutical industries (Castillo-Henríquez et al., 2020; Ngungeni et al., 2023). Recently, Torres-Arellano et al. (2023) utilized aqueous corncob ash extract to produce copper oxide (CuO) NPs. The resulting CuO NPs were characterized by agglomeration, porosity, rough edges, and the ability to generate twice as much H_2 as the traditional CuO catalysts. In another study, López-Alcántara et al. (2024) synthesized magnetic iron oxide NPs using sweet pepper extract as both reducing and capping agents. These NPs exhibited strong scavenging activity against DPPH (2, 2-diphenyl-1-picrylhydrazyl) with an IC50 of 128.1 μg/mL. Additionally, the stabilization of NPs with banana-based bioplastics effectively preserved grapes, resulting in a 27.4% lower weight loss rate after 144 h compared with the control group.

Several recent studies report the fractionation of various agriculture residue into silica NPs (also called silicon dioxide NPs or SiO2 NPs), which are useful in many areas, including energy, food, biomedicine, and personal care (Shahi et al., 2021; Shahbaz et al., 2023). A recent study by September et al. (2023) analyzes and summarizes the production of silica and Si NPs with potential applications in electronic devices.

It has also been shown that agricultural residues can be used as precursors for the extraction and synthesis of non-metallic NPs and hybrid-NCPs, such as carbon, pectin, nanocellulose, cellulose-metal-oxide polymer, metal-metal oxides, and metal-carbon dots with several conservative applications in food preservation, bioenergy production, and environmental remediation (Bhardwaj & Singh, 2023; Shalma et al., 2023). Thangadurai et al. (2023) recently synthesized carbon nanodots

Table 2.1 Successful examples of agricultural waste transformation into value-added bioproducts and biomaterials

Bioproducts or biomaterials	Agriculture residue	Valorization strategies	Outcomes	References
Bioactive metabolites and enzymes	Beetroot waste	Surfactant-assisted microwave extraction	Significant amount of pectin and bioactive compounds	del Amo-Mateos et al. (2023)
	Carrot bagasse	Solid-state fermentation (SSF) by Aspergillus niger with added nitrogen source	β-fructofuranosidase (90.82 U/mL)	Guerra et al. (2023)
	Citrus peels	Ultrasound-assisted and pulsed electric field-assisted extraction	Considerable amount of polyphenols, ascorbic acid, and carotenoids	Chatzimitakos et al. (2023)
	Mango seed waste	SSF with urea (nitrogen source), temperature 40 °C and pH 6.0	Cellulase (18 IU/gds filter paper), β-glucosidase (77 IU/gds), and endoglucanase (89 IU/gds)	Khan et al. (2024)
	Peanut meal	SSF with *Pseudomonas* aeruginosa FA1	Rhamnolipid (26.54 mg/g of substrate)	Zhao et al. (2023)
	Potato peels	Enzymatic hydrolysis	Isomaltose-oligosaccharide production	Shahapurkar et al. (2024)
	Pineapple peel and core	SSF with *Aspergillus niger* GH1	Increased yield of phenolic compounds (72%), β-glucosidase (29 U/g), and cellulose (4 U/g)	Paz-Arteaga et al. (2023)
	Raw rice bran and barley malt residue	SSF with *Gibberella fujikuroi*	Gibberellic acid (10.1 g/kg of substrates)	Pinheiro et al. (2024)
	Sugarcane bagasse and olive pits	Hydrothermal and dilute acid pretreatment with homo-fermentative *Bacillus coagulans* DSM2314 under non-sterile condition	Lactic acid (0.66 g/g from sugarcane bagasse and 0.71 g/g from olive pits)	Cox et al. (2023)
	Wheat bran, rice polishings and molasses	SSF anaerobically by *Clostridium tyrobutyricum* DSM	Butyric acid (5.63 mg/100 g of wheat bran, 5.5 mg/100 g of rice polishing and 4.26 mg/100 g of molasses)	Akhtar et al. (2020)

Biofuels and bioenergy	Corn cob, sugarcane bagasse, rice straw, and sweet sorghum bagasse	Fermentation with *Candida tropicalis*, *Saccharomyces cerevisiae*, and *Kluyveromyces marxianus*	Ethanol (0.38 g/g$Y_{eth/s}$), phenylacetylcarbinol (62.3 mM)	Nunta et al. (2023)
	Rice straw	Chemical pretreatments and fermentation with *Enterobacter aerogenes* and nano silica	Biohydrogen (178.3 mL/g of total solids)	Mojoodi et al. (2024)
	Rice straw	Anaerobic digestion, pyrolysis, and hydrodynamic cavitation	Methane (411 mL-CH_4/g-volatile solid)	Ezz et al. (2023)
	Rice straw	Dark fermentation with zinc molybdate nanomaterials	Biohydrogen (296.3 mL/g straw)	Zhou et al. (2023)
	Sugarcane bagasse	Fed-batch simultaneous saccharification, fermentation with *Kluyveromyces marxianus* JKH5 C60 and *Pichia stipitis* NCIM 3499	Ethanol yield (260.1 g/kg substrate)	Himanshi and Saini (2023)
	Wheat husk	Ionic liquid treatment (chlorochromate tethered [Py] HSO_4)	5-ethoxymethylfurfural (42% yield)	Naz and Uroos (2023)
Biofertilizer and biochar	Corncob, cassava rhizome, and rice husk	Controlled temperature biochar retort for slow pyrolysis process at 450–500 °C	Biochar with admirable properties for soil amendment and carbon sequestration	Wijitkosum (2022)
	Groundnut husk	Vermicomposting employing earthworms (*Eisenia fetida* and *Perionyx sansibaricus*)	Nitrogen-, phosphorous-, and potassium-rich vermifertilizer	Kumar et al. (2021)
	Rice straw	Delignification with laccase-assisted sodium chlorite, hydrolysis (cellulase cocktail), and fermentation by Saccharomyces cerevisiae NCIM	Extracted lignin showed biopesticidal activity and residual material exhibited biofertilizer properties	Saini et al. (2023)
Bioplastics and biopolymers	Apple waste	Delignification and reinforcement by nanocrystalline cellulose	Bioplastic with good storage modulus (0.42–4.23 MPa), hardness (5.22–0.51 kg), and biodegradability (58.1% after 4 weeks)	Loh et al. (2023)
	Banana peels	Chemical treatment with glycerine and hydrochloric acid	Bioplastic (biodegradability 82% after 7 days)	Karne et al. (2023)

(continued)

Table 2.1 (continued)

Bioproducts or biomaterials	Agriculture residue	Valorization strategies	Outcomes	References
	Cassava peel starch	Melt intercalation with additives (rice husk cellulose, polyvinyl alcohol, and zinc oxide)	Bioplastic with excellent tensile strength (37.38 MPa) and degradation capacity (71.02% after 11.62 days)	Liandi et al. (2023)
	Corn starch	Chemical treatment with glycerol (0.2 g to 0.8 g)	Bioplastic (1.52 MPa to 5.08 MPa tensile strength)	Rajesh et al. (2024)
	Corn husk	Delignification and chemical treatment with sorbitol	Bioplastic (50–70% yield) with excellent degradation frequency after 4 days	Setiawan et al. (2024)
	Orange, mango, banana, onion peels, and rice straw	Bioconversion with *Bacillus wiedmannii* AS-02 OK576278 under optimal pH (7.0), temperature (35 °C) and incubation period (72 h)	Polyhydroxybutyrates (423 mg/L from orange, 390 mg/L from mango, 249 mg/L from banana, 158 mg/L from onion peel and 144 mg/L from rice straw)	Danial et al. (2021)
	Sugarcane bagasse	Chemical pretreatment followed by acid-hydrolysis with additive (polyvinyl alcohol)	Composite films (43% biodegradability) with excellent removal capacity for methylene blue (70%) and crystal violet (64.5%)	Sarkhel et al. (2023)
Building materials	Cork, prickly pear, rice husk, and vine pruning residues	Bonded with a water-based acrylic resin in different proportions	Ceiling tiles with good thermal conductivity and acoustic properties	Maderuelo-Sanz et al. (2022)
	Pomegranate peel	Mixing with clay in substitution levels and combustion	Clay bricks with thermal conductivity (up to 21% at 900 °C), water absorption (22.6–29.1%), and shrinkage properties (4.3–5.4%)	Maafa et al. (2023)
	Rice straw	Blending with additive (gypsum) and foaming agents (aluminum sulphate and potassium alum)	Hollow gypsum blocks with excellent thermal insulation, acoustic and fire retardant properties	Singh et al. (2023a, b)
	Walnut shell	Mixing with clay (0–10 wt.%) and combustion (950 °C and 1100 °C)	Clay bricks with increased water absorption and specific heat capacity	Barnabas et al. (2023)

Feed and food additives	Corn stover and rice straw	Co-fermented with probiotics and enzymes on rumen fermentation	High concentration of volatile fatty acid, numbers of complex carbohydrate-decomposing, H_2-utilizing bacteria, and feed energy conversion efficiency	Hu et al. (2020)
	Groundnut shells, pigeon pea husk, and wheat straw	Microbial fermentation with *Colletotrichum* sp.	Improved amount of carbohydrates, crude protein, nitrogen, and low quantity of tannins and phytate	Patil et al. (2020)
	Potato waste	Fermentation using rumen fluid	High yield of volatile fatty acids and residual solid residues with high-protein and fat contents	Pourbayramian et al. (2021)
	Orange peel	Saccharification and aerobic fermentation	High in vitro organic matter digestibility with threefold protein content	Andrianou et al. (2023)
Nanoparticle and nanocomposites	Potato peelings and parsley stems	Hot water treatment and polyvinylpyrrolidone-silver nitrate solution	Silver nanoparticles with antimicrobial properties	Wolny-Kołladka et al. (2022)
	Peanut shell, rice husk and sugar cane bagasse	Mild hydrolysis and ultrasonic-assisted chemical synthesis	Silica nanoparticles and cellulose nanofibers with good crystallinity, and specific surface area	Shahi et al. (2021)
	Rice husks (sticky, red, and brown)	Magnesiothermic reduction	Silicon nanocrystals with excellent optical and textural characteristics	Sekar and Lee (2021)
	Rice husk	Sol-gel technique from crystalline nano silica using polyvinyl alcohol as a polymer binder	Crystalline nano silica (21–31 nm) with good mechanical and optical qualities	Kumari et al. (2023)
	Sweet pepper extract	Iron (III) chloride hexahydrate assisted chemical treatment	Magnetic iron oxide nanoparticles with possibilities in food preservation	López-Alcántara et al. (2024)

(CNDs) from rice husk biomass waste using a hydrothermal method. The synthesized CNDs exhibited a spherical morphology (size 25–30 nm) with 98% cell viability and remarkable anti-inflammatory activity (inhibition of lipoxygenase, cyclooxygenase, and nitric oxide) at 100 μL.

Chapter 6 in this book extensively deals with the production of nanoparticles and nanocomposites from agriculture waste and their applications

2.3.2 *High-Value Animal Feed and Food Additives*

Agricultural residues are a rich source of various nutritional components including carbohydrates, fibers, proteins, vitamins, and minerals. As a result, there is growing interest in utilizing agricultural wastes to produce high-value feed and food additives in the livestock industry. However, nutrient availability and digestion are hindered by antinutritional factors (ANFs) within agricultural residues, such as hydrogen cyanide, phytates, tannins, polyphenols, and others (Singh et al., 2023a, b). Recent studies have utilized various methods to transform agricultural waste into valuable animal feed and food ingredients, such as milling, and grinding, as well as chemical treatments such as alkaline and acid hydrolysis, and enzymatic hydrolysis. These techniques enhance nutrient accessibility and quality but are costly and labor-intensive (Mengqi et al., 2021). Also, the low nutritional content of most agro-industrial wastes, combined with the high demand for protein-rich foods and feed, has led to the development of strategies that allow for the efficient conversion of agricultural wastes into single-cell proteins (Tropea et al., 2022). Recently, extensive research has been conducted on SSF, a microbial fermentation process that reduces ANFs in agro-industrial wastes, thereby boosting nutrient digestibility and bioavailability (Chilakamarry et al., 2022). Several recent studies have confirmed the effectiveness of SSF in improving the nutrient content and quality of agricultural residues used as animal feed (Pourbayramian et al., 2021; Yafetto et al., 2023). Patil et al. (2020) studied the bioconversion of three different agro-residues using endophytic fungi (*Celastrus paniculatus*). Among the three agro-residues, groundnut shells exhibited the highest carbohydrate content (13.92 g/100 g), total crude protein (24.95 g/100 g), and total nitrogen content (15.53 g/100 g) after SSF biotransformation. Furthermore, significant reductions in tannin and phytate contents were observed in the processed waste. Additionally, the study revealed that up to 20% and 10% of biotransformed groundnut shells and pigeon pea waste, respectively, were added to the commercial poultry diet without any adverse effects. Recently, Andrianou et al. (2023) developed feedstuff by mixing the solid residue of orange peels that underwent saccharification under optimal conditions. The resulting feedstuff displayed a 23.11% increase in in vitro organic matter digestibility and a threefold increase in protein content. Similarly, a one-pot biotransformation process utilizing an in vitro coenzyme-free synthetic enzymatic pathway and baker's yeast can convert dilute sulfuric acid-pretreated corn stover to artificial starch and microbial proteins under aerobic conditions (Xu et al., 2023). Furthermore, animal trials have demonstrated that the digestion of artificial amylose produces small and slow changes in blood sugar levels, suggesting that it could be a promising component of healthy foods to prevent obesity and diabetes.

There are two chapters in this book (Chaps. 7 and 9) that describe in details valorization of agriculture waste into high-value animal feed and food additives.

2.3.3 *Biofertilizer and Biochar*

Agricultural residues possess high contents of organic matter and mineral nutrients, making them valuable inputs for the production of biofertilizers and biochar. Various conversion methods, including composting, anaerobic digestion, chemical hydrolysis, in-situ degradation, and direct burning, have been utilized in recent years to transform agricultural waste into biofertilizers and biochar. However, the sustainability of these processes depends on factors such as the feedstock composition, process efficiency, and end-product characteristics (Puglia et al., 2021). Recent studies have shown that biofertilizers produced from agro-industrial waste are rich in essential nutrients including nitrogen (N), phosphorus (P), potassium (K), micronutrients, and beneficial microorganisms, which enhance soil fertility, improve nutrient cycling, and promote plant growth (Vurukonda et al., 2024). In a recent study, Wang et al. (2021) developed a novel semi-interpenetrating polymer network fertilizer (CMCK-g-PAA/PDMUP) using wheat straw and linear polymers via solution polymerization and supplementation with polydihydroxymethyl-urea potassium phosphate (PDMUP). The prepared product exhibited excellent water absorbency (681.3 g/g) and increased the soil's water-retention capacity and cumulative release of N (56.1%), P (64.3%), and K (74.1%).

In addition, agro-waste-derived biochar has gained attention in the green energy sector due to its sustainability, low cost, and high supercapacitance (Khedulkar et al., 2023; Awogbemi & Von Kallon, 2023). It has also been demonstrated that biochar amendments improve soil structure, increase water retention, and improve plant nutrition (Hamidzadeh et al., 2023). Recent studies have shown that agro-waste-derived biochar can be used for wastewater treatment, contamination immobilization, and in situ carbon capture (Zhou et al., 2021; Wijitkosum, 2022).

Chapter 8 in this book extensively deals with the production of biofertilizers and biochar from agricultural waste and their applications.

2.3.4 *Bioplastics and Biopolymers*

In recent years, there has been growing interest in developing sustainable alternatives to conventional plastics by converting agro-industrial waste into bioplastics and biopolymers. Compared to conventional plastics, agro-waste-derived biopolymers provide several advantages such as biodegradability, renewable nature, and reduced carbon footprint (Díez-Pascual, 2019). Several recent studies have shown that bioplastics derived from agro-industrial waste exhibit a wide range of properties including mechanical strength, thermal stability, and barrier properties, making them suitable for diverse applications (Danial et al., 2021; Setiawan et al., 2024). For instance, biodegradable plastics have been synthesized from cassava peel starch through melt intercalation of rice husk cellulose, polyvinyl alcohol, and zinc oxide (Liandi et al., 2023). The synthesized bioplastic demonstrated excellent tensile strength (37.38 MPa) and degradation capacity (71.02% after 11.62 days). Similarly, apple waste has been converted into bioplastics by dignifying cellulose with nano-crystalline cellulose extracted from biowaste (Loh et al., 2023). The bioplastics displayed an improved storage modulus (0.42–4.23 MPa) and brittleness in terms of hardness (5.22–0.51 kg). Furthermore, it showed excellent biodegradability (58% after 4 weeks). In addition to bioplastic production, researchers have explored the potential use of agro-industrial waste-derived biopolymers in various applications, including the packaging, agriculture, biomedical, and automotive industries (Nanda et al., 2022; Gong et al., 2024). In a recent study, Sarkhel et al. (2023) reported that biodegradable composite films made from starch-based biopolymers derived from sugarcane bagasse can effectively remove methylene blue (70%) and crystal violet (64.5%).

There are two chapters in this book (Chaps. 10 and 11) that extensively deal with agriculture waste valorization into bioplastics, biopolymers, and packaging solutions

2.3.5 *Biofuels and Bioenergy*

Converting agricultural waste materials into renewable fuels offers a promising solution for reducing reliance on fossil fuels, mitigating greenhouse gas emissions, and promoting sustainability in the realms of energy, environment, and the economy. Bioethanol, a renewable fuel produced through the fermentation of sugars derived from biomass feedstock, has garnered significant research attention in recent years (Sharma et al., 2017; Joshi et al., 2020). Various pretreatment methods, enzymatic hydrolysis techniques, and fermentation processes using genetically engineered microorganisms and CPB approaches have been used to unlock the sugars present in diverse agro-waste biomasses and convert them into ethanol (Jayakumar et al., 2023; Kumar et al., 2023). In a recent study by Nuntra et al. (2023), the bioconversion efficacy of different substrates, including corn cob, sugarcane bagasse, rice straw, and sweet sorghum bagasse, for bioethanol and phenylacetylcarbinol (PAC), was examined using various yeast strains. The study reported an excellent ethanol yield of 0.38 g/g of product per substrate and 62.3 mM PAC yield when rice straw was cultivated with *Candida tropicalis*. Another study by Himanshi and Saini (2023) reported a significant ethanol yield of 260.1 g/kg substrate when sugarcane bagasse was subjected to fed-batch SSF in the presence of *Kluyveromyces marxianus* and *Pichia stipitis* strains.

In addition, agricultural residues have the potential to be utilized as feedstock for biohydrogen production through various fermentation pathways, including dark fermentation, photofermentation, and mixed culture fermentation (Zheng et al., 2022). Numerous recent studies have investigated different agricultural residues, microbial strains, reactor configurations, and process conditions to enhance H_2 production and efficiency (Zhou et al., 2023; Mojoodi et al., 2024). Moreover, agricultural waste can be digested by anaerobic microorganisms to produce biogas, which is primarily composed of methane (CH_4) and carbon dioxide. Biogas can be used directly as a fuel for heating and electricity generation, upgraded to biomethane for injection into natural gas pipelines, or used as a transportation fuel. In recent years, researchers have investigated various process parameters, substrate combinations, and digester configurations to enhance biogas yields and optimize biogas production systems (Dar et al., 2021; Sillero et al., 2022). In a recent study, anaerobic digestion and hydrodynamic cavitation of rice straw attained methane productivity of 411 mL-CH_4/g-volatile solid (Ezz et al., 2023).

Chapter 12 in this book extensively deals with the agriculture waste valorization for bioenergy production.

2.3.6 Construction or Building Materials

The use of agricultural waste for the production of construction materials is a rapidly growing and promising area with significant environmental and economic benefits. A variety of construction materials, including composite materials, insulation materials, panels, bricks, and blocks, can be produced using agricultural waste through methods such as compaction, molding, extrusion, and curing (Maraveas, 2020). These materials possess distinctive properties and provide effective thermal insulation, sound absorption, and fire resistance while reducing the carbon footprint of construction projects (Thomas et al., 2021). Recently, research efforts have concentrated on fine-tuning formulations to maximize material properties, such as strength, endurance, density, thermal conductivity, and moisture resistance (Sangmesh et al., 2023; Maafa et al., 2023). Furthermore, various recent studies have explored the use of additives, including binders, fibers, and fillers, to enhance material performance and functionality (Sathiparan et al., 2023; Barnabas et al., 2023). For instance, a recent study utilized rice straw and foaming agents such as aluminum sulfate and potassium alum to develop two-core hollow gypsum blocks (Singh et al., 2023a, b). The resulting hollow gypsum blocks exhibited exceptional thermal insulation and acoustic and fire-retardant properties. Another study investigated the bonding of rice husks, prickly pears, and vine pruning residues with a water-based acrylic resin in different proportions to create ceiling tiles with good thermal conductivity and acoustic properties (Maderuelo-Sanz et al., 2022).

Chapter 14 in this book describes in details the valorization of agriculture waste into sustainable building materials.

2.3.7 *Bioactive Metabolites and Enzymes*

Agricultural residues contain considerable amounts of various inherent bio-based molecules and phytochemicals, which can be efficiently extracted through innovative thermal (microwave-assisted, ultrasound-assisted, etc.) and non-thermal (i.e., pressurized liquid, supercritical fluid, deep eutectic solvent) extraction processes. These extraction techniques are more effective than conventional methods in terms of higher extraction efficiency, lower solvent consumption, shorter extraction time, and lower energy cost (Capanoglu et al., 2022). Several recent studies have reported the successful extraction of phytochemicals such as anthocyanins, carotenoids, glycoalkaloids, lutein, pectin, and polyphenols from various types of agricultural residues, with potential applications in functional foods, dietary supplements, pharmaceutical formulations, and cosmetic products (Sorrenti et al., 2023; Jimenez-Champi et al., 2023). Recently, remarkable developments have been made in the production of bioactive metabolites and enzymes through advances in chemical synthesis, biotechnology, and process engineering. Several recent studies have confirmed that solid-state fermentation (SSF) employing different engineered microorganisms can result in a significant yield of bioactive molecules and enzymes (Guerra et al., 2023; Khan et al., 2024).

Chapter 15 in this book describes in details the valorization of agriculture waste into useful products.

2.4 Conclusion

The valorization of agricultural residue presents a multifaceted opportunity to address environmental, economic, and societal challenges. As a result of recent advancements in various conversion processes, agricultural waste is now better capable of being converted into valuable products, such as biofuels, bioplastics, biopolymers, biochar, biofertilizers, metabolites, and nanocomposites. The integration of diverse valorizing strategies in a cascading manner can maximize resource efficiency and minimize waste generation. Building robust supply chains and scaling up these technologies are crucial for their widespread adoption and commercial viability to sustain a circular bioeconomy.

Acknowledgements PY and CY thankfully acknowledge the CSIR-UGC, New Delhi, India, for funding this work in the form of JRF vide NTA reference number 211610030312 and 221610048140 respectively.

References

Ab Rasid, N. S., Shamjuddin, A., Rahman, A. Z. A., & Amin, N. A. S. (2021). Recent advances in green pre-treatment methods of lignocellulosic biomass for enhanced biofuel production. *Journal of Cleaner Production, 321*, 129038.

Agapkin, A. M., Makhotina, I. A., Ibragimova, N. A., Goryunova, O. B., Izembayeva, A. K., & Kalachev, S. L. (2022). The problem of agricultural waste and ways to solve it. *IOP Conference Series: Earth and Environmental Science, 981*, 022009.

Akhtar, T., Hashmi, A. S., Tayyab, M., Anjum, A. A., Saeed, S., & Ali, S. (2020). Bioconversion of agricultural waste to butyric acid through solid state fermentation by *Clostridium tyrobutyricum*. *Waste and Biomass Valorization, 11*, 2067–2073.

Alcazar-Ruiz, A., Ortiz, M. L., Dorado, F., & Sanchez-Silva, L. (2022). Gasification versus fast pyrolysis bio-oil production: A life cycle assessment. *Journal of Cleaner Production, 336*, 130373.

Aldama, D. D., Grassauer, F., Zhu, Y., Ardestani-Jaafari, A., & Pelletier, N. (2023). Allocation methods in life cycle assessments (LCAs) of agri-food co-products and food waste valorization systems: Systematic review and recommendations. *Journal of Cleaner Production, 421*, 138488.

Alloun, W., & Calvio, C. (2024). Bio-driven sustainable extraction and AI-optimized recovery of functional compounds from plant waste: A comprehensive review. *Fermentation, 10*(3), 126.

Andrianou, C., Passadis, K., Malamis, D., Moustakas, K., Mai, S., & Barampouti, E. M. (2023). Upcycled animal feed: Sustainable solution to orange peels waste. *Sustainability, 15*(3), 2033.

Anonymous. (2024). *Global waste management outlook 2024*. https://doi.org/10.59117/20.500.11822/44939

Arora, J., Ramawat, K. G., & Mérillon, J. M. (2023). Disposal of agricultural waste and its effects on the environment, production of useful metabolites and energy: Potential and challenges. In K. Ramawat, J. M. Mérillon, & J. Arora (Eds.), *Agricultural waste: Environmental impact, useful metabolites and energy production* (pp. 3–20). Springer.

Awogbemi, O., & Von Kallon, D. V. (2023). Application of biochar derived from crops residues for biofuel production. *Fuel Communications, 15*, 100088.

Barnabas, A. A., Balogun, O. A., Akinwande, A. A., Ogbodo, J. F., Ademati, A. O., Dongo, E. I., & Romanovski, V. (2023). Reuse of walnut shell waste in the development of fired ceramic bricks. *Environmental Science and Pollution Research, 30*, 11823–11837.

Bhardwaj, K., & Singh, A. K. (2023). Bio-waste and natural resource mediated eco-friendly synthesis of zinc oxide nanoparticles and their photocatalytic application against dyes contaminated water. *Chemical Engineering Journal Advances, 16*, 100536.

Capanoglu, E., Nemli, E., & Tomas-Barberan, F. (2022). Novel approaches in the valorization of agricultural wastes and their applications. *Journal of Agricultural and Food Chemistry, 70*(23), 6787–6804.

Castillo-Henríquez, L., Alfaro-Aguilar, K., Ugalde-Álvarez, J., Vega-Fernández, L., Montes de Oca-Vásquez, G., & Vega-Baudrit, J. R. (2020). Green synthesis of gold and silver nanoparticles from plant extracts and their possible applications as antimicrobial agents in the agricultural area. *Nanomaterials, 10*(9), 1763.

Chatzimitakos, T., Athanasiadis, V., Kotsou, K., Bozinou, E., & Lalas, S. I. (2023). Response surface optimization for the enhancement of the extraction of bioactive compounds from citrus limon peel. *Antioxidants, 12*(8), 1605.

Chaudhary, G., Chaudhary, N., Saini, S., Gupta, Y., Vivekanand, V., & Panghal, A. (2024). Assessment of pretreatment strategies for valorization of lignocellulosic biomass: Path forwarding towards lignocellulosic biorefinery. *Waste and Biomass Valorization, 15*(1), 1–36.

Chilakamarry, C. R., Sakinah, A. M., Zularisam, A. W., Sirohi, R., Khilji, I. A., Ahmad, N., & Pandey, A. (2022). Advances in solid-state fermentation for bioconversion of agricultural wastes to value-added products: Opportunities and challenges. *Bioresource Technology, 343*, 126065.

Cox, R., Narisetty, V., Castro, E., Agrawal, D., Jacob, S., Kumar, G., & Kumar, V. (2023). Fermentative valorisation of xylose-rich hemicellulosic hydrolysates from agricultural waste residues for lactic acid production under non-sterile conditions. *Waste Management, 166*, 336–345.

da Silva, A. S. A., Espinheira, R. P., Teixeira, R. S. S., de Souza, M. F., Ferreira-Leitão, V., & Bon, E. P. (2020). Constraints and advances in high-solids enzymatic hydrolysis of lignocellulosic biomass: A critical review. *Biotechnology for Biofuels, 13*, 1–28.

Danial, A. W., Hamdy, S. M., Alrumman, S. A., Gad El-Rab, S. M., Shoreit, A. A., & Hesham, A. E. L. (2021). Bioplastic production by *Bacillus wiedmannii* AS-02 OK576278 using different agricultural wastes. *Microorganisms, 9*(11), 2395.

Dar, R. A., Parmar, M., Dar, E. A., Sani, R. K., & Phutela, U. G. (2021). Biomethanation of agricultural residues: Potential, limitations and possible solutions. *Renewable and Sustainable Energy Reviews, 135*, 110217.

Dastjerdi, B., Strezov, V., Rajaeifar, M. A., Kumar, R., & Behnia, M. (2021). A systematic review on life cycle assessment of different waste to energy valorization technologies. *Journal of Cleaner Production, 290*, 125747.

del Amo-Mateos, E., Fernández-Delgado, M., Lucas, S., López-Linares, J. C., García-Cubero, M. T., & Coca, M. (2023). Valorization of discarded red beetroot through the recovery of bioactive compounds and the production of pectin by surfactant-assisted microwave extraction. *Journal of Cleaner Production, 389*, 135995.

Dey, N., Kumar, G., Vickram, A. S., Mohan, M., Singhania, R. R., Patel, A. K., & Ponnusamy, V. K. (2022). Nanotechnology-assisted production of value-added biopotent energy-yielding products from lignocellulosic biomass refinery–a review. *Bioresource Technology, 344*, 126171.

Díez-Pascual, A. M. (2019). Synthesis and applications of biopolymer composites. *International Journal of Molecular Sciences, 20*(9), 2321.

Ezz, H., Ibrahim, M. G., Fujii, M., & Nasr, M. (2023). Dual biogas and biochar production from rice straw biomass: A techno-economic and sustainable development approach. *Biomass Conversion and Biorefinery, 13*(12), 10807–10821.

Fernandez, A., Ortiz, L. R., Asensio, D., Rodriguez, R., & Mazza, G. (2020). Kinetic analysis and thermodynamics properties of air/steam gasification of agricultural waste. *Journal of Environmental Chemical Engineering, 8*(4), 103829.

Fontaine, L., Legros, R., & Frayret, J. M. (2024). Solid waste generation prediction model framework using socioeconomic and demographic factors with real-time MSW collection data. *Waste Management and Research*. https://doi.org/10.1177/0734242X2412314

Ge, Y., Ding, S., Kong, X., Kantarelis, E., Engvall, K., & Pettersson, J. B. (2023). Online monitoring of alkali release during co-pyrolysis/gasification of forest and agricultural waste: Element migration and synergistic effects. *Biomass and Bioenergy, 172*, 106745.

Gong, L., Passari, A. K., Yin, C., Kumar Thakur, V., Newbold, J., Clark, W., & Gupta, V. K. (2024). Sustainable utilization of fruit and vegetable waste bioresources for bioplastics production. *Critical Reviews in Biotechnology, 44*(2), 236–254.

Guerra, L., Romanini, D., López, S., Castelli, V., & Clementz, A. (2023). Upcycling of carrot discards into prebiotics (fructooligosaccharides) as high value food ingredients. *Food and Bioproducts Processing, 138*, 172–180.

Hamdan, M. F., Mohd Noor, S. N., Abd-Aziz, N., Pua, T. L., & Tan, B. C. (2022). Green revolution to gene revolution: Technological advances in agriculture to feed the world. *Plants, 11*(10), 1297.

Hamidzadeh, Z., Ghorbannezhad, P., Ketabchi, M. R., & Yeganeh, B. (2023). Biomass-derived biochar and its application in agriculture. *Fuel, 341*, 127701.

Himanshi, & Saini, J. K. (2023). Enhanced cellulosic ethanol production via fed-batch simultaneous saccharification and fermentation of sequential dilute acid-alkali pretreated sugarcane bagasse. *Bioresource Technology, 372*, 128671.

Hong, T. B., Saruchi, S. A., Mustapha, A. A., Seng, J. L. L., Razap, A. N. A. A., Halisno, N., et al. (2023). Intelligent kitchen waste composting system via deep learning and internet-of-things (IoT). *Waste and Biomass Valorization*, 1–14. https://doi.org/10.1007/s12649-023-02341-y

Hu, Y., He, Y., Gao, S., Liao, Z., Lai, T., Zhou, H., & Lu, W. (2020). The effect of a diet based on rice straw co-fermented with probiotics and enzymes versus a fresh corn Stover-based diet on the rumen bacterial community and metabolites of beef cattle. *Scientific Reports, 10*(1), 10721.

Hussain, Z., Babe, M., Saravanan, S., Srimathy, G., Roopa, H., & Boopathi, S. (2023). Optimizing biomass-to-biofuel conversion: IoT and AI integration for enhanced efficiency and sustainability. In *Circular economy implementation for sustainability in the built environment* (pp. 191–214). IGI Global. https://doi.org/10.4018/978-1-6684-8238-4.ch009

Jayakumar, M., Gindaba, G. T., Gebeyehu, K. B., Periyasamy, S., Jabesa, A., Baskar, G., & Pugazhendhi, A. (2023). Bioethanol production from agricultural residues as lignocellulosic biomass feedstock's waste valorization approach: A comprehensive review. *Science of the Total Environment, 879*, 163158.

Jimenez-Champi, D., Romero-Orejon, F. L., Moran-Reyes, A., Muñoz, A. M., & Ramos-Escudero, F. (2023). Bioactive compounds in potato peels, extraction methods, and their applications in the food industry: A review. *CyTA-Journal of Food, 21*(1), 418–432.

Joshi, A., & Arora, J. (2023a). Recent trends and future prospects of nanotechnology for agro-waste valorization into biofuels. In K. Ramawat, J. M. Mérillon, & J. Arora (Eds.), *Agricultural waste: Environmental impact, useful metabolites and energy production* (pp. 559–582). Springer.

Joshi, A., & Arora, J. (2023b). Nanotechnology: Applications in biofuel production. In P. Moodley, R. C. Ray, & E. B. Gueguim Kana (Eds.), *Advances in lignocellulosic biofuel production systems* (pp. 305–328). Woodhead Publishing, Elsevier Inc.

Joshi, A., Kanthaliya, B., & Arora, J. (2020). Halophytes: The non-conventional crops as source of biofuel production. In M. N. Grigore (Ed.), *Handbook of halophytes: From molecules to ecosystems towards biosaline agriculture* (pp. 1–28). Springer.

Joshi, A., Kanthaliya, B., Meena, S., Khan, F., & Arora, J. (2021). Process consolidation approaches for cellulosic ethanol production. In R. C. Ray (Ed.), *Sustainable biofuels* (pp. 43–72). Academic Press, Elsevier BV.

Joshi, A., Verma, K. K., Rajput, D. V., Minkina, T., & Arora, J. (2022). Recent advances in metabolic engineering of microorganisms for advancing lignocellulose-derived biofuels. *Bioengineered, 13*(4), 8135–8163.

Joshi, A., Verma, K. K., Rajput, V. D., & Arora, J. (2023). Metabolic engineering of microorganisms in advancing biofuel production. In P. Moodley, R. C. Ray, & E. B. Gueguim Kana (Eds.), *Advances in lignocellulosic biofuel production systems* (pp. 181–202). Woodhead Publishing. Elsevier Inc.

Kanthasamy, R., Almatrafi, E., Ali, I., Sait, H. H., Zwawi, M., Abnisa, F., & Ayodele, B. V. (2023). Biochar production from valorization of agricultural wastes: Data-driven modelling using machine learning algorithms. *Fuel, 351*, 128948.

Karne, H. U., Gaydhane, P., Gohokar, V., Deshpande, K., Dunung, P., & Bendkule, G. (2023). Synthesis of biodegradable material from banana peel. *Materials Today: Proceedings*. In press. https://doi.org/10.1016/j.matpr.2023.05.157

Khan, F., Joshi, A., Devkota, H. P., Subramaniyan, V., Kumarasamy, V., & Arora, J. (2023). Dietary glucosinolates derived isothiocyanates: Chemical properties, metabolism and their potential in the prevention of Alzheimer's disease. *Frontiers in Pharmacology, 14*, 1214881.

Khan, P. A., Mohammad, A., Bansal, S. L., Lal, B., Singh, P., Singh, R., & O'Donovan, A. (2024). Biotransformation of raw mango seed waste into bacterial hydrolytic enzymes enhancement via solid state fermentation strategy. *Molecular Biotechnology*, 1–9.

Khanal, S. K., Tarafdar, A., & You, S. (2023). Artificial intelligence and machine learning for smart bioprocesses. *Bioresource Technology, 375*, 128826.

Khedulkar, A. P., Pandit, B., & Doong, R. A. (2023). Agricultural waste to real worth biochar as a sustainable material for supercapacitor. *Science of the Total Environment, 869*, 161441.

Kumar, S., Tripathi, G., & Mishra, G. V. (2021). A comparative study on agro-waste conversion into biofertilizer employing two earthworm species. *Applied Ecology and Environmental Sciences, 9*(2), 280–285.

Kumar, N., Yadav, A., Singh, G., Singh, A., Kumar, P., & Aggarwal, N. K. (2023). Comparative study of ethanol production from sodium hydroxide pretreated rice straw residue using *Saccharomyces cerevisiae* and *Zymomonas mobilis*. *Archives of Microbiology, 205*(4), 146.

Kumari, A., Singh, R. K., Kumar, N., Kumari, R., & Sharma, S. (2023). Green synthesis and physical properties of crystalline silica engineering nanomaterial from rice husk (agriculture waste) at different annealing temperatures for its varied applications. *Journal of the Indian Chemical Society, 100*(5), 100982.

Laksana, C., Sophiphun, O., & Chanprame, S. (2024). Lignin reduction in sugarcane by performing CRISPR/Cas9 site-direct mutation of SoLIM transcription factor. *Plant Science, 340*, 111987.

Liandi, A. R., Ahyar, H., Maladi, I., Azhari, A., & Cornelia, M. (2023). The influence of the filler addition of rice husk cellulose, polyvinyl alcohol, and zinc oxide on the characteristics of environmentally friendly cassava biodegradable plastic. *Case Studies in Chemical and Environmental Engineering, 8*, 100520.

Loh, N. Y. L., Pang, H. Y., Tee, W. T., Hiew, B. Y. Z., Hanson, S., Chong, S., & Lee, L. Y. (2023). Synthesis of renewable and cost-effective bioplastic from apple waste: Physicochemical and biodegradability studies. *Waste and Biomass Valorization, 14*(10), 3235–3252.

López-Alcántara, E. M., Colindres-Vásquez, G. M., Fodil, N., Sánchez-Barahona, M., Rivera-Flores, O., Romero, A., & Abdullah, J. A. A. (2024). Agro-waste sweet pepper extract-magnetic iron oxide nanoparticles for antioxidant enrichment and sustainable nanopackaging. *Polymers, 16*(4), 564.

Maafa, I. M., Abutaleb, A., Zouli, N., Zeyad, A. M., Yousef, A., & Ahmed, M. M. (2023). Effect of agricultural biomass wastes on thermal insulation and self-cleaning of fired bricks. *Journal of Materials Research and Technology, 24*, 4060–4073.

Maderuelo-Sanz, R., García-Cobos, F. J., Sánchez-Delgado, F. J., Mota-López, M. I., Meneses-Rodríguez, J. M., Romero-Casado, A., & López-Ramos, L. (2022). Mechanical, thermal and acoustical evaluation of biocomposites made of agricultural waste for ceiling tiles. *Applied Acoustics, 191*, 108689.

Magnus, I., Virte, M., Thienpont, H., & Smeesters, L. (2021). Combining optical spectroscopy and machine learning to improve food classification. *Food Control, 130*, 108342.

Maraveas, C. (2020). Production of sustainable construction materials using agro-wastes. *Materials, 13*(2), 262.

Meena, S., Kanthaliya, B., Joshi, A., Khan, F., & Arora, J. (2020). Biologiafutura: Medicinal plants-derived bioactive peptides in functional perspective—a review. *Biologia Futura, 71*, 195–208.

Mengqi, Z., Shi, A., Ajmal, M., Ye, L., & Awais, M. (2021). Comprehensive review on agricultural waste utilization and high-temperature fermentation and composting. *Biomass Conversion and Biorefinery*, 1–24.

Mojoodi, S., Zilouei, H., & Karimi, K. (2024). Improvement of biohydrogen production from rice straw hydrolysate by green-self-prepared nano-silica. *Fuel, 363*, 130887.

Nanda, S., Patra, B. R., Patel, R., Bakos, J., & Dalai, A. K. (2022). Innovations in applications and prospects of bioplastics and biopolymers: A review. *Environmental Chemistry Letters, 20*(1), 379–395.

Nath, P. C., Ojha, A., Debnath, S., Sharma, M., Sridhar, K., Nayak, P. K., & Inbaraj, B. S. (2023). Biogeneration of valuable nanomaterials from agro-wastes: A comprehensive review. *Agronomy, 13*(2), 561.

Naz, S., & Uroos, M. (2023). Synthesis of 5-ethoxymethylfurfural (5-EMF) from sugars and lignocellulosic crop residue using ionic liquids differing in their cationic structure. *Waste and Biomass Valorization, 14*(7), 2225–2236.

Ngungeni, Y., Aboyewa, J. A., Moabelo, K. L., Sibuyi, N. R., Meyer, S., Onani, M. O., & Madiehe, A. M. (2023). Anticancer, antioxidant, and catalytic activities of green synthesized gold nanoparticles using avocado seed aqueous extract. *ACS Omega, 8*(29), 26088–26101.

Nunta, R., Techapun, C., Sommanee, S., Mahakuntha, C., Porninta, K., Punyodom, W., & Leksawasdi, N. (2023). Valorization of rice straw, sugarcane bagasse and sweet sorghum bagasse for the production of bioethanol and phenylacetylcarbinol. *Scientific Reports, 13*(1), 727.

Oliveira, M., Ramos, A., Monteiro, E., & Rouboa, A. (2021). Modeling and simulation of a fixed bed gasification process for thermal treatment of municipal solid waste and agricultural residues. *Energy Reports, 7*, 256–269.

Patil, R. H., Patil, M. P., & Maheshwari, V. L. (2020). Microbial transformation of crop residues into a nutritionally enriched substrate and its potential application in livestock feed. *SN Applied Sciences, 2*(6), 1140.

Paz-Arteaga, S. L., Ascacio-Valdés, J. A., Aguilar, C. N., Cadena-Chamorro, E., Serna-Cock, L., Aguilar-González, M. A., & Torres-León, C. (2023). Bioprocessing of pineapple waste for sustainable production of bioactive compounds using solid-state fermentation. *Innovative Food Science and Emerging Technologies, 85*, 103313.

Pinheiro, U. V., Wancura, J. H., Brondani, M., da Silva, C. M., Mainardi, M. A., Gai, R. M., & Jahn, S. L. (2024). Production of gibberellic acid by solid-state fermentation using wastes from rice processing and brewing industry. *Applied Biochemistry and Biotechnology, 196*(3), 1493–1508.

Pourbayramian, R., Abdi-Benemar, H., Seifdavati, J., Greiner, R., Elghandour, M. M. M. Y., & Salem, A. Z. M. (2021). Bioconversion of potato waste by rumen fluid from slaughterhouses to produce a potential feed additive rich in volatile fatty acids for farm animals. *Journal of Cleaner Production, 280*, 124411.

Preethi, B., Karmegam, N., Manikandan, S., Vickram, S., Subbaiya, R., Rajeshkumar, S., & Govarthanan, M. (2024). Nanotechnology-powered innovations for agricultural and food waste valorization: A critical appraisal in the context of circular economy implementation in developing nations. *Process Safety and Environmental Protection, 184*, 477–491.

Puglia, D., Pezzolla, D., Gigliotti, G., Torre, L., Bartucca, M. L., & Del Buono, D. (2021). The opportunity of valorizing agricultural waste, through its conversion into biostimulants, biofertilizers, and biopolymers. *Sustainability, 13*(5), 2710.

Rahmati, S., Doherty, W., Dubal, D., Atanda, L., Moghaddam, L., Sonar, P., & Ostrikov, K. K. (2020). Pretreatment and fermentation of lignocellulosic biomass: Reaction mechanisms and process engineering. *Reaction Chemistry and Engineering, 5*(11), 2017–2047.

Rajesh, Y., Gautam, N., Saloni, P., Deore, V., Shivde, P., & Dabhade, G. (2024). Agricultural resources in focus: Eco-friendly bioplastic synthesis from corn starch. *Materials Today: Proceedings*. In press. https://doi.org/10.1016/j.matpr.2024.01.025

Roy, P., Mohanty, A. K., Dick, P., & Misra, M. (2023). A review on the challenges and choices for food waste valorization: Environmental and economic impacts. *ACS Environmental Au, 3*(2), 58–75.

Saini, S., Kuhad, R. C., & Sharma, K. K. (2023). Valorization of rice straw biomass for co-production of bioethanol, biopesticide and biofertilizer following an eco-friendly biorefinery process. *Process Safety and Environmental Protection, 173*, 823–836.

Sangmesh, B., Patil, N., Jaiswal, K. K., Gowrishankar, T. P., Selvakumar, K. K., Jyothi, M. S., & Kumar, S. (2023). Development of sustainable alternative materials for the construction of green buildings using agricultural residues: A review. *Construction and Building Materials, 368*, 130457.

Santos, A. G., de Albuquerque, T. L., Ribeiro, B. D., & Coelho, M. A. Z. (2021). In situ product recovery techniques aiming to obtain biotechnological products. A glance to current knowledge. *Biotechnology and Applied Biochemistry, 68*(5), 1044–1057.

Sarkhel, R., Ganguly, P., Das, P., Bhowal, A., & Sengupta, S. (2023). Synthesis of biodegradable PVA/cellulose polymer composites and their application in dye removal. *Environmental Quality Management, 32*(3), 313–323.

Sathiparan, N., Anburuvel, A., & Selvam, V. V. (2023). Utilization of agro-waste groundnut shell and its derivatives in sustainable construction and building materials–A review. *Journal of Building Engineering, 66*, 105866.

Sekar, S., & Lee, S. (2021). Derivation of luminescent mesoporous silicon nanocrystals from biomass rice husks by facile magnesiothermic reduction. *Nanomaterials, 11*(3), 613.

September, L. A., Kheswa, N., Seroka, N. S., & Khotseng, L. (2023). Green synthesis of silica and silicon from agricultural residue sugarcane bagasse ash–a mini-review. *RSC Advances, 13*(2), 1370–1380.

Setiawan, A., Mahfud, R. N., Mayangsari, N. E., Widiana, D. R., Iswara, A. P., & Dermawan, D. (2024). The potential of using sweet corn (*Zea mays* Saccharata) husk waste as a source for biodegradable plastics. *Industrial Crops and Products, 208*, 117760.

Shafiei, R., Hooper, M., McClellan, C., Oakey, H., Stephens, J., Lapierre, C., & Halpin, C. (2023). Downregulation of barley ferulate 5-hydroxylase dramatically alters straw lignin structure without impact on mechanical properties. *Frontiers in Plant Science, 13*, 1125003.

Shahapurkar, A., Joshi, S. M., Divyashree, T., Sasaki, K., & Jogaiah, S. (2024). Bioactive carbohydrate from potato peels: A sweet inducer that mediate resistance against powdery mildew disease of cucumber through phytohormonal alterations. *Biocatalysis and Agricultural Biotechnology, 103102*.

Shahbaz, A., Ayaz, M., Bin Khalid, U., & Liaqat, L. (2023). Sustainable synthesis of silica nanoparticles from agricultural waste and its utilization in modern technology: A review. *Energy Sources, Part A: Recovery, Utilization, and Environmental Effects, 45*(1), 1464–1484.

Shahi, N., Wang, P., Adhikari, S., Min, B., & Rangari, V. K. (2021). Biopolymers fractionation and synthesis of nanocellulose/silica nanoparticles from agricultural byproducts. *ACS Sustainable Chemistry and Engineering, 9*(18), 6284–6295.

Shalma, S., Shabbirahmed, A. M., Haldar, D., Patel, A. K., & Singhania, R. R. (2023). Influence of reaction conditions on synthesis and applications of lignin nanoparticles derived from agricultural wastes. *Environmental Technology and Innovation, 103163*.

Sharma, V., Joshi, A., Ramawat, K. G., & Arora, J. (2017). Bioethanol production from halophytes of Thar Desert: A "green gold". In *Environment at crossroads: Challenges, dynamics and solutions* (pp. 219–235). Haghshenass Publishing.

Sillero, L., Solera, R., & Perez, M. (2022). Biochemical assays of potential methane to test biogas production from dark fermentation of sewage sludge and agricultural residues. *International Journal of Hydrogen Energy, 47*(27), 13289–13299.

Singh, P., Pandey, V. K., Sultan, Z., Singh, R., & Dar, A. H. (2023a). Classification, benefits, and applications of various anti-nutritional factors present in edible crops. *Journal of Agriculture and Food Research, 14*, 100902.

Singh, S., Dalbehera, M. M., Kumar, A. A., Maiti, S., Balam, N. B., Bisht, R. S., & Panigrahi, S. K. (2023b). Elevated temperature and performance behaviour of rice straw as waste bio-mass based foamed gypsum hollow blocks. *Journal of Building Engineering, 69*, 106220.

Snoussi, Y., Sifaoui, I., El Garah, M., Khalil, A. M., Piñero, J. E., Jouini, M., & Chehimi, M. M. (2023). Green, zero-waste pathway to fabricate supported nanocatalysts and anti-kinetoplastid agents from sugarcane bagasse. *Waste Management, 155*, 179–191.

Sorrenti, V., Burò, I., Consoli, V., & Vanella, L. (2023). Recent advances in health benefits of bioactive compounds from food wastes and by-products: Biochemical aspects. *International Journal of Molecular Sciences, 24*(3), 2019.

Tao, J., Gu, Y., Hao, X., Liang, R., Wang, B., Cheng, Z., & Chen, G. (2023). Combination of hyperspectral imaging and machine learning models for fast characterization and classification of municipal solid waste. *Resources, Conservation and Recycling, 188*, 106731.

Thangadurai, T. D., Manjubaashini, N., Sowndarya, A., Subitha, A., Kausalya, G., Shanmugaraju, S., & Kalarikkal, N. (2023). Multipurpose biological applications of excitation-dependent fluorescent carbon nano dots emanated from biomass waste. *Materials Chemistry and Physics, 307*, 128113.

Thanigaivel, S., Priya, A. K., Dutta, K., Rajendran, S., Sekar, K., Jalil, A. A., & Soto-Moscoso, M. (2022). Role of nanotechnology for the conversion of lignocellulosic biomass into biopotent energy: A biorefinery approach for waste to value-added products. *Fuel, 322*, 124236.

Thomas, B. S., Yang, J., Mo, K. H., Abdalla, J. A., Hawileh, R. A., & Ariyachandra, E. (2021). Biomass ashes from agricultural wastes as supplementary cementitious materials or aggregate replacement in cement/geopolymer concrete: A comprehensive review. *Journal of Building Engineering, 40*, 102332.

Torres-Arellano, S., Torres-Martinez, L. M., Luévano-Hipólito, E., Aleman-Ramirez, J. L., & Sebastian, P. J. (2023). Biologically mediated synthesis of CuO nanoparticles using corn COB (*Zea mays*) ash for photocatalytic hydrogen production. *Materials Chemistry and Physics, 301*, 127640.

Tropea, A., Ferracane, A., Albergamo, A., Potortì, A. G., Lo Turco, V., & Di Bella, G. (2022). Single cell protein production through multi food-waste substrate fermentation. *Fermentation, 8*(3), 91.

Vurukonda, S. S. K. P., Fotopoulos, V., & Saeid, A. (2024). Production of a rich fertilizer base for plants from waste organic residues by microbial formulation technology. *Microorganisms, 12*(3), 541.

Wan, Z., Hu, H., Liu, K., Qiao, Y., Guo, F., Wang, C., & Jiang, M. (2024). Engineering industrial yeast for improved tolerance and robustness. *Critical Reviews in Biotechnology*, 1–17. https://doi.org/10.1080/07388551.2024.2326677

Wang, S., & Cheng, J. (2023). Co-generation and GHG emission from agricultural waste based on pyrolysis/gasification: Experimental and LCA approaches. *Energy Science and Engineering, 12*, 860–872.

Wang, W., Yang, S., Zhang, A., & Yang, Z. (2021). Synthesis of a slow-release fertilizer composite derived from waste straw that improves water retention and agricultural yield. *Science of the Total Environment, 768*, 144978.

Wijitkosum, S. (2022). Biochar derived from agricultural wastes and wood residues for sustainable agricultural and environmental applications. *International Soil and Water Conservation Research, 10*(2), 335–341.

Wolny-Koładka, K., Malina, D., Suder, A., Pluta, K., & Wzorek, Z. (2022). Bio-based synthesis of silver nanoparticles from waste agricultural biomass and its antimicrobial activity. *PRO, 10*(2), 389.

Woo, W. X., Tan, J. P., Wu, T. Y., Yeap, S. K., Indera Luthfi, A. A., Abdul Manaf, S. F., & Hui, Y. W. (2024). An overview on the factors affecting enzymatic saccharification of lignocellulosic biomass into fermentable sugars. *Reviews in Chemical Engineering, 40*(2), 279–303.

Xu, X., Zhang, W., You, C., Fan, C., Ji, W., Park, J. T., & Ma, Y. (2023). Biosynthesis of artificial starch and microbial protein from agricultural residue. *Science Bulletin, 68*(2), 214–223.

Yafetto, L., Odamtten, G. T., & Wiafe-Kwagyan, M. (2023). Valorization of agro-industrial wastes into animal feed through microbial fermentation: A review of the global and Ghanaian case. *Heliyon, 9*, e14814.

Zhao, F., Zheng, M., & Xu, X. (2023). Microbial conversion of agro-processing waste (peanut meal) to rhamnolipid by Pseudomonas aeruginosa: Solid-state fermentation, water extraction, medium optimization and potential applications. *Bioresource Technology, 369*, 128426.

Zheng, Y., Zhang, Q., Zhang, Z., Jing, Y., Hu, J., He, C., & Lu, C. (2022). A review on biological recycling in agricultural waste-based biohydrogen production: Recent developments. *Bioresource Technology, 347*, 126595.

Zhou, Y., Qin, S., Verma, S., Sar, T., Sarsaiya, S., Ravindran, B., & Awasthi, M. K. (2021). Production and beneficial impact of biochar for environmental application: A comprehensive review. *Bioresource Technology, 337*, 125451.

Zhou, Y., Wang, X., Huang, X., Deng, H., Hu, Y., & Lu, L. (2023). Tuning the isoelectric point of zinc molybdate nanomaterials to enhance the biohydrogen production of rice straws. *Chemical Engineering Journal, 452*, 139458.

Part II
Routes for Value Addition to Agricultural Waste

Chapter 3
Biochemical and Thermochemical Conversion Technologies for Agriculture Waste Transformation

Carlos J. Durán-Valle and Ignacio M. López-Coca

Abbreviations

AD	Anaerobic digestion
GHG	Greenhouse gas
IUPAC	International Union of Pure and Applied Chemistry
MACL	Moderated acid-catalyzed liquefaction
SCCD	Supercritical carbon dioxide
SCF	Supercritical fluids
SCW	Supercritical water

3.1 Introduction

The need to curtail carbon dioxide emissions, precipitated by the greenhouse effect primarily linked to fossil fuel consumption, has sparked a fervent quest for alternative energy sources and valuable compounds (Santos et al., 2020; Gao et al., 2022; Anu et al., 2023; Bala et al., 2023). Among the foremost contenders is biomass, offering a promising avenue for immediate action. Despite utilizing certain biomass reservoirs such as food, plant tissues, and medicinal flora, a considerable portion of commonly harvested vegetation remains untapped mainly, relegated to waste status. Yet, the prospect of deriving significant yields from these agricultural remnants persists. This potentiality has been unlocked through diligent research aimed at

C. J. Durán-Valle (✉)
IACYS, Faculty of Sciences, Universidad de Extremadura, Badajoz, Spain
e-mail: carlosdv@unex.es

I. M. López-Coca
INTERRA, School of Technology, Universidad de Extremadura, Cáceres, Spain

J. Arora et al. (eds.), *Transforming Agriculture Residues for Sustainable Development*, Waste as a Resource, https://doi.org/10.1007/978-3-031-61133-9_3

valorizing biomass residues, thereby extracting fuels and other valuable commodities from what was once deemed expendable. In doing so, potential pollutants are mitigated, and the prospect of economic returns emerges as a tangible incentive (Mishra et al., 2021; Sharma et al., 2023).

The global agricultural sector grapples with the substantial task of managing and disposing of extensive organic waste generated throughout cultivation and production processes. If left unaddressed, this waste presents significant environmental and economic hazards, encompassing soil and water contamination, greenhouse gas (GHG) emissions, and resource squandering. Nonetheless, these challenges are increasingly viewed as opportunities. Typically, agricultural waste exhibits consistent and well-defined properties, with its annual production coinciding with scheduled harvests. This waste can be efficiently gathered in concentrated forms owing to its human-derived origin, thereby reducing collection costs and potentially yielding additional economic benefits. Consequently, the exploration of agricultural waste utilization holds promise due to its dual impact on both environmental and economic fronts (Sharma et al., 2023; Yadav et al., 2023).

The chapter unfolds in two distinct sections. The initial segment delves into biochemical methodologies using microorganisms or specific natural compounds like enzymes. Conversely, the latter part elucidates thermochemical approaches predicated upon the application of heat to disintegrate biomass to acquire valuable outputs. Despite their disparate nature, these methods, albeit seldom employed in tandem, harbor the potential to yield akin outcomes, such as the production of biofuels or the facilitation of transesterification. These innovative technologies stand at the forefront of contemporary research endeavors, propelled by societal aspirations to curtail reliance on fossil fuels and foster the extraction of resources from renewable reservoirs.

3.2 Biochemical Technologies

Biochemical technologies offer innovative and promising solutions for harnessing the latent energy potential of biomass and agricultural waste. These technologies leverage the natural processes of microorganisms and enzymes to convert organic matter into valuable energy forms, such as biofuels, biogas, and electricity. By transforming waste materials into energy, we address waste management challenges and contribute to reducing our reliance on fossil fuels, mitigating GHG emissions, and promoting a circular economy (Gao et al., 2022).

The drawbacks found in the use of first-generation biofuels, that is, those derived from edible crops, have led to the study of the possibilities of obtaining bioenergy through the so-called second generation of biofuels, that is, those derived from biomass and agriculture wastes, formed mainly by lignocellulosic compounds (Nair et al., 2022). Lignocellulosic waste is often open-air incinerated, generating atmospheric pollution and a loss of value of carbon-rich material (Oanh et al., 2018). Moreover, the incorporation of lignocellulosic residues into the soil decreases the amount of nitrogen available to crops since the degradation of this waste by soil bacteria consumes more nitrogen than the waste itself delivers (Ocio et al., 1991). The transformation of this agricultural waste through biochemical technologies is more nascent than thermal technologies; however, an analysis of the published results shows more significant potential for lowering costs using these biochemical technologies instead of thermochemical technologies (Sims et al., 2010).

This exploration searches the diverse biochemical technologies available for converting biomass and agricultural waste into energy, highlighting their principles, applications, and the substantial environmental and economic benefits they bring (Chandra et al., 2012). It underscores the significance of these technologies in ushering in a greener and more sustainable era of energy production while mitigating the environmental impact of organic waste disposal (Lee et al., 2019).

3.2.1 Anaerobic Digestion

Anaerobic digestion (AD) has emerged as a robust and sustainable solution for transforming and valorizing agricultural waste. This process harnesses the natural power of methanogenic microorganisms to convert organic matter, in the absence of oxygen, into valuable biogas and nutrient-rich byproducts, offering multiple benefits to both agriculture and the environment (Weiland, 2010; Kainthola et al., 2019; Lee et al., 2019). These microorganisms break down complex organic materials, such as crop residues, animal manure, and food waste. The hallmark of this process is the production of biogas, a mixture of methane (50–75%), carbon dioxide (25–50%), and traces of other gases, such as hydrogen sulfide, which can be exploited as a renewable energy source. Furthermore, the residual materials, or digestate, are rich in nutrients and can be used as a valuable fertilizer for crop production. As a circular and sustainable approach, anaerobic digestion mitigates the environmental impact of agricultural waste and offers a source of renewable energy and valuable resources.

The significance of AD as a technology for agricultural waste management and valorization cannot be overstated. Agriculture remains one of the most significant contributors to global GHG emissions, primarily due to the release of methane during the decomposition of organic matter in landfills and lagoons. AD represents a potent tool for mitigating these emissions by capturing methane for productive use as an energy source, thereby reducing its release into the atmosphere.

One of the critical benefits of AD is its versatility in handling a wide range of agricultural waste streams. Many crop residues, including corn stalks, wheat straw, and sugarcane bagasse, are ideal substrates for anaerobic digestion due to their high organic content (Demirbas, 2010; Chandra et al., 2012; Bianco et al., 2021; Ketsub et al., 2021). Livestock operations can also benefit significantly from this technology (Chen et al., 2023). Furthermore, the food industry can utilize AD to manage food waste and byproducts, turning them into a valuable resource rather than contributing to landfill waste (Xu et al., 2018; Saravanan et al., 2023). This technology reduces odors associated with agricultural waste, benefiting both the environment and neighboring communities. The process also destroys harmful manure and food waste pathogens, enhancing public health and safety. Furthermore, the digestate produced as a byproduct of AD is an environmentally friendly alternative to chemical fertilizers, enriching the soil with organic matter and nutrients. This closed-loop nutrient cycling reduces the environmental impact of farming while improving soil quality and crop yields (Lee et al., 2019).

The utilization of AD is not a new concept; it has been practised for centuries as traditional, small-scale biogas digesters in various parts of the world. However, recent technological advances and a growing awareness of the need for sustainable waste management solutions have propelled anaerobic digestion into the mainstream (Lee et al., 2019). Modern anaerobic digestion facilities are equipped with state-of-the-art technology to optimize the process, maximize biogas production, and ensure environmental sustainability. These facilities vary in size, from small farm-based digesters to large industrial-scale installations, catering to diverse agricultural and waste management needs.

The adoption of AD is not without its challenges (Yadav et al., 2023). The initial investment required for setting up anaerobic digestion facilities can be substantial, and securing financing can be a barrier, especially for small-scale farmers. Moreover, the efficient operation of anaerobic digesters requires technical expertise and maintenance, which can pose challenges for some agricultural producers. The composition and quality of feedstock materials also play a crucial role in determining the success of anaerobic digestion, and variations in these factors can affect biogas yields and process stability. Future anaerobic digestion development should fill the gap between pilot-scale and profitable commercial-scale production. Upgradation focuses on optimizing production processes, including pretreatment and purifying the final product (biogas) (Mishra et al., 2021).

Pretreatment of Feedstock

The complex chemical structure of lignocellulosic waste requires pretreatment that facilitates the bacteria in charge of AD to carry out their work. This pretreatment, therefore, aims to expose cellulose and hemicellulose for subsequent anaerobic bacterial transformation and can be classified as physical, chemical, biological, or a combination of them (Baruah et al., 2018; Sims et al., 2010).

Physical pretreatment Several different physical pretreatment types can afford a larger accessible surface area and disrupt the crystalline, intricate structure of the biomass waste, hence facilitating the digestion of the feedstock. These treatments can be mechanical, such as milling or grinding, or use ultrasound or microwave irradiation; thermal treatment, such as heating at 60–200 °C, is another option (Khan et al., 2022; Ma et al., 2021; Naik et al., 2022). As a result of these pretreatment technologies, the polymeric crystalline structure of the lignocellulosic material is interrupted, which makes it easier for the bacteria to digest the feedstock (Abraham et al., 2020; Mishra et al., 2021; Sharma et al., 2023). To assess the performance of the different pretreatments, we must not forget that in addition to the improvement in the performance of biogas production, it is crucial to perform an accurate calculation of the costs associated with each type of method, such as energy fee or intrinsic cost of the equipment used (Baruah et al., 2018).

Chemical pretreatment The use of chemical compounds to pretreat lignocellulose residues has also been extensively studied; the goal is to degrade the feedstock into cellulose, hemicellulose, and lignin or even into the smaller molecules of which they are made and thus turn the biomass residue in a more easily digestible by anaerobic microorganisms (Mishra et al., 2021; Mussoline et al., 2013; Sharma et al., 2023).

Pretreatments with acids or bases have received considerable attention. When bases such as sodium or potassium hydroxide are applied, the lignin content is reduced, and a significant increase in methane production yield is achieved (Dai et al., 2018). For its part, the treatment with acids like sulfuric acid does not affect lignin similarly. Still, it can cause the hydrolysis of biopolymers, forming monosaccharides, which also increases the amount of biogas obtained (Sarto et al., 2019; Syaichurrozi et al., 2019; Xu et al., 2019; Zhou et al., 2014). Finally, promising results have been achieved when oxidizing agents, such as hydrogen peroxide, have been applied. Despite producing good results at an affordable cost, these chemical pretreatments suffer from a lack of environmental friendliness since they involve highly aggressive chemicals. That is why research has begun on using other chemical substances as pretreatment agents. These are, fundamentally, ionic liquids (Nargotra et al., 2019; Sharma et al., 2021a, b; Sharma et al., 2023) and deep eutectic solvents (Nargotra et al., 2022; Sharma et al., 2021a, b). Within each of these categories of neoteric solvents, there is the possibility of finding bio-friendly, low-toxicity compounds that can be considered green. These compounds facilitate the dissolution of cellulose and hemicellulose, thus facilitating access to enzymatic hydrolysis that degrades the polymers and improves the conditions for anaerobic digestion to produce biogas.

Biological pretreatment Biological pretreatment involves using microorganisms or enzymes before microbial AD. The goal is the enzymatic degradation of the cellulose and hemicellulose complex chains and cleavage of the aromatic rings in lignin within the biomass waste. This procedure facilitates the action of the microbes responsible for AD. As mentioned, microorganisms, one or more species or strains at a time, or enzymes can be used (Akyol et al., 2019; Liew et al., 2020; Zhang et al., 2020). This methodology is deemed greener and more biocompatible. Compared to no pretreatment, a 2.3-fold increase in biomethane yield was reported when pretreatment with a fungal mash rich in hydrolytic enzymes was used (Uçkun Kiran et al., 2015).

It is important to note that no pretreatment method guarantees satisfactory results in all cases. It has been proven that better results are obtained using physical, chemical, or biological technologies in certain instances. These hybrid methodologies have provided promising results, but further research is necessary (Bianco et al., 2021; Zhao et al., 2009). Finally, it must be considered that the costs associated with each pretreatment methodology influence its viability and sustainability (Mishra et al., 2021). However, an unequivocally cost-effective and environmentally friendly pretreatment method for lignocellulosic biomass has not been established yet (Baruah et al., 2018).

Production of Bioenergy

AD of pretreated biomass waste occurs through the participation of different microorganisms in four stages: hydrolysis, acidogenesis, acetogenesis, and methanogenesis (Kainthola et al., 2019). Hydrolysis is the rate-limiting step of the entire biogas production process through AD when treating lignocellulosic waste. Lipids, carbohydrates, or proteins decompose during hydrolysis by hydrolytic bacteria into constituent units, such as fatty acids, oligosaccharides, and amino acids. In cases where the feedstock is less recalcitrant to hydrolysis than the lignocellulosic biomass, the methanogenic microorganisms typically determine the process rate since they have the lowest growth rate among the participating bacteria (Kainthola et al., 2019).

Apart from the form of pretreatment, the performance obtained from AD depends on several factors. Not only does the composition and quality of the feedstock play a crucial role in the result, but also the nature of the microorganisms and operational variables, such as temperature, pH, feedstock loading speed, C/N ratio, or retention time, influence the performance of the biogas production process (Chandra et al., 2012; Sarker et al., 2019; Lee et al., 2019; Mishra et al., 2021; Sharma et al., 2023).

The feedstock for biogas production through AD can come from many sources, such as agricultural waste, municipal solid waste, industrial waste, and wastewater. Biomass waste has a high production potential due to its high content of polysaccharides, such as cellulose and hemicellulose, that biodegrade in days or weeks. However, lignin, which is difficult to degrade, makes the engineering of the process difficult (Sarker et al., 2019). Biogas yield, measured in cubic meters per kilogram, depending on the origin of the feedstock, has been reported to range from 0.15 to 0.39 with wheat straw, 0.24–0.27 for rice straw, and 0.27–0.31 using sugarcane waste (Mittal et al., 2019).

The anaerobic microorganisms fail to achieve satisfactory feedstock digestion and biogas production if the pH is not kept within an optimum operational range. By controlling other parameters, such as the organic loading rate, the pH can be held at 6.0–8.5 units; otherwise, it becomes toxic for the methanogenic bacteria. Temperature and sudden temperature changes also affect the digesting microbes. Methanogens are most active at 30–35 °C or 50–60 °C; they are inhibited between 40 °C and 50 °C. Most methanogens work best in the mesophilic range (Chandra et al., 2012; Gerardi, 2003).

Retention time is a critical parameter in biogas production because it directly influences the efficiency and effectiveness of the anaerobic digestion process. Adequate retention time ensures that microorganisms have sufficient time to carry out these biochemical reactions. A longer retention time contributes to higher methane content in the biogas by allowing the complete digestion of complex organic materials. This ensures that a wide range of substrates is utilized and converted into biogas. Also, sufficient retention time helps stabilize the digestate, reducing the potential for pathogens, although if the digestate is incinerated, the retention time may be shorter. Balancing the retention time is crucial for optimizing the biogas production rate (Gerardi, 2003; Yadvika et al., 2004).

The amount of volatile solids present should be around 10% for optimal gas yield through methanation, while a C/N ratio of 20–30 is considered the best for anaerobic digestion. If C/N is very high, the microorganisms will quickly consume the nitrogen they need, and there will be excess carbon. On the contrary, if there is a lot of nitrogen, it will be released as an ammonium ion, which will cause toxification due to an increase in the pH of the medium. For example, human excrement has a C/N ratio of 8, cereal straw can be up to 90, and cattle manure has an optimal C/N ratio of 24. An adequate C/N value can be achieved by co-digestion, that is, combining various organic feedstocks, hence maximizing biogas production (Chandra et al., 2012). By blending different organic materials, co-digestion maximizes biogas production and improves feedstock balancing and recycling.

3.2.2 *Composting*

Composting is a well-established biochemical technology that offers an environmentally responsible and economically viable solution to transform and valorize agricultural waste. By harnessing the natural decomposition processes, composting offers a sustainable solution to reduce waste, enhance soil fertility, and promote environmentally responsible farming practices (Harindintwali et al., 2020).

Composting is a natural biological process that mimics the decomposition of organic matter that occurs in nature. It relies on the activity of microorganisms, primarily bacteria, and fungi, which break down complex organic materials into simpler compounds. Adequate oxygen supply is essential for the aerobic decomposition process in composting; proper aeration and moisture levels of around 50–60% ensure the growth and activity of aerobic microorganisms. As with AD, achieving the right balance between carbon-rich (e.g., straw, leaves) and N-rich (e.g., manure, food scraps) materials is crucial. A C:N ratio of approximately 25–30:1 is considered ideal. Composting generates heat as a byproduct of microbial activity; proper temperature management, typically within 45–70 °C, optimizes decomposition (Harindintwali et al., 2020).

Aerobic composting is a valuable option for the treatment of lignocellulosic crop residues since it solves the environmental problems attributed to the treatment of lignocellulosic crop residues, such as open biomass incineration (Oanh et al., 2018); it also generates rich fertilizers that avoid the use of chemical fertilizers (Hubbe, 2014). The main shortcoming of conventional composting of lignocellulosic waste is its slowness, which hinders its application by farmers needing a faster process; this issue has led to the development of rapid composting methodologies, such as using cellulolytic nitrogen-fixing microorganisms (Harindintwali et al., 2020).

Table 3.1 Composition of the lignocellulosic crop wastes in % (Yang, 2007; Saini et al., 2015; Harindintwali et al., 2020)

Crop waste	Cellulose	Hemicellulose	Lignin	Other
Wheat straw	33–38	26–32	17–19	17
Bagasse	32–48	19–24	23–32	11
Rice straw	28–36	23–28	15–20	18
Corn stover	38–40	28–32	15–20	18
Barley straw	31–45	27–38	14–19	2–7
Sorghum straw	32	24	13	12

Composting plays a crucial role in various aspects of agricultural waste valorization:

- Crop residue management: Crop residues, such as corn stalks and wheat straw, can be composted to reduce waste and provide nutrient-rich organic matter for soil improvement.
- Manure handling: Livestock manure can be composted to reduce odors, pathogens, and weed seeds, transforming it into a valuable soil amendment.
- Food waste recycling: Food scraps from farms, markets, and restaurants can be composted, reducing landfill waste, and returning nutrients to the soil.
- Green waste: Yard trimmings, leaves, and pruning can be composted to create high-quality mulch and soil amendments for gardening and landscaping.
- Specialized composts: Composting can produce specialized composts, such as biochar or microbial inoculants, with specific benefits for crop production and soil health (Table 3.1).

Pretreatment of Feedstock

As mentioned, lignocellulosic waste is resistant to microbial degradation due to its complex and stable chemical structure. Therefore, it is necessary to pretreat this waste to facilitate and accelerate its composting. This pretreatment aims to partially decompose these biomolecules' highly linked polymer structure, making them more easily attackable by the microorganisms responsible for composting. Pretreatment can be physical, chemical, or biological (Harindintwali et al., 2020).

Physical pretreatment The most common mechanical methods of physical pretreatment, such as milling, grinding, or chipping, can be used before composting (Harindintwali et al., 2020). In all of them, a decrease in particle size and an increase in the surface accessible to bacteria are achieved (Putro et al., 2016). Recent advances aimed at accelerating the composting of lignocellulosic materials through physical pretreatment are the hydrothermal treatment technology (Nakhshiniev et al., 2014) and the hyperthermophilic pretreatment method (Cao et al., 2019). With hydrothermal treatment technology in relatively moderate process conditions (180 °C, MPa, 30 min), good results are achieved in accelerating the composting of rice straw. The hyperthermophilic pretreatment method uses an adjustable temperature reactor; it has been shown that 100 °C for 2 h yields good results before composting. Physical methods have the advantage of not producing compounds that can inhibit composting. However, they do not remove the lignin from the biomass, take a long time, and involve a high energy cost (Zheng et al., 2014).

Chemical pretreatment Lignocellulosic waste, such as agricultural residues (crop residues, straw), forestry residues, and other plant-based materials, comprises cellulose, hemicellulose, and lignin. Lignin, a complex and resistant polymer, significantly hinders the effective decomposition of lignocellulosic materials. The chemical pretreatment of lignocellulosic waste for composting involves the application of various chemical agents to modify the structure of lignocellulose and enhance its biodegradability during the composting process. Advanced chemical pretreatment methods aim to break down or modify lignin, disrupt the crystalline structure of cellulose, and solubilize hemicellulose, making the lignocellulosic material more accessible to microbial degradation during composting. Chemical pretreatment is very effective in facilitating the degradation of lignocellulosic waste; therefore, it is more used than other types of pretreatments. However, it presents some shortcomings, such as the potential environmental damage of the chemical used and the possible collateral formation of substances, such as some acids or phenols, that inhibit bacterial action in composting (Jönsson & Martín, 2016; Sheng et al., 2017; Harindintwali et al., 2020). Some essential chemical methods include alkali, acid, oxidant, and ionic liquid pretreatment. Alkali treatment is mainly based on hydroxides, such as NaOH, KOH, and $Ca(OH)_2$ (Karnchanawong et al., 2017; Zheng et al., 2018). These basic substances cause saponification of ester linkages in lignin, resulting in lignin breakdown. It also causes swelling of cellulose, increasing its susceptibility to enzymatic attack. Acid treatment, typically using sulfuric or hydrochloric acid, breaks down hemicellulose into simpler sugars and partially removes lignin, facilitating the exposure of cellulose (Zheng et al., 2018). Oxidative delignification entails the use of oxidizing agents, such as hydrogen peroxide or ozone, or Fenton's reagent, to selectively degrade lignin without significant cellulose and hemicellulose degradation (Kato et al., 2014; Sheng et al., 2017; Deng et al., 2019). Ionic liquids have also been used to disrupt the lignocellulosic structure (Carmichael et al., 2020; Ansari et al., 2021). These liquid salts penetrate the biomass, disrupting hydrogen bonding and breaking down and removing lignin.

Biological pretreatment Biological pretreatment of lignocellulosic waste is an advanced and environmentally friendly approach to enhancing the efficiency of composting processes. Biological pretreatment involves using microorganisms, such as fungi or bacteria, to selectively degrade the lignin and hemicellulose components of lignocellulosic materials. This process makes the remaining cellulose more accessible for subsequent microbial degradation during composting. One of the methods uses a combination of different microorganisms, or microbial consortiums, that target mainly hemicellulose and cellulose (Martínez-Gallardo et al., 2023). Fungi, on the other hand, hold the ability to break down lignin. White-rot fungi are more active in lignin hydrolysis than brown- and soft-rot fungi (Zheng et al.,

2014; Anu et al., 2023). Biological pretreatment of lignocellulosic waste holds promise for sustainable waste management practices, providing a means to efficiently convert complex biomass into valuable compost while minimizing environmental impacts (Martínez-Gallardo et al., 2023). Biological pretreatment is considered an environmentally friendly and cost-effective methodology. However, the long pretreatment periods required, well over 2 weeks, is a significant drawback. Continued research and development in this field aim to optimize pretreatment strategies for different types of lignocellulosic waste and improve overall composting performance (Harindintwali et al., 2020).

The choice of pretreatment method depends on the specific lignocellulosic material, desired outcomes, and the intended composting conditions. While chemical pretreatment can enhance lignocellulosic waste degradation, it's essential to consider the environmental and economic aspects associated with the chosen method. Applying these methods may also require optimization to balance enhanced biodegradability with potential environmental impacts.

Composting of Feedstock

Composting is a natural and environmentally friendly process that harnesses the activity of microorganisms to decompose organic matter, yielding a nutrient-rich soil amendment. Lignocellulosic materials, characterized by complex cellulose, hemicellulose, and lignin structures, present unique challenges for effective composting. Successful management of lignocellulosic waste requires understanding key parameters that influence the composting process, ensuring optimal microbial activity and nutrient cycling. In this context, several factors play crucial roles in shaping the dynamics of lignocellulosic waste composting, including the composition of the waste, environmental conditions, microbial activity, and process management. The main parameters that affect the composting of lignocellulosic waste are detailed below (Amuah et al., 2022; Alokika & Singh, 2023).

Carbon-to-nitrogen (C/N) ratio The C/N ratio is a critical parameter directly influencing microbial activity during composting. Lignocellulosic waste often has a high C/N ratio due to the abundance of carbon-rich components like cellulose and lignin. Microorganisms require a balanced C/N ratio for effective decomposition. The waste conversion will sag if the C/N ratio is below ten or above fifty. At a low level, this is so because the accumulation of substrates provokes nutrient deficiency in microorganisms. In contrast, ammonia and volatile fatty acids are formed at a high value, thus inhibiting the process. A C/N ratio of approximately 25–30:1 is generally considered optimal, ensuring a sufficient nitrogen supply for microbial growth and carbon for energy (Lin et al., 2019). Bulking agents are commonly added to the feedstock to improve its porosity, absorb moisture, decrease volatile compound emissions during composting, such as foul sulfur substances, and maintain an adequate C/N ratio during composting (Jain et al., 2019; Shao et al., 2014; Zhang & Sun, 2016).

Particle size and surface area The particle size of lignocellulosic waste affects the composting process by influencing microbial access to the substrate. Smaller particle size grants a larger surface area for microbial colonization and enzymatic activity. However, it can also produce compaction of the feedstock, thereby favoring unwanted anaerobic conditions. On the other hand, particles that are too large have less accessible surface area and generate less process temperature, thus slower composting (Ge et al., 2015; Zhang & Sun, 2016). According to the reported research on composting organic waste, the optimal particle size lies between 3.1 and 12.7 mm (Amuah et al., 2022).

Moisture content Adequate moisture is essential for microbial activity, enzyme function, and composting. Lignocellulosic waste tends to have low moisture content, and supplemental water may be required. The optimal moisture content for lignocellulosic waste composting typically falls within the range of 50–60%; nonetheless, according to separate reports, the range can be enlarged to 40–60%. Excessively dry or wet conditions can impede microbial activity and nutrient transformation (Amuah et al., 2022).

Aeration Proper aeration is crucial for supplying oxygen to aerobic microorganisms responsible for waste decomposition (Raza & Ahmad, 2016). Insufficient oxygen can lead to anaerobic conditions, resulting in the production of malodorous compounds. On the other hand, excessive aeration can slow the composting process due to significant losses of heat, water, and ammonia (Lin et al., 2019). Turning or aerating the compost pile helps maintain aerobic conditions, enhancing microbial efficiency and preventing the accumulation of anaerobic byproducts. In a study on paddy husk and chicken manure composting, it has been reported that the optimal oxygen level should be maintained between 15% and 20%, although there are studies that suggest that the ideal range for composting organic waste could be 5–15%. The nature of the raw material and that of the microorganisms involved in the process probably determines the variability of the degree of aeration described (Latifah et al., 2015; Harindintwali et al., 2020; Amuah et al., 2022).

pH level The pH of the composting material influences microbial activity and enzymatic processes. Lignocellulosic waste, especially materials rich in lignin, is typically more acidic than other organic wastes. A neutral to slightly basic pH range (7.0–8.0) is generally considered optimal for composting. However, different microorganisms can participate in this process, each adapted to different pH values. Hence, the optimal range can be as ample as 5–8 pH units (Torres-Climent et al., 2015). Adjustments, such as adding alkaline materials, may be necessary to maintain a favorable pH for microbial activity (Alokika & Singh, 2023).

Temperature Composting is a thermophilic process characterized by elevated temperatures resulting from microbial activity. The process occurs in three stages, defined by the pile's temperature (Waszkielis et al., 2013; Lin et al., 2019). Each of these stages favors the action of a specific type of microorganism. Thus, in the beginning, in the mesophilic stage, the temperature is low (25–40 °C), but the microbial action increases the temperature and causes the thermophilic stage to be reached (40–80 °C). In this stage, we can distinguish two ranges, the lower one reaching up to 60 °C, in which the rate of biodegradation processes of the raw material rises as the microbial population grows, provoking a temperature increase. Above 60 °C, many microorganisms die, including human pathogens, slowing the process and causing a temperature drop, leading to the cooling stage (Sudharsan Varma & Kalamdhad, 2015; Gerba & Pepper, 2015; Pandey et al., 2016). Lignocellulosic waste generally requires slightly higher temperatures to facilitate the breakdown of complex organic compounds. Regular turning of the compost pile helps sustain elevated temperatures. The optimum temperature range for composting animal manures is 40–65 °C, although depending on the nature of the feedstock, it can vary. Thermophilic temperatures (around 55–65 °C) promote the activity of heat-tolerant microorganisms, accelerating the decomposition of lignin and cellulose (Bernal et al., 2009).

Microbial inoculation Composting, in principle, does not require the addition of microbes; this, however, can improve the performance of the process in terms of both degradation time and quality of the compost. The microbial inoculum can consist of a single strain, a mixture of strains, or the addition of already matured compost. Introducing specific microbial inoculants, such as lignin-degrading fungi or cellulose-degrading bacteria, can enhance composting efficiency by accelerating the breakdown of lignocellulosic components. Well-established microbial consortia play a crucial role in lignocellulosic waste composting, and their composition can be influenced by factors like temperature, moisture, and nutrient availability (Zhao et al., 2016; Rastogi et al., 2020; Liu et al., 2023). The number of microbes for cellulose and lignin degradation during composting is limited. Among these are white-rot fungi and the bacteria *Cellulomonas*, *Pseudomonas*, *Bacillus* spp., and *Thermoactionmycetes* (Gautam et al., 2012; Awasthi et al., 2015; Rastogi et al., 2020). A study on organic municipal solid waste composting reported that process efficiency is maximum for multi-stage microbial inoculations (Xi et al., 2015).

Nutrient supplementation Lignocellulosic waste may have imbalances in essential nutrients. Supplementing the compost with nitrogen, phosphorus, and other micronutrients ensures that microorganisms have the necessary elements for growth and enzymatic activity. Nutrient supplementation helps maintain an optimal C/N ratio and supports microbial communities throughout composting (Sánchez et al., 2017).

Regularly monitoring critical parameters, such as temperature, moisture content, and pH, is essential for effectively controlling composting. Adjustments may be needed based on these monitoring results to maintain optimal conditions. Monitoring also involves assessing the progress of lignocellulosic decomposition, ensuring that the composting process is advancing effectively.

Methods of Composting

Various composting methods are employed, depending on the available resources and waste materials. Standard composting methods include:

Windrow composting Windrow composting is a large-scale, aerobic composting method where organic materials are piled into long, narrow rows, or "windrows." Typically used for agricultural or municipal composting, this method facilitates efficient decomposition by regularly turning the rows. Turning introduces oxygen, expedites the breakdown process, and helps maintain optimal temperatures. Windrow composting is well-suited for bulkier organic waste. It offers a systematic approach to managing more significant quantities of material, making it a practical choice for farms, waste management facilities, and composting operations on a larger scale (Hubbe, 2014).

Sheet composting Sheet composting is a straightforward and environmentally friendly method that does not require a traditional compost bin. Instead, it involves layering organic materials directly on the soil surface as mulch to create a nutrient-rich plant environment. Sheet composting is versatile and can be adapted to various scales, making it accessible for small-scale home gardens and larger agricultural applications. Garden debris and leaves are typical feedstocks for sheet composting (Amuah et al., 2022).

Vermicomposting It is an eco-friendly and space-efficient composting method that utilizes worms and microbes to break down organic waste into nutrient-rich fertilizer. In a controlled environment, such as in a worm bin, red worms (*Eisenia fetida*) consume kitchen scraps and other organic materials, producing nutrient-dense castings. The worms' natural activities aerate the compost, enhancing decomposition (Barthod et al., 2018; Liew et al., 2022).

In-vessel composting This method is an efficient and controlled process that accelerates the decomposition of organic waste within enclosed containers. Often used in commercial and municipal settings, this method provides optimal conditions for microbial activity, including temperature and aeration control. The enclosed vessels, ranging from drums to silos, effectively contain odors and hasten the composting process. In-vessel composting is a scalable solution, accommodating large volumes of organic waste, making it a preferred choice for waste management facilities, farms, and urban composting operations seeking controlled and accelerated decomposition (Amuah et al., 2022).

As a culmination, we can mention the advantages of composting:

- Waste reduction: Composting reduces the volume of organic waste, making it easier and more cost-effective to manage and dispose of residues.
- Soil enhancement: Finished compost is a valuable soil conditioner, enriching the soil with organic matter, improving structure, and enhancing nutrient content.

- Nutrient recycling: Compost returns essential nutrients, such as nitrogen, phosphorus, and potassium, to the soil, reducing the need for synthetic fertilizers.
- Improved crop yields: Compost-enhanced soils often result in higher crop yields and improved water retention, promoting sustainable farming practices.
- Carbon sequestration: Composting diverts organic waste from landfills, reducing methane emissions and mitigating climate change.
- Economic benefits: Composting reduces waste disposal costs and generates revenue by selling high-quality compost products.

3.2.3 Other Biochemical Technologies

Alcoholic Fermentation

Alcoholic fermentation stands as a remarkably adaptable biochemical process employing *Saccharomyces cerevisiae* and other microorganisms to convert fermentable sugars into bioethanol (Liu et al., 2022). These sugars, sourced from diverse agricultural residues like fruit pomace, sugarcane bagasse, or lignocellulosic feedstock (Baruah et al., 2018; García-Velásquez & Cardona, 2019), undergo pretreatment to render them amenable to fermentation. Lignocellulosic pretreatment, achieved through physical, chemical, or physicochemical means, such as milling, acid hydrolysis, or steam explosion, aims to liberate sugars from the biomass. However, caution must be exercised as acid hydrolysis may yield undesirable compounds like phenols and furfurals (Liu et al., 2022).

The process of alcoholic fermentation unfolds through several stages. Initially, agricultural waste is collected and pretreated to facilitate sugar release and microbial accessibility. The prepared feedstock and the chosen yeast strain are then introduced into a fermentation vessel. Within this vessel, yeast metabolizes the sugars, yielding ethanol and carbon dioxide as byproducts. Distillation, when necessary, separates ethanol from the fermentation broth, particularly for higher ethanol content. The resultant ethanol may be refined further to eliminate impurities and obtain the desired bioethanol fuel.

Alcoholic fermentation, occurring anaerobically, is conducive to sealed fermentation vessels. Although traditional fermentation typically occurs at temperatures ranging from ca. 20 °C to 30 °C (Liu et al., 2019), the utilization of thermotolerant yeast strains permits operation at higher temperatures, potentially reducing operational costs (Abdel-Banat et al., 2010). Solid-state fermentation technology has emerged as a viable alternative to conventional processing methods, prompting ongoing research efforts toward developing and operating large-scale, adaptable SSF bioreactors (Abdul Manan & Webb, 2017).

Photobiological Hydrogen Production

Photobiological hydrogen production, a cutting-edge biochemical technology, holds immense potential for transforming and valorizing agricultural waste (Xu et al., 2022). Photobiological hydrogen production harnesses the photosynthetic capabilities of certain microorganisms, such as green algae and cyanobacteria, to generate hydrogen gas (H_2) from organic matter (Arimbrathodi et al., 2023; Carbas et al., 2023). Hydrogenase enzymes are critical to photobiological hydrogen production. They catalyze the reversible conversion of protons (H^+) and electrons (e^-) into molecular hydrogen (H_2). Under specific conditions, photobiological organisms can redirect electrons from the photosynthetic electron transport chain toward hydrogenase enzymes, producing hydrogen gas. Photobiological hydrogen production involves several key processes. First, photosynthetic microorganisms are cultivated in controlled environments, such as bioreactors or open ponds, receiving sunlight, carbon dioxide, and nutrients. During photosynthesis, microorganisms capture solar energy, convert carbon dioxide into biomass, and release oxygen as a byproduct. To stimulate hydrogen production, researchers manipulate cultivation conditions by limiting sulfur availability, inducing nitrogen starvation, or adjusting light exposure. Generated hydrogen gas is typically collected and separated from the culture medium using various techniques, such as gas sparging or membrane separation (Kossalbayev et al., 2024). The advancement and integration of hydrogen energy represent a pivotal trajectory in contemporary energy research, poised as the most promising alternative to fossil fuels in the foreseeable future. In contrast to hydrogen production from fossil fuels, photobiological methods offer a distinctive advantage by enabling the utilization of biomass waste as raw material while concurrently mitigating the pollution associated with the process. Nonetheless, this technology faces the challenge of improving production capacity and cost to be competitive. Future endeavors should thus prioritize optimizing reaction conditions and production devices to enhance process efficiency (Xu et al., 2022).

Transesterification with Enzyme Catalysts

Transesterification with enzyme catalysts is a biochemical technology that converts agricultural waste into valuable biodiesel and other high-value products. Enzyme-catalyzed transesterification is a biodiesel production process that involves the conversion of triglycerides found in oils and fats into biodiesel and glycerol. In this process, enzymes, such as lipases or proteases, serve as biocatalysts to facilitate the reaction. Enzymes exhibit high specificity for the substrates they act upon; lipases, for example, selectively target the ester bonds in triglycerides. Enzymatic transesterification can be conducted under mild conditions, typically at lower temperatures and atmospheric pressure. The reaction results in the production of biodiesel (fatty acid methyl or ethyl esters) and glycerol. Agricultural waste materials, such as used cooking oil or animal fats, are collected and pretreated to remove impurities and water content. The pretreated feedstock is mixed with the enzyme catalyst and a methanol or ethanol source. The enzyme catalyzes the transesterification reaction, breaking ester bonds in triglycerides and forming biodiesel and glycerol (Xia et al., 2024). The mixture is then separated into layers, with biodiesel rising to the top and glycerol settling at the bottom. Biodiesel is purified to remove any remaining impurities, water, and excess alcohol. The predominant focus in biodiesel preparation lies within the realms of second- and third-generation feedstocks, comprising nonedible oils and waste cooking oil, respectively. Biodiesel constitutes an environmentally friendly alternative to diesel, due to its nontoxicity, renewability, and biodegradability (Malode et al., 2022). Even though the use of enzymes to catalyze transesterification usually involves a longer process and a higher cost, it also has advantages over other methods that must be considered, such as the fact that it does not cause the formation of foams with free fatty acids, has low energy and wastewater treatment demands, has no side reactions, uses mild conditions, and eases the purification process. Moreover, enzymes do not present the environmental challenges that can compromise the sustainability of the process that noneco-friendly catalysts, for example, HCl and NaOH, do (Leung et al., 2010; Kazemi Shariat Panahi et al., 2024; Xia et al., 2024).

One of the advances of this technology is the genetic modification of microbes to improve lipase production (Zhang et al., 2021; Joshi et al., 2022, 2023). Another active research field is the immobilization of enzymes to improve their performance and further facilitate their recovery and reuse; adsorption, encapsulation, covalent binding, or cross-linking are commonly used. Enzymes have been immobilized onto commercial materials such as silica; however, the Van der Waals forces implied are weak, and enzymatic activity is gradually lost over reutilization cycles. Immobilization onto nanomaterials, such as carbon nanotubes, electrospun nanofibers, and magnetic nanoparticles, efficiently increases reusability with a slight activity loss (Kazemi Shariat Panahi et al., 2024).

3.3 Thermochemical Technologies

Although almost as great a variety of products can be extracted from agricultural residues as from purpose-grown plants, thermochemical processes have been mainly oriented toward obtaining liquid, solid, or gaseous fuels (Jha et al., 2022). The most immediate form of utilization is the combustion of agricultural residues to produce heat. Still, in this case, we are not transforming agricultural residues into something more valuable, with a few exceptions. The combustion process has been known since the beginning of civilization. Another process that has been well-known for centuries is pyrolysis for producing charcoal. More recently, however, the need has arisen for energy products with other properties, such as being in the form of a liquid or gas or having a specific composition. An added advantage of using these methods over agricultural residues is that they do not lead to a net increase in GHG emissions as they are derived from biomass and are therefore GHG neutral (Yang et al., 2019).

Therefore, numerous thermochemical transformation methods have been developed, are still being researched, and will be discussed in this section.

Thermochemical processes involve using high temperatures to modify the chemical composition of the treated material. The classification of these processes is sometimes simplified by defining only three types: pyrolysis, liquefaction, and gasification, depending on whether the main product obtained is solid, liquid, or gas (Jha et al., 2022). However, other new methods must be added, such as hydrothermal treatments or modifications to the previous ones, such as using catalysts or chemical reactions like transesterification or hydrodeoxygenation. These methods are not only used with agricultural waste but also frequently with plastics and other types of waste.

The advantages of thermochemical methods include the following (Yang et al., 2019; Durán-Valle, 2020):

- They can be used with most biomass. Although some processes require the feedstock to be dry (e.g., pyrolysis), others allow a high degree of moisture (hydrothermal processes).
- High productivity is generally higher than with biochemical methods.
- Low waste generation.

Simple processes, as they do not usually require pretreatment, are carried out in short times, and the operating conditions do not require extreme precision to be effective.

Among their disadvantages are that they do not allow obtaining as many products with high added value as biochemical methods and that the working conditions are usually drastic, mainly due to high temperatures.

3.3.1 Biochar Production

Biochar is a solid, carbon-rich material obtained from biomass, including agricultural residues (Amalina et al., 2022). It is a material that has been known since ancient times and is produced in almost every country in the world. In most cases, it is produced in an artisanal way for local consumption. The primary use is for cooking and heating in developing countries. In more developed parts of the world, it is used for other applications. It is also used in cooking, but specifically in barbecues. It has been used in the metallurgy industry as it has a higher energy density than wood and can reach higher temperatures. But in this case, it is usually replaced by mineral coal, which is more polluting but cheaper. Another application is manufacturing activated carbon, an adsorbent widely used in many applications. And in recent decades it has begun to be applied in soil remediation and energy applications such as batteries and capacitors.

Pyrolysis

It is a term widely used to refer to the high-temperature treatment of biomass that produces its decomposition. However, the International Union of Pure and Applied Chemistry relates it to thermolysis in its Gold Book, indicating that this process is usually carried out at high temperatures and in an inert atmosphere. In this section, we will focus on processes carried out under vacuum or in the presence of inert gases and where no gassing agents or liquids are used, as these cases correspond to other techniques. Pyrolysis can be used to obtain gaseous, liquid, and solid products (Patra et al., 2021; Panwar & Pawar, 2022). Generally, the three types of products are obtained in different proportions depending on the raw material and the process conditions. Figure 3.1 explains the influence of temperature, heating rate, and residence time of the vapors formed on the proportion of the products obtained.

In manufacturing biochar, the aim is to favor the production of solid products, minimizing the production of liquids (tars) and gases as much as possible. In addition to obtaining a better yield, the production of pollutants is reduced, which must be managed appropriately. For this reason, slow pyrolysis methods are used, with residence times in the furnace ranging from several hours to several days. However, the traditional method, which is the most widely used worldwide, is not a true pyrolysis but a partial combustion that is carried out by limiting the amount of air that can enter the furnace. The combustion produces heat, which leads to pyrolysis. This method produces gases and liquids that are pollutants and are not always adequately processed. In modern industrial systems, heat can be supplied externally and the liquids and gases produced can be burned as fuels to generate heat to assist pyrolysis (Patra et al., 2021; Panwar & Pawar, 2022). In this way, liquid pollutant byproducts are eliminated and gaseous pollutants are transformed into less environmentally harmful substances (e.g., carbon dioxide instead of methane or carbon monoxide).

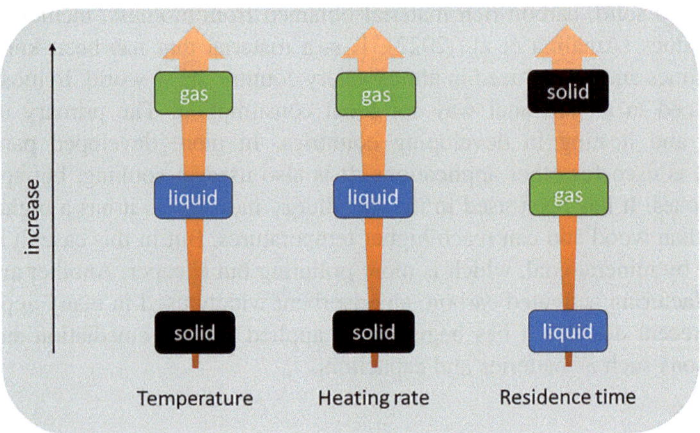

Fig. 3.1 Main products obtained according to pyrolysis conditions

A significant influence factor is the raw material used. In the case of agricultural residues, it can be assumed that most of them are lignocellulosic. A high lignin content will lead to a high yield as lignin reacts more slowly than cellulose and hemicellulose, resulting in more stable structures. The physical form of the material is also essential. As agricultural residues are not suitable heat transmitters, large pieces of agricultural residues will achieve a slower heat transmission to the inside so that the pyrolysis process can be very different on the inside and outside. This effect is more noticeable if short furnace residence times are used. Since in developing countries, it is not always possible to reduce the size of the biomass, kilns that are kept on for several days with a relatively low pyrolysis temperature must be used. In fast industrial processes, on the other hand, it is necessary to use small particles.

Hydrothermal Carbonization

When 100° C is exceeded in a closed container, the water is in an equilibrium between liquid and gaseous forms. Pressure increases with temperature, and both effects produce changes in its properties: lower dielectric constant, lower surface tension, and viscosity leading to better diffusion and higher reactivity (Martins-Vieira et al., 2022). The hydrothermal process can be used to obtain biochar (this section), and biofuels (Sect. 3.2.2) or to extract high-value-added compounds (Sect. 3.3.1).

Following the general principle of carbonization that milder conditions result in less gas and liquid production (Amalina et al., 2022), less drastic pressure and temperature conditions than for other hydrothermal processes are used to maximize the production of solid biochar. Generally, 250 °C and 2 MPa are not exceeded (Pavlovič et al., 2013) and the process lasts between 1 and 12 h. An advantage over pyrolysis is that the material does not need to be predried, translating into significant energy savings. If they are from agricultural waste, the materials used are usually lignocellulosic. In hydrothermal treatment, they undergo a series of parallel reactions (Pavlovič et al., 2013; Ponnusamy et al., 2020) such as hydrolysis, dehydration, or aromatization, but which are always preceded by hydrolysis of the existing macromolecules (cellulose, hemicellulose, and lignin). Although initially small molecules (e.g., short-chain organic acids, furfurans, aldehydes, phenols) are produced, they eventually condense with each other and undergo aromatization reactions to give rise to carbonaceous material. This material has been used as a catalyst, in soil remediation, energy applications, etc. (Jha et al., 2022). The aqueous liquid that remains a byproduct can be used as a fertilizer (Shan et al., 2023). Temperatures above the critical point of water (supercritical water) are not usually used to obtain biochar since, as indicated above, liquids and gases are preferentially obtained and the yield of the solid product is meager. One example of agricultural residues that have been used to obtain hydrothermal char is the use of rice husk and wheat straw (Shan et al., 2023). It is not uncommon for agricultural residues to be mixed with other types of waste such as sewage sludge (He et al., 2019) for carbonization. More examples of materials used and details on hydrothermal carbonization and its applications can be found in Titirici et al. (2015) and Durán-Valle et al. (2017).

Activation

Once either of the two previous methods has obtained the biochar, it can be activated to increase its specific surface area, develop its porosity, and thus obtain a material with greater added value that can be used as an adsorbent, catalyst, catalyst support, etc. The advantages of using forest residues are a decrease in the raw material cost and the use of renewable resources.There are two general methods of activation: physical activation and chemical activation (Tan et al., 2017; Danish & Ahmad, 2018).

Physical activation consists of heating the biochar with a moderately reactive gas (Panwar & Pawar, 2022). The most commonly used are air, carbon dioxide, and water vapor. The process can be considered as a limited gasification. The gases react with tars deposited on the coal's surface in the carbonization stage to produce a narrowing or plugging of the pores. As these tars are removed, porosity increases. But they can also react with the solid part of the biochar, changing its porous properties. A higher specific surface area, a larger pore volume, and more specifically micropores are obtained. Surface polarity is often reduced as functional groups with heteroatoms are usually more reactive than hydrocarbon chains or cycles.

Chemical activation involves a broader variety of reagents and a more comprehensive range of temperatures depending on the reagent used (Panwar & Pawar, 2022). Acids, alkalis, or oxidants may be used. The effect is similar to physical activation: removal of tars and limited reaction with the solid structure of the biochar. Although better end-products are usually achieved, it must be considered that chemical activation involves the use of pollutants. There are numerous examples in the scientific literature of the use of agricultural waste to obtain activated carbon (Ioannidou & Zabaniotou, 2007; Paraskeva et al., 2008; Yahya et al., 2015).

3.3.2 *Biofuel Production*

Biofuel production, excluding combustion, can be classified into three generations (Balagurumurthy et al., 2015; Sarwer et al., 2022). The first generation was obtained from human and animal food products such as grains, corn, sugarcane, or vegetable oils. Carbohydrates were fermented to produce alcohols and fats were transformed into biodiesel. Apart from having a negative impact on biodiversity and food prices in some parts of the world, the main problem is that only a tiny part of the plant was used. The use of large tracts of fertile land and the high consumption of water and fertilizers did not justify the low yields obtained from this methodology. As a result, it has been abandoned in much of the world. Second-generation biofuels have replaced them due to their greater cost-effectiveness. Not only economically but also because of their lower environmental impact and better use of resources (Joshi et al., 2020). This is because a higher percentage of the plant is used, producing more fuel with the same resources. Industrial waste, nonreusable plastics, wood, or plantations specially dedicated to biofuel production can be used. However, a significant part of the production is made from agricultural waste, so this subsection will focus mainly on second-generation biofuels. Of these residues, most are lignocellulosic materials. Finally, a third generation is currently under development, which is the production from algae (Chowdhury & Loganathan, 2019). These also make it possible to obtain high-value products other than fuels, neutral regarding the planet's greenhouse effect and allow using non-agricultural land and various hydrological resources. They still have problems, such as excessive phosphorus consumption and low energy yields. The algae obtained can also be used for thermochemical treatment. Several methodologies have been developed for biofuel production, which are described below.

Gasification

It consists of the reaction of a solid (agricultural waste) with a gas (generally low cost such as air, water vapor, carbon dioxide, or similar) at a high temperature (800–1300 °C) to produce a mixture of gases (Michailos et al., 2019). One of the most industrially used processes is the production of syngas, which is a mixture of gases in which dihydrogen and carbon monoxide predominate. One of these gases can be obtained by cryogenic separation. But it can also be used in the Fischer–Tropsch process in which, using catalysis and high pressure and temperature, this gas can be transformed into a mixture of hydrocarbons, including some oxygenates such as alcohols. The compounds obtained can be separated to produce fuels or feedstock for the chemical industry. Agricultural residues must be predried as removing moisture in the gasification process would consume a lot of energy. Heating involves both the decomposition of the feedstock into more labile and volatile molecules and the reaction with the gasifier.

Gasification, liquefaction, and pyrolysis all result in gaseous, liquid, and solid products. The methods employed seek to maximize one of these products, but inevitably, some amount of the other two is obtained. Sometimes, these byproducts may be of commercial interest, but more often they are burned to produce heat to assist in some stage of the process. Air is not usually used as a gasifying agent (Canabarro et al., 2013) as it produces compounds with a high oxygen content which, together with the dilution caused by the dinitrogen in the air, results in a lower calorific value and a lower amount of dihydrogen. The gas obtained can be burned but is not useful for the Fischer–Tropsch process. A higher quality product is obtained if pure dioxygen or water vapor is used. There are several gasification techniques. Some authors include pyrolysis or certain hydrothermal processes as a gasification method since combustible gases can also be obtained. But to be precise, gasification occurs when a reactive gas is used. Several methods sometimes differ only in the type of furnace used.

For example, fixed-bed reactors are used. They are simple and suitable for small installations and do not require any feedstock pretreatment other than a certain degree of drying. Fluidized bed reactors use small biomass particles suspended in the gas stream. They are easily adapted to different types of waste and even use mineral particles for heat transfer or to act as catalysts. They require cyclones to separate small solid particles entrained in the gas. More recent are plasma reactors (Ibrahimoglu et al., 2017). Highly reactive plasma acts on the biomass by breaking it into smaller molecules that become gaseous. A more detailed description of the reactors used can be found in (Durán-Valle, 2020).

Liquefaction and Hydrothermal Methods

This process aims to obtain liquid products that can be used mainly as liquid fuels. Some organic solvents or water are usually used, and a catalyst is also frequently used. The working conditions are milder than gasification but more drastic than hydrothermal carbonization (280–370 °C and 10–25 MPa; Pavlovič et al., 2013). The correlation can be maintained that at more drastic conditions (pressure, temperature, heating rate, reagents used), smaller and more volatile molecules are obtained. This explains why milder conditions are used since the aim is to obtain compounds of medium molecular mass. The milder conditions explain the desirability of using catalysts to increase the speed of the process. Both acid and alkaline catalysts are commonly used. Acid catalysts are more active but tend to deteriorate installations due to corrosion.

When water is used, and since the process is at medium temperature and pressure, it should be considered a hydrothermal process. In this case, it is not necessary to predry the waste used. Since most plant waste consists of lignocellulosic materials, the first reaction is the hydrolysis of ether bonds in cellulose, hemicellulose, and ester bonds in lignin. This results in oligomers of medium molecular mass which, as the process continues, eventually give rise to monomers. These can condense again to give larger molecules that, although oxygenated, form a hydrophobic mixture that can be separated from most water used. Sometimes, the process can be optimized to seek enrichment in a substance to give a simple composition. For example, the production of ethyl levulinate, a diesel fuel additive, from various agricultural residues (Zhuo et al., 2023). This technology can be combined with other technologies sequentially. For example, biodigestion (Klüpfel et al., 2023) of agricultural residues mixed with manure to obtain biogas, followed by hydrothermal liquefaction of the digestate obtained. Although this process is the most widely used industrially today, it has some limitations: low yield, the need for subsequent water removal, and low biofuel quality due mainly to higher oxygenation. For this reason, other options are being developed, such as moderate acid-catalyzed liquefaction (MACL; Jiang et al., 2018), which can be carried out at ambient pressure and lower temperatures.

Pyrolysis

This technique has already been mentioned in the production of biochar. It is advisable to increase the temperature and heating speed to obtain biofuels, as this results in a higher proportion of liquids and gases and a lower proportion of solids.

As in other techniques shown here, the water content of agricultural waste is decisive for whether or not energy and economic profitability can be obtained. This is mainly because evaporation is an endothermic process and requires more energy. In addition, if liquid fuels are obtained, they will contain a higher proportion of water, reducing their calorific value. Various raw materials have been used, and the use of plants cultivated for this purpose, such as timber trees, is common. Some agricultural residues have also been used, such as tree bark, olive husks, corn cobs, cotton stalks, and others (Mohan et al., 2006). There are several ways of supplying energy to the reactor. If the reactor is small, it can be heated externally, but this method is not effective if the size of the installation increases beyond a certain point. In this case, the heat must be supplied internally and this can be done with a carrier gas or solid that can be inert or act as a catalyst. Generally speaking, there are three types of pyrolysis (Jha et al., 2022):

Slow pyrolysis It has traditionally been used for biochar production and is still used in developing countries and semi-industrial kilns. A slow heating rate, moderate temperatures, and long processing times characterize it. Due to its wide use in many parts of the world, it is a process of great economic importance, but at the same time, it is often very polluting. The product obtained is mainly biochar.

Fast pyrolysis It is an industrial process carried out at a higher speed and the main product is liquid (biofuels). The very dry raw material is used in particles of a few millimeters to achieve this higher speed.

Flash pyrolysis It can be considered as an extremely fast variety of pyrolysis. The heating rates are very high (about 1000 °C/s) as well as the temperature and a very short reaction time, often less than 1 s. This requires that the agricultural residues are dried and prepared as a powder.

Many types of reactors can be used in fast and flash pyrolysis, resulting in different methodologies. Fixed bed reactors are more common in slow pyrolysis and are not commonly used. Some of these reactors are similar to those used in gasification, the difference between them is the gaseous atmosphere inside the reactor. A more detailed description of the reactor types can be found in Durán-Valle (2020).

Combustion

This is the simplest process for the energetic use of vegetable waste. Still, it cannot always be considered a transformation, as it does not usually produce a valuable end product and some residues (gases and ashes) are obtained. It consists of heating the raw material in the presence of air or another oxygen gas. As the process is usually exothermic, no external energy input is needed once the process has started. Current innovations are based on a reduction in the environmental impact (Najser et al., 2022) and a better use of the energy obtained (Isemin et al., 2010; Liang et al., 2022).

There are some cases where valuable products are obtained (Krishnamoorthy et al., 2015): using ashes as catalysts (see the section on transesterification of this chapter), fertilizers, or obtaining raw materials such as silicon compounds or alumina.

Catalytic Deoxygenation and Hydrodeoxygenation

In some of the processes shown so far, the resulting biofuel is not of sufficient quality to be used under the same conditions as fossil fuels. For this reason, other secondary processes must be carried out to improve its properties. One example (Filho et al., 2022) is the vapors from gasification or pyrolysis. These contain short-chain organic compounds with a high proportion of oxygen with low stability, low calorific value, low stability, and a tendency to polymerize and form carbonaceous deposits.

Two catalytic processes are used for this purpose: deoxygenation and hydrodeoxygenation (Lahijani et al., 2022). The former involves a condensation reaction between molecules with the loss of oxygen in the form of carbon oxides or water. In the latter, a dihydrogen-reducing atmosphere is used in addition to the catalyst so that oxygen is removed as water. This process is more expensive as it requires higher dihydrogen gas pressures and more sophisticated catalysts. But it has the advantage that no carbon atoms are lost and therefore better biofuel yield and quality are obtained.

In deoxygenation, a catalytic process is used (Filho et al., 2022; Ly et al., 2021) where larger molecules with lower oxygen content are obtained through condensation reactions. It is also possible to obtain high-value products (fine chemicals). For example, Mansur et al. (2014) performed pyrolysis of cocoa pod husks, and the oil obtained was treated with an iron oxide catalyst to selectively obtain aliphatic ketones, phenols, and alkylphenols.

Hydrodeoxygenation has been studied as an improvement application to industrial processes to obtain biofuels. Therefore, there are few examples of using agricultural residues treated with this technique. One example (Singh & Sarkar, 2023) is the roasting of bean stalks for pyrolysis and hydrodeoxygenation of the biofuel obtained. Preroasting was found to reduce the need for hydrodeoxygenation.

Transesterification with Acid/Base Catalyst

There is great interest in the production of biodiesel as it is a fuel obtained from renewable sources unlike diesel fuel which is obtained from petroleum (Odude et al., 2019). It is obtained from fats of biomass origin, such as vegetable oils, animal fats, microalgae, and food waste. The triglycerides in these fats react with alcohol to form esters in the presence of acidic or basic catalysts in a process called transesterification.

Most of the cost of biodiesel production is the feedstock (Atabani et al., 2012) so using low-cost waste is a major competitive advantage. However, most of the current production is made from animal fats, food fats, or plants grown specifically for biodiesel production, so agricultural waste is not very common. There is some research in this area, such as using rice bran, a material with a high oil content (Lourenço et al., 2021), and also *Annona Muricata* seed (Schroeder et al., 2018). However, it is more common to use oil already used in food, which, although of vegetable origin, should be considered food waste and not strictly agricultural waste (Dorado et al., 2004).

However, agricultural residues can play another role in this process: they can be the raw material for catalysts. For example, it has been tested with banana peel and cocoa pod husk (Atabani et al., 2012) that were calcined and whose ashes were used as catalysts. Calcined rice husk (rich in silicon) has also been used to support molybdenum and lanthanum oxides (Priya et al., 2020). There are numerous studies on preparing catalysts from agricultural waste to obtain heterogeneous catalysts. A review on this topic can be found in (Khan et al., 2021).

3.3.3 Subcritical and Supercritical Fluid Methods

These methods consist of (Martins-Vieira et al., 2022) in the use of a solvent at pressure and temperature conditions that are below (subcritical) or above (supercritical) the critical point of each solvent. In the first case, water is usually used, so it can be considered a hydrothermal process. Occasions where another solvent is used or the objective is biofuel production are discussed in Sect. 2.2.2. In the supercritical process, water or carbon dioxide is usually used, and less frequently other solvents.

Subcritical Fluid or Hydrothermal Method

One of the changes in the properties of water discussed above is the decrease in the dielectric constant with increasing temperature. This trend continues beyond the critical point. One consequence is that at lower temperatures, ionic compounds solubilize better, and as the temperature increases, covalent compounds dissolve better. There is also an increase in the ionic product until near critical temperatures when it starts to decrease. This can affect the reactivity of water, causing it to act as both an acid and an alkaline catalyst. In both cases, it increases the rate of hydrolysis of the polymers.

One example of use is the production of furfural and levoglucosan (Wang et al., 2020) from corn stalks, peanut shells, and rice stalks by a mild pyrolysis process followed by hydrothermal treatment. Another (Wu et al., 2021) in combination with alkaline catalysis is the hydrothermal hydrolysis of corn stalk lignin with subsequent treatment with propylene oxide to obtain a polyurethane precursor. An alternative means of delivering heat, such as microwaves, can also be used to obtain carbohydrates (Tsubaki et al., 2017).

Supercritical Fluid Methods

It consists of the use of supercritical fluids (SCFs) with low or zero environmental impact, such as supercritical water (SCW) or supercritical carbon dioxide (SCCD), which can also be reused for the most part. This technology stands out (Preetam et al., 2023) for being fast, with high reaction speed and high efficiency. As no high temperatures are used, no pollutants, such as NOx, dioxins, or soot particles, are produced.

SCFs are fluids that exist at temperatures and pressures above the critical point, which is specific to each substance. For example, for water, this is 374 °C y 22 MPa. Once the critical point is passed, the fluid obtained differs from gas and liquid. Its properties (density, viscosity, fluidity, etc.) are intermediate between both states of matter but closer to the liquid state (Okolie et al., 2020). Its density gives it a solvating capacity similar to a liquid's, but the increase in fluidity allows for better mass transfer and thus greater reactivity. For this reason, they stand out in extraction processes. And they are more useful in obtaining high-value products than in obtaining biofuels.

The method has been known for a long time. For example, in 1996, it was used to extract tea residues (Demirbas et al., 1996) using both SCW and supercritical acetone with considerably better results than extraction under subcritical conditions. It has also been used with peanut skin residues (Putra et al., 2018). In this case, the authors obtained oils, but could not extract bioactive compounds such as polyphenols. They explained this by the low polarity of the SCCD, as indicated in the previous section. The solution found was adding ethanol to increase the solvent's polarity (Putro et al., 2016).

As indicated above, not only SCW and SCCD are used as SCFs. For example, Demirbaş (2004) compared the use of supercritical water, acetone, ethanol, and methanol for the extraction of residues from wheat straw, corn cobs and stubble, tobacco stalks and leaves, olive residues, and nut and almond shells. The aim was to produce biofuels. The best results were achieved with water, although it must also be considered that in this work, it was the solvent with which the highest temperature was used. But if the temperature factor is not considered too important, this fact may explain why using solvents other than SCW and SCCD is rare. Other examples of extraction with SCFs are shown in Table 3.2.

SCFs are not always used for component extraction. The opposite process is also possible. Microcrystalline cellulose obtained from wheat straw was impregnated (Krivokapić et al., 2021) with ibuprofen dissolved in SCCD. The reason for using this method is that ibuprofen acid is poorly soluble in water, thus avoiding the use of more polluting organic solvents. Combining extraction with SCFs with other techniques, such as microwave digestion (Ateş et al., 2019) or hydrothermal processes (Zhao et al., 2009), is also possible.

Table 3.2 Valorization of plant wastes by extraction with SCFs

Agricultural waste	Target	SCF	References
Waste date palm	Waxes	SCCD	Al Bulushi et al. (2018)
Mandarin leaf waste	Polyphenols and flavonoids	SCCD	Ateş et al. (2019)
Stalks Wheat straw	Hexoses	SCW	Zhao et al. (2009)
Spearmint leaves	Flavonoids	SCCD	Bimakr et al. (2011)
Strawberry leaves	Phenols	SCCD mixed with ethanol, acetone, or water	Sato et al. (2019)
Olive mill wastewater	Oil	SCCD	Dali et al. (2022)

SCFs supercritical fluids, *SCCD* supercritical carbon dioxide, *SCW* supercritical water.

3.4 Conclusions and Future Perspectives

Human society is at a crossroads. On the one hand, there is a growing awareness of the impact of human activity on the environment. We are increasingly aware of the importance of preserving an environment in good conditions for the habitability of all living beings. But on the other hand, the increase in population and standard of living is forcing us to consume more resources. Many of these resources are limited or their extraction produces a strong environmental impact. Therefore, there are frequent dilemmas between using natural resources and respecting the environment.

The solution to these dilemmas often comes from scientific and technological development. One of the developing trends is the use of materials that were previously considered waste. A review of the properties of these wastes has allowed us to discover the possibility of obtaining energy and materials from them. Since they are resources of little or no value, their low cost is already an advantage for their use. Among these previously neglected materials are agricultural residues. Their production is inevitable since they are obtained as a byproduct of the production of foodstuffs, fibers, medicinal plants, wood, and other materials of biological origin. Moreover, as with the rest of the biomass, they can be considered neutral in the carbon cycle. And they are relatively easy to access because they come from human activity. Numerous techniques have been developed for this purpose. Some, such as alcoholic fermentation or combustion, have been known since ancient times. Others, such as enzymatic catalysis or the use of supercritical fluids, have been developed more recently. But we can say that today, there are methods for utilizing any agricultural waste that we can produce.

A different question is whether this utilization is economically viable. In many cases, the concentration of a valuable compound to be extracted or the energy density of a material is too low for the agricultural residue to be competitive with other more polluting sources. For this reason, political and fiscal measures are needed to evaluate not only the total cost but also the environmental impact of one or another technology or resource.

But the measures to be taken cannot stop here. It is also necessary to promote scientific and technological research to improve performance, reduce costs, and lower the environmental impact of the processes described in this chapter. Some of the ideas observed in recent years have been (a) improvement in heat recovery processes; (b) increasingly widespread use of catalysts; (c) use of hybrid techniques, mixing two or more of the technologies described to obtain greater efficiency; and (d) development of the biorefinery concept. Other measures that could be taken and would help to make better use of agricultural waste would be the transfer of technology to developing countries and the study of simple, small-scale on-farm systems. This would avoid the cost of transporting waste to distant facilities, which can be a disincentive.

In summary, we can say that we are at a time of solid development of waste recovery technologies. And that this has become an aspect of great social, economic, and environmental importance.

Acknowledgments This manuscript was supported by a grant from the European Regional Development Fund and the Junta de Extremadura (grant IB20026).

References

Abdel-Banat, B. M. A., Hoshida, H., Ano, A., Nonklang, S., & Akada, R. (2010). High-temperature fermentation: How can processes for ethanol production at high temperatures become superior to the traditional process using mesophilic yeast? *Applied Microbiology and Biotechnology, 85*(4), 861–867. https://doi.org/10.1007/s00253-009-2248-5

Abdul Manan, M., & Webb, C. (2017). Modern microbial solid state fermentation technology for future biorefineries for the production of added-value products. *Biofuel Research Journal, 4*(4), 730–740. https://doi.org/10.18331/BRJ2017.4.4.5

Abraham, A., Mathew, A. K., Park, H., Choi, O., Sindhu, R., Parameswaran, B., Pandey, A., Park, J. H., & Sang, B.-I. (2020). Pretreatment strategies for enhanced biogas production from lignocellulosic biomass. *Bioresource Technology, 301*, 122725. https://doi.org/10.1016/j.biortech.2019.122725

Akyol, Ç., Ince, O., Bozan, M., Ozbayram, E. G., & Ince, B. (2019). Biological pretreatment with Trametes versicolor to enhance methane production from lignocellulosic biomass: A metagenomic approach. *Industrial Crops and Products, 140*, 111659. https://doi.org/10.1016/j.indcrop.2019.111659

Al Bulushi, K., Attard, T. M., North, M., & Hunt, A. J. (2018). Optimisation and economic evaluation of the supercritical carbon dioxide extraction of waxes from waste date palm (*Phoenix dactylifera*) leaves. *Journal of Cleaner Production, 186*, 988–996. https://doi.org/10.1016/J.JCLEPRO.2018.03.117

Alokika, A., & Singh, B. (2023). Utilization of lignocellulosic plant residues for compost formation and its role in improving soil fertility. *Pedosphere, 33*(5), 700–716. https://doi.org/10.1016/j.pedsph.2022.11.008

Amalina, F., Razak, A. S. A., Krishnan, S., Zularisam, A. W., & Nasrullah, M. (2022). A comprehensive assessment of the method for producing biochar, its characterization, stability, and potential applications in regenerative economic sustainability—A review. *Cleaner Materials, 3*, 100045. https://doi.org/10.1016/J.CLEMA.2022.100045

Amuah, E. E. Y., Fei-Baffoe, B., Sackey, L. N. A., Douti, N. B., & Kazapoe, R. W. (2022). A review of the principles of composting: Understanding the processes, methods, merits, and demerits. *Organic Agriculture, 12*(4), 547–562. https://doi.org/10.1007/s13165-022-00408-z

Ansari, M., Zafar, U., Ejaz, U., Sohail, M., Pirzada, A., & Aman, A. (2021). Comparison of composting of chemically pretreated and fermented sugarcane bagasse for zero-waste biorefinery. *Journal of Material Cycles and Waste Management, 23*(3), 911–921. https://doi.org/10.1007/s10163-021-01176-w

Anu, A., Kuhad, R. C., Rapoport, A., Kumar, V., Singh, D., Kumar, V., Tiwari, S. K., Ahlawat, S., & Singh, B. (2023). Biological pretreatment of lignocellulosic biomass: An environment-benign and sustainable approach for conversion of solid waste into value-added products. *Critical Reviews in Environmental Science and Technology, 54*, 1–26. https://doi.org/10.1080/10643389.2023.2277670

Arimbrathodi, S. P., Javed, M. A., Hamouda, M. A., Aly Hassan, A., & Ahmed, M. E. (2023). Bio-H_2 production using microalgae: Highlights on recent advancements from a bibliometric analysis. *Water, 15*(1), 185. https://doi.org/10.3390/w15010185

Atabani, A. E., Silitonga, A. S., Badruddin, I. A., Mahlia, T. M. I., Masjuki, H. H., & Mekhilef, S. (2012). A comprehensive review on biodiesel as an alternative energy resource and its characteristics. *Renewable and Sustainable Energy Reviews, 16*(4), 2070–2093. https://doi.org/10.1016/J.RSER.2012.01.003

Ateş, F., Şahin, S., İlbay, Z., & Kırbaşlar, I. (2019). A green valorisation approach using microwaves and supercritical CO for high-added value ingredients from mandarin (*Citrus deliciosa* Tenore) leaf waste. *Waste and Biomass Valorization, 10*(3), 533–546. https://doi.org/10.1007/S12649-017-0074-Z/FIGURES/7

Awasthi, M. K., Pandey, A. K., Bundela, P. S., & Khan, J. (2015). Co-composting of organic fraction of municipal solid waste mixed with different bulking waste: Characterization of physicochemical parameters and microbial enzymatic dynamic. *Bioresource Technology, 182*, 200–207. https://doi.org/10.1016/j.biortech.2015.01.104

Bala, S., Garg, D., Sridhar, K., Inbaraj, B. S., Singh, R., Kamma, S., Tripathi, M., & Sharma, M. (2023). Transformation of agro-waste into value-added bioproducts and bioactive compounds: Micro/nano formulations and application in the agri-food-pharma sector. *Bioengineering, 10*(2), 152. https://doi.org/10.3390/bioengineering10020152

Balagurumurthy, B., Singh, R., & Bhaskar, T. (2015). Catalysts for thermochemical conversion of biomass. In *Recent advances in thermochemical conversion of biomass* (pp. 109–132). https://doi.org/10.1016/B978-0-444-63289-0.00004-1

Barthod, J., Rumpel, C., & Dignac, M.-F. (2018). Composting with additives to improve organic amendments. A review. *Agronomy for Sustainable Development, 38*(2), 17. https://doi.org/10.1007/s13593-018-0491-9

Baruah, J., Nath, B. K., Sharma, R., Kumar, S., Deka, R. C., Baruah, D. C., & Kalita, E. (2018). Recent trends in the pretreatment of lignocellulosic biomass for value-added products. *Frontiers in Energy Research, 6*. https://doi.org/10.3389/fenrg.2018.00141

Bernal, M. P., Alburquerque, J. A., & Moral, R. (2009). Composting of animal manures and chemical criteria for compost maturity assessment. A review. *Bioresource Technology, 100*(22), 5444–5453. https://doi.org/10.1016/j.biortech.2008.11.027

Bianco, F., Şenol, H., & Papirio, S. (2021). Enhanced lignocellulosic component removal and biomethane potential from chest nut shell by a combined hydrothermal–alkaline pretreatment. *Science of the Total Environment, 762*, 144178. https://doi.org/10.1016/j.scitotenv.2020.144178

Bimakr, M., Rahman, R. A., Taip, F. S., Ganjloo, A., Salleh, L. M., Selamat, J., Hamid, A., & Zaidul, I. S. M. (2011). Comparison of different extraction methods for the extraction of major bioactive flavonoid compounds from spearmint (*Mentha spicata* L.) leaves. *Food and Bioproducts Processing, 89*(1), 67–72. https://doi.org/10.1016/J.FBP.2010.03.002

Canabarro, N., Soares, J. F., Anchieta, C. G., Kelling, C. S., & Mazutti, M. A. (2013). Thermochemical processes for biofuel production from biomass. *Sustainable Chemical Processes, 1*(1), 1–10. https://doi.org/10.1186/2043-7129-1-22

Cao, Y., Wang, J., Huang, H., Sun, E., Butterly, C., Xu, Y., He, H., Zhang, J., & Chang, Z. (2019). Spectroscopic evidence for hyperthermophilic pretreatment intensifying humification during pig manure and rice straw composting. *Bioresource Technology, 294*, 122131. https://doi.org/10.1016/j.biortech.2019.122131

Carbas, B. B., Güler, M., Yücel, K., & Yildiz, H. B. (2023). Construction of novel cyanobacteria-based biological photovoltaic solar cells: Hydrogen and photocurrent generated via both photosynthesis and respiratory system. *Journal of Photochemistry and Photobiology A: Chemistry, 442*, 114764. https://doi.org/10.1016/j.jphotochem.2023.114764

Carmichael, E., Ghassemieh, E., & Lyons, G. (2020). Biorefining of lignocellulosic feedstock and waste materials using ionic liquid. *Materials Science and Engineering: B, 262*, 114741. https://doi.org/10.1016/j.mseb.2020.114741

Chandra, R., Takeuchi, H., & Hasegawa, T. (2012). Methane production from lignocellulosic agricultural crop wastes: A review in context to second generation of biofuel production. *Renewable and Sustainable Energy Reviews, 16*(3), 1462–1476. https://doi.org/10.1016/j.rser.2011.11.035

Chen, B., Azman, S., Dewil, R., & Appels, L. (2023). Alkaline anaerobic digestion of livestock manure: Unveiling mechanisms, applications, and perspective. *Chemical Engineering Journal, 477*, 146852. https://doi.org/10.1016/j.cej.2023.146852

Chowdhury, H., & Loganathan, B. (2019). Third-generation biofuels from microalgae: A review. *Current Opinion in Green and Sustainable Chemistry, 20*, 39–44. https://doi.org/10.1016/J.COGSC.2019.09.003

Dai, B., Guo, X., Yuan, D., & Xu, J. (2018). Comparison of different pretreatments of rice straw substrate to improve biogas production. *Waste and Biomass Valorization, 9*(9), 1503–1512. https://doi.org/10.1007/s12649-017-9950-9

Dali, I., Aydi, A., Stamenic, M., Kolsi, L., Ghachem, K., Zizovic, I., Manef, A., & Delgado, D. R. (2022). Extraction of lyophilized olive mill wastewater using supercritical CO2 processes. *Alexandria Engineering Journal, 61*(1), 237–246. https://doi.org/10.1016/J.AEJ.2021.04.088

Danish, M., & Ahmad, T. (2018). A review on utilization of wood biomass as a sustainable precursor for activated carbon production and application. *Renewable and Sustainable Energy Reviews, 87*, 1–21. https://doi.org/10.1016/J.RSER.2018.02.003

Demirbaş, A. (2004). Conversion of agricultural residues to fuel products via supercritical fluid extraction. *Energy Sources, 26*(12), 1095–1103. https://doi.org/10.1080/00908310390268029

Demirbas, A. (2010). Future fuels for internal combustion engines. *Energy Sources, Part A: Recovery, Utilization, and Environmental Effects, 32*(14), 1273–1281. https://doi.org/10.1080/15567030903060317

Demirbas, A., Caglar, A., Ayas, A., & Karslioglu, S. (1996). Supercritical and catalytic fluid extractions of tea wastes. *Fuel Science and Technology International, 14*(3), 395–404. https://doi.org/10.1080/08843759608947586

Deng, D., Duan, X.-X., Lu, J.-J., Qin, Y.-H., & Wang, C.-W. (2019). Electrogenerated alkaline hydrogen peroxide for rice straw pretreatment to enhance enzymatic hydrolysis. *Bioresource Technology, 292*, 122077. https://doi.org/10.1016/j.biortech.2019.122077

Dorado, M. P., Ballesteros, E., Mittelbach, M., & López, F. J. (2004). Kinetic parameters affecting the alkali-catalyzed transesterification process of used olive oil. *Energy and Fuels, 18*(5), 1457–1462. https://doi.org/10.1021/EF034088O

Durán-Valle, C. J. (2020). Thermochemical processes for the transformation of biomass into biofuels. In *Process systems engineering for biofuels development* (pp. 285–310). Wiley. https://doi.org/10.1002/9781119582694.CH11

Durán-Valle, C. J., Botet-Jiménez, A. B., & Omenat-Morán, D. (2017). Hydrothermal carbonisation: An eco-friendly method for the production of carbon adsorbents. In *Adsorption processes for water treatment and purification* (pp. 77–108). Springer. https://doi.org/10.1007/978-3-319-58136-1_4

Filho, R. M., Ranzi, E., Tognotti, L., Piazza, V., Lietti, L., & Beretta, A. (2022). Chemical engineering transactions catalytic upgrade of pyrolysis vapors: Biofuels production from acetic acid on Ru/TiO_2. *Catalysts*. https://doi.org/10.3303/CET2292055

Gao, Z., Alshehri, K., Li, Y., Qian, H., Sapsford, D., Cleall, P., & Harbottle, M. (2022). Advances in biological techniques for sustainable lignocellulosic waste utilization in biogas production. *Renewable and Sustainable Energy Reviews, 170*, 112995. https://doi.org/10.1016/j.rser.2022.112995

García-Velásquez, C. A., & Cardona, C. A. (2019). Comparison of the biochemical and thermochemical routes for bioenergy production: A techno-economic (TEA), energetic and environmental assessment. *Energy, 172*, 232–242. https://doi.org/10.1016/j.energy.2019.01.073

Gautam, S. P., Bundela, P. S., Pandey, A. K., Awasthi, M. K., & Sarsaiya, S. (2012). Diversity of cellulolytic microbes and the biodegradation of municipal solid waste by a potential strain. *International Journal of Microbiology*, 1–12. https://doi.org/10.1155/2012/325907

Ge, J., Huang, G., Huang, J., Zeng, J., & Han, L. (2015). Modeling of oxygen uptake rate evolution in pig manure–wheat straw aerobic composting process. *Chemical Engineering Journal, 276*, 29–36. https://doi.org/10.1016/j.cej.2015.04.067

Gerardi, M. H. (2003). *The microbiology of anaerobic digesters*. Wiley. https://doi.org/10.1002/0471468967

Gerba, C. P., & Pepper, I. L. (2015). Municipal wastewater treatment. In *Environmental microbiology* (pp. 583–606). Elsevier. https://doi.org/10.1016/B978-0-12-394626-3.00025-9

Harindintwali, J. D., Zhou, J., & Yu, X. (2020). Lignocellulosic crop residue composting by cellulolytic nitrogen-fixing bacteria: A novel tool for environmental sustainability. *Science of the Total Environment, 715*, 136912. https://doi.org/10.1016/j.scitotenv.2020.136912

He, C., Zhang, Z., Ge, C., Liu, W., Tang, Y., Zhuang, X., & Qiu, R. (2019). Synergistic effect of hydrothermal co-carbonization of sewage sludge with fruit and agricultural wastes on hydrochar fuel quality and combustion behavior. *Waste Management, 100*, 171–181. https://doi.org/10.1016/J.WASMAN.2019.09.018

Hubbe, M. A. (2014). Lignocellulose biodegradation in composting. In *Composting for sustainable agriculture* (pp. 43–66). https://doi.org/10.1007/978-3-319-08004-8_3

Ibrahimoglu, B., Cucen, A., & Yilmazoglu, M. Z. (2017). Numerical modeling of a downdraft plasma gasification reactor. *International Journal of Hydrogen Energy, 42*(4), 2583–2591. https://doi.org/10.1016/J.IJHYDENE.2016.06.224

Ioannidou, O., & Zabaniotou, A. (2007). Agricultural residues as precursors for activated carbon production - A review. *Renewable and Sustainable Energy Reviews, 11*, 1966–2005. https://doi.org/10.1016/j.rser.2006.03.013

Isemin, R., Konyakhin, V., Kuzmin, S., Zorin, A., Mikchalev, A., & Viryasov, D. (2010). Solution of some problems which can occur at production and combustion of straw pellets and pellets made of agricultural wastes. *Chemical Engineering Transactions, 19*, 55–60. https://doi.org/10.3303/CET1019010

Jain, M. S., Daga, M., & Kalamdhad, A. S. (2019). Variation in the key indicators during composting of municipal solid organic wastes. *Sustainable Environment Research, 29*(1), 9. https://doi.org/10.1186/s42834-019-0012-9

Jha, S., Okolie, J. A., Nanda, S., & Dalai, A. K. (2022). A review of biomass resources and thermochemical conversion technologies. *Chemical Engineering and Technology, 45*(5), 791–799. https://doi.org/10.1002/CEAT.202100503

Jiang, W., Kumar, A., & Adamopoulos, S. (2018). Liquefaction of lignocellulosic materials and its applications in wood adhesives—A review. *Industrial Crops and Products, 124*, 325–342. https://doi.org/10.1016/J.INDCROP.2018.07.053

Jönsson, L. J., & Martín, C. (2016). Pretreatment of lignocellulose: Formation of inhibitory by-products and strategies for minimizing their effects. *Bioresource Technology, 199*, 103–112. https://doi.org/10.1016/j.biortech.2015.10.009

Joshi, A., Kanthaliya, B., & Arora, J. (2020). Halophytes: The non-conventional crops as source of biofuel production. In M. N. Grigore (Ed.), *Handbook of halophytes* (pp. 1–28). Springer. https://doi.org/10.1007/978-3-030-17854-3_126-1

Joshi, A., Verma, K. K., Rajput, D. V., Minkina, T., & Arora, J. (2022). Recent advances in metabolic engineering of microorganisms for advancing lignocellulose-derived biofuels. *Bioengineered, 13*, 8135–8163. https://www.tandfonline.com/doi/full/10.1080/21655979.2022.2051856

Joshi, A., Verma, K. K., Rajput, V. D., & Arora, J. (2023). Metabolic engineering of microorganisms in advancing biofuel production. In P. Moodley, R. C. Ray, & E. B. G. Kana (Eds.), *Advances in lignocellulosic biofuel production systems* (pp. 181–195). Elsevier. ISBN: 978-0-323-91192-4 (print).

Kainthola, J., Kalamdhad, A. S., & Goud, V. V. (2019). A review on enhanced biogas production from anaerobic digestion of lignocellulosic biomass by different enhancement techniques. *Process Biochemistry, 84*, 81–90. https://doi.org/10.1016/j.procbio.2019.05.023

Karnchanawong, S., Mongkontep, T., & Praphunsri, K. (2017). Effect of green waste pretreatment by sodium hydroxide and biomass fly ash on composting process. *Journal of Cleaner Production, 146*, 14–19. https://doi.org/10.1016/j.jclepro.2016.07.126

Kato, D. M., Elía, N., Flythe, M., & Lynn, B. C. (2014). Pretreatment of lignocellulosic biomass using Fenton chemistry. *Bioresource Technology, 162*, 273–278. https://doi.org/10.1016/j.biortech.2014.03.151

Ketsub, N., Latif, A., Kent, G., Doherty, W. O. S., O'Hara, I. M., Zhang, Z., & Kaparaju, P. (2021). A systematic evaluation of biomethane production from sugarcane trash pretreated by different methods. *Bioresource Technology, 319*, 124137. https://doi.org/10.1016/j.biortech.2020.124137

Khan, H. M., Iqbal, T., Yasin, S., Ali, C. H., Abbas, M. M., Jamil, M. A., Hussain, A., Soudagar, M. E. M., & Rahman, M. M. (2021). Application of agricultural waste as heterogeneous catalysts for biodiesel production. *Catalysts, 11*(10), 1215. https://doi.org/10.3390/CATAL11101215

Khan, M. U., Usman, M., Ashraf, M. A., Dutta, N., Luo, G., & Zhang, S. (2022). A review of recent advancements in pretreatment techniques of lignocellulosic materials for biogas production: Opportunities and limitations. *Chemical Engineering Journal Advances, 10*, 100263. https://doi.org/10.1016/j.ceja.2022.100263

Klüpfel, C., Herklotz, B., & Biller, P. (2023). Influence of processing conditions and biochemical composition on the hydrothermal liquefaction of digested urban and agricultural wastes. *Fuel, 352*, 129016. https://doi.org/10.1016/J.FUEL.2023.129016

Kossalbayev, B. D., Yilmaz, G., Sadvakasova, A. K., Zayadan, B. K., Belkozhayev, A. M., Kamshybayeva, G. K., Sainova, G. A., Bozieva, A. M., Alharby, H. F., Tomo, T., & Allakhverdiev, S. I. (2024). Biotechnological production of hydrogen: Design features of photobioreactors and improvement of conditions for cultivating cyanobacteria. *International Journal of Hydrogen Energy, 49*, 413–432. https://doi.org/10.1016/j.ijhydene.2023.09.001

Krishnamoorthy, S., Iyer, N. R., & Murthy, A. R. (2015). Carbonization on combustion and biodegradation of agricultural waste as a possible source of silica. *Applied Biochemistry and Biotechnology, 175*(3), 1622–1632. https://doi.org/10.1007/S12010-014-1362-8

Krivokapić, J., Ivanović, J., Krkobabić, M., Arsenijević, J., & Ibrić, S. (2021). Supercritical fluid impregnation of microcrystalline cellulose derived from the agricultural waste with ibuprofen. *Sustainable Chemistry and Pharmacy, 21*, 100447. https://doi.org/10.1016/J.SCP.2021.100447

Lahijani, P., Mohammadi, M., Mohamed, A. R., Ismail, F., Lee, K. T., & Amini, G. (2022). Upgrading biomass-derived pyrolysis bio-oil to bio-jet fuel through catalytic cracking and hydrodeoxygenation: A review of recent progress. *Energy Conversion and Management, 268*, 115956. https://doi.org/10.1016/J.ENCONMAN.2022.115956

Latifah, O., Ahmed, O. H., Susilawati, K., & Majid, N. M. (2015). Compost maturity and nitrogen availability by co-composting of paddy husk and chicken manure amended with clinoptilolite zeolite. *Waste Management and Research: The Journal for a Sustainable Circular Economy, 33*(4), 322–331. https://doi.org/10.1177/0734242X15576771

Lee, S. Y., Sankaran, R., Chew, K. W., Tan, C. H., Krishnamoorthy, R., Chu, D.-T., & Show, P.-L. (2019). Waste to bioenergy: A review on the recent conversion technologies. *BMC Energy, 1*(1), 4. https://doi.org/10.1186/s42500-019-0004-7

Leung, D. Y. C., Wu, X., & Leung, M. K. H. (2010). A review on biodiesel production using catalyzed transesterification. *Applied Energy, 87*(4), 1083–1095. https://doi.org/10.1016/j.apenergy.2009.10.006

Liang, W., Jiang, C., Wang, G., Ning, X., Zhang, J., Guo, X., Xu, R., Wang, P., Ye, L., Li, J., & Wang, C. (2022). Research on the co-combustion characteristics and kinetics of agricultural waste hydrochar and anthracite. *Renewable Energy, 194*, 1119–1130. https://doi.org/10.1016/J.RENENE.2022.05.157

Liew, Y. X., Chan, Y. J., Manickam, S., Chong, M. F., Chong, S., Tiong, T. J., Lim, J. W., & Pan, G.-T. (2020). Enzymatic pretreatment to enhance anaerobic bioconversion of high strength wastewater to biogas: A review. *Science of the Total Environment, 713*, 136373. https://doi.org/10.1016/j.scitotenv.2019.136373

Liew, C. S., Yunus, N. M., Chidi, B. S., Lam, M. K., Goh, P. S., Mohamad, M., Sin, J. C., Lam, S. M., Lim, J. W., & Lam, S. S. (2022). A review on recent disposal of hazardous sewage sludge via anaerobic digestion and novel composting. *Journal of Hazardous Materials, 423*, 126995. https://doi.org/10.1016/j.jhazmat.2021.126995

Lin, L., Xu, F., Ge, X., & Li, Y. (2019). Biological treatment of organic materials for energy and nutrients production—Anaerobic digestion and composting. In *Advances in bioenergy* (pp. 121–181). Elsevier. https://doi.org/10.1016/bs.aibe.2019.04.002

Liu, C.-G., Li, K., Wen, Y., Geng, B.-Y., Liu, Q., & Lin, Y.-H. (2019). Bioethanol: New opportunities for an ancient product. In *Advances in bioenergy* (pp. 1–34). Elsevier. https://doi.org/10.1016/bs.aibe.2018.12.002

Liu, T., Miao, P., Shi, Y., Tang, K. H. D., & Yap, P.-S. (2022). Recent advances, current issues and future prospects of bioenergy production: A review. *Science of the Total Environment, 810*, 152181. https://doi.org/10.1016/j.scitotenv.2021.152181

Liu, H., Li, D., Huang, Y., Lin, Q., Huang, L., Cheng, S., Sun, S., & Zhu, Z. (2023). Addition of bacterial consortium produced high-quality sugarcane bagasse compost as an environmental-friendly fertilizer: Optimizing areca nut (*Areca catechu* L.) production, soil fertility and microbial community structure. *Applied Soil Ecology, 188*, 104920. https://doi.org/10.1016/j.apsoil.2023.104920

Lourenço, V. A., Nadaleti, W. C., Vieira, B. M., & Li, H. (2021). Investigation of ethyl biodiesel via transesterification of rice bran oil: Bioenergy from residual biomass in Pelotas, Rio Grande do Sul - Brazil. *Renewable and Sustainable Energy Reviews, 144*, 111016. https://doi.org/10.1016/J.RSER.2021.111016

Ly, H. V., Tran, Q. K., Kim, S. S., Kim, J., Choi, S. S., & Oh, C. (2021). Catalytic upgrade for pyrolysis of food waste in a bubbling fluidized-bed reactor. *Environmental Pollution, 275*, 116023. https://doi.org/10.1016/J.ENVPOL.2020.116023

Ma, S., Wang, H., Li, L., Gu, X., & Zhu, W. (2021). Enhanced biomethane production from corn straw by a novel anaerobic digestion strategy with mechanochemical pretreatment. *Renewable and Sustainable Energy Reviews, 146*, 111099. https://doi.org/10.1016/j.rser.2021.111099

Malode, S. J., Gaddi, S. A. M., Kamble, P. J., Nalwad, A. A., Muddapur, U. M., & Shetti, N. P. (2022). Recent evolutionary trends in the production of biofuels. *Materials Science for Energy Technologies, 5*, 262–277. https://doi.org/10.1016/j.mset.2022.04.001

Mansur, D., Tago, T., Masuda, T., & Abimanyu, H. (2014). Conversion of cacao pod husks by pyrolysis and catalytic reaction to produce useful chemicals. *Biomass and Bioenergy, 66*, 275–285. https://doi.org/10.1016/J.BIOMBIOE.2014.03.065

Martínez-Gallardo, M. R., Estrella-González, M. J., Suárez-Estrella, F., López-González, J. A., Jurado, M. M., Toribio, A. J., & López, M. J. (2023). Effect of upstream bioactivation of plant residues to accelerate the composting process and improve product quality. *Agronomy, 13*(6), 1638. https://doi.org/10.3390/agronomy13061638

Martins-Vieira, J. C., Torres-Mayanga, P. C., & Lachos-Perez, D. (2022). Hydrothermal processing of lignocellulosic biomass: An overview of subcritical and supercritical water hydrolysis. *BioEnergy Research, 16*(3), 1296–1317. https://doi.org/10.1007/S12155-022-10553-8

Michailos, S., Parker, D., & Webb, C. (2019). Design, sustainability analysis and multiobjective optimisation of ethanol production via syngas fermentation. *Waste and Biomass Valorization, 10*(4), 865–876. https://doi.org/10.1007/S12649-017-0151-3/FIGURES/8

Mishra, A., Kumar, M., Bolan, N. S., Kapley, A., Kumar, R., & Singh, L. (2021). Multidimensional approaches of biogas production and up-gradation: Opportunities and challenges. *Bioresource Technology, 338*, 125514. https://doi.org/10.1016/j.biortech.2021.125514

Mittal, S., Ahlgren, E. O., & Shukla, P. R. (2019). Future biogas resource potential in India: A bottom-up analysis. *Renewable Energy, 141*, 379–389. https://doi.org/10.1016/j.renene.2019.03.133

Mohan, D., Pittman, C. U., & Steele, P. H. (2006). Pyrolysis of wood/biomass for bio-oil: A critical review. *Energy and Fuels, 20*(3), 848–889. https://doi.org/10.1021/EF0502397

Mussoline, W., Esposito, G., Giordano, A., & Lens, P. (2013). The anaerobic digestion of rice straw: A review. *Critical Reviews in Environmental Science and Technology, 43*(9), 895–915. https://doi.org/10.1080/10643389.2011.627018

Naik, G. P., Poonia, A. K., & Chaudhari, P. K. (2022). Maximization of biogas by minimal microwave and alkaline pretreatment of rice straw. *Biomass Conversion and Biorefinery*. https://doi.org/10.1007/s13399-022-03539-1

Nair, L. G., Agrawal, K., & Verma, P. (2022). An insight into the principles of lignocellulosic biomass-based zero-waste biorefineries: A green leap towards imperishable energy-based future. *Biotechnology and Genetic Engineering Reviews, 38*(2), 288–338. https://doi.org/10.1080/02648725.2022.2082223

Najser, T., Gaze, B., Knutel, B., Verner, A., Najser, J., Mikeska, M., Chojnacki, J., & Němček, O. (2022). Analysis of the effect of catalytic additives in the agricultural waste combustion process. *Materials, 15*(10), 3526. https://doi.org/10.3390/MA15103526

Nakhshiniev, B., Biddinika, M. K., Gonzales, H. B., Sumida, H., & Yoshikawa, K. (2014). Evaluation of hydrothermal treatment in enhancing rice straw compost stability and maturity. *Bioresource Technology, 151*, 306–313. https://doi.org/10.1016/j.biortech.2013.10.083

Nargotra, P., Sharma, V., & Bajaj, B. K. (2019). Consolidated bioprocessing of surfactant-assisted ionic liquid-pretreated *Parthenium hysterophorus* L. biomass for bioethanol production. *Bioresource Technology, 289*, 121611. https://doi.org/10.1016/j.biortech.2019.121611

Nargotra, P., Sharma, V., Sharma, S., Kapoor, N., & Bajaj, B. K. (2022). Development of consolidated bioprocess for biofuel-ethanol production from ultrasound-assisted deep eutectic solvent pretreated *Parthenium hysterophorus* biomass. *Biomass Conversion and Biorefinery, 12*(12), 5767–5782. https://doi.org/10.1007/s13399-020-01017-0

Oanh, N. T. K., Permadi, D. A., Hopke, P. K., Smith, K. R., Dong, N. P., & Dang, A. N. (2018). Annual emissions of air toxics emitted from crop residue open burning in Southeast Asia over the period of 2010–2015. *Atmospheric Environment, 187*, 163–173.

Ocio, J. A., Brookes, P. C., & Jenkinson, D. S. (1991). Field incorporation of straw and its effects on soil microbial biomass and soil inorganic N. *Soil Biology and Biochemistry, 23*(2), 171–176. https://doi.org/10.1016/0038-0717(91)90131-3

Odude, V. O., Adesina, A. J., Oyetunde, O. O., Adeyemi, O. O., Ishola, N. B., Etim, A. O., & Betiku, E. (2019). Application of agricultural waste-based catalysts to transesterification of esterified palm kernel oil into biodiesel: A case of banana fruit peel versus cocoa pod husk. *Waste and Biomass Valorization, 10*(4), 877–888. https://doi.org/10.1007/S12649-017-0152-2/TABLES/10

Okolie, J. A., Nanda, S., Dalai, A. K., Berruti, F., & Kozinski, J. A. (2020). A review on subcritical and supercritical water gasification of biogenic, polymeric and petroleum wastes to hydrogen-rich synthesis gas. *Renewable and Sustainable Energy Reviews, 119*, 109546. https://doi.org/10.1016/J.RSER.2019.109546

Panahi, K. S. H., Hosseinzadeh-Bandbafha, H., Dehhaghi, M., Orooji, Y., Mahian, O., Shahbeik, H., Kiehbadroudinezhad, M., Kalam, M. A., Karimi-Maleh, H., Salehi Jouzani, G., Mei, C., Guillemin, G. G., Nizami, A.-S., Wang, Y., Gupta, V. K., Lam, S. S., Pan, J., Kim, K.-H., Peng,

W., et al. (2024). Nanotechnology applications in biodiesel processing and production: A comprehensive review. *Renewable and Sustainable Energy Reviews, 192*, 114219. https://doi.org/10.1016/j.rser.2023.114219

Pandey, P. K., Cao, W., Biswas, S., & Vaddella, V. (2016). A new closed loop heating system for composting of green and food wastes. *Journal of Cleaner Production, 133*, 1252–1259. https://doi.org/10.1016/j.jclepro.2016.05.114

Panwar, N. L., & Pawar, A. (2022). Influence of activation conditions on the physicochemical properties of activated biochar: A review. *Biomass Conversion and Biorefinery, 12*(3), 925–947. https://doi.org/10.1007/S13399-020-00870-3/FIGURES/4

Paraskeva, P., Kalderis, D., & Diamadopoulos, E. (2008). Production of activated carbon from agricultural by-products. *Journal of Chemical Technology and Biotechnology, 83*(5), 581–592. https://doi.org/10.1002/JCTB.1847

Patra, B. R., Nanda, S., Dalai, A. K., & Meda, V. (2021). Slow pyrolysis of agro-food wastes and physicochemical characterization of biofuel products. *Chemosphere, 285*, 131431. https://doi.org/10.1016/J.CHEMOSPHERE.2021.131431

Pavlovič, I., Knez, Ž., & Škerget, M. (2013). Hydrothermal reactions of agricultural and food processing wastes in sub- and supercritical water: A review of fundamentals, mechanisms, and state of research. *Journal of Agricultural and Food Chemistry, 61*(34), 8003–8025. https://doi.org/10.1021/JF401008A/ASSET/IMAGES/MEDIUM/JF-2013-01008A_0003.GIF

Ponnusamy, V. K., Nagappan, S., Bhosale, R. R., Lay, C. H., Duc Nguyen, D., Pugazhendhi, A., Chang, S. W., & Kumar, G. (2020). Review on sustainable production of biochar through hydrothermal liquefaction: Physico-chemical properties and applications. *Bioresource Technology, 310*, 123414. https://doi.org/10.1016/J.BIORTECH.2020.123414

Preetam, A., Jadhao, P. R., Naik, S. N., Pant, K. K., & Kumar, V. (2023). Supercritical fluid technology - an eco-friendly approach for resource recovery from e-waste and plastic waste: A review. *Separation and Purification Technology, 304*, 122314. https://doi.org/10.1016/J.SEPPUR.2022.122314

Priya, A. S., Devi, K. R. S., Karthik, K., & Sugunan, S. (2020). Modified rice husk silica from biowaste: An efficient catalyst for transesterification of diethyl malonate and benzyl alcohol. *Waste and Biomass Valorization, 11*(9), 4809–4819. https://doi.org/10.1007/S12649-019-00808-5/TABLES/6

Putra, N. R., Yian, L. N., Nasir, H. M., Binti Idham, Z., & Yunus, M. A. C. (2018). Effects of process parameters on peanut skins extract and CO_2 diffusivity by supercritical fluid extraction. *IOP Conference Series: Materials Science and Engineering, 334*(1), 012057. https://doi.org/10.1088/1757-899X/334/1/012057

Putro, J. N., Soetaredjo, F. E., Lin, S.-Y., Ju, Y.-H., & Ismadji, S. (2016). Pretreatment and conversion of lignocellulose biomass into valuable chemicals. *RSC Advances, 6*(52), 46834–46852. https://doi.org/10.1039/C6RA09851G

Rastogi, M., Nandal, M., & Khosla, B. (2020). Microbes as vital additives for solid waste composting. *Heliyon, 6*(2), e03343. https://doi.org/10.1016/j.heliyon.2020.e03343

Raza, S., & Ahmad, J. (2016). Composting process: A review. *International Journal of Biological Research, 4*(2), 102. https://doi.org/10.14419/ijbr.v4i2.6354

Saini, J. K., Saini, R., & Tewari, L. (2015). Lignocellulosic agriculture wastes as biomass feedstocks for second-generation bioethanol production: Concepts and recent developments. *3. Biotech, 5*(4), 337–353. https://doi.org/10.1007/s13205-014-0246-5

Sánchez, Ó. J., Ospina, D. A., & Montoya, S. (2017). Compost supplementation with nutrients and microorganisms in composting process. *Waste Management, 69*, 136–153. https://doi.org/10.1016/j.wasman.2017.08.012

Santos, J., Ouadi, M., Jahangiri, H., & Hornung, A. (2020). Thermochemical conversion of agricultural wastes applying different reforming temperatures. *Fuel Processing Technology, 203*, 106402. https://doi.org/10.1016/j.fuproc.2020.106402

Saravanan, A., Senthil Kumar, P., Rangasamy, G., Hariharan, R., Hemavathy, R. V., Deepika, P. D., Anand, K., & Karthika, S. (2023). Strategies for enhancing the efficacy of anaerobic digestion of food industry wastewater: An insight into bioreactor types, challenges, and future scope. *Chemosphere, 310*, 136856. https://doi.org/10.1016/j.chemosphere.2022.136856

Sarker, S., Lamb, J. J., Hjelme, D. R., & Lien, K. M. (2019). A review of the role of critical parameters in the design and operation of biogas production plants. *Applied Sciences, 9*(9), 1915. https://doi.org/10.3390/app9091915

Sarto, S., Hildayati, R., & Syaichurrozi, I. (2019). Effect of chemical pretreatment using sulfuric acid on biogas production from water hyacinth and kinetics. *Renewable Energy, 132*, 335–350. https://doi.org/10.1016/j.renene.2018.07.121

Sarwer, A., Hussain, M., Al-Muhtaseb, A. H., Inayat, A., Rafiq, S., Khurram, M. S., Ul-Haq, N., Shah, N. S., Alaud Din, A., Ahmad, I., & Jamil, F. (2022). Suitability of biofuels production on commercial scale from various feedstocks: A critical review. *ChemBioEng Reviews, 9*(5), 423–441. https://doi.org/10.1002/CBEN.202100049

Sato, T., Ikeya, Y., Ichi, A. S., Yagasaki, K., Ichi, N. K., & Itoh, N. (2019). Extraction of strawberry leaves with supercritical carbon dioxide and entrainers: Antioxidant capacity, total phenolic content, and inhibitory effect on uric acid production of the extract. *Food and Bioproducts Processing, 117*, 160–169. https://doi.org/10.1016/J.FBP.2019.07.003

Schroeder, P., dos Santos Barreto, M., Romeiro, G. A., & Figueiredo, M. K. K. (2018). Development of energetic alternatives to use of waste of *Annona muricata* L. *Waste and Biomass Valorization, 9*(8), 1459–1467. https://doi.org/10.1007/S12649-017-9869-1/TABLES/7

Shan, G., Li, W., Bao, S., Li, Y., & Tan, W. (2023). Co-hydrothermal carbonization of agricultural waste and sewage sludge for product quality improvement: Fuel properties of hydrochar and fertilizer quality of aqueous phase. *Journal of Environmental Management, 326*, 116781. https://doi.org/10.1016/J.JENVMAN.2022.116781

Shao, L.-M., Zhang, C.-Y., Wu, D., Lü, F., Li, T.-S., & He, P.-J. (2014). Effects of bulking agent addition on odorous compounds emissions during composting of OFMSW. *Waste Management, 34*(8), 1381–1390. https://doi.org/10.1016/j.wasman.2014.04.016

Sharma, S., Nargotra, P., Sharma, V., Bangotra, R., Kaur, M., Kapoor, N., Paul, S., & Bajaj, B. K. (2021a). Nanobiocatalysts for efficacious bioconversion of ionic liquid pretreated sugarcane tops biomass to biofuel. *Bioresource Technology, 333*, 125191. https://doi.org/10.1016/j.biortech.2021.125191

Sharma, V., Nargotra, P., Sharma, S., & Bajaj, B. K. (2021b). Efficacy and functional mechanisms of a novel combinatorial pretreatment approach based on deep eutectic solvent and ultrasonic waves for bioconversion of sugarcane bagasse. *Renewable Energy, 163*, 1910–1922. https://doi.org/10.1016/j.renene.2020.10.101

Sharma, V., Sharma, D., Tsai, M.-L., Ortizo, R. G. G., Yadav, A., Nargotra, P., Chen, C.-W., Sun, P.-P., & Dong, C.-D. (2023). Insights into the recent advances of agro-industrial waste valorization for sustainable biogas production. *Bioresource Technology, 390*, 129829. https://doi.org/10.1016/j.biortech.2023.129829

Sheng, T., Zhao, L., Liu, W.-Z., Gao, L., & Wang, A.-J. (2017). Fenton pre-treatment of rice straw with citric acid as an iron chelate reagent for enhancing saccharification. *RSC Advances, 7*(51), 32076–32086. https://doi.org/10.1039/C7RA04329E

Sims, R. E. H., Mabee, W., Saddler, J. N., & Taylor, M. (2010). An overview of second generation biofuel technologies. *Bioresource Technology, 101*(6), 1570–1580. https://doi.org/10.1016/j.biortech.2009.11.046

Singh, R. K., & Sarkar, A. (2023). Pyrolysis of torrefied crop residue: Optimization using response surface methodology and benefits of bio-char in co-combustion with coal. *Industrial Crops and Products, 199*, 116786. https://doi.org/10.1016/J.INDCROP.2023.116786

Sudharsan Varma, V., & Kalamdhad, A. S. (2015). Evolution of chemical and biological characterization during thermophilic composting of vegetable waste using rotary drum composter. *International journal of Environmental Science and Technology, 12*(6), 2015–2024. https://doi.org/10.1007/s13762-014-0582-3

Syaichurrozi, I., Villta, P. K., Nabilah, N., & Rusdi, R. (2019). Effect of sulfuric acid pretreatment on biogas production from *Salvinia molesta*. *Journal of Environmental Chemical Engineering, 7*(1), 102857. https://doi.org/10.1016/j.jece.2018.102857

Tan, X. F., Liu, B. S., Liu, G. Y., Gu, L. Y., Zeng, M. G., Hu, J. X., Wang, X., Liu, H. S., & Jiang, H. L. (2017). Biochar as potential sustainable precursors for activated carbon production: Multiple applications in environmental protection and energy storage. *Bioresource Technology, 227*, 359–372. https://doi.org/10.1016/J.BIORTECH.2016.12.083

Titirici, M. M., Funke, A., & Kruse, A. (2015). Hydrothermal carbonization of biomass. In *Recent advances in thermochemical conversion of biomass* (pp. 325–352). https://doi.org/10.1016/B978-0-444-63289-0.00012-0

Torres-Climent, A., Martin-Mata, J., Marhuenda-Egea, F., Moral, R., Barber, X., Perez-Murcia, M. D., & Paredes, C. (2015). Composting of the solid phase of digestate from biogas production: Optimization of the moisture, C/N Ratio, and pH Conditions. *Communications in Soil Science and Plant Analysis, 46*(sup1), 197–207. https://doi.org/10.1080/00103624.2014.988591

Tsubaki, S., Onda, A., Hiraoka, M., Fujii, S., Azuma, J., & Wada, Y. (2017). Microwave-assisted water extraction of carbohydrates from unutilized biomass. In *Water extraction of bioactive compounds: From plants to drug development* (pp. 199–219). Elsevier. https://doi.org/10.1016/B978-0-12-809380-1.00007-3

Uçkun Kiran, E., Trzcinski, A. P., & Liu, Y. (2015). Enhancing the hydrolysis and methane production potential of mixed food waste by an effective enzymatic pretreatment. *Bioresource Technology, 183*, 47–52. https://doi.org/10.1016/j.biortech.2015.02.033

Wang, X., Liu, Y., Cui, X., Xiao, J., Lin, G., Chen, Y., Yang, H., & Chen, H. (2020). Production of furfural and levoglucosan from typical agricultural wastes via pyrolysis coupled with hydrothermal conversion: Influence of temperature and raw materials. *Waste Management, 114*, 43–52. https://doi.org/10.1016/J.WASMAN.2020.06.045

Waszkielis, K. M., Wronowski, R., Chlebus, W., Białobrzewski, I., Dach, J., Pilarski, K., & Janczak, D. (2013). The effect of temperature, composition and phase of the composting process on the thermal conductivity of the substrate. *Ecological Engineering, 61*, 354–357. https://doi.org/10.1016/j.ecoleng.2013.09.024

Weiland, P. (2010). Biogas production: Current state and perspectives. *Applied Microbiology and Biotechnology, 85*(4), 849–860. https://doi.org/10.1007/s00253-009-2246-7

Wu, M., Peng, J., Dong, Y., Pang, J., & Zhang, X. (2021). Extraction and oxypropylation of lignin by an efficient and mild integration process from agricultural waste. *Industrial Crops and Products, 172*, 114013. https://doi.org/10.1016/J.INDCROP.2021.114013

Xi, B., He, X., Dang, Q., Yang, T., Li, M., Wang, X., Li, D., & Tang, J. (2015). Effect of multi-stage inoculation on the bacterial and fungal community structure during organic municipal solid wastes composting. *Bioresource Technology, 196*, 399–405. https://doi.org/10.1016/j.biortech.2015.07.069

Xia, S., Lin, J., Sayanjali, S., Shen, C., & Cheong, L. (2024). Lipase-catalyzed production of biodiesel: A critical review on feedstock, enzyme carrier and process factors. *Biofuels, Bioproducts and Biorefining, 18*(1), 291–309. https://doi.org/10.1002/bbb.2561

Xu, F., Li, Y., Ge, X., Yang, L., & Li, Y. (2018). Anaerobic digestion of food waste—Challenges and opportunities. *Bioresource Technology, 247*, 1047–1058. https://doi.org/10.1016/j.biortech.2017.09.020

Xu, N., Liu, S., Xin, F., Zhou, J., Jia, H., Xu, J., Jiang, M., & Dong, W. (2019). Biomethane production from lignocellulose: Biomass recalcitrance and its impacts on anaerobic digestion. *Frontiers in Bioengineering and Biotechnology, 7*. https://doi.org/10.3389/fbioe.2019.00191

Xu, X., Zhou, Q., & Yu, D. (2022). The future of hydrogen energy: Bio-hydrogen production technology. *International Journal of Hydrogen Energy, 47*(79), 33677–33698. https://doi.org/10.1016/j.ijhydene.2022.07.261

Yadav, A., Sharma, V., Tsai, M.-L., Chen, C.-W., Sun, P.-P., Nargotra, P., Wang, J.-X., & Dong, C.-D. (2023). Development of lignocellulosic biorefineries for the sustainable production of biofuels: Towards circular bioeconomy. *Bioresource Technology, 381*, 129145. https://doi.org/10.1016/j.biortech.2023.129145

Yadvika, S., Sreekrishnan, T. R., Kohli, S., & Rana, V. (2004). Enhancement of biogas production from solid substrates using different techniques—a review. *Bioresource Technology, 95*(1), 1–10. https://doi.org/10.1016/j.biortech.2004.02.010

Yahya, M. A., Al-Qodah, Z., & Ngah, C. W. Z. (2015). Agricultural bio-waste materials as potential sustainable precursors used for activated carbon production: A review. *Renewable and Sustainable Energy Reviews, 46*, 218–235. https://doi.org/10.1016/J.RSER.2015.02.051

Yang, S.-T. (2007). Bioprocessing—from biotechnology to biorefinery. In *Bioprocessing for value-added products from renewable resources* (pp. 1–24). Elsevier. https://doi.org/10.1016/B978-044452114-9/50002-5

Yang, Z., Wu, Y., Zhang, Z., Li, H., Li, X., Egorov, R. I., Strizhak, P. A., & Gao, X. (2019). Recent advances in co-thermochemical conversions of biomass with fossil fuels focusing on the synergistic effects. *Renewable and Sustainable Energy Reviews, 103*, 384–398. https://doi.org/10.1016/J.RSER.2018.12.047

Zhang, L., & Sun, X. (2016). Influence of bulking agents on physical, chemical, and microbiological properties during the two-stage composting of green waste. *Waste Management, 48*, 115–126. https://doi.org/10.1016/j.wasman.2015.11.032

Zhang, S., Zou, L., Wan, Y., Ye, M., Ye, J., Li, Y.-Y., & Liu, J. (2020). Using an expanded granular sludge bed reactor for advanced anaerobic digestion of food waste pretreated with enzyme: The feasibility and its performance. *Bioresource Technology, 311*, 123504. https://doi.org/10.1016/j.biortech.2020.123504

Zhang, J., Chen, X., Lv, P., Luo, W., Wang, Z., Xu, J., & Wang, Z. (2021). Bionic-immobilized recombinant lipase obtained via bio-silicification and its catalytic performance in biodiesel production. *Fuel, 304*, 121594. https://doi.org/10.1016/j.fuel.2021.121594

Zhao, Y., Lu, W. J., Wang, H. T., & Yang, J. L. (2009). Fermentable hexose production from corn stalks and wheat straw with combined supercritical and subcritical hydrothermal technology. *Bioresource Technology, 100*(23), 5884–5889. https://doi.org/10.1016/J.BIORTECH.2009.06.079

Zhao, Y., Lu, Q., Wei, Y., Cui, H., Zhang, X., Wang, X., Shan, S., & Wei, Z. (2016). Effect of actinobacteria agent inoculation methods on cellulose degradation during composting based on redundancy analysis. *Bioresource Technology, 219*, 196–203. https://doi.org/10.1016/j.biortech.2016.07.117

Zheng, Y., Zhao, J., Xu, F., & Li, Y. (2014). Pretreatment of lignocellulosic biomass for enhanced biogas production. *Progress in Energy and Combustion Science, 42*, 35–53. https://doi.org/10.1016/j.pecs.2014.01.001

Zheng, Q., Zhou, T., Wang, Y., Cao, X., Wu, S., Zhao, M., Wang, H., Xu, M., Zheng, B., Zheng, J., & Guan, X. (2018). Pretreatment of wheat straw leads to structural changes and improved enzymatic hydrolysis. *Scientific Reports, 8*(1), 1321. https://doi.org/10.1038/s41598-018-19517-5

Zhou, X., Ma, J., Ji, Z., Zhang, X., Ramaswamy, S., Xu, F., & Sun, R. (2014). Dilute acid pretreatment differentially affects the compositional and architectural features of *Pinus bungeana*Zucc. compression and opposite wood tracheid walls. *Industrial Crops and Products, 62*, 196–203. https://doi.org/10.1016/j.indcrop.2014.08.035

Zhuo, C., Xueqin, L., Zhiwei, W., Yantao, Y., Tanglei, S., Taoli, H., Peng, L., Yanling, L., Youqing, W., Tingzhou, L., & Jingshen, Q. (2023). Techno-economic and whole life cycle assessment of ester fuels production from agricultural waste via hydrothermal liquefaction. *Industrial Crops and Products, 192*, 116096. https://doi.org/10.1016/J.INDCROP.2022.116096

Chapter 4
Biotechnological Approaches for Agriculture Waste Transformation

Cristina Soares, Olena Dorosh, Filipe Fernandes, Clara Grosso, Manuela Moreira, and Cristina Delerue-Matos

Abbreviations

3D	Three-dimensional
CCoAOMT	Caffeoyl-CoA O-methyltransferase
COMT	Caffeic acid O-methyltransferase
CRISPR	Clustered Regularly Interspaced Short Palindromic Repeats
CRISPR-Cas	Clustered Regularly Interspaced Short Palindromic Repeats-CRISPR-associated proteins (Cas)
CSE	Caffeoyl shikimate esterase
DNA	Deoxyribonucleic acid
DSBs	Double-strand breaks
EMS	Ethyl Methane Sulphonate
F5H	Ferulate 5-hydroxylase
GBSSI	Granule-bound starch synthase I
GMOs	Genetically modified organisms
gRNA	Guide RNA
Hv	Hordeum vulgare
KK3	Khon Kaen 3 sugarcane variety
M2	Second-generation plants that have been subjected to mutagenesis

The original version of the chapter has been revised. A correction to this chapter can be found at https://doi.org/10.1007/978-3-031-61133-9_18

C. Soares (✉) · O. Dorosh · F. Fernandes · C. Grosso · M. Moreira · C. Delerue-Matos
REQUIMTE/LAQV, Instituto Superior de Engenharia do Porto, Instituto Politécnico do Porto, Rua Dr. António Bernardino de Almeida, Porto, Portugal
e-mail: cds@isep.ipp.pt; olena.dorosh@graq.isep.ipp.pt; filipe.fernandes@graq.isep.ipp.pt; claragrosso@graq.isep.ipp.pt; manuela.moreira@graq.isep.ipp.pt; cmm@isep.ipp.pt

J. Arora et al. (eds.), *Transforming Agriculture Residues for Sustainable Development*, Waste as a Resource, https://doi.org/10.1007/978-3-031-61133-9_4

MAE	Microwave-assisted extraction
mito-TALEN	mitochondrial-TALEN
NPs	Nanoparticles
PBS	Primer binding site
PE	Prime editor
pegRNA	Prime editing guide RNA
RNA	Ribonucleic acid
RNAi	RNA interference
RT	Reverse transcription
S/G	Syringyl/guaiacyl
SDGs	Sustainable development goals
SoLIM	Limonene synthase gene
SWE	Subcritical water extraction
TALENs	Transcription activator-like effector nucleases
TILLING	Targeting Induced Local Lesions IN Genomes
TLPs	Trait landing pads
UAE	Ultrasonic-assisted extraction
ZFNs	Zinc finger nucleases
ZFPs	Zinc finger proteins

4.1 Introduction

As the world's population grows, so does the demand for agricultural products, increasing agricultural waste. The absolute volume of this waste generated is staggering. For instance, the global agricultural sector is estimated to produce 1.3 billion tons of waste annually, including crop residues, livestock manure, and agro-industrial byproducts (Babla et al., 2022).

Environmentally, the improper management of agricultural waste has far-reaching impacts. It contributes to major environmental problems such as water pollution, soil degradation, and air pollution (Abbasi et al., 2014). Nutrient surplus from agricultural waste, particularly nitrogen and phosphorus, leads to the eutrophication of water bodies, adversely affecting aquatic life and water quality (Tiwari & Pal, 2022). Additionally, the open burning of crop residues, a common practice in many regions, releases harmful pollutants into the atmosphere, contributing to air quality issues and respiratory problems in humans (Lan et al., 2022). Agricultural waste also significantly contributes to global greenhouse gas emissions. The decomposition of organic waste, especially under anaerobic conditions, releases methane, a potent greenhouse gas (Sánchez et al., 2015). This contributes to climate change, exacerbating extreme weather conditions and impacting global ecosystems. The mismanagement of agricultural waste also represents a lost opportunity in terms of resource recovery. Much of this waste contains valuable organic material that could be repurposed into bioenergy, biofertilizers, and other beneficial products, promoting a circular economy (Awasthi et al., 2022; Babla et al., 2022). However, this potential remains largely unexploited due to technological, logistical, and economic barriers (Batista et al., 2021).

Therefore, addressing the challenges posed by agricultural waste is critical for environmental sustainability, climate change mitigation, and the efficient use of natural resources (Çakmakçı et al., 2023). It necessitates innovative and sustainable management strategies to turn this expanding problem into an opportunity for green growth and sustainable development (Awasthi et al., 2022; Koul et al., 2022).

Historically, agricultural waste management primarily involved open burning or landfill disposal, leading to environmental hazards and resource wastage (Ferronato & Torretta, 2019; Ma & Liu, 2019; Koul et al., 2022). With growing environmental awareness, practices evolved to advanced composting and manure recycling, yet challenges like nutrient runoff and greenhouse gas emissions persist (Koul et al., 2022).

Biotechnology offers a groundbreaking solution in agricultural waste management, providing efficient, environmentally friendly alternatives to traditional methods. Biotechnology efficiently processes various agricultural wastes using microorganisms, enzymes, and genetic engineering, converting them into valuable products (Verstraete, 2002; Degli Esposti et al., 2021; Pandya et al., 2023). This reduces the environmental footprint of waste disposal and facilitates the creation of circular economies in agriculture (Yang et al., 2023). By transforming waste into resources, biotechnological methods mitigate environmental concerns and generate new economic opportunities, contributing to a more sustainable future (Ma & Liu, 2019). These approaches markedly reduce greenhouse gas emissions and pollution, aligning with global sustainability goals.

Biotechnology in agricultural waste management transforms waste into valuable commodities like biofuels and bioplastics, creating economic growth and job opportunities in the green economy (Verstraete, 2002; Awasthi et al., 2022; Yang et al., 2023; Pandya et al., 2023). This supports circular economy principles by repurposing waste into valuable inputs while being more energy-efficient and requiring less space than traditional methods. Biotechnological byproducts like biofertilizers enhance soil health, promoting sustainable agriculture (Ma & Liu, 2019). However, challenges like technological complexity, high initial costs, and the need for supportive policies exist (Batista et al., 2021; Koul et al., 2022). Overcoming these through research, public-private partnerships and regulatory frameworks are crucial for realizing the potential of biotechnology in sustainable waste management and environmental protection (Batista et al., 2021).

Transitioning into the biotechnological approaches in agricultural waste management, it is explored how it potentially revolutionizes waste processing. This includes tailoring microbial capabilities and manipulating genetic material for enhanced waste degradation and byproduct production. Techniques like extraction, composting, digestion, and fermentation, alongside biosorption and phytoremediation, will be examined for their roles in waste transformation. A discussion on current challenges and prospects will highlight biotechnology's potential in sustainable waste management.

4.2 Concept of Biotechnology in Waste Management

The concept of biotechnology in waste management represents a paradigm shift from traditional waste-handling methods to innovative and efficient solutions for converting agricultural waste into valuable products (Verstraete, 2002). At its core, biotechnology leverages the natural abilities of microorganisms, such as bacteria and fungi, to decompose organic waste (Agrawal et al., 2023). This process is enhanced through genetic manipulation, where specific strains are selected and genetically engineered to improve efficiency and adaptability to different environmental conditions (Yi et al., 2022; Agrawal et al., 2023).

Genetic engineering plays a pivotal role in optimizing microorganisms for waste management. By modifying genes, these organisms can more effectively break down complex materials like lignocellulosic biomass in crop residues (Agrawal et al., 2023). This includes the production of enzymes crucial for decomposing tough agricultural waste, such as the ones rich in cellulosic materials, and making microorganisms resilient to adverse conditions (Yadav, 2021; Anand et al., 2022; Agrawal et al., 2023).

Biotechnology also involves the creation of microbial consortia that work synergistically to decompose various waste components efficiently (Ghosh et al., 2016; Bhatia et al., 2018). Modern biotechnological tools allow for real-time monitoring and control of microbial activity, ensuring optimal degradation conditions. Furthermore, biotechnology tailor's microbial processes to produce valuable byproducts, such as biofuels, organic acids, and bioplastics, by altering metabolic pathways and enzymes to optimize waste conversion (Verstraete, 2002; Ma & Liu, 2019; Degli Esposti et al., 2021).

4.3 Biotechnological Tools for Agriculture Waste Transformation

In biotechnology, the transformation of agricultural waste into valuable resources has been significantly enhanced by the arrival of advanced biotechnological tools. These tools, summarized in Table 4.1, include clustered regularly interspaced short palindromic repeats (CRISPR) and CRISPR-associated proteins (CRISPR-Cas), Prime editing, Transcription Activator-Like Effector Nucleases (TALENs), Zinc Finger Nucleases (ZFNs), RNA Interference (RNAi), and Targeting Induced Local Lesions IN Genomes (TILLING), which have revolutionized how we approach agricultural waste management and increase their value (Hamdan et al., 2022). Agricultural technology has significantly evolved to enhance crop yields and quality, addressing the growing global food demand. Traditional methods like seed saving, selective breeding, and mutation breeding contributed to a surge in crop production during the "Green Revolution" in the 1990s (Hamdan et al., 2022). Despite these advancements, challenges like shrinking arable land, climate change, and rising food demands continue to pressure agricultural production and threaten food security. Entering the "Gene Revolution" era, biotechnological innovations offer new

Table 4.1 Overview of biotechnological tools and their applications in agricultural waste management

Biotechnological tool	Description	Applications in agricultural waste management
CRISPR-Cas9	A gene-editing technology that allows for precise modifications in the DNA of living organisms	Modification of crop genetics for enhanced biomass Lignin content reduction in plants for improved biofuel production
TALENs	Transcription Activator-Like Effector Nucleases, another gene-editing technique that facilitates targeted changes in the genome	Crop residue management for increased digestibility Development of pest-resistant crops to reduce waste
Prime editing	Prime editing is a genome editing technique enabling precise DNA sequence alterations without inducing double-strand breaks	Edition of rice and wheat genome
ZFNs	Zinc Finger Nucleases are custom-designed proteins that can cut specific DNA sequences	Engineering microbial strains for better fermentation processes Modification of plants for better growth and yield
RNAi	RNA interference is a method used to silence gene expression	Reducing enzymes involved in lignin synthesis to facilitate biomass conversion Enhancing plant resistance to pests and diseases
TILLING	A reverse genetics technique that combines chemical mutagenesis with high-throughput molecular screening	Identification of mutations for improved plant traits such as higher biomass yield Development of crops with modified starch content for bioethanol production

strategies to improve crop yield, quality, and resilience to biotic and abiotic stresses (Jung et al., 2012; Bhatia et al., 2017; Iqbal et al., 2020). The adoption and future of these modern tools depend on differing regulatory frameworks across countries, which govern the development and production of genetically engineered products (Hamdan et al., 2022).

4.3.1 CRISPR-Cas9 Systems (Clustered Regularly Interspaced Short Palindromic Repeats-CRISPR-Associated Protein 9)

CRISPR-Cas9, a revolutionary genome editing tool, has transformed genetic engineering due to its precision, simplicity, and versatility (Kaur et al., 2021). Derived from a bacterial defence system, CRISPR-Cas9 allows for precise modifications of DNA sequences in living organisms. It comprises a guide RNA (gRNA) to match a desirable gene sequence and the Cas9 enzyme, an endonuclease that causes a double-stranded DNA break. This break prompts the cell's natural repair mechanisms, enabling the insertion, deletion, or alteration of specific DNA sequences (Kaur et al., 2021). CRISPR-Cas9 has been utilized in agriculture waste transformation to enhance crop residues' biochemical composition, making them more suitable for biofuel production. For instance, modifying lignin content in plant cell walls can significantly ease the conversion of plant biomass into biofuels (Jung et al., 2012; Lee et al., 2021; Kaur et al., 2021; Hemalatha et al., 2023).

4.3.2 Prime Editing

The prime editing system is a relatively new and highly precise genome editing technique that allows for the direct alteration of specific DNA sequences within the genome without the requirement for creating double-strand breaks (DSBs) as in traditional methods like CRISPR-Cas9. It comprises two components: an engineered prime editing guide RNA (pegRNA) and a prime editor (PE). PegRNA contains a spacer sequence that is complementary to one strand of the target DNA, carries a reverse transcription (RT) template and primer binding site (PBS), and PE has a mutant Cas9 protein that can only cut one strand of DNA (Cas9n). Thus, pegRNA guides the complex to the target DNA sequence, and Cas9n beaks single DNA strands, allowing for the incorporation of the desired edit in one strand of the DNA (Mahmudul & TuskanGerald, 2020). Plant prime editing was recently applied to edit the rice and wheat genome (Tang et al., 2020; Lin et al., 2021).

4.3.3 TALENs (Transcription Activator-Like Effector Nucleases)

TALENs are another set of engineered proteins used for genome editing (Iqbal et al., 2020). A TALEN consists of distinct segments: an N-terminal segment, a central repeat domain, a C-terminal segment, and a FokI catalytic domain (nuclease). The central repeat domain contains consecutive tandem repeats, each approximately 33–35 amino acids in length, capable of recognizing a single nucleotide. When two TALENs form a heterodimer in a tail-to-tail configuration at the target site, the FokI catalytic domains induce a site-specific DNA double-strand break (Sun & Zhao, 2013). The precision of TALENs in targeting specific DNA sequences allows for the editing of genes with remarkable accuracy (Bhushan et al., 2016). This technology has found application in modifying crop plants to increase their tolerance to stress and disease, thereby reducing agricultural waste (Bhardwaj & Nain, 2021). Moreover, TALENs have been used to enhance the characteristics of certain crops, making them more amenable to processing into biofuels or other value-added products (Jung & Altpeter, 2016; Kiran & Abdin, 2020; Iqbal et al., 2020).

4.3.4 ZFNs (Zinc Finger Nucleases)

ZFNs are artificial endonucleases that result from the linking of the DNA-binding domain of a versatile class of eukaryotic transcription factors—zinc finger proteins (ZFPs)—with the nuclease domain of the FokI restriction enzyme (Urnov et al., 2010). Heterodimeric ZFNs can be designed to target specific DNA sequences, facilitating targeted gene disruptions or insertions (Davies et al., 2017). In agricultural waste, ZFNs have been employed to genetically modify plants to improve their growth and yield, directly impacting the amount and quality of agricultural waste for further processing. Additionally, ZFNs have been instrumental in developing crops with altered lignocellulosic properties for more efficient bioconversion (Davies et al., 2017).

4.3.5 RNAi (RNA Interference)

RNAi is a natural cellular process of gene silencing mediated by small RNA molecules. This technology has been harnessed to downregulate specific genes in plants, offering a way to alter plant traits without changing the DNA sequence (Bharathi et al., 2023). In agricultural waste management, RNAi has been used to reduce the expression of certain enzymes involved in lignin synthesis in plants (Bewg et al., 2016). This reduction leads to lower lignin content in the biomass, facilitating its processing into biofuels and other bioproducts. RNAi has also been employed to enhance plant resistance to pests and diseases, thereby reducing crop loss and waste (Bharathi et al., 2023).

4.3.6 TILLING (Targeting Induced Local Lesions IN Genomes)

TILLING is a reverse genetics approach that combines traditional mutagenesis with high-throughput molecular screening (Singh et al., 2023). It enables the discovery of mutations in specific genes and the study of their functions in an organism. TILLING does not involve direct gene modification but exploits naturally occurring or induced mutations (Singh et al., 2023). In agricultural applications, TILLING has been utilized to identify and select plant varieties with desirable traits, such as increased biomass yield or modified composition, which are crucial for improving the efficiency of waste valorization (Slade et al., 2005; Reddy et al., 2012).

Integrating these advanced biotechnological tools in agriculture has opened new avenues for transforming agricultural waste into valuable products. By enabling precise genetic modifications, these technologies enhance the efficiency of waste conversion processes, contribute to the sustainable management of agricultural residues, and pave the way for innovative approaches in bioeconomy (Usmani et al., 2020; Jimoh et al., 2021; Aggarwal et al., 2023). As research and applications evolve, these tools will undoubtedly play a critical role in meeting the growing demands for sustainable agricultural practices and waste management solutions.

4.4 Applications of Biotechnologies for Waste Transformation

In sustainable agriculture, biotechnological methods stand the lead in transforming agricultural waste into valuable resources (Verstraete, 2002). The utilization of biotechnologies in transforming agricultural waste represents a significant shift toward sustainable practices. With the arrival of advanced biotechnological tools like CRISPR-Cas9, Prime Editing, TALENs, ZFNs, RNAi, and TILLING, the potential for converting agricultural waste into valuable resources has vastly expanded. This section explores various applications of these technologies in agricultural waste transformation, aligning with sustainability principles and the circular economy. These applications are summarized in Table 4.2.

4.4.1 Genetic Engineering in Agriculture Waste Transformation

Genetic engineering has revolutionized agricultural waste management approaches, particularly through CRISPR-Cas9 and TALENs (Pazhany & Henry, 2019). By altering the genetic makeup of crops, these technologies have enabled the development of plants with reduced lignin content, facilitating easier biomass conversion to biofuels and other value-added products (Pazhany & Henry, 2019). Biotechnological tools are key to addressing urgent global challenges such as climate change, unsustainable food systems, energy crises, and environmental degradation, offering faster, eco-friendly, and cost-effective solutions compared to traditional methods (Hamdan et al., 2022). They do this by promoting the development of eco-friendly, bio-based products through the valorization of plant and food waste, aligning with a circular bioeconomy and the sustainable development goals (SDGs). A key focus is using the CRISPR-Cas9 system, a groundbreaking tool in genome editing known for its simplicity and efficiency across various organisms. CRISPR-Cas9 can contribute to the bioeconomy by reducing food loss and waste across the entire food supply chain, from farm to table (Kaur et al., 2021). This includes the utilization of food waste by genome-edited microorganisms for food valorization, the efficient microbial conversion of low-cost substrates into biofuels, the valorization of agro-industrial wastes, the reduction of greenhouse gas emissions through forestry and plantation crops, and the protection of ecosystems and the environment (Hemalatha

Table 4.2 Genetic engineering in agricultural waste valorization

Method	Crop	Key genetic modification	Change in waste composition	Bioproduct yield increase	Additional notes	References
CRISPR/Cas9	Hybrid poplar	Reduced lignin content and altered xylem vessel shapes	High increase in saccharification efficiency	Direct use in biorefineries	No significant change in growth	Jang et al. (2021)
	Barley	Reduced lignin content	14% reduction in lignin and a 34% improvement in fermentable glucose recovery	34% higher bioethanol concentration and a 12% increase in yield	No effect on barley's growth or appearance	Lee et al. (2021)
	Sugarcane	Reduced lignin content	Decreased cell wall recalcitrance	Enhanced increase in ethanol production	Enhanced fermentable sugar recovery	Laksana et al. (2024)
TALENs	Sugarcane	Reduced lignin content	Lower syringyl subunit and higher hemicellulose contents	Potential use of lignocellulosic ethanol	No significant change in growth	Jung and Altpeter (2016)
	Sugarcane	Reduced lignin content	43.8% increase in saccharification efficiency	Increased efficiency in biofuel production	No significant difference in biomass production and agronomic performance	Kannan et al. (2018)
RNAi	Sugarcane	Reduced lignin content	Altering lignin deposition and composition and improving the release of glucose after enzymatic hydrolysis	Increased sucrose levels	Improved quality of bagasse for biofuels	Bewg et al. (2016)
	Barley	Reduced lignin syringyl/guaiacyl (S/G) ratio in straw	No apparent impact on saccharification	Improved lignocellulosic biomass as a feedstock for green technologies	Lignin content, straw mechanical properties, plant growth, and grain characteristics all remained unaffected	Shafiei et al. (2023)
TILLING	Wheat	Reduced amylose production	Higher fermentation efficiency	More efficient conversion to ethanol, potentially reducing processing times and costs	Crop improvement	Slade et al. (2005)
	Tobacco	Enhanced leaf yield	Improved leaf yield and broader leaves	Potential increase of oil accumulation for biofuel production	Agronomically superior traits such as more and broader leaves	Reddy et al. (2012)

et al., 2023). Genome editing, a powerful technique for specific DNA modifications in any organism, has evolved significantly with designer gene-manipulating tools like ZFNs, TALENs, and CRISPR/Cas9 (Gaj et al., 2016). While CRISPR-Cas9 has gained substantial attention, TALENs and ZFNs have unique advantages. TALENs, for instance, can target mitochondrial DNA (mito-TALEN), a task challenging for CRISPR gRNA (Bhardwaj & Nain, 2021).

CRISPR-Cas9 in Lignin Modification

CRISPR-Cas9 has been pivotal in modifying the lignin biosynthesis pathway in plants. By targeting specific genes involved in lignin production, researchers have been able to produce plants with altered lignin content and composition (Lee et al., 2021). This modification significantly enhances the efficiency of converting plant biomass into biofuels, reducing pretreatment costs and energy consumption.

Barley (*Hordeum vulgare* (Hv)), a versatile cereal crop, has been genetically modified to enhance its potential as a biofuel feedstock (Lee et al., 2021). Researchers have enhanced barley for bioethanol production by using CRISPR-Cas9 to target the Hv-Caffeic acid O-methyltransferase (HvCOMT1) gene involved in lignin synthesis. The modified barley showed a 14% reduction in lignin and a 34% improvement in fermentable glucose recovery. This resulted in a 34% higher bioethanol concentration and a 12% increase in yield, without affecting the barley's growth or appearance, making it a viable option for more efficient biofuel production (Lee et al., 2021).

In another study, researchers have developed a method to reduce lignin content in sugarcane (*Saccharum* spp. hybrids) stalks for improved bioethanol production using CRISPR-Cas 9 technology (Laksana et al., 2024). Genetic editing of the sugarcane variety KK3, targeting the limonene synthase gene (SoLIM) gene via Agrobacterium-mediated transformation, resulted in reduced lignin content by 9.74–51.46% and altered expression of lignin-related genes. This modification increased the syringyl/guaiacyl (S/G) ratio, confirmed by histological and microscopic analysis, making the edited sugarcane a more efficient substrate for second-generation bioethanol production (Laksana et al., 2024).

Caffeoyl shikimate esterase (CSE) plays a crucial role in plant lignin biosynthesis, positioning it as a key target for improving lignocellulosic biomass in biofuel production. CRISPR-Cas9 was used to edit CSE genes in hybrid poplar (*Populus alba* × *P. glandulosa*), leading to specific knockouts that resulted in a 29.1% reduction in lignin content and altered xylem vessel shapes without affecting growth or morphology compared to wild-type poplars (Jang et al., 2021). These transgenic poplars, with mutations in either one or both CSE genes, showed up to 25% increased saccharification efficiency, demonstrating the potential of targeted gene editing in improving lignocellulosic biomass for biofuel production without compromising plant growth (Jang et al., 2021).

TALENs in Crop Residue Management

TALENs have been employed to develop crop varieties with enhanced resistance to pests and diseases, leading to reduced crop waste (Bhardwaj & Nain, 2021). Additionally, TALENs have been used to modify the cellulose and hemicellulose content in plant cell walls, improving agricultural residues'digestibility and fermentation efficiency. Sugarcane, known for its efficiency in bioethanol production, combines high biomass production and significant sucrose accumulation in stems. Integrating first- and second-generation ethanol conversion processes can boost biofuel yields and use sucrose and cell wall-bound sugars (Jung & Altpeter, 2016).

RNAi suppression of the lignin biosynthetic gene COMT has shown promise in enhancing bioethanol production from lignocellulosic biomass. However, genome editing in sugarcane is complex due to its highly polyploid genome (Kannan et al., 2018). A recent study employed TALEN technology for multi-allelic mutagenesis, targeting a conserved region of COMT to alter lignin biosynthesis in sugarcane (Kannan et al., 2018). Field-grown TALEN-mediated COMT mutants exhibited up to 19.7% less lignin and a significantly lower S/G ratio, leading to a remarkable 43.8% increase in saccharification efficiency. Importantly, these COMT mutants maintained biomass production comparable to the original cultivar under field conditions. These results stress the potential of co-mutagenesis of numerous target alleles in crops with complex genomes for biofuel improvement (Kannan et al., 2018).

Sugarcane stands out as a key crop for biofuel production, with advanced technology exploiting sucrose in stems and cell wall-bound sugars for ethanol production (Jung & Altpeter, 2016). Targeted TALEN-induced mutations in the COMT gene, crucial for lignin synthesis, successfully reduced lignin content in sugarcane by 29–32%, despite its complex polyploidy and gene redundancy. Mutations were found in up to 74% of lines, directly correlating with lignin reduction and resulting in altered cell wall properties, including lower syringyl subunit and higher hemicellulose contents. This genome editing approach demonstrates the potential for improving lignocellulosic ethanol production in crops with complex genomes, with mutation effects consistently passed to vegetative progenies (Jung & Altpeter, 2016).

4.4.2 Metabolic Engineering for Waste Valorization

Metabolic engineering, using tools like ZFNs and RNAi, has enabled the modification of microbial pathways to enhance the conversion of agricultural waste into biofuels, bioplastics, and other chemicals (Joshi et al., 2022, 2023). Plant genome modification has advanced significantly in the last two decades with technologies like ZFNs (Davies et al., 2017).

RNAi technology, on the other hand, allows for the specific down-regulation of genes without affecting the expression of others (Bharathi et al., 2023). It has become a preferred method for studying gene functions and silencing genes in various organisms. Despite its benefits, challenges in RNAi technology, such as selecting appropriate gene candidates, stabilizing trigger molecules, and choosing target species and crops, remain. However, in the past decade, significant progress has been made in identifying crop target genes for improvement against biotic and abiotic stresses like drought, salinity, and temperature (Bharathi et al., 2023).

ZFNs in Crop Engineering

ZFNs have been used to engineer microbial strains for improved fermentation processes. These engineered microbes can efficiently convert agricultural residues into ethanol, biodiesel, and other biofuels. Applying ZFNs in microbial engineering also extends to producing high-value compounds like bioplastics and pharmaceuticals from agricultural waste (Davies et al., 2017).

Developing crops with multiple traits is challenging due to the random integration of transgenes. A versatile and broadly applicable solution has been developed: high-efficiency targeted genome editing with ZFNs combined with modular "trait landing pads" (TLPs). These TLPs facilitate on-demand transgene integration and trait stacking in crops. This method's effectiveness is demonstrated in maize by stacking herbicide resistance traits. Initially, an herbicide resistance gene, pat, flanked by a TLP (including ZFN target sites and sequences for incoming DNA), was integrated into the maize genome using WHISKERS™-mediated transformation. Subsequently, a targeted transgene integration method was established using microparticle bombardment to precisely deliver a second herbicide resistance gene, aad1, into the TLP. This was achieved by co-transformation with donor DNA containing aad1 and homologous sequences to the integrated TLP, along with a corresponding ZFN expression construct. This method allowed the co-segregation of both traits, showcasing the potential of ZFN-mediated integration for simplifying trait stacking in various crops (Davies et al., 2017).

RNAi in Plant Waste Management

RNAi technology has facilitated the reduction in specific enzymes in plants responsible for recalcitrant components like lignin. This reduction in lignin makes the biomass more amenable to bioconversion and improves the overall yield and quality of the agricultural waste for various applications (Bewg et al., 2016).

Sugarcane, a substantial biomass-producing subtropical crop, is crucial for sugar production and has the potential for biofuel production using residual bagasse. However, lignin in sugarcane poses a barrier to accessing its polysaccharides, making it a target for genetic modification. This study applied RNAi to suppress three key genes in the lignin biosynthesis pathway: caffeoyl-CoA O-methyltransferase (CCoAOMT), ferulate 5-hydroxylase (F5H), and COMT affecting lignin content and composition. RNAi analysis in sugarcane revealed differential impacts on lignin content and glucose release. While CCoAOMT lines showed minimal lignin reduction, one-line enhanced glucose release. F5H lines, without reducing lignin, also had a line with improved glucose release. Notably, a COMT line reduced lignin and increased glucose release. Two lines from F5H and COMT exhibited lower S/G ratios, indicating altered lignin composition, suggesting that genetic modification can boost biofuel potential without compromising, and in some cases enhancing, sugar yield (Bewg et al., 2016).

In another study, barley, a significant cereal crop in temperate climates and a genetic model for polyploid wheat presents an attractive feedstock source for green technologies due to its cereal straw biomass. The research showed that instead of reducing lignin content, altering its composition via downregulating the HvF5H1 gene, which affects lignin's S/G ratio, preserves plant traits, and improves straw properties for bioconversion. This approach maintains lignin content, mechanical properties, growth, and grain characteristics, while metabolic profiling revealed changes in 173 features, highlighting barley's potential in optimizing biomass for green technologies (Shafiei et al., 2023).

4.4.3 TILLING for Enhanced Biomass Production

TILLING has been instrumental in identifying plant varieties with desirable traits for biomass production. By exploiting natural genetic variations, TILLING has enabled the selection of plant varieties with higher biomass yield, improved digestibility, and modified composition, suitable for efficient waste valorization (Reddy et al., 2012).

Tobacco, primarily known for its use in cigarettes, has been explored for alternative applications, particularly focusing on leaf yield, a key economic factor. TILLING was used as an Ethyl Methane Sulphonate (EMS) mutant population to identify mutants enhancing leaf yield and related metabolic traits. This involved preparing DNA and seed libraries from 7800 second-generation plants (M_2) subjected to mutagenesis. Using high-throughput sequencing and three-dimensional pooling, the study screened mutations in 3072 tobacco plants, identifying seven mutations affecting leaf biomass and secondary metabolism. This resulted in a mutation density of one per 1.423 million bases. Notably, a mutation in Ntfw2.2 led to a premature stop codon, which was used to create a DNA marker. Mutants with increased leaf number and size were identified for further breeding, showing potential for tobacco crop improvement through genetic screening methods like TILLING (Reddy et al., 2012).

Another study highlighted the application of TILLING to enhance a quality characteristic in a polyploid crop. Waxy starches, predominantly made of amylopectin, are known for their unique physicochemical properties. In wheat, only one or two active waxy genes (granule-bound starch synthase I, or GBSSI) result in starch with intermediate amylopectin levels. Through TILLING, 246 alleles of the waxy genes were identified in 1920 allohexaploid and allotetraploid wheat plants. Using TILLING, a bread wheat line with almost no waxy trait was developed, featuring homozygous mutations in two waxy homoeologs and a deletion in the third. This approach, revealing more genetic diversity than seen in the past 25 years, shows significant promise for crop improvement (Slade et al., 2005).

4.4.4 *Practical Applications*

Enhancing Extraction Techniques Using Genetic Tools

Agricultural wastes can be used to recover natural bioactive compounds, particularly phenolic compounds, rather than accumulating in landfills or being incinerated (Azeez et al., 2017; Dorosh et al., 2022; Vieira et al., 2022; Bala et al., 2023; Soares et al., 2023).

Advanced biotechnological tools can play a pivotal role in optimizing extraction processes. CRISPR-Cas9 and TALENs, for instance, can be used to genetically modify plants to produce higher concentrations of extractable bioactive compounds (Niazian, 2019; Manchanda et al., 2022), thus making processes such as subcritical water extraction (SWE), microwave-assisted extraction (MAE), and ultrasonic-assisted extraction (UAE) more efficient and yield higher-quality products (Quitério et al., 2022; Soares et al., 2023).

Genetic Engineering in Composting and Vermicomposting

Microbial composting is a process that uses the natural decomposition abilities of bacteria and fungi (Jurado et al., 2014). In this biologically rich process, microorganisms consume organic waste, breaking it down into simpler substances. This decomposition is a natural phenomenon, but when used and controlled, it becomes a powerful tool for transforming waste into valuable resources (Jurado et al., 2014; Ribeiro et al., 2020).

Genetic tools like RNAi and ZFNs can be employed to engineer microbial strains used in composting. These engineered microbes can accelerate the decomposition process, resulting in a more efficient conversion of organic waste into nutrient-rich compost (Pant et al., 2021; Aggarwal et al., 2023).

Optimizing Anaerobic Digestion Through Metabolic Engineering

Anaerobic digestion, a key innovation in waste management and energy production, transforms organic waste into renewable energy, aligning with sustainability principles (Li et al., 2020). This biotechnological process produces biogas, primarily methane and carbon dioxide, through a multi-stage biochemical transformation without oxygen (Li et al., 2020).

CRISPR-Cas9 and ZFNs can be instrumental in genetically modifying microorganisms used in anaerobic digestion. These modifications can lead to increased biogas production or allow for the digestion of more recalcitrant forms of agricultural waste. RNAi technology could silence specific genes in microbes, optimizing their metabolic pathways for improved biogas yield (Christy et al., 2014).

Advanced Genetic Tools in Fermentation

Fermentation plays a crucial role in sustainable waste management by transforming agricultural waste into biofuels like ethanol and biodiesel through the metabolic processes of microorganisms. This approach not only reduces reliance on fossil fuels but also utilizes resources such as sugarcane, corn, and recycled oils for energy production (Adewuyi, 2020).

The fermentation process can be significantly enhanced by using genetically engineered yeasts and bacteria developed using tools like CRISPR-Cas and TALENs. These organisms can be designed to have increased tolerance to inhibitors present in agricultural waste or enhanced capabilities to convert waste into biofuels like ethanol and biodiesel (Xia et al., 2023).

Biochar Production and Genetic Optimization

Agricultural waste is converted into biochar through thermal processing in an inert environment, producing a carbon-rich material with a vast pore structure and large surface area (Fernandes et al., 2019; Kamali et al., 2022; Wang et al., 2022). Biochar is effective for adsorption, catalysis, environmental cleanup, and soil improvement, also aiding in long-term carbon sequestration and climate change mitigation (Zhang et al., 2012; Puga et al., 2022; Wang et al., 2022; Soares et al., 2022).

While biochar production primarily relies on physical and chemical processes, genetic engineering tools can be used to develop crop varieties that produce biomass more suitable for biochar production (Panahi et al., 2020; Kwon et al., 2020; Seow et al., 2022; Yadav et al., 2023). This could involve modifying plant genetics using tools like CRISPR-Cas9 to alter the lignin content, resulting in biochar with improved properties (Pazhany & Henry, 2019; Lee et al., 2021).

Enhancing Biosorption and Phytoremediation

Biosorption and phytoremediation are eco-friendly techniques for detoxifying polluted environments using microbes and plants (Karić et al., 2022; Priya et al., 2023). Biosorption employs microorganisms like bacteria, fungi, and algae to remove pollutants from water and soil through mechanisms such as adsorption and ion exchange, offering a cost-effective solution for heavy metal and organic contaminant removal (Vijayaraghavan & Yun, 2008; Zulfiqar et al., 2023). Phytoremediation uses plants to absorb, stabilize, or degrade contaminants, providing a noninvasive method to clean sites polluted with heavy metals and other toxins while preserving natural landscapes and potentially generating bioenergy (Nedjimi, 2021; Park & Oh, 2023).

Advanced genetic tools can enhance the capabilities of microbes and plants used in biosorption and phytoremediation. Genetically modified plants with enhanced heavy metal uptake capabilities, developed using TALENs or ZFNs, could significantly improve the efficiency of phytoremediation processes (Yadav et al., 2010; Venegas-Rioseco et al., 2021). Similarly, RNAi can silence genes in plants that interfere with pollutant uptake (Li et al., 2007).

Synthesis of Nanoparticles and Genetic Engineering

The integration of nanotechnology with agricultural waste management presents innovative and sustainable solutions, including the creation of nanoparticles (NPs) from waste for applications in soil improvement, pest control, and crop health, aligning with green chemistry and circular economy principles (Abdelbasir et al., 2020; Joshi & Arora, 2023).

While the synthesis of NPs from agricultural waste is primarily a chemical process, biotechnological tools can assist in this area. For example, genetically modified bacteria or plants developed using CRISPR-Cas9 or TALENs could be used to produce specific enzymes or metabolites that facilitate the synthesis of NPs (Bahrulolum et al., 2021).

The integration of advanced biotechnological tools like CRISPR-Cas9, TALENs, ZFNs, RNAi, and TILLING in agricultural waste management opens new frontiers for innovation. These tools enhance existing processes and define the way for developing novel approaches to transform agricultural waste into valuable resources, aligning with sustainability and the circular economy. As research in these areas progresses, a significant impact on the efficiency and effectiveness of agricultural waste transformation techniques can be expected.

4.5 Challenges and Future Perspectives

The integration of biotechnology into sustainable waste management is a promising approach, yet it encounters a range of challenges and offers significant potential. The key limitations and hurdles in implementing biotechnological strategies include economic feasibility, technological complexities, policy restrictions, and public acceptance. The next points outline the most significant current challenges in this field:

1. Technological limitations: One of the primary challenges is the technological complexity and cost associated with biotechnological processes. Advanced bioreactors and genetically modified organisms (GMOs) require substantial investment in research and development. The scalability of laboratory successes in industrial-scale applications remains a significant hurdle (Awasthi et al., 2022).
2. Regulatory and ethical concerns: The use of GMOs in waste management raises ethical and regulatory questions. There is a need for stringent regulations to ensure environmental safety and public health, particularly concerning the release of modified organisms into the environment (Harfouche et al., 2021).

3. Public perception and acceptance: Public skepticism towards GMOs and biotechnological solutions can impede implementation. Misinformation and lack of awareness about biotechnological processes can result in resistance from communities and stakeholders (Harfouche et al., 2021).
4. Economic viability: The economic feasibility of biotechnological solutions compared to traditional waste management methods is a key concern. High initial costs and uncertainty about long-term financial returns can deter investment (Kumar & Agrawal, 2020; Awasthi et al., 2022).

The future trends and opportunities in the field of biotechnology for agricultural waste transformation are vast and evolving. The focus on potential advancements in biotechnological techniques and emerging opportunities presents a promising outlook for the sustainable management and valorization of agricultural waste as presented below:

1. Advanced genetic engineering: With rapid advancements in genetic engineering, particularly CRISPR-Cas9 technology, there is potential for creating more efficient microorganisms tailored for specific waste types. This precision could significantly enhance waste processing and resource recovery (Hemalatha et al., 2023).
2. Circular economy integration: Biotechnology is pivotal in transitioning to a circular economy. By transforming waste into valuable resources like biofuels, bioplastics, and fertilizers, biotechnological methods can reduce reliance on nonrenewable resources and minimize environmental impact (Awasthi et al., 2022).
3. Collaborative ventures: Partnerships between governments, academic institutions, and industries can accelerate the development and adoption of biotechnological solutions. Collaborative efforts can pool resources, share risks, and foster innovation (Ma & Liu, 2019).
4. Sustainability and environmental protection: Biotechnology in waste management aligns with global sustainability goals. It offers solutions for reducing greenhouse gas emissions, minimizing landfill use, and conserving natural resources (Ma & Liu, 2019).

While biotechnology in waste management faces several challenges, its integration is crucial for sustainable development. Continued innovation, collaboration, and supportive policies can unlock its full potential, leading to a more sustainable and resilient future.

4.6 Conclusion

The comprehensive exploration of biotechnological applications in agricultural waste management elucidates a transformative approach towards sustainability and environmental preservation. This description, stretching across the discussions of biotechnological advancements, practical applications, and the challenges within, underscores the potential of these technologies to revolutionize waste management practices globally. By harnessing advanced genetic engineering, metabolic engineering, and bioprocessing techniques, biotechnology offers viable solutions for converting agricultural waste into biofuels, bioplastics, organic fertilizers, and other value-added products. This addresses the pressing issue of waste accumulation and contributes significantly to creating a circular economy, thereby promoting resource efficiency and environmental sustainability.

However, the journey towards fully integrating biotechnology in waste management is fraught with challenges, including economic constraints, technological complexities, regulatory barriers, and the need for broader public acceptance. The potential for future advancements, such as the application of CRISPR-Cas9 technology and artificial intelligence, alongside collaborative efforts across governmental, industrial, and academic sectors, presents a hopeful outlook. These advancements are crucial for overcoming existing hurdles and realizing the full potential of biotechnological applications in waste management.

In conclusion, the role of biotechnology in agricultural waste management is pivotal for advancing sustainable development goals. It embodies an innovative approach to waste valorization that mitigates environmental impacts and fosters economic growth and resource conservation. Moving forward, continuous innovation, supportive policy frameworks, and increased public awareness must converge to unlock the transformative potential of biotechnology. This integration will pave the way for more efficient, environmentally friendly, and economically viable waste management practices, setting a new standard in the pursuit of global sustainability.

Acknowledgments This work received support and help from FCT/MCTES (LA/P/0008/2020 DOI 10.54499/LA/P/0008/2020, UIDP/50006/2020 DOI 10.54499/UIDP/50006/2020 and UIDB/50006/2020 DOI 10.54499/UIDB/50006/2020), through national funds. Olena Dorosh (2021.05322.BD) and Filipe Fernandes (2021.06806.BD) are thankful for their Ph.D. grant financed by FCT/MCTES. Manuela M. Moreira (CEECIND/02702/2017) is thankful for her contract financed by FCT/MCTES—CEEC Individual Program Contract and to REQUIMTE/LAQV.

References

Abbasi, A., Sajid, A., Haq, N., et al. (2014). Agricultural pollution: An emerging issue. In P. Ahmad, M. Wani, M. Azooz, & L. S. Tran (Eds.), *Improvement of crops in the era of climatic changes* (pp. 347–387). Springer.

Abdelbasir, S. M., McCourt, K. M., Lee, C. M., & Vanegas, D. C. (2020). Waste-derived nanoparticles: Synthesis approaches, environmental applications, and sustainability considerations. *Frontiers in Chemistry, 8*, 782.

Adewuyi, A. (2020). Challenges and prospects of renewable energy in Nigeria: A case of bioethanol and biodiesel production. *Energy Reports, 6*, 77–88.

Aggarwal, N., Pham, H. L., Ranjan, B., Saini, M., Liang, Y., Hossain, G. S., et al. (2023). Microbial engineering strategies to utilize waste feedstock for sustainable bioproduction. *Nature Reviews Bioengineering, 2*, 1–20.

Agrawal, K., Nair, L. G., Chaturvedi, V., & Verma, P. (2023). Designing microbial cellulases using genetic engineering approach: A promising strategy towards zero-waste cellulosic biorefinery. *Biocatalysis and Agricultural Biotechnology, 102830*.

Anand, U., Vaishnav, A., Sharma, S. K., et al. (2022). Current advances and research prospects for agricultural and industrial uses of microbial strains available in world collections. *Science of the Total Environment, 842*, 156641. https://doi.org/10.1016/j.scitotenv.2022.156641

Awasthi, M. K., Singh, E., Binod, P., Sindhu, R., Sarsaiya, S., Kumar, A., et al. (2022). Biotechnological strategies for bio-transforming biosolid into resources toward circular bio-economy: A review. *Renewable and Sustainable Energy Reviews, 156*, 111987.

Azeez, S., Narayana, C. K., & Oberoi, H. S. (2017). Extraction and utilisation of bioactive compounds from agricultural waste. In *Utilisation of bioactive compounds from agricultural and food production waste* (pp. 127–158). CRC Press.

Babla, M., Katwal, U., Yong, M.-T., et al. (2022). Value-added products as soil conditioners for sustainable agriculture. *Resources, Conservation and Recycling, 178*, 106079. https://doi.org/10.1016/j.resconrec.2021.106079

Bahrulolum, H., Nooraei, S., Javanshir, N., et al. (2021). Green synthesis of metal nanoparticles using microorganisms and their application in the agrifood sector. *Journal of Nanobiotechnology, 19*, 86. https://doi.org/10.1186/s12951-021-00834-3

Bala, S., Garg, D., Sridhar, K., Inbaraj, B. S., Singh, R., Kamma, S., et al. (2023). Transformation of agro-waste into value-added bioproducts and bioactive compounds: Micro/nano formulations and application in the agri-food-pharma sector. *Bioengineering, 10*(2), 152.

Batista, M., Caiado, R. G. G., Quelhas, O. L. G., Lima, G. B. A., Leal Filho, W., & Yparraguirre, I. T. R. (2021). A framework for sustainable and integrated municipal solid waste management: Barriers and critical factors to developing countries. *Journal of Cleaner Production, 312*, 127516.

Bewg, W. P., Poovaiah, C., Lan, W., Ralph, J., & Coleman, H. D. (2016). RNAi downregulation of three key lignin genes in sugarcane improves glucose release without reduction in sugar production. *Biotechnology for Biofuels, 9*, 1–13.

Bharathi, J. K., Anandan, R., Benjamin, L. K., Muneer, S., & Prakash, M. A. S. (2023). Recent trends and advances of RNA interference (RNAi) to improve agricultural crops and enhance their resilience to biotic and abiotic stresses. *Plant Physiology and Biochemistry, 194*, 600–618.

Bhardwaj, A., & Nain, V. (2021). TALENs—an indispensable tool in the era of CRISPR: A mini review. *Journal of Genetic Engineering and Biotechnology, 19*(1), 125.

Bhatia, R., Gallagher, J. A., Gomez, L. D., & Bosch, M. (2017). Genetic engineering of grass cell wall polysaccharides for biorefining. *Plant Biotechnology Journal, 15*, 1071–1092. https://doi.org/10.1111/pbi.12764

Bhatia, S. K., Bhatia, R. K., Choi, Y. K., Kan, E., Kim, Y. G., & Yang, Y. H. (2018). Biotechnological potential of microbial consortia and future perspectives. *Critical Reviews in Biotechnology, 38*(8), 1209–1229.

Bhushan, K., Pratap, D., & Sharma, P. K. (2016). Transcription activator-like effector nucleases (TALENs): An efficient tool for plant genome editing. *Engineering in Life Sciences, 16*(4), 330–337.

Çakmakçı, R., Salık, M. A., & Çakmakçı, S. (2023). Assessment and principles of environmentally sustainable food and agriculture systems. *Agriculture, 13*(5), 1073.

Christy, P. M., Gopinath, L. R., & Divya, D. (2014). A review on anaerobic decomposition and enhancement of biogas production through enzymes and microorganisms. *Renewable and Sustainable Energy Reviews, 34*, 167–173.

Davies, J. P., Kumar, S., & Sastry-Dent, L. (2017). Use of zinc-finger nucleases for crop improvement. *Progress in Molecular Biology and Translational Science, 149*, 47–63.

Degli Esposti, M., Morselli, D., Fava, F., Bertin, L., Cavani, F., Viaggi, D., & Fabbri, P. (2021). The role of biotechnology in the transition from plastics to bioplastics: An opportunity to reconnect global growth with sustainability. *FEBS Open Bio, 11*(4), 967–983.

Dorosh, O., Rodrigues, F., Delerue-Matos, C., & Moreira, M. M. (2022). Increasing the added value of vine-canes as a sustainable source of phenolic compounds: A review. *Science of the Total Environment, 830*, 154600.

Fernandes, M. J., Moreira, M. M., Paíga, P., Dias, D., Bernardo, M., Carvalho, M., et al. (2019). Evaluation of the adsorption potential of biochars prepared from forest and agri-food wastes for the removal of fluoxetine. *Bioresource Technology, 292*, 121973.

Ferronato, N., & Torretta, V. (2019). Waste mismanagement in developing countries: A review of global issues. *International Journal of Environmental Research and Public Health, 16*(6), 1060.

Gaj, T., Sirk, S. J., Shui, S. L., & Liu, J. (2016). Genome-editing technologies: Principles and applications. *Cold Spring Harbor Perspectives in Biology, 8*(12), a023754.

Ghosh, S., Chowdhury, R., & Bhattacharya, P. (2016). Mixed consortia in bioprocesses: Role of microbial interactions. *Applied Microbiology and Biotechnology, 100*, 4283–4295.

Hamdan, M. F., Mohd Noor, S. N., Abd-Aziz, N., Pua, T. L., & Tan, B. C. (2022). Green revolution to gene revolution: Technological advances in agriculture to feed the world. *Plants, 11*(10), 1297.

Harfouche, A. L., Petousi, V., Meilan, R., Sweet, J., Twardowski, T., & Altman, A. (2021). Promoting ethically responsible use of agricultural biotechnology. *Trends in Plant Science, 26*(6), 546–559.

Hemalatha, P., Abda, E. M., Shah, S., Prabhu, S. V., Jayakumar, M., Karmegam, N., et al. (2023). Multi-faceted CRISPR-Cas9 strategy to reduce plant based food loss and waste for sustainable bio-economy–a review. *Journal of Environmental Management, 332*, 117382.

Iqbal, Z., Iqbal, M. S., Ahmad, A., Memon, A. G., & Ansari, M. I. (2020). New prospects on the horizon: Genome editing to engineer plants for desirable traits. *Current Plant Biology, 24*, 100171.

Jang, H. A., Bae, E. K., Kim, M. H., Park, S. J., Choi, N. Y., Pyo, S. W., et al. (2021). CRISPR-knockout of CSE gene improves saccharification efficiency by reducing lignin content in hybrid poplar. *International Journal of Molecular Sciences, 22*(18), 9750.

Jimoh, A. A., Senbadejo, T. Y., Adeleke, R., & Lin, J. (2021). Development and genetic engineering of hyper-producing microbial strains for improved synthesis of biosurfactants. *Molecular Biotechnology, 63*, 267–288.

Joshi, A., & Arora, J. (2023). Recent trends and future prospects of nanotechnology for agro-waste valorization into biofuels. In K. G. Ramawat, J. M. Mérillon, & J. Arora (Eds.), *Agricultural waste: Environmental impact, useful metabolites and energy production* (pp. 559–582). Springer Nature.

Joshi, A., Verma, K. K., Rajput, D. V., Minkina, T., & Arora, J. (2022). Recent advances in metabolic engineering of microorganisms for advancing lignocellulose-derived biofuels. *Bioengineered, 13*, 8135–8163. https://www.tandfonline.com/doi/full/10.1080/21655979.2022.2051856

Joshi, A., Verma, K. K., Rajput, V. D., & Arora, J. (2023). Metabolic engineering of microorganisms in advancing biofuel production. In P. Moodley, R. C. Ray, & E. B. G. Kana (Eds.), *Advances in lignocellulosic biofuel production systems* (pp. 181–195). Elsevier. ISBN: 978-0-323-91192-4 (print).

Jung, J. H., & Altpeter, F. (2016). TALEN mediated targeted mutagenesis of the caffeic acid O-methyltransferase in highly polyploid sugarcane improves cell wall composition for production of bioethanol. *Plant Molecular Biology, 92*, 131–142.

Jung, H. J. G., Samac, D. A., & Sarath, G. (2012). Modifying crops to increase cell wall digestibility. *Plant Science, 185*, 65–77.

Jurado, M., López, M. J., Suárez-Estrella, F., Vargas-García, M. C., López-González, J. A., & Moreno, J. (2014). Exploiting composting biodiversity: Study of the persistent and biotechnologically relevant microorganisms from lignocellulose-based composting. *Bioresource Technology, 162*, 283–293.

Kamali, M., Sweygers, N., Al-Salem, S., Appels, L., Aminabhavi, T. M., & Dewil, R. (2022). Biochar for soil applications-sustainability aspects, challenges and future prospects. *Chemical Engineering Journal, 428*, 131189.

Kannan, B., Jung, J. H., Moxley, G. W., Lee, S. M., & Altpeter, F. (2018). TALEN-mediated targeted mutagenesis of more than 100 COMT copies/alleles in highly polyploid sugarcane improves saccharification efficiency without compromising biomass yield. *Plant Biotechnology Journal, 16*(4), 856–866.

Karić, N., Maia, A. S., Teodorović, A., Atanasova, N., Langergraber, G., Crini, G., et al. (2022). Bio-waste valorisation: Agricultural wastes as biosorbents for removal of (in) organic pollutants in wastewater treatment. *Chemical Engineering Journal Advances, 9*, 100239.

Kaur, H., Pandey, D. K., Goutam, U., & Kumar, V. (2021). CRISPR/Cas9-mediated genome editing is revolutionizing the improvement of horticultural crops: Recent advances and future prospects. *Scientia Horticulturae, 289*, 110476.

Kiran, U., & Abdin, M. Z. (2020). Genome editing technologies for value-added traits in plants. In *Transgenic technology based value addition in plant biotechnology* (pp. 51–67). Academic Press.

Koul, B., Yakoob, M., & Shah, M. P. (2022). Agricultural waste management strategies for environmental sustainability. *Environmental Research, 206*, 112285.

Kumar, A., & Agrawal, A. (2020). Recent trends in solid waste management status, challenges, and potential for the future Indian cities–A review. *Current Research in Environmental Sustainability, 2*, 100011.

Kwon, G., Bhatnagar, A., Wang, H., Kwon, E. E., & Song, H. (2020). A review of recent advancements in utilization of biomass and industrial wastes into engineered biochar. *Journal of Hazardous Materials, 400*, 123242.

Laksana, C., Sophiphun, O., & Chanprame, S. (2024). Lignin reduction in sugarcane by performing CRISPR/Cas9 site-direct mutation of SoLIM transcription factor. *Plant Science, 340*, 111987.

Lan, R., Eastham, S. D., Liu, T., Norford, L. K., & Barrett, S. R. (2022). Air quality impacts of crop residue burning in India and mitigation alternatives. *Nature Communications, 13*(1), 6537.

Lee, J. H., Won, H. J., Hoang Nguyen Tran, P., Lee, S. M., Kim, H. Y., & Jung, J. H. (2021). Improving lignocellulosic biofuel production by CRISPR/Cas9-mediated lignin modification in barley. *GCB Bioenergy, 13*(4), 742–752.

Li, J. C., Guo, J. B., Xu, W. Z., & Ma, M. (2007). RNA interference-mediated silencing of phytochelatin synthase gene reduce cadmium accumulation in rice seeds. *Journal of Integrative Plant Biology, 49*(7), 1032–1037.

Li, Y., Han, Y., Zhang, Y., Luo, W., & Li, G. (2020). Anaerobic digestion of different agricultural wastes: A techno-economic assessment. *Bioresource Technology, 315*, 123836.

Lin, Q., Jin, S., Zong, Y., Yu, H., Zhu, Z., Liu, G., et al. (2021). High-efficiency prime editing with optimized, paired pegRNAs in plants. *Nature Biotechnology, 39*(8), 923–927.

Ma, Y., & Liu, Y. (2019). Turning food waste to energy and resources towards a great environmental and economic sustainability: An innovative integrated biological approach. *Biotechnology Advances, 37*(7), 107414.

Mahmudul, H., & TuskanGerald, A. (2020). Prime editing technology and its prospects for future applications in plant biology research. *BioDesign Research*. Published online 2020 Jun 26. https://doi.org/10.34133/2020/9350905

Manchanda, P., Sharma, D., Kaur, A., & Singh, A. (2022). Improvement of genetic variation for nutrients and bioactive food components in cereal crops. In *Functional cereals and cereal foods: Properties, functionality and applications* (pp. 51–80). Springer International Publishing.

Nedjimi, B. (2021). Phytoremediation: A sustainable environmental technology for heavy metals decontamination. *SN Applied Sciences, 3*(3), 286.

Niazian, M. (2019). Application of genetics and biotechnology for improving medicinal plants. *Planta, 249*, 953–973.

Panahi, H. K. S., Dehhaghi, M., Ok, Y. S., Nizami, A. S., Khoshnevisan, B., Mussatto, S. I., et al. (2020). A comprehensive review of engineered biochar: Production, characteristics, and environmental applications. *Journal of Cleaner Production, 270*, 122462.

Pandya, R. S., Kaur, T., Bhattacharya, R., Bose, D., & Saraf, D. (2023). Harnessing microorganisms for bioenergy with microbial fuel cells: Powering the future. *Water-Energy Nexus, 7*, 1–12.

Pant, G., Garlapati, D., Agrawal, U., Prasuna, R. G., Mathimani, T., & Pugazhendhi, A. (2021). Biological approaches practised using genetically engineered microbes for a sustainable environment: A review. *Journal of Hazardous Materials, 405*, 124631.

Park, J. K., & Oh, K. (2023). Advancements in phytoremediation research for soil and water resources: Harnessing plant power for environmental clean up. *Sustainability, 15*(18), 13901.

Pazhany, A. S., & Henry, R. J. (2019). Genetic modification of biomass to alter lignin content and structure. *Industrial and Engineering Chemistry Research, 58*(35), 16190–16203.

Priya, A. K., Muruganandam, M., Ali, S. S., & Kornaros, M. (2023). Clean-up of heavy metals from contaminated soil by phytoremediation: A multidisciplinary and eco-friendly approach. *Toxics, 11*(5), 422.

Puga, A., Moreira, M. M., Pazos, M., Figueiredo, S. A., Sanromán, M. Á., Delerue-Matos, C., & Rosales, E. (2022). Continuous adsorption studies of pharmaceuticals in multicomponent mixtures by agroforestry biochar. *Journal of Environmental Chemical Engineering, 10*(1), 106977.

Quitério, E., Grosso, C., Ferraz, R., Delerue-Matos, C., & Soares, C. (2022). A critical comparison of the advanced extraction techniques applied to obtain health-promoting compounds from seaweeds. *Marine Drugs, 20*(11), 677.

Reddy, T. V., Dwivedi, S., & Sharma, N. K. (2012). Development of TILLING by sequencing platform towards enhanced leaf yield in tobacco. *Industrial Crops and Products, 40*, 324–335.

Ribeiro, I. C., Teixeira, R. A., Rabello, L. L., Bringhenti, J. R., & Nicolau Korres, A. M. (2020). Composting and anaerobic digestion as biotechnological alternatives for the valorization of used coffee ground in University Campus. In *Universities as living labs for sustainable development: Supporting the implementation of the sustainable development goals* (pp. 789–802). Springer.

Sánchez, A., Artola, A., Font, X., Gea, T., Barrena, R., Gabriel, D., et al. (2015). Greenhouse gas emissions from organic waste composting. *Environmental Chemistry Letters, 13*, 223–238.

Seow, Y. X., Tan, Y. H., Mubarak, N. M., Kansedo, J., Khalid, M., Ibrahim, M. L., & Ghasemi, M. (2022). A review on biochar production from different biomass wastes by recent carbonization technologies and its sustainable applications. *Journal of Environmental Chemical Engineering, 10*(1), 107017.

Shafiei, R., Hooper, M., McClellan, C., Stephens, J., Goeminne, G., Boerjan, W., et al. (2023). Downregulation of barley ferulate 5-hydroxylase dramatically alters straw lignin structure without impact on mechanical properties. *Frontiers in Plant Science, 13*, 1125003.

Singh, D., Chaudhary, P., Taunk, J., Singh, C. K., Chinnusamy, V., Sevanthi, A. M., et al. (2023). Targeting Induced Local Lesions in Genomes (TILLING): Advances and opportunities for fast tracking crop breeding. *Critical Reviews in Biotechnology*, July, 1–20. https://doi.org/10.1080/07388551.2023.2231630

Slade, A. J., Fuerstenberg, S. I., Loeffler, D., Steine, M. N., & Facciotti, D. (2005). A reverse genetic, nontransgenic approach to wheat crop improvement by TILLING. *Nature Biotechnology, 23*(1), 75–81.

Soares, C., Correia-Sá, L., Paíga, P., Barbosa, C., Remor, P., Freitas, O. M., et al. (2022). Removal of diclofenac and sulfamethoxazole from aqueous solutions and wastewaters using a three-dimensional electrochemical process. *Journal of Environmental Chemical Engineering, 10*(5), 108419.

Soares, C., Moreira, M. M., Ramos, S., Ramalhosa, M. J., Correia, M., Svarc-Gajić, J., et al. (2023). A critical assessment of extraction methodologies for the valorization of agricultural wastes: Polyphenolic profile and bioactivity. *Processes, 11*(6), 1767.

Sun, N., & Zhao, H. (2013). Transcription activator-like effector nucleases (TALENs): A highly efficient and versatile tool for genome editing. *Biotechnology and Bioengineering, 110*(7), 1811–1821.

Tang, X., Sretenovic, S., Ren, Q., Jia, X., Li, M., Fan, T., et al. (2020). Plant prime editors enable precise gene editing in rice cells. *Molecular Plant, 13*(5), 667–670.

Tiwari, A. K., & Pal, D. B. (2022). Nutrients contamination and eutrophication in the river ecosystem. In *Ecological significance of river ecosystems* (pp. 203–216). Elsevier.

Urnov, F. D., Rebar, E. J., Holmes, M. C., Zhang, H. S., & Gregory, P. D. (2010). Genome editing with engineered zinc finger nucleases. *Nature Reviews Genetics, 11*(9), 636–646.

Usmani, Z., Sharma, M., Sudheer, S., Gupta, V. K., & Bhat, R. (2020). Engineered microbes for pigment production using waste biomass. *Current Genomics, 21*(2), 80–95.

Venegas-Rioseco, J., Ginocchio, R., & Ortiz-Calderón, C. (2021). Increase in phytoextraction potential by genome editing and transformation: A review. *Plants, 11*(1), 86.

Verstraete, W. (2002). Environmental biotechnology for sustainability. *Journal of Biotechnology, 94*(1), 93–100.

Vieira, E. F., Souza, S., Moreira, M. M., Cruz, R., Silva, A. B. D., Casal, S., & Delerue-Matos, C. (2022). Valorization of phenolic and carotenoid compounds of *Sechium edule* (Jacq. Swartz) leaves: Comparison between conventional, ultrasound-and microwave-assisted extraction approaches. *Molecules, 27*(21), 7193.

Vijayaraghavan, K., & Yun, Y. S. (2008). Bacterial biosorbents and biosorption. *Biotechnology Advances, 26*(3), 266–291.

Wang, C., Luo, D., Zhang, X., Huang, R., Cao, Y., Liu, G., et al. (2022). Biochar-based slow-release of fertilizers for sustainable agriculture: A mini-review. *Environmental Science and Ecotechnology, 10*, 100167.

Xia, Y., Li, Y., Shen, W., Yang, H., & Chen, X. (2023). CRISPR-Cas technology for bioengineering conventional and non-conventional yeasts: Progress and new challenges. *International Journal of Molecular Sciences, 24*(20), 15310.

Yadav, A. N. (2021). Biodiversity and bioprospecting of extremophilic microbiomes for agro-environmental sustainability. *Journal of Applied Biology and Biotechnology, 9*(3), 1–6.

Yadav, R., Arora, P., Kumar, S., & Chaudhury, A. (2010). Perspectives for genetic engineering of poplars for enhanced phytoremediation abilities. *Ecotoxicology, 19*, 1574–1588.

Yadav, S. P. S., Bhandari, S., Bhatta, D., Poudel, A., Bhattarai, S., Yadav, P., et al. (2023). Biochar application: A sustainable approach to improve soil health. *Journal of Agriculture and Food Research, 11*, 100498.

Yang, M., Chen, L., Wang, J., Msigwa, G., Osman, A. I., Fawzy, S., et al. (2023). Circular economy strategies for combating climate change and other environmental issues. *Environmental Chemistry Letters, 21*(1), 55–80.

Yi, W., Ziyu, Z., Shujun, Y., Hao, L., Noman, A., & Zhang, S. H. (2022). Cellulose degradation microorganisms and environmental-friendly solution to the agricultural waste management. In *Beneficial microorganisms in agriculture* (pp. 307–328). Springer Nature Singapore.

Zhang, M., Gao, B., Yao, Y., Xue, Y., & Inyang, M. (2012). Synthesis of porous MgO-biochar nanocomposites for removal of phosphate and nitrate from aqueous solutions. *Chemical Engineering Journal, 210*, 26–32.

Zulfiqar, U., Haider, F. U., Maqsood, M. F., Mohy-Ud-Din, W., Shabaan, M., Ahmad, M., et al. (2023). Recent advances in microbial-assisted remediation of cadmium-contaminated soil. *Plants, 12*(17), 3147.

Chapter 5
Emerging Technologies for Efficient Extraction of Value-Added Compounds from Agriculture Waste

Megha and Nisha Singh

Abbreviations

AI	Artificial intelligence
ANN	Artificial neural network
DNN	Deep neural network
EAE	Enzyme-assisted extraction
HD	Hydro-distillation
ILs	Ionic liquids (ILs)
IoT	Internet of Thing
LLE	Liquid-liquid extraction
MAE	Microwave-assisted extraction (MAE)
ML	Machine learning
PA	Precision agriculture
PLE	Pressurized liquid extraction
SCW	Subcritical water extraction
SE	Solvent extraction
SFE	Supercritical fluid extraction
UAE	Ultrasound-assisted extraction (UAE)
UAV	Unmanned aerial vehicles

Megha · N. Singh (✉)
Gujarat Biotechnology University (GBU), Gandhinagar, Gujarat, India
e-mail: nisha.singh@gbu.edu.in

J. Arora et al. (eds.), *Transforming Agriculture Residues for Sustainable Development*, Waste as a Resource, https://doi.org/10.1007/978-3-031-61133-9_5

5.1 Introduction

Agricultural and food wastes pose significant challenges globally, affecting the environment, economy, and society. The increasing global population, along with changing food consumption patterns, contributes to the generation of substantial amounts of agricultural and food waste. This waste not only represents a loss of valuable resources but also has negative environmental impacts, including greenhouse gas emissions and soil contamination. The world generates a vast number of agricultural raw resources, which are primarily used for human and animal sustenance as well as energy generation. Appropriate organic byproduct management can provide environmental and economic benefits by decreasing food waste (Arora & Ramawat, 2023; Pattnaik et al., 2021). The United Nations' global agenda for 2030 has defined several goals for sustainable development. It aims at a collaborative effort among nations to prevent food loss and waste generation in SDG (Sustainable Development Goals)-12, responsible consumption, and production. However, it is believed that up to 50% of the raw materials are lost during harvest, post-harvest, slaughter, transport, processing, storage, and consumption (Koul et al., 2022; Spang et al., 2019). Furthermore, considerable food is lost after harvest when the product of interest is removed from the soil or separated from the mother plant, and degradation begins (Gomes et al., 2021). Hence, substantial interventions are necessary to ensure the sustainable utilization of agricultural waste. From agro-waste, high-value components including proteins, polysaccharides, fibers, flavor compounds, and phytochemicals can be reused for nutraceuticals, functional ingredients, and value-added compounds (Table 5.1) (Pateiro et al., 2020; Dora et al., 2020; Ray, 2022, 2024). This process involves isolating these valuable compounds from a source material, often a plant or natural product, to utilize them for their potential health benefits or functional properties. This approach aligns with the broader trend in the food and pharmaceutical industries to harness the nutritional and bioactive potential of natural substances for various health and wellness applications (Sawicka, 2019; Sadh et al., 2018). Different extraction procedures have been selected depending on the nature of the raw material and the type of compounds extracted (Zainal et al., 2017; Dorta & Sogi, 2017; Sooch et al., 2022). In recent years, this new reality has prompted a model of sustainable development that asks for significant changes in conventional agricultural production practices and waste utilization. Furthermore, the majority of non-conventional extraction procedures can be used with modern, cutting-edge analytical equipment. In general, agricultural residues and waste contain a wide variety of biochemicals. Their availability is determined by the extraction processes used (Bala et al., 2023).

Natural bioactive compounds are actively being explored for the treatment and prevention of human diseases. These compounds demonstrate effective interactions with proteins, DNA, and other biological molecules, yielding desired outcomes. This capability makes them valuable candidates for the design of natural therapeutic agents, offering potential avenues for developing treatments derived from natural sources that can interact with and modulate biological processes in beneficial ways. The fourth agricultural revolution integrates cross-industry technology to increase

Table 5.1 The functional qualities of bioactive substances derived from plant residues

Source	Method	Bioactive compounds	Residual part	Functionality of processed product	Reference
Tomato	Solid-liquid extraction	Lycopene	Peel	Antioxidants, food colorants	Pandya et al. (2017)
Potato	Solid-liquid extraction	Polyphenols (gallic, chlorogenic, caffeic, ferulic, syringic, and p coumaric acids); and steroidal alkaloids (α-solanine, α-chaconine, aglycone solanidine)	Peel/skin	Antioxidants, thickener	Martinez (2019), Cardullo et al. (2021)
Pumpkin	Ultrasound-assisted three phases partitioning extraction	p-hydroxybenzoic acid, trans-p-coumaric acid, p-hydroxybenzaldehyde, caffeic acid	Seeds, shells	Antioxidants, fiber-rich component	Leichtweiset al. (2022)
Apple	Supercritical fluid extraction	Anthocyanins, cinnamic acid, dihydrochalcones (phloretin), flavan-3-ol (epicatechin), flavonol (quercitin glycosides)	Pomace, peel	Antioxidant activity (ROS and RNS), food additive (natural alternative to synthetic antioxidants and anti-microbial)	Vodnar et al. (2017)
Citrus fruits	Sequential extraction with a combination of microwave-assisted extraction and hydrodistillation and microwave-assisted extraction	Naringin, eriocitrin, hesperidin, narirutin, limonin	Lemon peel, mandarin peel, lemon seeds, Valencia orange seeds	Thickening and gelling agent, stabilizer, food additive	Singanusong et al. (2015), Mahato et al. (2019), Jagannath and Biradar (2019)
Logan/ Litchi	Microwave-assisted extraction with different conditions	Epicatechin-3-gallate, malvidin-3-glucoside, procyanidin B2, gallic acid, procyanidin B4, anthocyanins, naringin, isoscopoletin, coumaric acid	Seeds, Pericarp	Antioxidants, food colorants, food additives	Bai et al. (2019), Capanoglu et al. (2022)
Papaya	Ultrasound-assisted extraction/conventional organic solvent extraction	Carpaine, glucotropaeolin, benzyl isothiocyanate, benzyl thiourea, benzyl glucosinolate, sitosterol, hentriacontane, papain, caffeic acid, chlorogenic acid, p-coumaric acid	Seeds, leaf/peel	Rich in digestive enzymes	Rodrigues et al. (2019)

precision agriculture (PA) output and efficiency. Technologies such as artificial intelligence (AI), blockchain, machine learning (ML), unmanned aerial vehicles (UAV), and artificial neural network (ANN) are transforming agricultural activities on digital platforms. Subsets of AI have found application in decision-making processes across various stages of the food chain ecosystem. Notably, AI has ushered in significant advancements in agriculture, impacting crop production, harvesting, and marketing. Serving as a modernized technology, AI can effectively substitute human intelligence in troubleshooting and decision-making tasks. When integrated with the Internet of Things (IoT), AI becomes a powerful tool across the entire food system, extending its influence from agricultural practices to the management of food waste (Goel et al., 2022). This synergy allows for enhanced efficiency, precision, and automation in processes related to food production, distribution, and waste management. This chapter aims to discuss techniques for extracting valuable compounds from agricultural sources in order to provide a useful resource for medical scientists, pharmaceutical and food industries, and others interested in achieving high yields of valuable organic compounds from agricultural sources, residues, and waste to meet market demand. Sustainable agriculture is crucial to preserving the environment and biodiversity.

5.2 Extraction Procedure of Bioactive compounds

Agro-waste cannot be utilized without proper pretreatment since modified agro-waste is more effective and useful (Dey et al., 2021). Before being used, agricultural waste is subjected to some treatment techniques. Agro-waste residues are composed of lignocellulosic components that require chemical, physical, and biological treatments to break down the complexity (Capanoglu et al., 2022, Sooch et al., 2022). A wide range of extraction procedures are now employed to recover valuable organic molecules from agricultural crop residue/waste. Agro-food waste is often composed of polysaccharides (cellulose, lignin, hemicellulose), sugars (glucose, mannose, galactose), and phenolic chemicals (organic acids). The extraction of bioactive chemicals from agricultural waste is accomplished in three steps:

- Preparation,
- Extraction and purification
- Storage (De Souza & Shivakumar, 2021)

5.2.1 Pretreatment

The process involves the particular preparation or transformation of agro-food waste or byproducts to facilitate the release of desired components and eliminate insoluble, macromolecules, and microorganisms. This stage could be followed by several processes such as physical, chemical, biological, or enzymatic. Particle size reduction by milling and grinding and compound hydrolysis such as steam explosion and hydrothermal are all examples of physical preparation. Chemical pretreatment, on the other hand, comprises alkaline hydrolysis, which aids in the degradation of polysaccharides or polymeric substances. The biological pretreatment approach is based on the use of microorganisms such as fungi and bacteria to aid in polysaccharide breakdown (Fierascu et al., 2019).

5.2.2 *Extraction*

The process is a critical stage that is carried out utilizing various solvents, acids, alkalis, steam diffusion, and hydro-distillation (HD). Liquid-liquid extraction (LLE) is one of the most successful extraction procedures. LLE is determined by the volatility and solubility of the chemicals in the chosen solvent. Carotenoids, for example, are fat soluble in polar aprotic solvents such as acetone, ethanol, and acetic acid, as well as nonpolar solvents such as ethyl acetate and benzene (Nguyen et al., 2023).

5.2.3 *Purification*

The process includes at least one selective technology, such as chromatography or a membrane system capable of separating molecules of interest while eliminating contaminants (Gullón et al., 2020). Byproducts of extraction techniques are usually not useful enough for nutraceuticals and medicines, thus additional purification steps are frequently used, employing adsorbents, membranes, or integrated procedures combining several methodologies (Dzah et al., 2020; Mahato et al., 2019). The resulting fractions are rich in certain phytochemicals with similar properties and bioactivities that can be used to make nutritional supplements, nutraceuticals, or medicines (Yammine et al., 2018; Vidal-Casanella et al., 2021).

The extraction method and process operating parameters are always chosen depending on the overall performance of bioactive substances. However, to optimize the operating conditions, it is required to establish the family of compounds or the individual compounds to be extracted (Zhang et al., 2020; Koul et al., 2022).

5.3 Extraction Technique Used for the Production of Bioactive Compounds

Various approaches can be used to extract bioactive components found in agro-industrial waste. The availability of these approaches allows for the best use of any of them for the recovery of specific chemicals. According to a literature review, the following extraction methods are used for extracting bioactive compounds (Sooch et al., 2022):

- Solvent extraction (SE).
- Supercritical fluid extraction (SFE)
- Subcritical water extraction (SCW) and enzyme-assisted extraction (EAE),
- Ultrasound-assisted extraction (UAE)
- Microwave-assisted extraction (MAE)

Given the wide variety of bioactive compounds found in different fruit and vegetable waste, it is critical to design a standardized and effective approach for screening out health-beneficial substances for mankind (Table 5.2). The major extraction methods discussed in this section are listed below:

5.3.1 Solvent Extraction (SE)

This extraction method exposes the appropriately sized raw material to various organic solvents, which absorb the soluble components of interest as well as additional flavoring and coloring compounds, such as the anti-inflammatory and anti-cancer anthocyanins. After samples are typically filtered and centrifuged to eliminate solid residue, the extract can be utilized as a food supplement, additive, or to make functional foods (Nguyen et al., 2023). In addition to approaches that employ compressed fluids as extracting agents and sophisticated auxiliaries, it is crucial to identify suitable solvents with low toxicity, reduced environmental impact, increased extraction efficiency, and biodegradability (Hashemi et al., 2022). In addition to other environmentally friendly solvents like hot water or traditional organic liquids, ionic liquids (ILs) are recognized as innovative solvents for phenol extraction. Salts that exist as liquids at room temperature and with melting or glass transition temperatures lower than 100 °C are referred to as ILs. Because of their higher solubility power, ILs make superior solvents than water. Because of their distinct chemical characteristics, ILs have a lot to offer in terms of breaking down cell walls and ensnaring their targets through chemical selectivity, which increases the extract's effectiveness quickly (De Hoyos-Martínez et al., 2019; Xiao et al., 2018). When compared to other techniques, solvent Extraction is advantageous because of its low processing costs and simplicity of use. Nevertheless, this process necessitates the use of hazardous solvents, necessitates an evaporation/concentration phase for recovery, and typically requires a considerable amount of solvent to be used over an extended period of time. Furthermore, because of the high temperatures of the solvents during the lengthy extraction processes, it is impossible to rule out the possibility of thermal degradation of naturally occurring bioactive components. In order to get higher yields, solvent extraction has been enhanced by various techniques such as Soxhlet's, ultrasound or microwave extraction, and SFE (Del Pilar Sánchez-Camargo et al., 2017).

Table 5.2 Different methods for extraction of bioactive compounds

Plant waste	Extraction methods	Compounds	Benefits of the extracted compounds	Reference
Pomegranate peels, Pouteria sapota seeds	Solvent extraction	Carotenoids; punicalagins and ellagic acids	Provitamin A carotenoids can be converted into vitamin A, which is essential for growth, immune system function, and eye health	Goula et al. (2017)
Tobacco waste (midrib, dust, scrap), pomegranate peels, jackfruit peel	Ultrasound-assisted extraction	Chlorogenic, carotenoids, pectin	Helps in regulating glucose and lipid metabolism, improving insulin resistance, and reducing the risk of type 2 diabetes and cardiovascular diseases	Banožić et al. (2019), Goula et al. (2017), Moorthy et al. (2017)
Blackberry, blueberry, and Grumichama wastes	Ultrasound-assisted extraction and pressurized liquid extraction	Anthocyanin	Reduction in blood pressure protection and improvement in brain function	Machado et al. (2017), Mattioli et al. (2020)
Vegetable peel wastes (sweet potato, tomato, apricot, peach) and citrus peels and seeds	Supercritical fluid extraction	Carotenoids, total phenolic content	Helps in reducing microbial infections, oxidative stress, inflammation, diabetes, cancers, neurodegenerative disorders, cardiac disorders, etc.	Ferrentino et al. (2018), Singh et al. (2016)
Pistachio hulls	Subcritical water extraction	Gallotannin and flavonols	Bioactivities involve antioxidant, anti-inflammatory, enzyme inhibition, and protein binding agents	Erşan et al. (2018), He (2022)
Vine prune residues	Microwave-assisted extraction	Total phenolic content	Phenolic compounds and their catabolites also have health benefits in rheumatoid arthritis, depression, eye diseases, and cardiovascular diseases	Jesus et al. (2019), Matsumura et al. (2023)

5.3.2 *Supercritical Fluid Extraction (SFE)*

Supercritical fluid extraction, which uses carbon dioxide as a renewable and environmentally friendly solvent, is a popular green extraction technique for obtaining bioactive chemicals from various source materials. This technique is a fantastic way to extract bioactive compounds from agro-industrial leftovers because it is inexpensive, doesn't require high extraction temperatures, and doesn't include carbon dioxide, which is harmful (Ahangari et al., 2021; Tita et al., 2021). SF behaves as a single phase with both gas and liquid qualities. SF has no surface tension because it lacks a phase boundary between gas and liquid. The most significant benefit of SF is its adaptability. Even minor changes in temperature and pressure have a large impact on density around the critical point. Depending on the solubility of the material, SF can be changed to a more gas-like or liquid-like mobile phase (Amran et al., 2021). For high-value food applications, this extraction procedure employs supercritical CO_2. In contrast to many organic solvents, CO_2 is nontoxic, allows for greater extraction of nonpolar or partially polar compounds in supercritical CO_2, has a high solubility of oxygenated organic compounds of medium molecular weight, and is nonexplosive (Bubalo et al., 2018).

Jiao extracted haskap berries with supercritical (SC) CO_2 and water as a co-solvent. According to the findings, employing SC CO_2 and water as a co-solvent resulted in a higher extraction yield than utilizing water or ethanol as a solvent, 52.7% and 38.3%, respectively (Jiao, 2018). According to B. Wang and Lao's research, high-pressure CO_2-H_2O (107 Pa, 60 °C, 15–20 min) extraction yielded more colorants from red cabbage and purple sweet potato than typical extraction procedures using water and 70% acidified EtOH (Wang et al., 2022; Lao et al., 2020). Because of the greater extract yields, both studies demonstrated the feasibility of the $SCCO_2$-H_2O extraction process as a viable future technology. Despite this, this technique has yet to be implemented in industry and is not widely used due to the high cost of the equipment, massive technological investments, and complicated control parameters.

5.3.3 Ultrasound-Assisted Extraction (UAE)

For the extraction of bioactive chemicals from natural goods, ultrasound-assisted extraction is thought to be a simpler and more successful process than standard extraction methods. Ultrasound causes higher solvent diffusion into cellular materials, boosting mass transfer, and also breaks cell walls, allowing bioactive components to be released. UAE improves extraction and reduces extraction time by utilizing the cavitation and strong shear force created by ultrasound at frequencies ranging from 20 kHz to 100 MHz (Belwal et al., 2019). UAE is a low-cost, high-value technology that does not require complicated equipment. The extraction of chemicals from leaves, stalks, fruit, and berries such as strawberries, and plant seeds has been widely used. Ultrasound power, temperature, a solvent with low vapor pressure, extraction time, and sonication mode (continuous or pulsed) are the most important factors during extraction. These components determine the production and content of the extracts (Nguyen et al., 2023). In the Pieiro project, ultrasound-assisted extraction was optimized and validated for the fast extraction of stilbenes from grape canes. Stilbenes in grape canes were extracted for 10 min using this method, with an extraction temperature of 75 °C and ethanol (60%) as the extraction solvent. Grape cane byproducts were determined to be prospective sources of bioactive chemicals of relevance to the pharmaceutical and food sectors (Piñeiro et al., 2016). Another study looked at the influence of ultrasound technology on extracting water-soluble polysaccharides from dried and milled Agaricus bisporus byproducts. Glucan was extracted from mushroom byproducts in amounts of 1.01 and 0.98 g/100 g dry mass, with particle diameters of 355–250 and 150–125 m, respectively. The greatest extraction yield of 4.7% was obtained using a 15-min extraction time, an amplitude of 100 m, and 1 h of precipitation in 80% ethanol (Aguiló-Aguayo et al., 2017). This finding demonstrated that ultrasound could be used as a substituent energy source.

5.3.4 Enzyme-Assisted Extraction (EAE)

In recent years, there has been an upsurge in the generation or biotransformation of bioactive chemicals from agro-industrial leftovers by fermentation or enzymatic processes. Solid-state fermentation is a bioprocess in which enzymes generated by a microbe (bacteria, fungi, or yeast) act as biocatalysts, allowing for improved release and extraction of bioactive substances from various substrates. Enzymatic processes are more complex, more expensive, and have limitations such as limited stability, recovery, and temperature sensitivity, which can be mitigated by enzyme immobilization (Sharma et al., 2021). Enzymes including cellulase, −glucosidase, xylanase, −gluconase, and pectinase serve to deconstruct cell wall structure and depolymerize plant cell wall polysaccharides, allowing related chemicals to be released (Singh et al., 2016). Reshmitha et al. (2017) extracted lycopene-rich extracts from tomato peel using cellulase (20 units/g) and pectinase (30 units/g) at 50 °C for 60 min. These findings imply that the release of bioactive substances from plant cells via cell disruption and extraction can be optimized by utilizing enzyme preparations alone or in combination. Enzyme-assisted extraction is a viable replacement for traditional solvent-based extraction procedures. It is based on enzymes' ability to catalyze reactions in aqueous solutions under mild processing conditions (Nguyen et al., 2023).

5.3.5 Microwave-Assisted Extraction (MAE)

It is a microwave-heated solid-liquid extraction. Microwaves are electromagnetic waves with frequencies ranging from 300 MHz to 300 GHz. Microwave heating is caused by the samples' ability to absorb microwave energy and convert it to heat. Microwave energy is converted into heat via dipolar and ionic processes. Because of the dipolar nature of water, the presence of polar molecules causes microwave heating. The microwave creates an oscillating electric field, and the permanently polarized dipolar molecules strive to realign in the direction of the electric field, resulting in internal friction and heating of the material (Guo et al., 2017; Araújo et al., 2020; Sarfarazi et al., 2020). Kulkarni and Rathod (2016) used microwave-assisted extraction conditions and water as a solvent to extract mangiferin from *Mangifera indica* leaves. The greatest extraction yield of 55 mg/g was obtained after 5 min of extraction, a solid-to-solvent ratio of 1:20, and a microwave power of 272 W. In comparison to sequential batch extraction and Soxhlet extraction, MAE boosted extraction yield in a short period while also reducing solvent need. MAE has lately been used to extract plants such as purple sweet potato, red cabbage, blackberry, cranberry, and sour cherry. However, the use of MAE and UAE technologies should be used with caution, as these methods might disrupt the structure of flavonoids and tannins by subjecting the samples to high temperatures and extreme vibration, respectively (Hashemi et al., 2022). Smiderle et al. (2017) investigated MAE and pressurized liquid extraction (PLE) as advanced techniques for extracting polysaccharides (particularly biologically active-glucans) from *Pleurotus ostreatus* and *Ganoderma lucidum* fruiting bodies and found α- and β-glucans, as well as heteropolysaccharides, in all extracts. As a result, it can be stated that the microwave-assisted approach has numerous advantages over other methods due to its increased extraction efficiency, shorter extraction time, less labor, and high extraction selectivity, making it a preferred method for bioactive component extraction.

Radial innovation, relatively fast growth, coherence, tremendous impact, and uncertainty and ambiguity are the five characteristics that technology must have to be termed emergent. So that its use can be sustained over time. Aside from the extraction techniques outlined above, AI, ML, and IoT were used to improve agricultural systems, including agricultural waste (Fig. 5.1). Several scientific research have been conducted on the management of food waste. Precision agriculture and waste management systems that use AI integrate cross-industry technology to boost agricultural waste output and efficiency (Boopathi et al., 2022a, b; Boopathi & Myilsamy, 2021). Precision agriculture is being used expressly to contribute to potential increases in food demand while also achieving sustainability. Smart bins with RFID (radio-frequency identification) tags and global positioning system (GPS) modules are used for agricultural waste management. When the bins are a quarter full, a signal is sent to the smart vehicle. The agricultural stage, collection, and storage, pretreatment and treatment stage, and products and delivery stage are the major components of such a system. The quantification of garbage available for collection is critical for starting an effective management loop. Sensors and the Internet of Things (IoT) have enabled real-time design and feedback loops for trash management (Heideker et al., 2020; Shafi et al., 2019). The use of RFID-tagged smart bins and automatic weight sensors at all stages can help to prevent illicit agricultural waste transportation outside of the control of pretreatment and treatment units (Babu et al., 2022; Jeevanantham et al., 2022). This technology attempts to enhance the amount as well as the quality of agricultural waste. IoT sensors can provide information about the agricultural sector and take action based on user input (Koul et al., 2022).

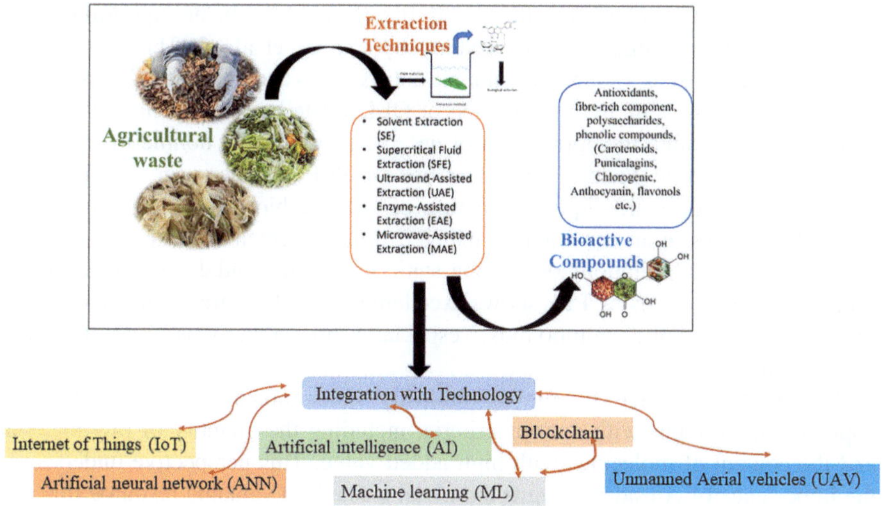

Fig. 5.1 Representation of Integration of emerging extraction techniques with technology for efficient extraction of bioactive compounds from agricultural waste

Precision agriculture (PA) is being introduced specifically to contribute to the potential growth in food demand and to achieve sustainability. To support farmers in tracking the status of their crops, PA combines various sensing technologies with communication protocols. "Smart" bins with RFID tags and GPS modules are placed. When the bins are a quarter full, signals will be sent to "smart" trucks.

5.4 Extraction of Compounds from Agro-Waste in Production Models Using AI, ML, and IoT

The need for approaches that can handle complex and ambiguous situations becomes increasingly crucial with the advancement of science and technology in addressing challenges in daily life. Soft computing, a growing collection of techniques, employs a variety of statistical, stochastic, and optimization tools to analyze datasets, categorize data, discover new patterns, and forecast future developments quickly. The primary goal of soft computing is to achieve a balance between addressing poor accuracy, dealing with ambiguity, and handling unclear truths in various applications (Ahmad et al., 2021). The integration of emerging remote-sensing technologies and artificial intelligence applications has become indispensable for addressing challenges such as new virus diseases, extreme weather conditions, climate change, environmental depletion, and the increasing demand for food. A notable development in the twenty-first century is the application of artificial intelligence techniques in agriculture and related fields. Agricultural waste, containing valuable components like soluble sugars, fiber, vitamins, and minerals, presents an opportunity for recycling by living organisms. In a study by Pappas et al. (2021), olive leaves (OLL) are identified as a source of valuable antioxidants and agricultural byproducts. The research focuses on the valorization of OLL waste by maximizing polyphenol concentration through the application of an emerging "green" nonthermal technology, Pulsed Electric Field (PEF). The use of PEF enhances the efficiency of conventional static solid–liquid extraction of specific bioactive compounds from fresh olive leaves in an eco-friendly manner, utilizing green solvents. While there may be industrial limitations due to the static nature of standalone extraction, the proposed technology using PEF shows excellent potential for green selective extraction of polyphenolic compounds, especially in continuous-flow industrial applications.

The waste generated by the agroindustry holds potential for a biorefinery approach, as it aligns with various criteria, including the quantity of raw material and the content of molecules with high added value. This perspective underscores the importance of considering agro-industrial waste as a valuable resource that can be efficiently utilized through biorefinery processes to extract and derive high-value compounds, contributing to sustainable and eco-friendly practices in resource management (Panda et al., 2016; Zuin & Ramin, 2018; El Sheikha & Ray, 2022). Spectroscopic techniques coupled with chemometrics offer a viable and environmentally friendly alternative to conventional methods for determining bioactive compounds in foods and food waste. The use of spectroscopic techniques provides several advantages over conventional methods, including simplified sample preparation procedures and shortened time for data collection and analysis. These advancements contribute to more efficient and streamlined processes, making spectroscopy coupled with chemometrics an attractive green solution for analyzing bioactive compounds in various food-related contexts (Lucarini et al., 2018). Moreover, AI holds the potential to drive the automation of processes, integrating into cyber-physical systems and advancing modeling for improved agriculture more efficiently than previous technological stages in this sector. Agricultural communities and businesses are urged to adopt a strategic approach for continuous improvement in production processes by implementing faster, safer, and more cost-effective plans through the utilization of data analytics and the concept of farming as a service (FaaS). This underscores the transformative role AI can play in enhancing the productivity and sustainability of agriculture (Yaqot & Menezes, 2021). In another study, it was discovered that radioactive substances are emitted during the recycling of agricultural waste materials, posing potential risks to both the environment and human health. To assess these risks, the research employed deep neural networks (DNN) with various network architectures. The Levenberg–Marquardt backpropagation algorithm was utilized for training, and the neural network was constructed with a single target variable, the hazard index, along with three input variables: 226Ra, 232Th, and 40K. The findings indicated that all evaluated agricultural byproducts presented no potential indoor or outdoor risks. The employed neural network structures demonstrated high accuracy in predicting the hazard indexes of agricultural byproducts (Oyebisi et al., 2023). ML approaches and optimization algorithms are crucial for making data-driven decisions, ensuring system safety and reliability, and identifying cost-effective strategies. In the context of agriculture, residues are predominantly composed of lignocellulosic compounds bound by covalent C-H bonds to form a complex crystalline polymer (Schmitt et al., 2020). Lignin, a component of biomass, plays a protective role by shielding the cellulose and hemicellulose composition against biochemical and thermal activities that could lead to degradation during utilization processes (Chen et al., 2021). Recently, ML has been employed to comprehend multi-dimensional correlations in biomass pretreatment for BioH2 production. ML can independently learn and recognize patterns from large datasets without explicit programming. By assessing the significance of each input characteristic, ML enhances the understanding of the system under consideration, aiding in the design of experiments based on model interpretation. ML applications in biomass conversion and upgrading processes, including torrefaction, pyrolysis, hydrothermal treatment, and fermentation, have been explored in the literature. Support vector machines (SVM), random forests (RF), and gradient boosting (GB) are among the most commonly used algorithms in ML for biomass research (Sharma et al., 2022a, b). Certainly, the exploration of advanced machine learning techniques, including deep learning and multi-view learning, is encouraged for further research. Integrating theory-driven models into machine learning models can enhance interpretability and prediction accuracy. This interdisciplinary approach, combining domain knowledge with the power of machine learning algorithms, holds promise for addressing complex challenges and advancing our understanding in various fields.

5.5 Significance of Extracted Compounds from Food Waste

Agricultural waste contains high levels of polysaccharides, dietary fibers, oils, phenolics, carotenoids, and other pigments (Kotsanopoulos et al., 2022; Sooch et al., 2022), it is regarded as an important source of complex carbohydrates, proteins, lipids, and phytochemicals. Therefore, the high concentration of physiologically active chemicals in wasted food is necessary for any potential health advantages (Ravindran & Jaiswal, 2016; Dueñas & García, 2020). Food waste can be used as a substrate to recover the value-added products, which include fine chemicals, nutraceuticals, antioxidants, bioactive, biopolymers, industrial enzymes, polysaccharides, organic acids, pigments, sugar, wax, esters, and xanthan gum (Kavitha et al., 2020; Kotsanopoulos et al., 2022). Several valuable bioproducts can be extracted from agricultural waste biomass using potential bioconversion pretreatment approaches, such as in agriculture (vermicomposting, biofertilizers, biochar, wood vinegar), bioenergy (biofuels), and pharma (antimicrobials, antioxidants, and antibiotic production). Available techniques for managing food waste that are more sustainable and profitable emerge as alternative possibilities for obtaining value-added products. Furthermore, unique compounds derived from food waste, ranging from solvents to antioxidants, are required for nutraceutical and biomaterial applications (Xiong et al., 2019). Combining biological, chemical, and physical techniques should be used to separate potentially marketable molecules contained in food wastes and byproducts to selectively extract and modify the desirable components and convert them to higher-value food products and additives. These methods should be used with caution to avoid microbiological dangers and to guarantee that the final goods are fit for consumer taste and are created following food laws. Through the effective exploitation of agricultural and forestry residues, aquatic biomass, and various waste streams, industrial development will be sustainable through specialty product extraction, conversion by green chemical or biotechnological processes, integrated biorefining, industrial symbiosis, cascade processing, and on-site processing of seasonal waste streams (Plazzotta & Manzocco, 2019; More et al., 2022).

Countries are currently experimenting with using food waste as fuel. A project including Nordic countries, for example, focuses on strategies to increase the use of food waste and the development of new technologies to convert these wastes into transport fuels. Denmark, Finland, and Sweden are currently producing biodiesel, bioethanol, and biogas from food wastes such as fruit and vegetable wastes, animal-based wastes, bakery wastes, and biowastes from households, agricultural leftovers, industrial and commercial origins (Nordic Energy Research, 2019). Direct or indirect transesterification produces biodiesel, which is a fatty acid methyl ester, from a variety of plant oils, including soybean, rapeseed, and canola (Joshi & Arora, 2023; Kannah et al., 2020). Karmee and Lin's research demonstrated the use of low-cost food waste in biodiesel manufacturing. The lipids were acquired through fungal hydrolysis of food wastes, and these lipids were trans-esterified to make biodiesel. Recently, a new solid-state micro-anaerobic digestion technique for degrading food waste has been created. This method compacts anaerobic digestion, requiring little water and energy (Degueurce et al., 2022). Biomaterials with high value are utilized in the cosmetics, pharmaceutical, chemical, food, and beverage industries (Nayak & Bhushan, 2019). Enzymes, which are important elements in various goods and processes, are extremely important in the business because they display selectivity against substrate and product, mild reaction conditions, minimal byproduct generation, and high yield. Raw material expenses can account for up to 30% of overall enzyme production costs. As a result, using food wastes and byproducts to reduce raw material prices for enzyme manufacturing is a viable option. It also reduces waste and mitigates its negative environmental implications (Ravindran & Jaiswal, 2016; Liu et al., 2023). These health-promoting bioactive chemicals in agricultural waste are critical to their anticancer, antimutagenic, antiviral, antioxidant, and antitumor actions, as well as their capacity to reduce the risk of cardiometabolic disorders. Furthermore, new phytochemical studies indicate that the significant antioxidant qualities of polyphenols derived from plant sources, such as skins, seeds, pulp, or pomace, make polyphenols one of the key phytochemicals crucial for health (McClements & Öztürk, 2021; Thakur & Borah, 2021; Castro et al., 2022).

5.6 Conclusion and Future Direction

The evolving nature of emerging technologies has prompted both researchers and companies to explore their potential advantages over traditional extraction processes. These modern methods demonstrate effectiveness in successfully extracting various compound types, including both organic and inorganic substances while preserving their properties and enhancing their quality. The development of efficient bioprocesses for recovering bioactive compounds from food waste not only contributes to sustainable resource utilization but also fosters economic opportunities for farmers and reduces the environmental footprint of the food industry. Leveraging this waste as a source of bioactive compounds not only offers environmental benefits by minimizing waste but also presents an opportunity to improve farmers' financial standing. Improvements in extraction methods that minimize or eliminate solvent usage are pivotal for sustainable bioprocessing. The organic waste generated by industries poses environmental hazards but also represents a potential bioresource for extracting bioactive components. This chapter explores how various technologies can facilitate the extraction of chemicals suitable for use as nutraceuticals and dietary supplements. Each extraction method has its strengths and limitations in recovering specific chemicals from plants. Depending on the plant matrix and the desired compounds, the most appropriate methodology can be selected with the assistance of technologies such as artificial intelligence and blockchain designs. Additionally, machine learning can optimize extraction parameters such as duration, temperature, pressure, reactor types, and raw material quality during methodological execution. By leveraging advancements in technology and innovation, the future of extraction processes holds the potential for enhanced efficiency, reduced environmental impact, and increased utilization of organic waste as valuable bioresources. Moreover, the integration of such technical solutions can foster collaboration and innovation among stakeholders, including farmers, researchers, industry players, and policymakers. By promoting interdisciplinary approaches and knowledge exchange, we can unlock new opportunities for value creation and sustainable development in the agriculture sector and beyond.

References

Aguiló-Aguayo, I., Walton, J., Viñas, I., & Tiwari, B. K. (2017). Ultrasound-assisted extraction of polysaccharides from mushroom by-products. *LWT, 77*, 92–99.

Ahangari, H., King, J. W., Ehsani, A., & Yousefi, M. (2021). Supercritical fluid extraction of seed oils–A short review of current trends. *Trends in Food Science and Technology, 111*, 249–260.

Ahmad, A., Ostrowski, K. A., Maślak, M., Farooq, F., Mehmood, I., & Nafees, A. (2021). Comparative study of supervised machine learning algorithms for predicting the compressive strength of concrete at high temperature. *Materials, 14*(15), 4222.

Amran, M. A., Palaniveloo, K., Fauzi, R., Satar, N. M., Mohidin, T. B. M., Mohan, G., et al. (2021). Value-added metabolites from agricultural waste and application of green extraction techniques. *Sustainability, 13*(20), 11432.

Araújo, R. G., Rodríguez-Jasso, R. M., Ruiz, H. A., Govea-Salas, M., Rosas-Flores, W., Aguilar-González, M. A., et al. (2020). Hydrothermal–microwave processing for starch extraction from Mexican avocado seeds: Operational conditions and characterization. *PRO, 8*(7), 759.

Arora, J., Ramawat, K.G., Mérillon, JM. (2023). Disposal of agricultural waste and its effects on the environment, production of useful metabolites and energy: Potential and challenges. In: Ramawat, K., Mérillon, JM., Arora, J. (eds) Agricultural waste: Environmental impact, useful metabolites and energy production. Sustainable development and biodiversity, Springer, 31, pp. 3–20. https://doi.org/10.1007/978-981-19-8774-8_1

Babu, B. S., Kamalakannan, J., Meenatchi, N., Karthik, S., & Boopathi, S. (2022, December). Economic impacts and reliability evaluation of battery by adopting electric vehicle. In *2022 International Conference on Power, Energy, Control and Transmission Systems (ICPECTS)* (pp. 1–6). IEEE.

Bai, X., Pan, R., Li, M., Li, X., & Zhang, H. (2019). HPLC profile of longan (cv. Shixia) pericarp-sourced phenolics and their antioxidant and cytotoxic effects. *Molecules, 24*(3), 619.

Bala, S., Garg, D., Sridhar, K., Inbaraj, B. S., Singh, R., Kamma, S., et al. (2023). Transformation of agro-waste into value-added bioproducts and bioactive compounds: Micro/nanoformulations and application in the agri-food-pharma sector. *Bioengineering, 10*(2), 152.

Banožić, M., Banjari, I., Jakovljević, M., Šubarić, D., Tomas, S., Babić, J., & Jokić, S. (2019). Optimization of ultrasound-assisted extraction of some bioactive compounds from tobacco waste. *Molecules, 24*(8), 1611.

Belwal, T., Huang, H., Li, L., Duan, Z., Zhang, X., Aalim, H., & Luo, Z. (2019). Optimization model for ultrasonic-assisted and scale-up extraction of anthocyanins from *Pyrus communis* 'Starkrimson' fruit peel. *Food Chemistry, 297*, 124993.

Boopathi, S., & Myilsamy, S. (2021). Material removal rate and surface roughness study on near-dry wire electrical discharge machining process. *Materials Today: Proceedings, 45*, 8149–8156.

Boopathi, S., Arigela, S. H., Raman, R., Indhumathi, C., Kavitha, V., & Bhatt, B. C. (2022a). Prominent rule control-based Internet of Things: Poultry farm management system. In *2022 International Conference on Power, Energy, Control and Transmission Systems (ICPECTS)* (pp. 1–6). IEEE.

Boopathi, S., Thillaivanan, A., Pandian, M., Subbiah, R., & Shanmugam, P. (2022b). Friction stir processing of boron carbide reinforced aluminum surface (Al-B4C) composite: Mechanical characteristics analysis. *Materials Today: Proceedings, 50*, 2430–2435.

Bubalo, M. C., Vidović, S., Redovniković, I. R., & Jokić, S. (2018). New perspective in the extraction of plant biologically active compounds by green solvents. *Food and Bioproducts Processing, 109*, 52–73.

Capanoglu, E., Nemli, E., & Tomas-Barberan, F. (2022). Novel approaches in the valorization of agricultural wastes and their applications. *Journal of Agricultural and Food Chemistry, 70*(23), 6787–6804.

Cardullo, N., Leanza, M., Muccilli, V., & Tringali, C. (2021). Valorization of agri-food waste from pistachio hard shells: Extraction of polyphenols as natural antioxidants. *Resources, 10*(5), 45.

Castro-Muñoz, R., Díaz-Montes, E., Gontarek-Castro, E., Boczkaj, G., & Galanakis, C. M. (2022). A comprehensive review of current and emerging technologies toward the valorization of bio-based wastes and by-products from foods. *Comprehensive Reviews in Food Science and Food Safety, 21*(1), 46–105.

Chen, W. H., Eng, C. F., Lin, Y. Y., Bach, Q. V., Ashokkumar, V., & Show, P. L. (2021). Two-step thermodegradation kinetics of cellulose, hemicelluloses, and lignin under isothermal torrefaction analyzed by particle swarm optimization. *Energy Conversion and Management, 238*, 114116.

De Hoyos-Martínez, P. L., Merle, J., Labidi, J., & Charrier–El Bouhtoury, F. (2019). Tannins extraction: A key point for their valorization and cleaner production. *Journal of Cleaner Production, 206*, 1138–1155.

De Souza, L., & Shivakumar, S. (2021). Conversion of agro-industrial wastes for the manufacture of bio-based plastics. In *Bioplastics for sustainable development* (pp. 177–204).

Degueurce, A., Dabert, P., Argence, V., Blondel, L., Le Bihan, A., Lebreton, M., et al. (2022). An innovative solid-state micro-anaerobic digestion process to valorize food waste: Technical development constraints and consequences on biological performances. *Waste and Biomass Valorization, 13*(1), 617–630.

Del Pilar Sánchez-Camargo, A., Pleite, N., Herrero, M., Cifuentes, A., Ibáñez, E., & Gilbert-López, B. (2017). New approaches for the selective extraction of bioactive compounds employing bio-based solvents and pressurized green processes. *The Journal of Supercritical Fluids, 128*, 112–120.

Dey, T., Bhattacharjee, T., Nag, P., Ghati, A., & Kuila, A. (2021). Valorization of agro-waste into value-added products for sustainable development. *Bioresource Technology Reports, 16*, 100834.

Dora, M., Wesana, J., Gellynck, X., Seth, N., Dey, B., & De Steur, H. (2020). Importance of sustainable operations in food loss: Evidence from the Belgian food processing industry. *Annals of Operations Research, 290*, 47–72.

Dorta, E., & Sogi, D. S. (2017). Value-added processing and utilization of pineapple by-products. In M. G. Lobo & R. E. Paull (Eds.), *Handbook of pineapple technology: Production, postharvest science, processing and nutrition* (pp. 196–220). Wiley.

Dueñas, M., & García-Estévez, I. (2020). Agricultural and food waste: Analysis, characterization and extraction of bioactive compounds and their possible utilization. *Food, 9*(6), 817.

Dzah, C. S., Duan, Y., Zhang, H., Boateng, N. A. S., & Ma, H. (2020). Latest developments in polyphenol recovery and purification from plant by-products: A review. *Trends in Food Science and Technology, 99*, 375–388.

El Sheikha, A. F., & Ray, R. C. (2022). Bioprocessing of horticultural wastes by solid-state fermentation into value-added/innovative bioproducts: A review. *Food Reviews International*. https://doi.org/10.1080/87559129.2021.2004161

Erşan, S., Üstündağ, Ö. G., Carle, R., & Schweiggert, R. M. (2018). Subcritical water extraction of phenolic and antioxidant constituents from pistachio (*Pistacia vera* L.) hulls. *Food Chemistry, 253*, 46–54.

Ferrentino, G., Morozova, K., Mosibo, O. K., Ramezani, M., & Scampicchio, M. (2018). Biorecovery of antioxidants from apple pomace by supercritical fluid extraction. *Journal of Cleaner Production, 186*, 253–261.

Fierascu, R. C., Fierascu, I., Avramescu, S. M., & Sieniawska, E. (2019). Recovery of natural antioxidants from agro-industrial side streams through advanced extraction techniques. *Molecules, 24*(23), 4212.

Goel, N., Kumar, Y., Kaur, S., Sharma, M., & Sharma, P. (2022). Machine learning-based remote monitoring and predictive analytics system for monitoring and livestock monitoring. In *Application of machine learning in agriculture* (pp. 47–67). Academic Press.

Gomes-Araújo, R., Martínez-Vázquez, D. G., Charles-Rodríguez, A. V., Rangel-Ortega, S., & Robledo-Olivo, A. (2021). Bioactive compounds from agricultural residues, their obtaining techniques, and the antimicrobial effect as postharvest additives. *International Journal of Food Science*. https://doi.org/10.1155/2021/9936722

Goula, A. M., Ververi, M., Adamopoulou, A., & Kaderides, K. (2017). Green ultrasound-assisted extraction of carotenoids from pomegranate wastes using vegetable oils. *Ultrasonics Sonochemistry, 34*, 821–830.

Gullón, P., Gullón, B., Romaní, A., Rocchetti, G., & Lorenzo, J. M. (2020). Smart advanced solvents for bioactive compounds recovery from agri-food by-products: A review. *Trends in Food Science and Technology, 101*, 182–197.

Guo, Q., Sun, D. W., Cheng, J. H., & Han, Z. (2017). Microwave processing techniques and their recent applications in the food industry. *Trends in Food Science and Technology, 67*, 236–247.

Hashemi, B., Shiri, F., Švec, F., & Nováková, L. (2022). Green solvents and approaches recently applied for extraction of natural bioactive compounds. *TrAC Trends in Analytical Chemistry, 157*, 116732.

He, H. F. (2022). Recognition of gallotannins and the physiological activities: From chemical view. *Frontiers in Nutrition, 9*, 888892.

Heideker, A., Ottolini, D., Zyrianoff, I., Neto, A. T., Cinotti, T. S., & Kamienski, C. (2020, November). IoT-based measurement for smart agriculture. In *2020 IEEE international workshop on metrology for agriculture and forestry (MetroAgriFor)* (pp. 68–72). IEEE.

Jagannath, A., & Biradar, R. (2019). Comparative evaluation of soxhlet and ultrasonics on the structural morphology and extraction of bioactive compounds of lemon (*Citrus limon* L.) peel. *Journal of Food Chemistry and Nanotechnology, 5*(3), 56–64.

Jeevanantham, Y. A., Saravanan, A., Vanitha, V., Boopathi, S., & Kumar, D. P. (2022, December). Implementation of Internet-of Things (IoT) in soil irrigation system. In *2022 International Conference on Power, Energy, Control and Transmission Systems (ICPECTS)* (pp. 1–5). IEEE.

Jesus, M. S., Genisheva, Z., Romaní, A., Pereira, R. N., Teixeira, J. A., & Domingues, L. (2019). Bioactive compounds recovery optimization from vine pruning residues using conventional heating and microwave-assisted extraction methods. *Industrial Crops and Products, 132*, 99–110.

Jiao, G. (2018). Extraction of anthocyanins from haskap berry pulp using supercritical carbon dioxide: Influence of co-solvent composition and pretreatment. *Lwt, 98*, 237–244.

Joshi, A., & Arora, J. (2023). Recent trends and future prospects of nanotechnology for agro-waste valorization into biofuels. In K. G. Ramawat, J. M. Mérillon, & J. Arora (Eds.), *Agricultural waste: Environmental impact, useful metabolites and energy production* (pp. 559–582). Springer Nature Singapore. https://doi.org/10.1007/978-981-19-8774-8_21

Kannah, R. Y., Merrylin, J., Devi, T. P., Kavitha, S., Sivashanmugam, P., Kumar, G., & Banu, J. R. (2020). Food waste valorization: Biofuels and value-added product recovery. *Bioresource Technology Reports, 11*, 100524.

Kavitha, S., Kannah, R. Y., Kumar, G., Gunasekaran, M., & Banu, J. R. (2020). Introduction: Sources and characterization of food waste and food industry wastes. In R. Banu, G. Kumar, M. Gunasekaran, & S. Kavitha (Eds.), *Food waste to valuable resources* (pp. 1–13). Academic Press.

Kotsanopoulos, K., Behera, S. S., & Ray, R. C. (2022). Microbial production of polysaccharides, oligosaccharides and sugar alcohols from vegetables and fruit wastes. In R. C. Ray (Ed.), *Fruits and vegetable wastes: Valorization to bioproducts and platform chemicals (.)* (pp. 343–364). Springer.

Koul, B., Yakoob, M., & Shah, M. P. (2022). Agricultural waste management strategies for environmental sustainability. *Environmental Research, 206*, 112285.

Kulkarni, V., & Rathod, V. (2016). Green process for extraction of mangiferin from *Mangifera indica* leaves. *Journal of Biologically Active Products from Nature, 6*(5–6), 406–411.

Lao, F., Cheng, H., Wang, Q., Wang, X., Liao, X., & Xu, Z. (2020). Enhanced water extraction with high-pressure carbon dioxide on purple sweet potato pigments: Comparison to traditional aqueous and ethanolic extraction. *Journal of CO2 Utilization, 40*, 101188.

Leichtweis, M. G., Molina, A. K., Pires, T. C., Dias, M. I., Calhelha, R., Bachari, K., et al. (2022). Biological activity of pumpkin byproducts: Antimicrobial and antioxidant properties. *Molecules, 27*(23), 8366.

Liu, Z., de Souza, T. S., Holland, B., Dunshea, F., Barrow, C., & Suleria, H. A. (2023). Valorization of food waste to produce value-added products based on its bioactive compounds. *PRO, 11*(3), 840.

Lucarini, M., Durazzo, A., Romani, A., Campo, M., Lombardi-Boccia, G., & Cecchini, F. (2018). Bio-based compounds from grape seeds: A biorefinery approach. *Molecules, 23*(8), 1888.

Machado, A. P. D. F., Pereira, A. L. D., Barbero, G. F., & Martínez, J. (2017). Recovery of anthocyanins from residues of Rubus fruticosus, *Vaccinium myrtillus,* and *Eugenia brasiliensis* by ultrasound-assisted extraction, pressurized liquid extraction and their combination. *Food Chemistry, 231*, 1–10.

Mahato, N., Sinha, M., Sharma, K., Koteswararao, R., & Cho, M. H. (2019). Modern extraction and purification techniques for obtaining high-purity food-grade bioactive compounds and value-added co-products from citrus wastes. *Food, 8*(11), 523.

Martinez, J. (2019). Recovering valuable bioactive compounds from potato peels via sequential hydrothermal extraction. In *41st symposium on biotechnology for fuels and chemicals*. SIMB.

Matsumura, Y., Kitabatake, M., Kayano, S. I., & Ito, T. (2023). Dietary phenolic compounds: Their health benefits and association with the gut microbiota. *Antioxidants, 12*(4), 880.

Mattioli, R., Francioso, A., Mosca, L., & Silva, P. (2020). Anthocyanins: A comprehensive review of their chemical properties and health effects on cardiovascular and neurodegenerative diseases. *Molecules, 25*(17), 3809.

McClements, D. J., & Öztürk, B. (2021). Utilization of nanotechnology to improve the application and bioavailability of phytochemicals derived from waste streams. *Journal of Agricultural and Food Chemistry, 70*(23), 6884–6900.

Moorthy, I. G., Maran, J. P., Ilakya, S., Anitha, S. L., Sabarima, S. P., & Priya, B. (2017). Ultrasound-assisted extraction of pectin from waste *Artocarpus heterophyllus* fruit peel. *Ultrasonics Sonochemistry, 34*, 525–530.

More, P. R., Jambrak, A. R., & Arya, S. S. (2022). Green, environment-friendly and sustainable techniques for extraction of food bioactive compounds and waste valorization. *Trends in Food Science and Technology, 128*, 296–315.

Nayak, A., & Bhushan, B. (2019). An overview of the recent trends on the waste valorization techniques for food wastes. *Journal of Environmental Management, 233*, 352–370.

Nguyen, T. L., Ora, A., Häkkinen, S. T., Ritala, A., Räisänen, R., Kallioinen-Mänttäri, M., & Melin, K. (2023). Innovative extraction technologies of bioactive compounds from plant by-products for textile colorants and antimicrobial agents. *Biomass Conversion and Biorefinery*, 1–30.

Nordic Energy Research (2019). *Report on Food waste to Biofuels*. https://www.nordicenergy.org/wordpress/wpcontent/uploads/2019/04/Food-Waste-to-Biofuels_FINAL.pdf

Oyebisi, S., Igba, T., Olutoge, F., & Ede, A. (2023). Application of artificial intelligence in the hazard indexes of recycled agricultural waste materials. In *Multiscale and multidisciplinary modeling, experiments and design* (pp. 1–12). Springer.

Panda, S. K., Mishra, S. S., Kayitesi, E., & Ray, R. C. (2016). Microbial-processing of fruit and vegetable wastes for production of vital enzymes and organic acids: Biotechnology and scopes. *Environmental Research (Elsevier), 146*, 161–172.

Pandya, D., Akbari, S., Bhatt, H., Joshi, D. C., & Darji, V. (2017). Standardization of solvent extraction process for lycopene extraction from tomato pomace. *Journal of Applied Biology & Biotechnology, 2*(1), 12–16.

Pappas, V. M., Lakka, A., Palaiogiannis, D., Athanasiadis, V., Bozinou, E., Ntourtoglou, G., et al. (2021). Optimization of pulsed electric field as a standalone "green" extraction procedure for the recovery of high value-added compounds from fresh olive leaves. *Antioxidants, 10*(10), 1554.

Pateiro, M., Munekata, P. E., Tsatsanis, C., Domínguez, R., Zhang, W., Barba, F. J., & Lorenzo, J. M. (2020). Evaluation of the protein and bioactive compound bioaccessibility/bioavailability and cytotoxicity of the extracts obtained from aquaculture and fisheries by-products. *Advances in Food and Nutrition Research, 92*, 97–125.

Pattnaik, M., Pandey, P., Martin, G. J., Mishra, H. N., & Ashokkumar, M. (2021). Innovative technologies for extraction and microencapsulation of bioactives from plant-based food waste and their applications in functional food development. *Food, 10*(2), 279.

Piñeiro, Z., Marrufo-Curtido, A., Serrano, M. J., & Palma, M. (2016). Ultrasound-assisted extraction of stilbenes from grape canes. *Molecules, 21*(6), 784.

Plazzotta, S., & Manzocco, L. (2019). Food waste valorization. In C. M. Galanakis (Ed.), *Saving food* (pp. 279–313). Academic Press.

Ravindran, R., & Jaiswal, A. K. (2016). Exploitation of food industry waste for high-value products. *Trends in Biotechnology, 34*(1), 58–69.

Ray, R. C. (2022). *Fruits and vegetable wastes: Valorization into bioproducts and platform chemicals* (p. 448). Springer.

Ray, R. C. (2024). *Roots, tubers and bulb crop wastes; management by biorefinery approaches* (p. 374). Springer.

Reshmitha, T. R., Thomas, S., Geethanjali, S., Arun, K. B., & Nisha, P. (2017). DNA and mitochondrial protective effect of lycopene-rich tomato (*Solanum lycopersicum* L.) peel extract prepared by enzyme-assisted extraction against H_2O_2 induced oxidative damage in L6 myoblasts. *Journal of Functional Foods, 28*, 147–156.

Rodrigues, L. G. G., Mazzutti, S., Vitali, L., Micke, G. A., & Ferreira, S. R. S. (2019). Recovery of bioactive phenolic compounds from papaya seeds agro-industrial residue using subcritical water extraction. *Biocatalysis and Agricultural Biotechnology, 22*, 101367.

Sadh, P. K., Duhan, S., & Duhan, J. S. (2018). Agro-industrial wastes and their utilization using solid-state fermentation: A review. *Bioresources and Bioprocessing, 5*(1), 1–15.

Sarfarazi, M., Jafari, S. M., Rajabzadeh, G., & Galanakis, C. M. (2020). Evaluation of microwave-assisted extraction technology for separation of bioactive components of saffron (*Crocus sativus* L.). *Industrial Crops and Products, 145*, 111978.

Sawicka, B. (2019). Post-harvest losses of agricultural produce (No. Chapter 40-1). *Sustainable Development, 1*, 1–16.

Schmitt, C. C., Moreira, R., Neves, R. C., Richter, D., Funke, A., Raffelt, K., et al. (2020). From agriculture residue to upgraded product: The thermochemical conversion of sugarcane bagasse for fuel and chemical products. *Fuel Processing Technology, 197*, 106199.

Shafi, U., Mumtaz, R., García-Nieto, J., Hassan, S. A., Zaidi, S. A. R., & Iqbal, N. (2019). Precision agriculture techniques and practices: From considerations to applications. *Sensors, 19*(17), 3796.

Sharma, P., Gaur, V. K., Sirohi, R., Varjani, S., Kim, S. H., & Wong, J. W. (2021). Sustainable processing of food waste for the production of bio-based products for circular bioeconomy. *Bioresource Technology, 325*, 124684.

Sharma, A. K., Ghodke, P. K., Goyal, N., Nethaji, S., & Chen, W. H. (2022a). Machine learning technology in biohydrogen production from agriculture waste: Recent advances and future perspectives. *Bioresource Technology, 364*, 128076.

Sharma, P., Vimal, A., Vishvakarma, R., Kumar, P., porto de Souza Vandenberghe, L., Gaur, V. K., & Varjani, S. (2022b). Deciphering the black box of omics approaches and artificial intelligence in food waste transformation and mitigation. *International Journal of Food Microbiology, 372*, 109691.

Singanusong, R., Nipornram, S., Tochampa, W., & Rattanatraiwong, P. (2015). Low power ultrasound-assisted extraction of phenolic compounds from mandarin (*Citrus reticulata* Blanco cv. Sainampueng) and lime (*Citrus aurantifolia*) peels and the antioxidant. *Food Analytical Methods, 8*, 1112–1123.

Singh, G., Verma, A. K., & Kumar, V. (2016). Catalytic properties, functional attributes and industrial applications of β-glucosidases. *3 Biotech, 6*, 1–14.

Smiderle, F. R., Morales, D., Gil-Ramírez, A., de Jesus, L. I., Gilbert-López, B., Iacomini, M., & Soler-Rivas, C. (2017). Evaluation of microwave-assisted and pressurized liquid extractions to obtain β-d-glucans from mushrooms. *Carbohydrate Polymers, 156*, 165–174.

Sooch, B. S., Mann, M. K., Sharma, P., & Ray, R. C. (2022). Phenolics and other bioactive compounds from vegetables and fruit wastes: Extraction methods and their possible utilizations. In R. C. Ray (Ed.), *Fruits and vegetable wastes: Valorization to bioproducts and platform chemicals (.)* (pp. 81–114). Springer.

Spang, E. S., Moreno, L. C., Pace, S. A., Achmon, Y., Donis-Gonzalez, I., Gosliner, W. A., et al. (2019). Food loss and waste: Measurement, drivers, and solutions. *Annual Review of Environment and Resources, 44*, 117–156.

Thakur, J., & Borah, A. (2021). Microcapsules of bioactive compounds from fruits and vegetable waste and their utilization: A review. *Pharma Innovation, 10*(5), 151–157.

Tita, G. J., Navarrete, A., Martin, A., & Cocero, M. J. (2021). Model-assisted supercritical fluid extraction and fractionation of added-value products from tobacco scrap. *The Journal of Supercritical Fluids, 167*, 105046.

Vidal-Casanella, O., Núñez, O., Granados, M., Saurina, J., & Sentellas, S. (2021). Analytical methods for exploring nutraceuticals based on phenolic acids and polyphenols. *Applied Sciences, 11*(18), 8276.

Vodnar, D. C., Călinoiu, L. F., Dulf, F. V., Ştefănescu, B. E., Crişan, G., & Socaciu, C. (2017). Identification of the bioactive compounds and antioxidant, antimutagenic, and antimicrobial activities of thermally processed agro-industrial waste. *Food Chemistry, 231*, 131–140.

Wang, B., Yang, S., Xu, L., Wang, X., Mi, L., Wang, K., et al. (2022). Evaluation study on extraction of anthocyanins from red cabbage using high-pressure CO_2+ H_2O: A fuzzy logic model and metabolomic analysis. *Sustainability, 14*(3), 1369.

Xiao, J., Chen, G., & Li, N. (2018). Ionic liquid solutions as a green tool for the extraction and isolation of natural products. *Molecules, 23*(7), 1765.

Xiong, X., Iris, K. M., Tsang, D. C., Bolan, N. S., Ok, Y. S., Igalavithana, A. D., et al. (2019). Value-added chemicals from food supply chain wastes: State-of-the-art review and future prospects. *Chemical Engineering Journal, 375*, 121983.

Yammine, S., Brianceau, S., Manteau, S., Turk, M., Ghidossi, R., Vorobiev, E., & Mietton-Peuchot, M. (2018). Extraction and purification of high added value compounds from by-products of the winemaking chain using alternative/nonconventional processes/technologies. *Critical Reviews in Food Science and Nutrition, 58*(8), 1375–1390.

Yaqot, M., & Menezes, B. C. (2021). Unmanned aerial vehicle (UAV) in precision agriculture: Business information technology towards farming as a service. In *2021 1st international conference on emerging smart technologies and applications (eSmarTA)* (pp. 1–7). IEEE.

Zainal-Abidin, M. H., Hayyan, M., Hayyan, A., & Jayakumar, N. S. (2017). New horizons in the extraction of bioactive compounds using deep eutectic solvents: A review. *Analytica Chimica Acta, 979*, 1–23.

Zhang, J., Wen, C., Zhang, H., Duan, Y., & Ma, H. (2020). Recent advances in the extraction of bioactive compounds with subcritical water: A review. *Trends in Food Science and Technology, 95*, 183–195.

Zuin, V. G., & Ramin, L. Z. (2018). Green and sustainable separation of natural products from agro-industrial waste: Challenges, potentialities, and perspectives on emerging approaches. In *Chemistry and chemical technologies in waste valorization* (pp. 229–282). Springer.

Chapter 6
Production of Metallic Nanoparticles From Agriculture Waste and Their Applications

Neetika Kimta, Rajni Dhalaria, Kamil Kuča, Richard Cimler, Vandana Guleria, Shivani Guleria, and Harsh Kumar

Abbreviations

Ag NPS	Silver nanoparticles
Au-NPS	Aureum (gold) nanoparticle
DLS	Dynamic light scattering
EDX	Energy dispersive X-ray
FTIR	Fourier transform infrared
FVWs	Fruits and vegetable wastes
HepG2	Hepatoblastoma cancer cells
MNPs	Metallic nanoparticles,
MONPs	Metal oxide nanoparticles

N. Kimta · R. Dhalaria
School of Biological and Environmental Sciences, Shoolini University of Biotechnology and Management Sciences, Solan, India

K. Kuča
Department of Chemistry, Faculty of Science, University of Hradec Kralove, Hradec Kralove, Czech Republic
e-mail: kamil.kuca@uhk.cz

R. Cimler · H. Kumar (✉)
Centre of Advanced Technologies, Faculty of Science, University of Hradec Kralove, Hradec Kralove, Czech Republic
e-mail: richard.cimler@uhk.cz

V. Guleria
School of Bioengineering and Food Technology, Shoolini University of Biotechnology and Management Sciences, Solan, India

S. Guleria
Department of Biotechnology, TIFAC-Centre of Relevance and Excellence in Agro and Industrial Biotechnology (CORE), Thapar Institute of Engineering and Technology, Patiala, India

J. Arora et al. (eds.), *Transforming Agriculture Residues for Sustainable Development*, Waste as a Resource, https://doi.org/10.1007/978-3-031-61133-9_6

NP	Nanoparticles
PdNPS	Palladium nanoparticles
Pt	Platinum nanoparticles
SEM-EDX	Scanning electron microscopy with energy dispersive X-ray spectroscopy
TEM	Transmission electron microscopy
TiO_2	Titanium oxide
UV-Vis	Ultraviolet-visible spectrophotometer
XPS	X-ray photoelectron spectroscopy
XRD	X-ray diffraction

6.1 Introduction

Waste organization has become an indigenous and global challenge that affects society, animals, and the environment. Garbage generation in emerging countries has expanded in parallel with increasing economic growth and the amount of rubbish produced per person as a result of population increase. In the absence of an all-encompassing and effective solid waste management strategy, garbage produced by various anthropological activities can cause health hazards and have a negative impact on the environment (Tun et al., 2018). When contrasting waste that decomposes and that doesn't, both have downsides and negative impacts that can have a devastating impact on society if waste is not adequately managed or disposed of. Biodegradable waste includes both plant and animal waste. Because increasing garbage generation rates are closely correlated with population density and per capita garbage output, the urban population in the majority of developing Asian nations may increase by more than 50% by 2025. Every year, India produces over 500 million tons of crop residue (Siwal et al., 2021). Nowadays, a lot of biodegradable litter is disposed of in dangerous ways, like burning, careless dumping, or spilling into waterways. Moreover, a variety of bioresources, including wood, agricultural residue, and cakes made from animal excrement, are frequently used as cooking fuel, which contaminates the air within. Bacteria and other decomposers can decompose household trash such as leftover tea leaves, vegetable peels, and fruit peels. An abundance of biodegradable waste in the ecosystem can contaminate it. In the presence of rubbish, they produce a tremendous volume of microbial flora. These bacteria have the latent to generate a multitude of infectious diseases in humans, plants, and animals. Certain gases produced during the burning of these wastes give them an unpleasant smell, and burning them can additionally emit greenhouse gases like methane and carbon dioxide. Disease carriers find refuge in waste dumps. A multitude of diseases are transmitted by rodents and mosquitoes, which proliferate in waste dumps and serve as carriers and vectors of disease. As proof of climate change's consequences grows more pervasive, scientists are still looking for ways to lessen the harm caused by hazardous manufacturing practices (Huston et al.,

2021). Rice is the most important farmed food crop, with a silica concentration of 91–93% and rice husk serves as a major agro-waste (Ding et al., 2005). An additional byproduct is bagasse of sugarcane juice crushing and extraction (Norsuraya et al., 2016). With an anticipated yield of 419.25 million tons in 2021–2022, including 91 million tons of bagasse waste, India is the second-largest producer of sugarcane in the world (Munagala et al., 2022). Wheat straw is produced when plants are harvested for their wheat and paddy (Munagala et al., 2022). Approximately 760 million tons of wheat straw waste, with cellulose content ranging from 30% to 40%, are generated annually. Fruit wastes containing phenolic chemicals are another viable and sustainable option (Yan et al., 2014). Additional waste materials including microbes, various plant parts, eggshells, coconut shells, walnut shells, and so on are also sought (Mohan & Panneerselvam, 2021). Waste utilization and recycling are in high demand because they promote global economic growth, reduce pollution, and eliminate waste (Razak et al., 2022). One application for biomass or biodegradable garbage is as a raw source for nanoparticle creation via green synthesis. It is a heterogeneous renewable basis that may be chemically mixed, and it is great because it is readily available for the production of essentials, affordable, abundant, and sustainable (Siwal et al., 2021).

In the past few years, nanotechnology has drawn a lot of interest. Nanotechnology refers to materials with dimensions between one and one hundred nm. Their applications span nearly all sectors of the economy, including electronics, agriculture, and medicine (Huston et al., 2021). The usage of nanoparticles aids in the improvement of thermal, mechanical, and barrier properties (Aswathi et al., 2023). A vast array of applications and potentially infinite potential for technological improvement are made possible by the fine-tuning of many reaction conditions to produce nanoparticles with different morphologies, such as spheres, rods, quantum dots, and particles. Classical synthesis methods, such as the top-down approach, were employed to create nanosized materials, which required both carcinogenic chemicals and a substantial amount of energy. Because the standard synthesis method produces pollutants, environmentally friendly synthesis techniques must be developed (Aswathi et al., 2023). Therefore, one way to lessen and reuse waste that pollutes our environment is to synthesize nanoparticles from biodegradable wastes. In green material synthesis techniques, natural biological systems are exploited to generate nanomaterials (Huston et al., 2021).

Green nanotechnology is a cutting-edge method for converting agricultural waste into nanodimensional, biodegradable, biocompatible, and renewable nanoparticles (Bello et al., 2015). Utilizing crop wastes such as rice husk, sugarcane bagasse, maize stalk, and maize, a high yield of MNPs was produced. A redox mechanism involving metabolites produces green nanomaterials, namely zero valent MNPs. Secondary metabolites identified in bio-waste include phenols, terpenes, alkaloids, and flavonoids (Kuppusamy et al., 2016). Few plant species produce secondary metabolites, while all plant groups produce principal metabolites, which are essential to the regular development, growth, and reproduction of all plant groups (Pichersky & Gang, 2000). Virtually all agro-industrial and food residues contain certain functional groups that improve reduction and stabilization. Fruits and vegetables encompass phenols, which can act as free radical scavengers and avert oxidation events in foodstuffs (Bancal & Ray, 2022). They contain other bioactive constituents, such as carotenoids, vitamins, lipids, and enzymes. Fruit residual biomass waste contains a range of flavonoids that can bind and convert metal ions into MNPs. As a result, they are utilized in the manufacture of nanoparticles. This chapter is primarily useful for learning about how the generation of MNPs from various agro-removing garbage contributes to maintaining a cleaner environment. The uses of MNPs in numerous evolving sectors, moderately than just one came to light to demonstrate that, in addition to inventive creation, they have a high commercial usage and limitless potential.

6.2 Source of Agro-Waste

Agricultural waste is the discarded material from farming deeds, namely harvesting, dispensation, and manufacturing crops, meat, poultry, fruits, and dairy products (Pattanaik et al., 2019). Agricultural leftovers are high in bioactive and beneficial chemicals (Lou et al., 2023). These leftovers have multiple industrial applications as raw materials after purification. The utilization of agro-residue can be beneficial and successful two goals are to employ natural and affordable resources for a variety of enterprises to stop its buildup in the environment and lessen the pollution it causes (Sud et al., 2008; Koul et al., 2022). According to Sadh et al. (2018), Sarkar et al. (2012), and Nguyen et al. (2019), there are two types of agricultural wastes: those that come straight from agricultural goods, such as bagasse, straw, and husks, and those that come from farm products utilized in many sectors, such as fruit cores and peels (Fig. 6.1). Certain components of agricultural waste will be discussed in the following sections.

6.2.1 Crop Residues

The bulk of waste residues at the field level that are derived directly from agricultural production are crop leftovers, including leaves, straws, stovers, and seed pods. These crop deposits are recognized as the most affordable and plentiful biological wastes that may be easily converted into an array of goods with additional value (Sadh et al., 2018). The main agricultural wastes are rice and wheat straws. Slightly more than half of these crops are used to produce biofuel; the rest are generally available all year round or as feed, while the remainder is burned, posing serious environmental concerns (Sadh et al., 2018). Of these three main agricultural waste products, rice straw is one of the most valuable and promising biomasses worldwide, producing an estimated 731 million tons annually, with Asia being the world's top producer (Afolalu et al., 2019). The majority of wheat waste gathered is wheat straw, which has 1–3 tons per acre as the annual yield, and an estimated 354.34 million tons of

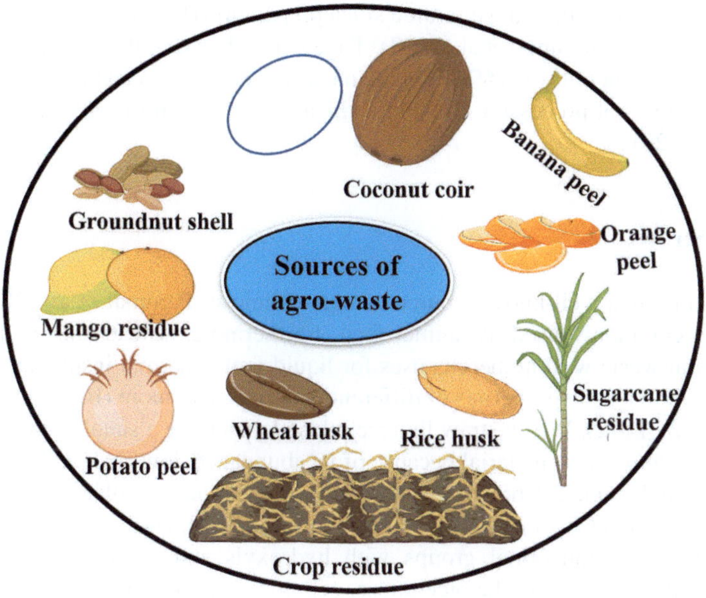

Fig. 6.1 Sources of agro-waste

production worldwide (Sarkar et al., 2012). The corn stover exhibits great potential agricultural leftovers for lingo-cellulosic ethanol, with an estimated production of 4.0 tons per acre and a yearly yield of over 128 million tons worldwide. Additionally adding to agro-wastes are crop residues from barley, sorghum, and oats (Jahirul et al., 2012).

Rice Husk

Worldwide, more than 600 million tons of rice are produced annually, with rice husk making almost 20% of the total weight. In other words, 0.23 tons of rice husk are expected to be formed for each ton of rice produced (Abbas & Ansumali, 2010; Kwan & Wong, 2020). One of the advantages of rice byproduct rice husk is that it is very biodegradable, has a low density, and is simple to gather (Aridi et al., 2017). It's also relatively cheap and high in cellulose, lignin, and silica (Yi et al., 2016). Rice husk's chemical composition differs among samples because of differences in climatic and terrestrial conditions, the kind of paddy, and so on. Rice husk has an ash mass fraction of 13–29%, which pollutes the environment. SiO_2 is the most abundant constituent of rice husk ash, accounting for up to 97.3% of the total (Renuka et al., 2015). When rice husk was seared in the open air, it produced slag with colors ranging from gray, based on the ratio of unburned carbon to inorganics, to black (Das et al., 2019; Gehlen et al., 2023). The rice husk ash is the utmost cost-effective basis for nanostructured silica preparation (Haider et al., 2022; Aprilia et al., 2023; Kumchompoo et al., 2023). Together with nano-SiO_2, rice husk ash also includes trace amounts (5–14%) of alkali metal oxides, which can be eliminated with the right chemical processing to yield a silica concentration of >99 weight percent (Lee et al., 2017).

Wheat Straw

Wheat straw is a well-known source of lignocellulose. Perumal et al. (2018) state that it is commonly used as animal feed. Furthermore, wheat stalk is a type of agricultural waste with numerous uses for liquid fuels, thermal insulating material, enhancing the tensile properties of different materials, and more (Binici et al., 2020; Gajera et al., 2022). Wheat straw lignocellulose looks to be a suitable carbonaceous microwave absorption material because of its abundance, low cost, and high carbon content (Singh et al., 2016; Zhou et al., 2018). The unique vascular bundle structure of wheat straw adds more surface area for chemical changes. It also has a high concentration of functional groups with hydroxyls and complex materials like protein, fatty acids, lignin, hemicellulose, cellulose, and pectin (Tian et al., 2011).

6.2.2 Agricultural Industry Wastes

Agricultural industrial processing waste is the second type of agro-waste. This comprises leftovers and byproducts from the meat and egg processing and abattoir industries, as well as animal fat and chicken skin, as well as fruit peels and vegetables, fruit marc, and sugarcane bagasse after oil or juice extraction (Rahman et al., 2008) De-oiled seed cakes made from edible oil and starch leftovers from the starch-manufacturing industries manufacturing industries, molasses/black treacle obtained from sugarcane. The most frequent agro-industrial waste is sugarcane bagasse. The dry, pulpy, fibrous residue leftover after sugarcane juice is removed is called bagasse. Sugarcane bagasse availability is expected to be over 180 million tons globally (Putri et al., 2020). Fruit leftovers from cider, fruit juice, and other food manufacturing operations, as well as orange peel and apple pomace, are examples of agro-industry wastes. In addition, some non-food-based agro-industrial wastes come from de-oiled seed cakes that are made from non-edible oil plants like *Jatropha curcas* and *Pongamia pinnata* (Maiti et al., 2016).

Wheat Bran

Wheat bran is a significant byproduct of the foodstuff commerce (Cheng et al., 2022; Glaser et al., 2023; Wayenbergh et al., 2023). Wheat bran's great availability and low cost make it useful not just in food applications but also in applications such as the synthesis of more expensive adsorbents (Curti et al., 2013). Water, ash, fat, protein, and carbohydrates such as cellulose, starch, glucan, arabinoxylans, and phenols make up wheat bran (Apprich et al., 2014). The wheat bran's chemical configuration diversity makes it an intriguing biomass with numerous applications. One of the applications of wheat bran is the removal of certain harmful ions. In other words, these materials absorb a wide range of hazardous ions quite well (Wen et al., 2014).

Bagasse

Another natural pore structure-containing renewable biomass source is bagasse. Cane stalk fiber remains from crushed cane and is present in this agricultural waste. It should be emphasized that bagasse (about 54 million dry tons) is produced globally annually (Mkhwanazi et al., 2023). Bagasse is used for a range of tenacities and consists of basic ingredients for a variety of building materials, crops for pulp and paper, particular chemicals, as well as fuel for sugar plant boilers (Sun et al., 2003). Bagasse has been extensively researched for its ability to remove harmful ions. It has a variety of advantages, including biodegradability, economic efficacy, modifiability, non-toxicity, and renewability, (Bhatnagar et al., 2015; Son et al., 2018). Bagasse is also high in lignin (23%), hemicellulose (27%), and cellulose (46%) (Doke et al., 2019). As a result, it may be a suitable choice for the manufacture of carbonaceous materials.

Eggshells

Eggshells are a recent agricultural waste that has received major attention. According to Sagar et al. (2018), millions of eggshells are produced as bio-waste every day worldwide. The majority of eggshell garbage is generated by residences, eateries, and bakeries. Eggshells are mostly of pure calcium carbonate with very little porosity, which can be made into valuable products. The sol-gel process can be utilized to create nano-calcium oxide from leftover eggshells (Siddiqui et al., 2014). The membrane of an eggshell is employed for the fabrication of magnetic $CuFe_2O_4$ nanoparticles with multifunctional features, namely its porous and hierarchical structure, which confers antibacterial and catalytic properties used in industrial water treatment (Santo et al., 2012).

6.2.3 *Livestock Wastes*

Worldwide, millions of farming households depend on the production of animals and crops for their livelihood. In certain regions, producing biogas from cattle manure or just making cakes for burning is a common technique that makes use of this renewable energy source (Canete-Rodriguez et al., 2016). Animal slurry, digested, burned chicken litter, and large amounts of straw are fuel sources for the cogeneration of heat and power (Jayasuriya et al., 2003). However, in many situations, these initiatives can only be made profitable if there is a market for the heat they generate (for use at adjacent enterprises) and byproducts (fertilizers for farms) can be established. Waste heat can be reticulated and piped for use on the farm or in nearby settlements, and power can also be used on the farm (Jayasuriya et al., 2003).

6.2.4 *Fruit and Vegetable Wastes*

Fruits and vegetable wastes (FVWs) are a type of agro-waste that include unprocessed vegetables and fruits, namely bananas, oranges, mangos, jack fruit, pineapple, and tomatoes (Fig. 6.1). FVWs created by food processing companies and wholesale markets are massive (Ruanglek et al., 2006; Ray, 2022). Furthermore, because organic waste is highly perishable, it is a major environmental contaminant (Pattanaik et al., 2019). Statistics show that the amount of waste produced in the United States, China, the Philippines, and India from the processing, packing, distribution, and consumption of fruits and vegetables is estimated to be 15, 32, 6.5, and 1.8 million tons, respectively (Pattanaik et al., 2019). Currently, India produces approximately 5.6 million tons of FVWs each year, which are disposed of via dumping on the city outskirts (Petruccioli et al., 2011).

6.3 Composition of Agro-Wastes

A lot of research is done on the lingo-cellulosic content of feedstock (cellulose, hemicellulose, and lignin) as well as the biochemical content (total carbohydrates, protein, lipids, and so forth). Crop residues have a cellulose content of 30–50% and a hemicellulose content of 20–38%. Contrarily, crop lignin contents vary from 7% to 21%. Depending on where they come from, agro-processing wastes can range in cellulose content, hemicellulose ranges from 15% to 33%, and lignin ranges from 5% to 24%. Rice bran waste is a good feedstock for industrial applications while containing only 5% lignin. High lignin sugarcane bagasse needs to be processed to remove lignin (Sun & Cheng, 2002) In industrial waste, the total solids consist of 40–85% carbohydrate molecules. Similarly, the quantity of cattle dung (Møller et al., 2004) combined with its 50–60% carbohydrate content makes it a good fuel for biogas production. Analogously, a feedstock's capacity for producing biogas is influenced by its protein and lipid composition. In the realm of antimicrobial research, nano-biotechnology has been used to probe into the recesses of scientific intricacies confined inside the confines of medical practice. It has been established that nanoparticles have excellent antibacterial characteristics, allowing them to tackle antibiotic-resistant superbugs. Scientists harnessed the synergistic effect induced by the metal as a precursor and plants as a capping and reducing agent interaction to create MNPs able to eradicate fungus and bacteria that are resistant to multiple drugs.

6.4 Green Synthesis of Metallic Nanoparticles by Agro-Waste

Conventional methods used a top-down approach to manufacture nanoparticles. The MNPs are created by breaking down bulk material into smaller fragments using various physical, chemical, and mechanical processes (Alparslan et al., 2017). It includes a wide range of procedures, such as laser ablation, mechanical milling, and sputtering (Etxabide et al., 2017). A bottom-up approach to nanoparticle biogenesis was established. This method uses a biological system to generate MNPs at room pressure and temperature without the use of hazardous substances or reagents, with the primary reaction being oxidation-reduction (Alparslan et al., 2017). Atoms serve as the beginning material for the production of NPs in this case. It covers procedures that are solid-state (physical and chemical vapor deposition), liquid-state (spray pyrolysis, laser ablation, etc.), and gas-phase (sol-gel and hydrothermal processes, among others) (Etxabide et al., 2017). The typical technique for the synthesis of MNPs from agro-wastes includes the following steps:

1. Agricultural waste acquisition
2. Agro-waste pretreatment
3. Intended MNP output
4. Characterization (Fig. 6.2)

Harvested from a nearby market or mill, agricultural waste is crushed, cleaned with distilled water, and then dried in an oven. At higher temperatures, certain wastes are calcined to go through a physiochemical transition. The trash is assorted with specific chemicals to produce the needed MNPs, which are then characterized using a variety of approaches. Information regarding the size, diameter, and shape of nanoparticles can be obtained by microscopy (TEM/SEM). MNP's highest absorption peak at a certain wavelength is determined via UV spectroscopy. The MNPs' % composition is investigated using energy dispersive X-ray (EDX). The crystalline structure of the MNPs is investigated using X-ray diffraction (XRD) and the functional groups were identified using Fourier transform infrared (FTIR). The charge and thermal stability of MNPs can be determined by dynamic light scattering (DLS). Although biowaste can be used to create a variety of nanomaterials, in this work, we will concentrate on metal and metal oxide nanoparticles.

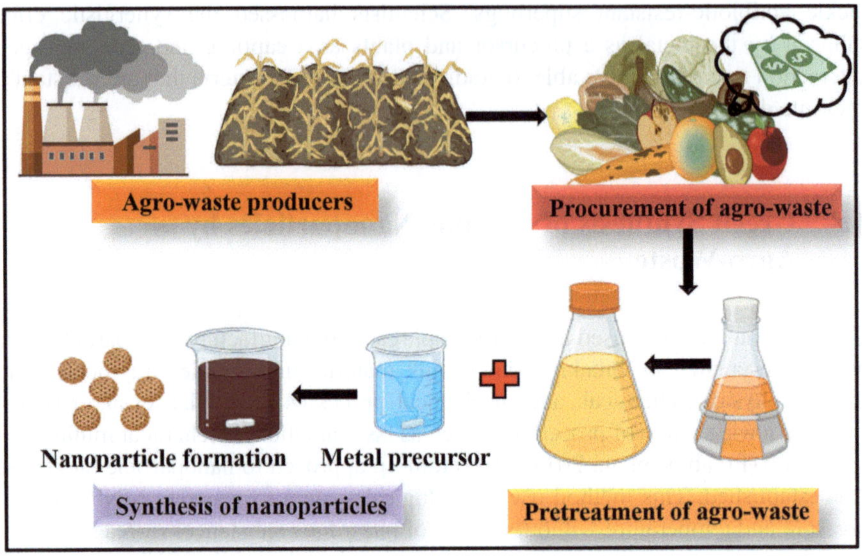

Fig. 6.2 Diagrammatic representation of the synthesis of agro-waste mediated nanoparticles

6.4.1 Metal and Metal Oxide Nanoparticles

MNPs are made up of an inorganic metal or metal oxide core encased in an organic, inorganic, or metal oxide shell. MNP shape is a key factor that determines biological response. Metallic rods, spherical, cylinders, triangular, ellipsoids, hexagonal, and other shapes are among them. NP geometry is claimed to influence MNPs' ability to cross biological barriers, the method of MNP entry into cells, the length of the cycle, and targeted impact (Jiang et al., 2008). MNPs can exhibit a wide range of characteristics, like the adsorption of particles onto surfaces the interactions of water with natural organic matter, and aggregation or agglomeration. These behaviors are caused in part by physicochemical properties (Khan, 2020). The influence of solubility affects the properties of MNPs as well. MNPs are intended to operate as a source of metal ions within cells, continuously releasing metal ions into the cytoplasm. The metal ion's solubility and rate of dissolution determine its release. Even though some MNPs are poorly soluble, they are extremely hazardous in physiological media (Khan, 2020). Noble metal nanoparticles have received the majority of attention in metal nanoparticle research. At temperatures below 100 °C, metal oxide nanoparticles (MONPs) are commonly generated during the hydrolysis of metal salts (Facure et al., 2021). Metal oxide nanoparticles can have remarkable chemical and physical characteristics because of their compact size and high density of edge or corner surface spots. Beneficial bioactive components of fruits and vegetables, including proteins, polysaccharides, and secondary metabolites among others, function as reducing agents in the creation of MNPs (Phang et al., 2021; Rathnasamy et al., 2017). MONPs have been reported to be synthesized from fruit peels, including banana, jackfruit, lemon, mango, pomegranate, and tangerine (Jain et al., 2009). The manufacturing of MNPs can employ agricultural waste. Weeds serve as bioreactors for the synthesis of MNPs and MONPs, which can be produced from them. *Lantana camara* can be used to make copper nanoparticles, *Ipomoea carnea* can make silver nanoparticles, and *Gloriosa superba* L can make copper oxide nanoparticles (Zamare et al., 2016). For the intra- and extracellular synthesis of MNPs and MONPs, bacteria, yeast, and fungi serve as nanofactories. For the large-scale production of MNPs and MONPs, this method was thought to be a promising option due to its eco-friendliness, economy, and affordability.

Gold Nanoparticles (Au-NPs)

Despite its historical fascination, gold now has a much wider range of uses because of developments in nanotechnology. Au-NPs, in particular, are benign and non-toxic. Gold is suitable for long-term usage in biological applications since it does not oxidize. Au-NPs have been found to improve proteins and antibodies have medicinal and sensing properties. Applications for Au-NPs include bio-imaging, bio-sensing, bio-labeling, bio-antibacterial, thrombolytic, photo-thermal, fluorescent, and anticancer properties. Several physical and chemical approaches have been used to successfully manufacture precise Au-NPs, but they are costly and employ toxic ingredients. Furthermore, the chemical manufacturing of Au-NPs may induce toxic chemicals to stick to the particles' surfaces, which could be harmful in medical applications (Thangadurai et al., 2021). It may therefore be beneficial to produce NPs using microbes or substances derived from plants. Biomolecules derived from biomass act as reducing and capping agents, resulting in more biocompatible NPs and eliminating the need for abrasive chemical treatments. Watermelon rind extract was used to produce and characterize Au-NPs (Tassoni et al., 2020). The Au-NPs generated showed significant synergistic effects and antibacterial efficiency against food-borne diseases. It also exhibited potent antioxidant and proteasome inhibitory action. Therefore, it is suggested that using agro-wastes to produce Au-NPs is an environmentally benign process, which makes them a good option for use in medication delivery, food, cosmetic, pharmaceutical, and biological applications (Muniyappan et al., 2021).

Palladium Nanoparticles (Pd-NPs)

Due to the huge surface area, Pd-NPs are efficient catalysts for chemical reactions. One distinguishing aspect of these catalysts is that as their size lowers, their relative surface area increases, and hence their activity rises. The use of an adequate stabilizing agent is therefore crucial not only for obtaining Pd-NPs of a suitable size to form highly active catalysts but also for stabilizing the surface and reducing leaching. Pd-NPs were synthesized using peel extract from *Annona squamosa*. The combined mixture was heated to 60 °C after adding 10 mL of aqueous extract and 80 mL of 1 mM palladium acetate for 4 h. The particles were characterized using XRD, UV-Vis (Ultraviolet-Visible) spectrophotometer, and TEM (transmission electron microscopy). It was proven that Pd-NPs with a spherical shape and an average size of 100 nm were present (Lakshmipathy et al., 2015). To cap and reduce Pd-NPs, watermelon rind aqueous extract was used. The polyhydroxy groups in the extract were found to be important in the creation of Pd-NPs. Pd-NPs with a mean size of 96.4 nm and a polydispersity index of 0.243 demonstrated potential for industrial catalysis by Suzuki coupling aryl halide with phenylboronic acid at room temperature (Roopan et al., 2012a, b).

Platinum Nanoparticles (Pt-NPs)

Pt-NPs are gaining popularity as unique nano-sized tools for the automobile industry, chemical industry, and biomedical device manufacturing (Shi et al., 2015). The automotive and chemical industries are collaborating to create more efficient materials based on Pt-NPs for catalytic converters and high-value chemicals. Dauthal and Mukhopadhyay (2015) utilized *Punica granatum* peel extract to make Pt-NPs with dimensions 16–23 nm; the particles were crystalline and spherical. The negative zeta potential of nanoparticles' indicates that they were extremely stable. The peel extracts polyphenolic components and the FTIR (Fourier transform infrared) study indicates that the capping and stability of the generated particles are attributed to quinones, which are composed of hydroxyl and carbonyl functional groups. The particles demonstrated excellent catalytic activity when the man-made impurity 3-nitrophenol was removed by sodium tetrahydridoborate (Roopan et al., 2012a, b).

Silver Nanoparticles (AgNPs)

Many individuals are interested in nanotechnology because of its many applications in various technological fields. AgNPs are frequently employed in their unique form, size, optical, electrical, and magnetic capabilities, coated materials, sensing materials, electrical components, and antimicrobial applications (Salem & Fouda, 2021). Due to their large volume ratio to surface area ratio, Ag-MNPs have superior physiochemical properties and numerous uses, including the treatment of industrial wastes and antibacterial and antioxidant properties (Salem & Fouda, 2021). Saad et al. (2021) synthesized AgNPs using the residue of *Citrullus lanatus* with great antibacterial potential. The residue of *Citrus limon* was utilized to synthesize silver nanoparticles with a size range of 9 nm that possess biological efficacy (Ahmed et al., 2018).

6.4.2 *Composites*

Nanocomposites are made up of numerous nanoparticles that are entrapped within a bulk material and can include a soft and a hard nanomaterial, two soft nanomaterials, or two hard nanomaterials (Falcaro et al., 2016). Nanocomposites are a type of nanomaterial in which one or more nano-sized phases (zero, one, and two dimensions) are incorporated in a ceramic, metal, or polymer material. These can be created at the molecular level by inorganic or organic components to achieve novel features (Shifrina et al., 2019). Because of its crucial uses, such as catalysis, nanocomposites have attracted a lot of attention.

Zinc and Nickel Oxide Nanoparticles

Rambutan peels were used to make biogenic zinc oxide nanocrystals. Rambutan peel extract and zinc nitrate hexahydrate were reacted for 2 h at 80 °C to form the zinc ellagate complex. Zinc oxide nanocrystals were then produced by calcining the zinc ellagate complex at 450 °C in a muffle furnace after it had been oven-dried for 8 h at 40 °C. Standard characterization techniques like XRD, SEM-EDX (Scanning electron microscopy with energy dispersive X-ray spectroscopy), XPS (X-ray photoelectron spectroscopy), and FTIR were used to show that zinc oxide nanocrystals' p-track conjugation activity boosted polyphenol production. Cotton fabric treated with zinc oxide nanocrystals showed impressive antibacterial activity against *Escherichia coli* (18.5 mm) and *Saccharomyces aureus* (23 mm) due to increased ZnO nanoparticle adsorption on the cotton surface (Yuvakumar et al., 2014a, b). Yuvakkumar et al. (2014a, b) successfully synthesized nickel-oxide (NiO) nanocrystals from rambutan peels for the first time to produce nickel-ellagate complexes in another work. The polyphenols in the extract aided the transformation of nickel nitrate hexahydrate into NiO nanocrystals. Synthetic NiO crystals applied to cotton fabric showed strong antibacterial activity of 25 and 35 mm against *Saccharomyces aureus* and *Escherichia coli*, respectively.

Iron Oxide (Fe_3O_4) Magnetic Nanoparticles

Watermelon rind extract was used to create Fe_3O_4 magnetic nanoparticles (Prasad et al., 2016). By utilizing XRD, DLS (Dynamic Light Scattering), FTIR, TEM, and UV-Vis spectrophotometer the spherical-shaped biogenic particles-whose sizes varied from 2 to 20 nm and showed a notable magnetic response were characterized. The polyphenol particles generated by the extract's polyphenols showed good catalytic activity in the manufacture of a 2-oxo-1,2,3,4-tetrahydropyrimidine derivative with extraordinarily high yields of 94%. Similarly, Venkateswarlu et al. (2013) recounted the synthesis of FeONPs using peel extract. The nanomaterials were detected using standard analytical procedures. The particles were sphere-shaped and equally distributed. TEM determined their size to be less than 50 nm. The bio-produced Fe_3O_4 magnetic nanoparticles exhibited excellent magnetic characteristics and had an inundation magnetization of 15.8 emu/g (Adelere & Lateef, 2016).

6.5 Biological Applications

This section investigates the novel synthetic methodologies used in the synthesis of MNPs. It will be a great delight to learn how mixing agro-waste into miniature MNPs benefits society and the environment.

6.5.1 Antibacterial Activity

Prolonged antibiotic usage in water and agriculture pollutes the environment. Iron oxide MNPs are commonly utilized for antibiotic adsorption due to their increased reusability, adsorption capacity, surface area/volume ratio, and stability. MNPs and MONPs (metal oxide nanoparticles) antibacterial effect is frequently mediated by the generation of reactive oxidative species (ROS). One kind of ROS is singlet oxygen (O_2), followed by hydroxyl radicals (OH) and superoxide radicals (O_2). It inhibits the electron transport chain, transcription, translation, and enzymatic activity, as well as causes protein inactivation and DNA degradation, which can result in bacterial cell death (Shahzadi et al., 2018). MNPs and MONPs have shown promise in inhibiting bacterial growth and combating antibiotic resistance. Researchers used an aqueous extract of *Musa balbisiana*, *Colla pseudostem* bio-waste, to produce ZnO nanoparticles (MNPs) with diameters ranging from 45 to 65 nm. It demonstrated substantial antibacterial action against gram-positive and gram-negative microorganisms, such as *Escherichia coli*, *Staphylococcus aureus*, *Bacillus subtilis*, and *Pseudomonas aeruginosa*; through the release of Zn^{2+} ions from ZnO MNPs, they attach themselves to the membrane of the bacterial cell and release ROS, which denaturizes DNA, lipids, and proteins and impair cellular activity. ZnO MNPs can also demonstrate anti-biofilm activity against *P. aeruginosa*, making them useful as a water disinfectant in the pharmaceutical industry and food preservation. By identifying the breakdown of bacterial cell membranes and cytoplasmic elements including proteins, reducing sugars, and nucleic acid, the mechanism of the antibacterial response was examined. Ag MNPs were shown to have antibacterial capability and synergistic efficacy when used with standard antibiotics in opposition to harmful bacterial infections in humans. This study indicated grape's potential as a novel source of Ag MNP biosynthesis, which could open up new pathways in nanomedicine (Saratale et al., 2020). Several researchers found that ZnO nanocrystals derived from rambutan peel waste extract have antibacterial action against pathogenic microorganisms and are also suited for biomedical nanotechnology (Table 6.1). Ag MNPs may also be made from rambutan peel waste extract,

Table 6.1 Antibacterial potential of agro-waste derived nanoparticles

Botanical name	Agro-waste	Nanoparticles	Size (nm)	Strains	Application	References
Artocarpus heterophyllus	Jackfruit residue	Silver	15	*Escherichia coli*	100% inhibition	Rajput et al. (2022)
Carthamus tinctorius	Safflower residue	Silver	2–50	*Pseudomonas* sp. and *Staphylococcus* sp.	100% inhibition	Rodríguez-Felix et al. (2021)
Citrus limon	Lemon residue	Silver	9	*S. aureus* and *E. coli*	ZOI 8 and 7 mm	Ahmed et al. (2018)
Citrus maxima	Pomelo peel	Silver	–	*Bacillus cereus* and *E. coli*	ZOI 8–16 mm and 7–10 mm	Barbhuiya et al. (2022)
Citrus sinensis	Orange residue	Silver	46	*S. aureus* and *E. coli*	ZOI 8 and 7 mm	Ahmed et al. (2018)
Cocos nucifera	Coconut shell	Silver	14.2–22.96	*S. aureus* and *E. coli*	ZOI 15 and 13 mm	Sinsinwar et al. (2018)
Mangifera indica	Mango peel	Silver	25–70	*E. coli* and *S. aureus*	MIC 8.3 and 6.22 mg/L	Xing et al. (2021)
Oryza sativa	Rice husk	Gold	15	*S. aureus*	ZOI 16 mm	Harby et al. (2022)
Punica granatum	Pomegranate residue	Silver	25	*E. coli*	100% inhibition	Rajput et al. (2022)

(– Not reported)

which has antibacterial activity against *Salmonella paratyphi* with a 4 mm inhibitory zone. This suggests that *Salmonella paratyphi* can cause enteric fever, also known as paratyphoid fever (Lestari et al., 2018).

6.5.2 Anticancerous

Cancer is a class of disorders distinguished by unrestrained cell division (Matthews et al., 2022). Cancer claims the lives of 8.2 million people each year. There are about 200 different types of cancer. Immunology and nanotechnology are being integrated to create nano-immune chemotherapy, an effective cancer treatment. Various MNPs and MONPs display cytotoxicity against malignant cells while not affecting normal cells, making them effective anticancer agents. Das et al. created Ag MNPs that included discarded pineapple peel extract that was cytotoxic to hepatoblastoma (HepG2) cancer cells, antioxidative, and diabetic-preventive. Along with helping to create drugs to cure conditions like diabetes and cancer, it is helpful in the treatment of acute ailments. It can also be used to patch wounds and cure bacterial infections. *Sargassum wightii*, a brown seaweed, was a source of aqueous extracts that were used to generate MgO-MNPs. They include vitamins, polysaccharides, carotenoids, amino acids, and polyphenols, which function as reducing and capping agents to help generate MgO-MNPs. By boosting ROS formation, these biosynthesized MgO-MNPs induce apoptosis in malignant cells, indicating potential lethal efficacy against the lung cancer cell line A549 (Singh et al., 2021) (Table 6.2). Another study used *Trapa natans* peel extract to create bimetallic nanoparticles. *Trapa* peel extract was used to create gold (Au) (25 nm) and silver (Ag) NPs with encouraging cytotoxic action against certain cancer cells. Since the characteristics of Ag and Au NPs are combined in bimetallic NPs, they are more favorable than uni-metallic nanoparticles. According to the findings, bimetallic composite nanoparticles can trigger apoptosis and mitochondrial stress, which cause ROS-mediated p53-independent cell death in cancer cells. Bimetallic composite NPs may be used in the future in nanomedicine to treat cancer since it is a straightforward, effective, and affordable method (Ahmad et al., 2019).

6.5.3 Dye Degradation

Dyes were widely used in industries, resulting in increased biochemical oxygen demand and chemical oxygen demand levels, neurological diseases, cancer, respiratory problems, paralysis, and death. The selection of adsorbents can result in a higher percentage of dye removal from aqueous solution. Changing to MNPs instead of wasting massive volumes of agricultural waste would keep the environment clean and human life healthy (Hu et al., 2018). MNPs remediated methyl orange (91.1%) and phenol red dyes (92.6%) under optimal environmental parameters such as pH

Table 6.2 Anticancer potential of agro-waste derived nanoparticles

Botanical name	Agro-waste	Nanoparticles	Size (nm)	Cell line	Application	References
Ananas comosus	Pineapple residue	Gold	20.7	–	Less harmful to healthy cells and more biocompatible	Pechyen et al. (2021)
Cocos nucifera	Coconut shell	Silver	–	HepG2	IC_{50} value of 15.28 μg/mL against cancer cells; tiny size allows NPs to enter cancer cells easily, leading to cell death	Das et al. (2021)
Musa spp.	Banana peel	Gold	10	Adherent mouse fibroblast (L929) cells	(>100 g/mL) reduced mouse fibroblast development and had no adverse effects on healthy organs	Liu et al. (2020)
Passiflora edulis	Passion peels	Gold	18.6	–	Biocompatible and less harmful to healthy cells	Pechyen et al. (2021)

(– Not reported)

(3 and 5), adsorbent dosage (0.4 g/L), temperature (343 K), and beginning dye concentration (10 ppm). MNPs demonstrated a high potential for color removal from wastewater. Hossain et al. discovered that MNPs might be more promising as wastewater adsorbents after rice husk was modified with potassium hydroxide and AgNPs (Hossain et al., 2022). Within 10 min, the MNPs demonstrated catalytic capabilities for complete congo red deprivation. These nanoparticles also boosted biogas production via anaerobic digestion of Enteromorpha. The synthesis of iron oxide MNPs was thought to be both pollution-free and economical. Rhodamine B was catalyzed to degrade by 90–95 nm-sized iron oxide MNPs that were created when coconut pulp was hydrothermally treated with a ferric chloride ($FeCl_3$) solution (Rajendran et al., 2021). Simonetti et al. created titanium dioxide (TiO_2)-C composites with a methylene blue dye elimination rate of 0.006/min. The carbon recovered from the coconut shell performed well as an adsorbent (Simonetti et al., 2018). In order to photodegrade methylene blue, Yunarti et al. (2022) employed graphene oxide (GO) generated from recycled coconuts coated with titanium dioxide (TiO_2) MNPs. In 120 min, the co-precipitation method converted coconut husk ash into ZnO NPs with significant photocatalytic efficacy for the remediation of 97% MB and 68% methyl orange (MO) dyes (Wary et al., 2022). When making high-yielding ZnO NPs, the CH served as a precipitating agent. The MNPs increased dye degradation efficiency due to their large surface area and excellent quenching effects (Dayanidhi et al., 2020). For the 97–98% dyes degradation, an agro-waste extract

Table 6.3 Dye degradation activity of agro-waste derived nanoparticles

Agro-waste	Nanoparticles	Size (nm)	Dye used	Dye degradation	References
Bahera fruit	Silver	–	Methylene blue	–	Sharma (2021)
Banana peel	Silver	10–40	Methyl orange	97.96%	Abdelghaffar (2021)
Coconut husk	Zinc	9–14	Methylene blue, methyl orange	97%, 68%	Wary et al. (2022)
Coconut shell	Titanium dioxide	–	Methylene blue	Removal rate 0.006/min	Simonetti et al. (2018)
Coconut shell	Titanium dioxide	10–11	Methylene blue	51.30%	Yunarti et al. (2022)
Eggshell	Iron oxide	21	Congo red, Rhodamine B	100%, 92%	Rajendran et al. (2021)
Eggshell	Titanium dioxide	78.73	Methylene blue, rhodamine	–	Singh et al. (2017)
Eggshell membrane	Gold	–	Nitrophenol	Reduction of 4-nitrophenol to 4-aminophenol	Jeon et al. (2021)
Eggshell membrane	CuO–ZnO	100	Congo red, nitrophenol	Adsorption capacity 775.2 mg/g, 98%	He et al. (2019)
Pineapple	ZnO	10–60	–	99.6%	Mirgane et al. (2021)

(– Not reported)

comprising several phytochemicals worked as a reducing agent in the conversion of stabilized zirconium (Zr^{4+}) ions to ZrO_2 NPs (Narasaiah et al., 2022a, b) (Table 6.3). Stannum dioxide (SnO_2) NPs were formed after 3 h of centrifugation, cleaning, and calcination after constant stirring of tin chloride dihydrate and cotton peel aqueous extract at 80 °C. Within 30–100 min, the dyes MB and MO were destroyed by the MNPs' functional photocatalytic capabilities, which destroyed 96–98% of the dyes (Narasaiah et al., 2022a, b). SnO_2 NPs are non-toxic and environmentally safe, and they can be reused five times, with degrading efficiency remaining at 95.82 and 92.1%, respectively.

6.6 Conclusion

Recycling agricultural waste and its connected applications are essential for both economic growth and scientific study. Utilizing agricultural waste as a multipurpose material produced reusable, biocompatible, non-toxic, and environmentally beneficial nanoparticles (MNPs) of varied sizes, shapes, and surface area. According to a study conducted over the preceding two decades, the rate of biowaste-assisted nanomaterial synthesis has increased exponentially due to its ease of availability. Finally, in terms of zero contamination, simple procedures, low toxicity, high stability, and cost-effectiveness in the synthesis of MNPs, the utilization of biowaste is assessed to be more favorable than other raw materials. Finally, in terms of zero contamination, simple procedures, low toxicity, high stability, and cost-effectiveness in the synthesis of MNPs, the utilization of biowaste is assessed to be more favorable than other raw materials. On the other hand, further study is required to increase the environmentally benign uses of nanomaterials in energy, biosensing, biomedicine, and battery storage. This covers the following: functionalization, cytotoxicity, bulk industrial production, shape-selective synthesis of nanoparticles, the interaction between the metal precursor and biowaste, the mechanism of interaction in metal precursor reduction, and techniques for isolating or purifying the interesting component from biowaste. The manufacture of new MNPs with plentiful pore volume, high pore area, and functional groupings, on the other hand, remains a difficulty. Furthermore, it is critical to continue investigating the potential of MNPs in the fields of cosmetics and optoelectronics. Improving MNPs' demand and supply can be increased by industrial production. Because MNPs agro-waste combination usage can solve a range of challenges, value-added product commercialization will boost net productivity and sustainability. As a result, advancements in nano-devices will benefit the agricultural waste business.

References

Abbas, A., & Ansumali, S. (2010). Global potential of rice husk as a renewable feedstock for ethanol biofuel production. *Bio Energy Research, 3*, 328–334.

Abdelghaffar, F. (2021). Biosorption of anionic dye using nanocomposite derived from chitosan and silver nanoparticles synthesized *via* cellulosic banana peel bio-waste. *Environmental Technology and Innovation, 24*, 101852. https://doi.org/10.1016/j.eti.2021.101852

Adelere, I. A., & Lateef, A. (2016). A novel approach to the green synthesis of metallic nanoparticles: The use of agro-wastes, enzymes, and pigments. *Nanotechnology Reviews, 5*(6), 567–587.

Afolalu, S. A., Oladipupo, S., Bose, M. E., Abioye, A. A., Adejuyigbe, S. B., Ajayi, O. O., & Ongbali, S. O. (2019). Agro waste a sustainable source for steel reinforcement-review. In: *Journal of Physics: Conference Series* (Vol. 1378, No. 3, p. 032032). IOP Publishing.

Ahmad, N., Sharma, A. K., Sharma, S., Khan, I., Sharma, D. K., Shamsi, A., Santhosh Kumar, T. R., & Seervi, M. (2019). Biosynthesized composites of Au-Ag nanoparticles using Trapa peel extract induced ROS-mediated p53 independent apoptosis in cancer cells. *Drug and Chemical Toxicology, 42*(1), 43–53. https://doi.org/10.1080/01480545.2018.1463241

Ahmed, S., Kaur, G., Sharma, P., Singh, S., & Ikram, S. (2018). Fruit waste (peel) as bio-reductant to synthesize silver nanoparticles with antimicrobial, antioxidant and cytotoxic activities. *Journal of Applied Biomedicine, 16*(3), 221–231. https://doi.org/10.1016/j.jab.2018.02.002

Alparslan, Y., Metin, C., Yapıcı, H. H., Baygar, T., Günlü, A., & Baygar, T. (2017). Combined effect of orange peel essential oil and gelatin coating on the quality and shelf life of shrimps. *Journal of Food Safety and Food Quality, 68*, 69–78.

Apprich, S., Tirpanalan, Ö., Hell, J., Reisinger, M., Böhmdorfer, S., Siebenhandl-Ehn, S., Novalin, S., & Kneifel, W. (2014). Wheat bran-based biorefinery 2: Valorization of products. *LWT-Food Science and Technology, 56*(2), 222–231.

Aprilia, S., Rosnelly, C. M., Fitriani, F., Akbar, E. H., Raqib, M., Rahmah, K., Amin, A., & Baity, R. A. (2023). Synthesis of amorphous silica from rice husk ash using the sol–gel method: Effect of alkaline and alkaline concentration. *Materials Today: Proceeding*. https://doi.org/10.1016/j.matpr.2023.02.403

Aridi, N., Sapuan, S., Zainudin, E., & Al-Oqla, F. (2017). A review of rice husk bio-based composites. *Current Organic Synthesis, 14*(2), 263–271.

Aswathi, V. P., Meera, S., Maria, C. A., & Nidhin, M. (2023). Green synthesis of nanoparticles from biodegradable waste extracts and their applications: A critical review. *Nanotechnology For Environmental Engineering, 8*(2), 377–397.

Bancal, V., & Ray, R. C. (2022). Overview of food loss and waste in fruits and vegetables: From issue to resources. In *Fruits and vegetable wastes: Valorization to bioproducts and platform chemicals* (pp. 3–29). Springer Nature Singapore.

Barbhuiya, R. I., Singha, P., Asaithambi, N., & Singh, S. K. (2022). Ultrasound-assisted rapid biological synthesis and characterization of silver nanoparticles using pomelo peel waste. *Food Chemistry, 385*, 132602. https://doi.org/10.1016/j.foodchem.2022.132602

Bello, S. A., Agunsoye, J. O., & Hassan, S. B. (2015). Synthesis of coconut shell nanoparticles via a top down approach: Assessment of milling duration on the particle sizes and morphologies of coconut shell nanoparticles. *Materials Letters, 159*, 514–519. https://doi.org/10.1016/j.matlet.2015.07.063

Bhatnagar, A., Sillanpää, M., & Witek-Krowiak, A. (2015). Agricultural waste peels as versatile biomass for water purification–A review. *Chemical Engineering Journal, 270*, 244–271.

Binici, H., Aksogan, O., Dıncer, A., Luga, E., Eken, M., & Isikaltun, O. (2020). The possibility of vermiculite, sunflower stalk and wheat stalk using for thermal insulation material production. *Thermal Science and Engineering Progress, 18*, 100567.

Canete-Rodriguez, A. M., Santos-Duenas, I. M., Jimenez-Hornero, J. E., Ehrenreich, A., Liebl, W., & Garcia-Garcia, I. (2016). Gluconic acid: Properties, production methods and applications—An excellent opportunity for agro-industrial byproducts and waste bio-valorization. *Process Biochemistry, 51*(12), 1891–1903.

Cheng, W., Sun, Y., Fan, M., Li, Y., Wang, L., & Qian, H. (2022). Wheat bran, as the resource of dietary fiber: A review. *Critical Reviews in Food Science and Nutrition, 62*(26), 7269–7281.

Curti, E., Carini, E., Bonacini, G., Tribuzio, G., & Vittadini, E. (2013). Effect of the addition of bran fractions on bread properties. *Journal of Cereal Science, 57*(3), 325–332.

Das, G., Patra, J. K., Basavegowda, N., Vishnuprasad, C. N., & Shin, H. S. (2019). Comparative study on antidiabetic, cytotoxicity, antioxidant and antibacterial properties of biosynthesized silver nanoparticles using outer peels of two varieties of *Ipomoea batatas* (L.) Lam. *International Journal of Nanomedicine*, 4741–4754. https://doi.org/10.2147/IJN.S210517

Das, G., Shin, H. S., Kumar, A., Vishnuprasad, C. N., & Patra, J. K. (2021). Photo-mediated optimized synthesis of silver nanoparticles using the extracts of outer shell fibre of *Cocos nucifera* L. fruit and detection of its antioxidant, cytotoxicity and antibacterial potential. *Saudi Journal of Biological Sciences, 28*(1), 980–987. https://doi.org/10.1016/j.sjbs.2020.11.022

Dauthal, P., & Mukhopadhyay, M. (2015). Biofabrication, characterization, and possible bio-reduction mechanism of platinum nanoparticles mediated by agro-industrial waste and their catalytic activity. *Journal of Industrial Engineering, 22*, 185–192. https://doi.org/10.1016/j.jiec.2014.07.009

Dayanidhi, K., Vadivel, P., Jothi, S., & Eusuff, N. S. (2020). Facile synthesis of silver@ eggshell nanocomposite: A heterogeneous catalyst for the removal of heavy metal ions, toxic dyes and microbial contaminants from water. *Journal of Environmental Management, 271*, 110962. https://doi.org/10.1016/j.jenvman.2020.110962

Ding, T. P., Ma, G. R., Shui, M. X., Wan, D. F., & Li, R. H. (2005). Silicon isotope study on rice plants from the Zhejiang province, China. *Chemical Geology, 218*(1–2), 41–50. https://doi.org/10.1016/j.chemgeo.2005.01.018

do Espírito Santo, A. P., Cartolano, N. S., Silva, T. F., Soares, F. A., Gioielli, L. A., Perego, P., Converti, A., & Oliveira, M. N. (2012). Fibers from fruit by-products enhance probiotic viability and fatty acid profile and increase CLA content in yoghurts. *International Journal of Food Microbiology, 154*(3), 135–144. https://doi.org/10.1016/j.ijfoodmicro.2011.12.025

Doke, D. S., Advani, J. H., Naikwadi, D. R., Gawande, M. B., Walke, P., Umbarkar, S. B., & Biradar, A. V. (2019). Utilization of waste biomass for the synthesis of functionalizable support for covalent anchoring of active organo catalyst. *ACS Sustainable Chemistry & Engineering, 7*(3), 3018–3026.

Etxabide, A., Urdanpilleta, M., Gómez-Arriaran, I., De La Caba, K., & Guerrero, P. (2017). Effect of pH and lactose on cross-linking extension and structure of fish gelatin films. *Reactive and Functional Polymers, 117*, 140–146.

Facure, M. H. M., Braunger, M. L., Mercante, L. A., Paterno, L. G., Riul, A., & Correa, D. S. (2021). Electrical impedance-based electronic tongues: Principles, sensing materials, fabrication techniques and applications. In *Reference module in biomedical sciences*. Elsevier. https://doi.org/10.1016/B978-0-12-822548-6.00091-1

Falcaro, P., Ricco, R., Yazdi, A., Imaz, I., Furukawa, S., Maspoch, D., Ameloot, R., Evans, J. D., & Doonan, C. J. (2016). Application of metal and metal oxide nanoparticles@ MOFs. *Coordination Chemistry Reviews, 307*, 237–254.

Gajera, B., Tyagi, U., Sarma, A. K., & Jha, M. K. (2022). Impact of torrefaction on thermal behavior of wheat straw and groundnut stalk biomass: Kinetic and thermodynamic study. *Fuel Communications, 12*, 100073.

Gehlen, G. S., Nogueira, A. P., Carlevaris, D., Barros, L. Y., Poletto, J. C., Lasch, G., Straffelini, G., Ferreira, N. F., & Neis, P. D. (2023). Tribological assessment of rice husk ash in eco-friendly brake friction materials. *Wear, 516*, 204613.

Glaser, S. J., Al-Rudainy, B., Hatti-Kaul, R., & Galbe, M. (2023). Wheat bran fractionation: Effect of steam explosion and hydrotropic extraction conditions on the recovery of sugars and lignin. *Industrial Crops and Products, 195*, 116405.

Haider, J. B., Haque, M. I., Hoque, M., Hossen, M. M., Mottakin, M., Khaleque, M. A., Johir, M. A., Zhou, J. L., Ahmed, M. B., & Zargar, M. (2022). Efficient extraction of silica from openly burned rice husk ash as adsorbent for dye removal. *Journal of Cleaner Production, 380*, 135121.

Harby, A. G., El-Borady, O. M., & El-Kemary, M. (2022). The exploitation of rice husk biomass for the bio-inspired synthesis of gold nanoparticles as a multifunctional material for various biological and photocatalytic applications. *Bioprocess and Biosystems Engineering*, 1–4. https://doi.org/10.1007/s00449-021-02639-y

He, X., Yang, D. P., Zhang, X., Liu, M., Kang, Z., Lin, C., Jia, N., & Luque, R. (2019). Waste eggshell membrane-templated CuO-ZnO nanocomposites with enhanced adsorption, catalysis and antibacterial properties for water purification. *Chemical Engineering Journal, 369*, 621–633. https://doi.org/10.1016/j.cej.2019.03.047

Hossain, N., Nizamuddin, S., Selvakannan, P., Griffin, G., Madapusi, S., & Shah, K. (2022). The effect of KOH activation and Ag nanoparticle incorporation on rice husk-based porous materials for wastewater treatment. *Chemosphere, 291*, 132760. https://doi.org/10.1016/j.chemosphere.2021.132760

Hu, Y., Liu, Z., & He, Y. (2018). Effects of SiO_2 nanoparticles on pool boiling heat transfer characteristics of water based nanofluids in a cylindrical vessel. *Powder Technology, 327*, 79–88.

Huston, M., DeBella, M., DiBella, M., & Gupta, A. (2021). Green synthesis of nanomaterials. *Nanomater, 11*(8), 2130. https://doi.org/10.3390/nano11082130

Jahirul, M. I., Rasul, M. G., Chowdhury, A. A., & Ashwath, N. (2012). Biofuels production through biomass pyrolysis—a technological review. *Energies, 5*(12), 4952–5001.

Jain, D., Daima, H. K., Kachhwaha, S., & Kothari, S. L. (2009). Synthesis of plant-mediated silver nanoparticles using papaya fruit extract and evaluation of their anti-microbial activities. *Digest Journal of Nanomaterials and Biostructures, 4*(3), 557–563. https://doi.org/10.1016/j.jfoodeng.2014.06.014

Jayasuriya, H. P., Soni, P., & Steele, P. (2003). *AGS divisional project 214A1 'Small Farmer Livelihoods'*.

Jeon, J., Choi, M., Kim, S. B., Seo, T. H., Ku, B. C., Ryu, S., Park, J. H., & Kim, Y. K. (2021). Eggshell membrane hydrolysate as a multi-functional agent for synthesis of functionalized graphene analogue and its catalytic nanocomposites. *Journal of Industrial and Engineering Chemistry, 102*, 233–240. https://doi.org/10.1016/j.jiec.2021.07.010

Jiang, W., Kim, B. Y., Rutka, J. T., & Chan, W. C. (2008). Nanoparticle-mediated cellular response is size-dependent. *Nature Nanotechnology*, 145–150. https://doi.org/10.1038/nnano.2008.30

Khan, S. A. (2020). Metal nanoparticles toxicity: Role of physicochemical aspects. In *Metal nanoparticles for drug delivery and diagnostic applications* (pp. 1–11). Elsevier. https://doi.org/10.1016/B978-0-12-816960-5.00001-X

Koul, B., Yakoob, M., & Shah, M. P. (2022). Agricultural waste management strategies for environmental sustainability. *Environmental Research, 206*, 112285.

Kumchompoo, J., Kunthadee, P., Laorodphan, N., Kidkhunthod, P., Kuimalee, S., Tangkuaram, T., & Puntharod, R. (2023). The solid-state reaction facilitated by a microwave-assisted method for lithium vanadium silicon oxide synthesis by incorporating pure silica and rice husk ash for the application as anode material in lithium-ion battery. *Radiation Physics and Chemistry, 207*, 110863.

Kuppusamy, P., Yusoff, M. M., Maniam, G. P., & Govindan, N. (2016). Biosynthesis of metallic nanoparticles using plant derivatives and their new avenues in pharmacological applications–An updated report. *Saudi Pharmaceutical Journal, 24*(4), 473–484. https://doi.org/10.1016/j.jsps.2014.11.013

Kwan, W. H., & Wong, Y. S. (2020). Acid leached rice husk ash (ARHA) in concrete: A review. *Materials Science for Energy Technologies, 3*, 501–507.

Lakshmipathy, R., Palakshi Reddy, B., Sarada, N. C., Chidambaram, K., & Khadeer Pasha, S. K. (2015). Watermelon rind-mediated green synthesis of noble palladium nanoparticles: Catalytic application. *Applied Nanoscience, 5*, 223–228.

Lee, J. H., Kwon, J. H., Lee, J. W., Lee, H. S., Chang, J. H., & Sang, B. I. (2017). Preparation of high purity silica originated from rice husks by chemically removing metallic impurities. *Journal of Industrial and Engineering Chemistry, 50*, 79–85.

Lestari, P., Pratiwi, I., & Juliani, A. (2018). Green synthesis of silver nanoparticle using rambutan (*Nephelium lappaceum* L.) peel extract and its antibacterial activity against *Salmonella parathypi* A. In *MATEC web of conferences* (Vol. 154, p. 01024). EDP Sciences. https://doi.org/10.1051/matecconf/201815401024

Liu, Y., Song, X., Cao, F., Li, F., Wang, M., Yang, Y., Liu, M., Liu, A., Xin, H., & Wang, X. (2020). Banana peel-derived dendrite-shaped Au nanomaterials with dual inhibition toward tumor growth and migration. *International Journal of Nanomedicine, 15*, 2315–2322.

Lou, J., Babadi, M. R., Otadi, M., Tarahomi, M., Van Le, Q., Khonakdar, H. A., & Li, C. (2023). Agricultural waste valorization towards (nano) catalysts for the production of chemicals and materials. *Fuel, 351*, 128935.

Maiti, S., Sarma, S. J., Brar, S. K., Le Bihan, Y., Drogui, P., Buelna, G., & Verma, M. (2016). Agro-industrial wastes as feedstock for sustainable bio-production of butanol by *Clostridium beijerinckii*. *Food and Bioproducts Processing, 98*, 217–226.

Matthews, H. K., Bertoli, C., & de Bruin, R. A. (2022). Cell cycle control in cancer. *Nature Reviews Molecular Cell Biology, 23*(1), 74–88. https://doi.org/10.1038/s41580-021-00404-3

Mirgane, N. A., Shivankar, V. S., Kotwal, S. B., Wadhawa, G. C., & Sonawale, M. C. (2021). Waste pericarp of *Ananas comosus* in green synthesis zinc oxide nanoparticles and their application in wastewater treatment. *Materials Today: Proceedings, 37*, 886–889. https://doi.org/10.1016/j.matpr.2020.06.045

Mkhwanazi, Z., Isa, Y. M., & Vallabh, S. T. (2023). Production of biocoal from wastewater sludge and sugarcane bagasse: A review. *Atmosphere, 14*(1), 184.

Mohan, S., & Panneerselvam, K. (2021). An investigation on antibacterial filler property of silver nanoparticles generated from Walnut shell powder by *in situ* process. *Materials Today: Proceedings, 39*, 368–372.

Møller, H. B., Sommer, S. G., & Ahring, B. K. (2004). Methane productivity of manure, straw and solid fractions of manure. *Biomass and Bioenergy, 26*(5), 485–495.

Munagala, M., Shastri, Y., Nagarajan, S., & Ranade, V. (2022). Production of Bio-CNG from sugarcane bagasse: Commercialization potential assessment in Indian context. *Industrial Crops and Products, 188*, 115590. https://doi.org/10.1016/j.indcrop.2022.115590

Muniyappan, N., Pandeeswaran, M., & Amalraj, A. (2021). Green synthesis of gold nanoparticles using *Curcuma pseudomontana* isolated curcumin: Its characterization, antimicrobial, antioxidant and anti-inflammatory activities. *Environmental Chemistry and Ecotoxicology, 3*, 117–124.

Narasaiah, B. P., Banoth, P., Sohan, A., Mandal, B. K., Bustamante Dominguez, A. G., De Los Santos Valladares, L., & Kollu, P. (2022a). Green biosynthesis of tin oxide nanomaterials mediated by agro-waste cotton boll peel extracts for the remediation of environmental pollutant dyes. *ACS Omega, 7*(18), 15423–15438. https://doi.org/10.1021/acsomega.1c07099

Narasaiah, B. P., Koppala, S., Kar, P., Lokesh, B., & Mandal, B. K. (2022b). Photocatalytic and antioxidant studies of bioinspired ZrO_2 nanoparticles using agriculture waste durva grass aqueous extracts. *Journal of Hazardous Materials Advances, 7*, 100112. https://doi.org/10.1016/j.hazadv.2022.100112

Nguyen, H., Jamali Moghadam, M., & Moayedi, H. (2019). Agricultural wastes preparation, management, and applications in civil engineering: A review. *Journal of Material Cycles and Waste Management, 21*, 1039–1051.

Norsuraya, S., Fazlena, H., & Norhasyimi, R. (2016). Sugarcane bagasse as a renewable source of silica to synthesize Santa Barbara Amorphous-15 (SBA-15). *Procedia Engineering, 148*, 839–846.

Pattanaik, L., Pattnaik, F., Saxena, D. K., & Naik, S. N. (2019). Biofuels from agricultural wastes. In *Second and third generation of feedstocks* (pp. 103–142). Elsevier.

Pechyen, C., Ponsanti, K., Tangnorawich, B., & Ngernyuang, N. (2021). Waste fruit peel–mediated green synthesis of biocompatible gold nanoparticles. *Journal of Materials Research and Technology, 14*, 2982–2991. https://doi.org/10.1016/j.jmrt.2021.08.111

Perumal, A. B., Sellamuthu, P. S., Nambiar, R. B., & Sadiku, E. R. (2018). Development of polyvinyl alcohol/chitosan bio-nanocomposite films reinforced with cellulose nanocrystals isolated from rice straw. *Applied Surface Science, 449*, 591–602.

Petruccioli, M., Raviv, M., Di Silvestro, R., & Dinelli, G. (2011). Agriculture and agro-industrial wastes, byproducts, and wastewaters: Origin, characteristics, and potential in bio-based-compounds production. In *Environmental biotechnology and safety* (pp. 531–545). Elsevier.

Phang, Y. K., Aminuzzaman, M., Akhtaruzzaman, M., Muhammad, G., Ogawa, S., Watanabe, A., & Tey, L. H. (2021). Green synthesis and characterization of CuO nanoparticles derived from papaya peel extract for the photocatalytic degradation of palm oil mill effluent (POME). *Sustainability, 13*(2), 796. https://doi.org/10.3390/su13020796

Pichersky, E., & Gang, D. R. (2000). Genetics and biochemistry of secondary metabolites in plants: An evolutionary perspective. *Trends in Plant Science, 5*(10), 439–445. https://doi.org/10.1016/S1360-1385(00)01741-6

Prasad, C., Gangadhara, S., & Venkateswarlu, P. (2016). Bio-inspired green synthesis of Fe_3O_4 magnetic nanoparticles using watermelon rinds and their catalytic activity. *Applied Nanoscience, 6*, 797–802.

Putri, D. N., Sahlan, M., Montastruc, L., Meyer, M., Negny, S., & Hermansyah, H. (2020). Progress of fermentation methods for bio-succinic acid production using agro-industrial waste by *Actinobacillus succinogenes*. *Energy Reports, 6*, 234–239.

Rahman, S. M., Islam, M. A., Rahman, M. M., & Oh, D. H. (2008). Effect of cattle slurry on growth, biomass yield and chemical composition of maize fodder. *Asian-Australasian Journal of Animal Sciences, 21*(11), 1592–1598.

Rajendran, A., Alsawalha, M., & Alomayri, T. (2021). Biogenic synthesis of husked rice-shaped iron oxide nanoparticles using coconut pulp (*Cocos nucifera* L.) extract for photocatalytic degradation of Rhodamine B dye and their in vitro antibacterial and anticancer activity. *Journal of Saudi Chemical Society, 25*(9), 101307. https://doi.org/10.1016/j.jscs.2021.101307

Rajput, H., Kedia, A., Shah, D., Gamit, H. A., & Amaresan, N. (2022). Sustainable synthesis of silver nanoparticles using fruit waste and its antibacterial activity. *Materials Today: Proceedings, 68*, 299–305. https://doi.org/10.1016/j.matpr.2022.05.181

Rathnasamy, R., Thangasamy, P., Thangamuthu, R., Sampath, S., & Alagan, V. (2017). Green synthesis of ZnO nanoparticles using *Carica papaya* leaf extracts for photocatalytic and photovoltaic applications. *Journal of Materials Science: Materials in Electronics, 28*, 10374–10381. https://doi.org/10.1007/s10854-017-6807-8

Ray, R. C. (2022). *Fruits and vegetable wastes: Valorization to bioproducts and platform chemicals* (p. 448). Springer Nature.

Razak, N. A., Othman, N. H., Shayuti, M. S., Jumahat, A., Sapiai, N., & Lau, W. J. (2022). Agricultural and industrial waste-derived mesoporous silica nanoparticles: A review on chemical synthesis route. *Journal of Environmental Chemical Engineering, 10*(2), 107322. https://doi.org/10.1016/j.jece.2022.107322

Renuka, N. K., Anas, K., & Aniz, C. U. (2015). Synthesis, characterisation and activity of SBA-16 supported oxidation catalysts for CO conversion. *Chinese Journal of Catalysis, 36*(8), 1237–1241.

Rodríguez-Félix, F., López-Cota, A. G., Moreno-Vásquez, M. J., Graciano-Verdugo, A. Z., Quintero-Reyes, I. E., Del-Toro-Sánchez, C. L., & Tapia-Hernández, J. A. (2021). Sustainable-green synthesis of silver nanoparticles using safflower (Carthamus tinctorius L.) waste extract and its antibacterial activity. *Heliyon, 7*(4).

Roopan, S. M., Bharathi, A., Kumar, R., Khanna, V. G., & Prabhakarn, A. (2012a). Acaricidal, insecticidal, and larvicidal efficacy of aqueous extract of Annona squamosa L peel as biomaterial for the reduction of palladium salts into nanoparticles. *Colloids and Surfaces. B, Biointerfaces, 92*, 209–212.

Roopan, S. M., Bharathi, A., Prabhakarn, A., Rahuman, A. A., Velayutham, K., Rajakumar, G., Padmaja, R. D., Lekshmi, M., & Madhumitha, G. (2012b). Efficient phyto-synthesis and structural characterization of rutile TiO_2 nanoparticles using *Annona squamosa* peel extract. *Spectrochimica Acta Part A: Molecular and Biomolecular Spectroscopy, 98*, 86–90.

Ruanglek, V., Maneewatthana, D., & Tripetchkul, S. (2006). Evaluation of Thai agro-industrial wastes for bio-ethanol production by *Zymomonas mobilis*. *Process Biochemistry, 41*(6), 1432–1437.

Saad, A. M., El-Saadony, M. T., El-Tahan, A. M., Sayed, S., Moustafa, M. A., Taha, A. E., Taha, T. F., & Ramadan, M. M. (2021). Polyphenolic extracts from pomegranate and watermelon wastes as substrate to fabricate sustainable silver nanoparticles with larvicidal effect against *Spodoptera littoralis Saudi. Journal of Biological Sciences, 28*(10), 5674–5683. https://doi.org/10.1016/j.sjbs.2021.06.011

Sadh, P. K., Duhan, S., & Duhan, J. S. (2018). Agro-industrial wastes and their utilization using solid state fermentation: A review. *Bioresources and Bioprocessing, 5*(1), 1–5.

Sagar, N. A., Pareek, S., Sharma, S., Yahia, E. M., & Lobo, M. G. (2018). Fruit and vegetable waste: Bioactive compounds, their extraction, and possible utilization. *Comprehensive Reviews in Food Science and Food Safety, 17*(3), 512–531. https://doi.org/10.1111/1541-4337.12330

Salem, S. S., & Fouda, A. (2021). Green synthesis of metallic nanoparticles and their prospective biotechnological applications: an overview. *Biological trace element research, 199*(1), 344–370.

Saratale, G. D., Saratale, R. G., Kim, D. S., Kim, D. Y., & Shin, H. S. (2020). Exploiting fruit waste grape pomace for silver nanoparticles synthesis, assessing their antioxidant, antidiabetic potential and antibacterial activity against human pathogens: A novel approach. *Nanomater, 10*(8), 1457. https://doi.org/10.3390/nano10081457

Sarkar, N., Ghosh, S. K., Bannerjee, S., & Aikat, K. (2012). Bioethanol production from agricultural wastes: An overview. *Renewable Energy, 37*(1), 19–27.

Shahzadi, S., Zafar, N., & Sharif, R. (2018). Antibacterial activity of metallic nanoparticles. In *Bacterial pathogenesis and antibacterial control* (Vol. 51). https://doi.org/10.5772/intechopen.72526

Sharma, R. (2021). Synthesis of *Terminalia bellirica* fruit extract mediated silver nanoparticles and application in photocatalytic degradation of wastewater from textile industries. *Materials Today Proceedings, 44*, 1995–1998. https://doi.org/10.1016/j.matpr.2020.12.118

Shi, Y., Lin, M., Jiang, X., & Liang, S. (2015). Recent advances in FePt nanoparticles for biomedicine. *Journal of Nanomaterials, 2015*, 2.

Shifrina, Z. B., Matveeva, V. G., & Bronstein, L. M. (2019). Role of polymer structures in catalysis by transition metal and metal oxide nanoparticle composites. *Chemical Reviews, 120*(2), 1350–1396.

Siddiqui, A., Salahuddin, T., & Riaz, A. (2014). Production of amylase from *Bacillus* sp. AY3 using fruit peels as substrate. *FUUAST Journal of Biology, 4*(2), 213–215.

Simonetti, E. A., de Simone, C. L., da Silva Fonseca, B. C., de Freitas, A. P., dos Reis, C. A., & Thim, G. P. (2018). TiO_2 Carbon composite using coconut waste as carbon source: Sonocatalysis and adsorption evaluation. *Surfaces and Interfaces, 12*, 124–134. https://doi.org/10.1016/j.surfin.2018.04.008

Singh, R., Srivastava, M., & Shukla, A. (2016). Environmental sustainability of bioethanol production from rice straw in India: A review. *Renewable and Sustainable Energy Reviews, 54*, 202–216.

Singh, R., Kumari, P., Chavan, P. D., Datta, S., & Dutta, S. (2017). Synthesis of solvothermal derived TiO2 nanocrystals supported on ground nano eggshell waste and its utilization for the photocatalytic dye degradation. *Optical Materials, 73*, 377–383. https://doi.org/10.1016/j.optmat.2017.08.040

Singh, K. R., Nayak, V., Singh, J., Singh, A. K., & Singh, R. P. (2021). Potentialities of bioinspired metal and metal oxide nanoparticles in biomedical sciences. *RSC Advances, 11*(40), 24722–24746. https://doi.org/10.1039/D1RA04273D

Sinsinwar, S., Sarkar, M. K., Suriya, K. R., Nithyanand, P., & Vadivel, V. (2018). Use of agricultural waste (coconut shell) for the synthesis of silver nanoparticles and evaluation of their antibacterial activity against selected human pathogens. *Microbial Pathogenesis, 124*, 30–37. https://doi.org/10.1016/j.micpath.2018.08.025

Siwal, S. S., Zhang, Q., Devi, N., Saini, A. K., Saini, V., Pareek, B., Gaidukovs, S., & Thakur, V. K. (2021). Recovery processes of sustainable energy using different biomass and wastes. *Renewable and Sustainable Energy Reviews, 150*, 111483.

Son, E. B., Poo, K. M., Chang, J. S., & Chae, K. J. (2018). Heavy metal removal from aqueous solutions using engineered magnetic biochars derived from waste marine macro-algal biomass. *Science of the Total Environment, 615*, 161–168.

Sud, D., Mahajan, G., & Kaur, M. P. (2008). Agricultural waste material as potential adsorbent for sequestering heavy metal ions from aqueous solutions–A review. *Bioresource Technology, 99*(14), 6017–6027.

Sun, Y., & Cheng, J. (2002). Hydrolysis of lignocellulosic materials for ethanol production: A review. *Bioresource Technology, 83*(1), 1–1.

Sun, X. F., Sun, R. C., Tomkinson, J., & Baird, M. S. (2003). Preparation of sugarcane bagasse hemicellulosic succinates using NBS as a catalyst. *Carbohydrate Polymers, 53*(4), 483–495.

Tassoni, A., Tedeschi, T., Zurlini, C., Cigognini, I. M., Petrusan, J. I., Rodríguez, Ó., Neri, S., Celli, A., Sisti, L., Cinelli, P., & Signori, F. (2020). State-of-the-art production chains for peas, beans and chickpeas—valorization of agro-industrial residues and applications of derived extracts. *Molecules, 25*(6), 1383.

Thangadurai, D., Naik, J., Sangeetha, J., Al-Tawaha, A. R., Adetunji, C. O., Islam, S., David, M., Shettar, A. K., & Adetunji, J. B. (2021). Nanomaterials from agrowastes: Past, present, and the future. In *Handbook of nanomaterials and nanocomposites for energy and environmental applications* (pp. 471–487). Springer.

Tian, Y., Wu, M., Lin, X., Huang, P., & Huang, Y. (2011). Synthesis of magnetic wheat straw for arsenic adsorption. *Journal of Hazardous Materials, 193*, 10–16.

Tun, M. M., Juchelková, D., Raclavská, H., & Sassmanová, V. (2018). Utilization of biodegradable wastes as a clean energy source in the developing countries: A case study in Myanmar. *Energies, 11*(11), 3183. https://doi.org/10.3390/en11113183

Venkateswarlu, S., Rao, Y. S., Balaji, T., Prathima, B., & Jyothi, N. V. (2013). Biogenic synthesis of Fe_3O_4 magnetic nanoparticles using plantain peel extract. *Materials Letters, 100*, 241–244.

Wary, R. R., Baglari, S., Brahma, D., Gautam, U. K., Kalita, P., & Baruah, M. B. (2022). Synthesis, characterization, and photocatalytic activity of ZnO nanoparticles using water extract of waste coconut husk. *Environmental Science and Pollution Research, 29*(28), 42837–42848. https://doi.org/10.1007/s11356-022-18832-9

Wayenbergh, E., Langenaeken, N. A., Struyf, N., Goos, P., Foubert, I., & Courtin, C. M. (2023). Stabilisation of vitamin A by wheat bran is affected by wheat bran antioxidants, bound lipids and endogenous lipase activity. *Food Research International, 169*, 112911.

Wen, X., Chen, X., Tian, N., Gong, J., Liu, J., Rümmeli, M. H., Chu, P. K., Mijiwska, E., & Tang, T. (2014). Nanosized carbon black combined with Ni_2O_3 as "universal" catalysts for synergistically catalyzing carbonization of polyolefin wastes to synthesize carbon nanotubes and application for supercapacitors. *Environmental Science & Technology, 48*(7), 4048–4055.

Xing, Y., Liao, X., Liu, X., Li, W., Huang, R., Tang, J., Xu, Q., Li, X., & Yu, J. (2021). Characterization and antimicrobial activity of silver nanoparticles synthesized with the peel extract of mango. *Materials, 14*(19), 5878. https://doi.org/10.3390/ma14195878

Yan, D., Zhang, H., Chen, L., Zhu, G., Wang, Z., Xu, H., & Yu, A. (2014). Super capacitive properties of Mn_3O_4 nanoparticles bio-synthesized from banana peel extract. *RSC Advances, 4*(45), 23649–23652. https://doi.org/10.1039/c4ra02603a

Yi, Y., Lee, B. D., Kim, S. K., Jung, D. H., Jung, E. M., Hwang, S. M., Choi, S. Y., & Peck, D. H. (2016). Performance of porous carbon as catalyst support for anode from rice husk in a direct methanol fuel cell. *International Journal of Electrochemical Science, 11*, 5909–5923.

Yunarti, R. T., Safitri, T. N., Dimonti, L. C., Aulia, G., Khalil, M., & Ridwan, M. (2022). Facile synthesis of composite between titania nanoparticles with highly exposed (001) facet and coconut shell-derived graphene oxide for photodegradation of methylene blue. *Journal of Physics and Chemistry of Solids, 160*. https://doi.org/10.1016/j.jpcs.2021.110357

Yuvakkumar, R., Suresh, J., Nathanael, A. J., Sundrarajan, M., & Hong, S. I. (2014a). Novel green synthetic strategy to prepare ZnO nanocrystals using rambutan (*Nephelium lappaceum* L.) peel extract and its antibacterial applications. *Materials Science and Engineering: C, 41*, 17–27.

Yuvakkumar, R., Suresh, J., Nathanael, A. J., Sundrarajan, M., & Hong, S. I. (2014b). Rambutan (Nephelium lappaceum L.) peel extract assisted biomimetic synthesis of nickel oxide nanocrystals. *Materials Letters, 128*, 170–174.

Zamare, D., Vutukuru, S. S., & Babu, R. (2016). Biosynthesis of nanoparticles from agro-waste: A sustainable approach. *International Journal of Engineering and Applied Science & Technology, 1*(12), 85–92.

Zhou, X., Zhou, X., Tang, X., & Xu, Y. (2018). Process for calcium xylonate production as a concrete admixture derived from in-situ fermentation of wheat straw pre-hydrolysate. *Bioresource Technology, 261*, 288–293.

Part III
Recent Trends in Valorization of Agricultural Waste

Chapter 7
Agriculture and Other Waste Substrates for Single-Cell Protein Production

S. Fatemeh S. Reihani and Kianoush Khosravi-Darani

Abbreviations

BOD	Biological oxygen demand
COD	Chemical oxygen demand
DPLJ	Deproteinized leaf juice
LMs	Lignocellulosic material
SCP	Single cell protein

7.1 Introduction

Accounting for about 18% of the global protein supply, animal-based protein is certainly an extensively resource-consuming food product (Alexander et al., 2017; Piercy et al., 2023; Zhao et al., 2010). The global increase in demand for meat protein to 9.7 billion by 2050 leads to rising environmental concerns (Poore & Nemecek, 2018; Rischer et al., 2020; Piercy et al., 2023). Moreover, global food and agricultural supply chains are extremely affected by the COVID-19 pandemic at multiple levels, threatening food security with an estimation of 909 million undernutrition people by 2030 (Wen et al., 2019; Upcraft et al., 2021; Ušča & Tisenkopfs, 2023). Therefore, exploiting novel biotechnological methods of safe and sustainable protein production is becoming more critical. On the other hand, according to the statistics and estimations, about 33% of globally produced food is underused due to logistics reasons of supply and demand. This trend which is equivalent to 1.3 billion

S. F. S. Reihani · K. Khosravi-Darani (✉)
Research Department of Food Technology, National Nutrition and Food Technology Research Institute, Faculty of Nutrition Sciences and Food Technology, ShahidBeheshti University of Medical Sciences, Tehran, Iran
e-mail: k.khosravi@sbmu.ac.ir

J. Arora et al. (eds.), *Transforming Agriculture Residues for Sustainable Development*, Waste as a Resource, https://doi.org/10.1007/978-3-031-61133-9_7

tons of wasted food (by which 2 billion people can be adequately fed) is evident in both developed regions with over-nutrition and less developed countries with increasing rates of under-nutrition (Kaza et al., 2018; Reihani et al., 2023). As documented by the United Nations Environment Program (UNEP), almost 650 million tons of untreated raw materials are annually released by the agricultural sector, which costs about 800 billion euros (UNEP, 2021). Today, by exploiting innovative biotechnological techniques, high-added-value products can be driven from biodegradable waste of which a large proportion is being directly burned or ultimately processed into rather low-added-value products such as biogas, bioenergy, and biofuels. Single-cell protein (SCP), as one of these sustainable valorization products, is mainly derived from microalgae, fungi, yeast, or bacteria and can be used in animal and human nutrition (Spalvins et al., 2018; Reihani & Khosravi-Darani, 2019b). It is known as a health-beneficial source of protein compared to plant or animal proteins since SCP contains a wide range of essential amino acids as well as cell wall biomass which is rich in fiber (Hashempour-Baltork et al., 2020a, b; Bajić et al., 2022; Molfetta et al., 2022). In addition, produced SCP from agro-waste residues and byproducts could also be incorporated as powder or wet into the formulation of local foods for all age categories as a functional supplement.

For example, Khan et al. enriched wheat flour bread at different concentrations with their SCP produced from fruit and vegetable residues (Khan et al., 2022; Salazar-López et al., 2022). SCP also has a rapid growth rate and a wide range of raw materials can be used as substrate including low-value agro-industrial residues and byproducts as well as municipal and biodegradable industrial waste. Moreover, SCP production requires rather a small land area for batch or continuous bioprocesses regardless of climate variations. Another advantage of using SCP is that they acquire essential amino acid nutrition recommended by the FAO/WHO for humans and can be genetically modified more easily than plants and animals (Bajić et al., 2022; Voutilainen et al., 2021). SCP obtained from agro-waste byproducts and residues could also be incorporated into various edible products as a functional supplement. For example, Khan et al. enriched wheat flour bread at different concentrations with their SCP produced from fruit and vegetable residues (Khan et al., 2022). Considering agro-industrial waste composition and their extensive amount, they can play a major role in the sustainable bio-production of SCP. Moreover, any industrial residue that is not further used can also be exploited for SCP production. Since waste from factories, industries, mills, and mining operations may include undesirable residues such as chemical solvents, pigments, dyes, metal processing waste, radioactive waste, etc., only biodegradable industrial wastes such as sludge, paper waste, and production residues, specific industrial and chemical byproducts and waste gases can be used for microbial fermentation by using a range of sustainable technologies. As such, nutrients and organic compounds present in the contaminant-free waste can be extracted or converted to 40–70% cell biomass, which can be used as food- or feed-grade protein (Ferreira et al., 2016; Reihani & Khosravi-Darani, 2019b; Piercy et al., 2023). By using waste products in the production of SCP, these residues are treated and disposed of for the benefit of the economy and environment, owing to significantly lower production costs of protein production. For waste sources to be a favorable substrate, availability, low cost of pretreatment, low cost of transportation, and high protein yield in the final microbial biomass are among the important factors (Ivanovs et al., 2018; Spalvins et al., 2018; Bajić et al., 2022). Nevertheless, the characteristics of different waste products and their regional distribution are limited and for both research and industrial applications, the availability of waste products is a subject of high interest depending on local economies (Amran et al., 2021).

This chapter aims to discuss and categorize various agro-industrial and municipal wastes that can be used as worthy substrates for the production of SCP as food-grade or animal feed-grade protein through sustainable bioconversion technologies. These categories include crop cultivation residues such as stalks and stubble (stems), and straws, as well as forestry waste including wood chips, bark, and sawdust; food industry waste including liquid and solid streams; and organic gas generated from processing and manufacturing within the food and drink sector. These waste streams offer considerable potential for resource recovery and protein production due to the high concentrations of nutrients, degradable organic compounds, and the absence of pathogens, toxic metals, and other hazardous contaminants.

7.2 Waste Types as Substrates for SCP Production

The type of substrate utilized and the composition of the cultivation media determine how much SCP is produced, keeping in mind that selecting an appropriate substrate has a direct impact on the effectiveness of the bioprocess (Bajić et al., 2022; Reihani & Khosravi-Darani, 2019a). A variety of substrates have been utilized to produce SCP and can be categorized as high-energy resources (gas oil, natural gas, ethanol, methanol, n-alkanes, and acetic acid), renewable plant resources (starch, sugar, and cellulose), various wastes (sulfite waste liquor, molasses, whey, milk, and fruit waste), and carbon dioxide (Raziq et al., 2020). Agro-industrial residues, byproducts, and waste encompass agricultural residues (leaves, stalks, seed pods, stems, straw, molasses, husks, bagasse, seeds, shell, pulp, stubble, peel, roots, etc.), which can be further divided into field residues (present in the field after the process of crop harvesting), process residues (present after the crop is processed), and industrial residues (produced in different branches of food processing industries such as juice, chips, meat, confectionary, and fruit industries, and including potato peel, orange peel, cassava peel, coconut oil cake, soybean oil cake, etc.) (Bajić et al., 2022; Sadh et al., 2018). The scheme of various categories of waste substrates is given in Fig. 7.1.

The substrate used for bioconversion purposes must be nutritious, particularly in terms of carbon and nitrogen, and also attainable for microorganisms (Sharif et al., 2021). The substrate with a higher ratio of simple and fermentable carbohydrates leads to higher SCP yield (Khan et al., 2022). Furthermore, compounds with antimicrobial attributes such as limonene, terpenes, and camphene in the citrus peel can suppress microorganisms and consequently, hinder SCP growth. Thus, it

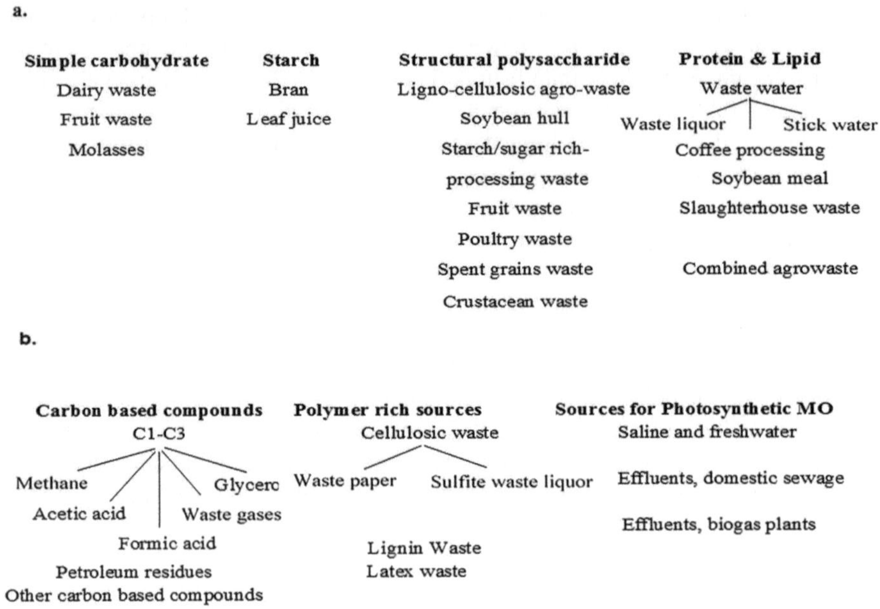

Fig. 7.1 (**a**) Agro-industrial; (**b**) industrial and municipal waste used for SCP production

is critical to prevent such interferences by applying pretreatments before cultivation (Khan et al., 2022). It is advised to regulate the carbon source and its conversion capacity by the addition of volatile fatty acids in the fermentation liquid waste, which in turn produces higher yields of biomass with the aid of photosynthetic bacteria. This occurs due to the activation of certain metabolic pathways that convert volatile fatty acids into tricarboxylic acids and/or acetyl-CoA and both are critical pathways in SCP generation (Zhu et al., 2022). It should be noted that in some cases, depending on the used microorganisms and bioconversion conditions, the multi-complex substrate can lead to higher yield biomass, protein, and productivity of the end-product which could be used as human food or animal feed. For instance, Tropea et al. exploited animal (fish) and agro-food (pineapple, apple, banana, and citrus peels) waste together as a multi-complex substrate to produce SCP by *Saccharomyces cerevisiae* and reported a final valorized product rich in protein (40%) suitable for animal feed (Tropea et al., 2022). Ivanovs et al. categorized agricultural wastes that can be used for SCP production as sources rich in simple carbohydrates, starch, structural polysaccharides, and protein or lipid (Ivanovs et al., 2018). The agricultural wastes for SCP production are discussed by various researchers (Tropea et al., 2022). According to the definition of industrial wastes, several agricultural wastes can also be categorized as industrial ones because they come from industrial processing plants; however, in the context of this review, they were categorized as agricultural because the main resource used in the waste generation was originally grown on agricultural lands. By reviewing available literature on the use of industrial waste products suitable for SCP cultivation, it was possible to categorize industrial wastes into three groups of byproducts: polymer-rich sources (industrial waste), carbon compounds, sources for photosynthetic MO.

7.3 Agro-Industrial Waste

Agro-industrial waste resources applicable for SCP production reviewed in this chapter are summarized in Table 7.1.

7.3.1 Simple Carbohydrate-Rich Sources

Dairy Waste

The wastewater released from dairy industries and its insufficient treatment is indeed one of the main sources of environmental pollution. According to statistics and reports, annually, more than 139 million tons of whey are released only from cheese production and dairy industries, of which half of this amount is directly discharged into sewage or local waters (Yadav et al., 2014a, b). As such, whey and other organic wastes should be effectively processed and removed from these streams (Ivanovs et al., 2018; Suman et al., 2017). Dairy waste can have high levels of either lactose or protein (Aggelopoulos et al., 2014; Yadav et al., 2014a, b), depending on milk processing purposes. If lactose-rich, it is considered a suitable substrate for the production of SCP by choosing the right MO. Presence of lactose as the main compound of dairy wastewater leads to high measures of BOD (biological oxygen demand) and COD (chemical oxygen demand). In another study, the physiochemical and biological properties of both untreated and treated dairy wastewater were evaluated. Furthermore, both treated and untreated wastewater were used as substrates for SCP production and the end-products were characterized and compared (Suman et al., 2017). As a result, untreated dairy waste has constituents such as fat, carbohydrates, proteins, and lactose that enhance culture growth, and the dry cell mass obtained in powder form is thus further used as SCP.

Fruit Waste

A variety of research has been conducted on SCP production from fruit waste as carbon substrates. Depending on the type and part of the used fruit, the mass content of fruit processing waste will differ (Ivanovs et al., 2018). Bananas and figs are among those fruits which due to regulations and incorrect storage and transport in some cases are considered deteriorated and the whole fruit is discharged (Hashem et al., 2014; Salazar-López et al., 2022).

Table 7.1 Agro-industrial waste used as substrate for the production of single-cell protein (SCP) and biomass yield

No.	Substrates	Microorganisms	Yield	References
1	Cheese whey (Lactose)	*K. lactis* *K. marxianus* *S. fragilis* *K. lactis* *S. cerevisiae*	12.43 g/L 12.68 g/L 9.62 g/L 11.98 g/L 10.30 g/L	Anvari and Khayati (2011)
2	Cheese whey filtrate	*T. harzianum*	34.2%	Şişman et al. (2013)
3	Cheese whey	*K. marxianus* *C. krusei* *K. marxianus*	Batch: *K. marxianus*: 0.26 g/g Mixed 0.31 g/g 0.12–0.19 g/g	Yadav et al. (2014a, b)
4	Thin stillage (an ethanol production facility mostly based on wheat)	*N. intermedia* *Rhizopus* sp. *A. oryzae* *F. venenatum* *M. purpureus*	*N. intermedia:* 16 g/L 0.17 g/g biomass (56% w/w protein), *A. oryzae:* 19 g/L, 0.2 g/g TS. (48% crude protein), *Rhizopus* sp.*:* 15 g/L (55%), *F. venenatum:* 14 g/L (56%), *M. purpureus:* 12 g/L	Ferreira et al. (2014)
5	Banana skin, mango waste, sweet orange peel, pomegranate rind, apple waste	*S. cerevisiae*	Banana peel 58.62%, pomegranate rind 54.28% apple waste 50.86%, mango waste 39.98%, sweet orange peel 26.26%	Khan et al. (2010)
6	Banana peel, citrus peel, carrot pomace, potato peel	*S. cerevisiae*	Up to 12 g/100 g of waste biomass 47.7% protein content	Khan et al. (2022)
7	Beet pulp hydrolysate supp. with molasses and glucose	Mixed culture of *C. utilis* and, *B. lactofermentum*	0.49–0.56 g/g substrate (Max: 0.56 with corn steep liquor conc. 30 g/L)	Rajoka et al. (2012)
8	Vinasse	*S. cerevisiae*	73.2 g/L per day 53.3%	Pires et al. (2016)
9	Soy molasses	*C. tropicalis*	56.4%	Gao et al. (2012)
10	Waste capsicum powder	*C. utilis*	48.2%	Zhao et al. (2010)
11	Date juice	*F. venenatum*	Protein 46.48%, Biomass: 4.84 g/L	Hosseini et al. (2011)
12	Date waste	*F. venenatum*	7.2 g/L, 0.2 g/g	Reihani and Khosravi-Darani (2019a, b)
13	Spoiled date palm fruits	*H. uvarum*, *Z. rouxii*	48.9%	Hashem et al. (2014)
	Rice bran	*A. flavus* *A. ochraceus*	11.5% 10%	Valentino et al. (2016)

(continued)

Table 7.1 (continued)

No.	Substrates	Microorganisms	Yield	References
14	Leaf juice	*S. cerevisiae*	45.6%	Chanda and Chakrabarti (1996)
15	Prawn shell wastes	Unspecified, marine yeast	61–70%	Rhishipal and Philip (1998)
16	Bagasse hemicellulose, (84–100% Xylose)	*P. janthinellum*	28 g/L biomass	Rao et al. (2010)
17	Ram horn hydrolysate	*B. cereus*, *B. subtilis* and *E. coli*	Protein: *B. cereus* 66%, *B. subtilis* 68%, *E. coli* 71%. *B. cereus:* highest yield: 7.3 g/L	Kurbanoglu and Algur (2002)
18	Wheat straw	*P. florida*	62.8%	Ahmadi et al. (2010)
19	Combined agricultural waste	*S. cerevisiae* *K. marxianus*	38.5% 33.7%	Aggelopoulos et al. (2014)
20	Soybean hull Soybean meal	*B. subtilis*	12.3% —	Wongputtisin et al. (2012, 2014)
21	Potato starch processing	*B. licheniformis*	38.2%	Liu et al. (2014)
22	Cellulose-rich waste	*B. pumilus*, *Candida*	46.1%	Liu et al. (2013)
23	Poultry litter (fiber-rich)	*C. utilis*	29%	Jalasutram et al. (2013)
24	Stick-water	*A. niger*	48.7%	Kam et al. (2012)
25	Sugar industry wastewater	*R. faecalis*	2.5 g/L per day 51.5%	Saejung and Salasook (2020)
26	Brewery wastewater	*Rhizospheric diazotrophs*	>55%	Lee et al. (2015)

Abbreviations: *A.: Aspergillus, B.: Bacillus, C.: Candida, E.: Escherichia, F.: Fusarium, H.: Hanseniaspora, K.: Kluyveromyces, M.: Monascus, N.: Neurospora, P. (Janthinellum): Penicillum, P.(florida): Pleurotus, R.: Rhodopseudomonas, S.: Saccharomyces, T.: Trichoderma, Z.: Zygosaccharomyces*

Molasses

As a rich source of simple carbohydrates, molasses can be used as a suitable substrate for SCP production with no pretreatment demand. It is a byproduct of the sugar cane and sugar beet industries as well as a byproduct of soy protein concentrate which the later has become more accessible in recent years. Utilization of molasses (45–60% fermentable carbohydrates) in SCP production is evaluated by its accessibility, cost, composition, and growth-limiting impurities as well as toxic compounds (Aggelopoulos et al., 2014).

7.3.2 *Starch-Rich Sources*

As one of the major agricultural byproducts, starch-rich residues such as grains and tuber waste can be widely used in SCP production. Nevertheless, these substrates require chemical or enzymatic pretreatments for the starch to break into simple carbohydrates before cultivation. This may cause an increase in the production cost. The use of microorganisms with amylolytic activities together with SCP synthesizing microorganisms can solve this issue. As such, the first amylolytic MO hydrolyzes starch, and hence, SCP-generating microorganisms can ingest their essential substrate, that is, simple carbohydrates. Examples are brans and leaf juices. As a byproduct of grain processing, bran is utilized in oil extraction, food additive, or used as animal feed (Valentino et al., 2016). Another starch-rich product is deproteinized leaf juice (DPLJ) with high BOD and COD. Similar to dairy wastewater, discharging untreated DPLJ with high levels of organic compounds leads to environmental degradation. Thus, microbiological pretreatment of this type of waste is favorable when integrated with SCP production, thus providing the whole process a good economic perspective (Ivanovs et al., 2018).

7.3.3 *Structural Polysaccharides-Rich Sources*

Lignocellulosic Residues

Lignocellulosic materials (LMs) are the most abundant constituent of agricultural residues. According to the statistics and reports, in 2015, 737, 984, and 474 million tons of wheat, maize, and rice, respectively, were harvested worldwide. The proportion of straw to grain for rice, wheat, and corn differs from 0.7 to 1.7 (Ivanovs et al., 2018). This data suggests that the volume of straw and dried stalks and leaves are almost equal or even greater than that of actually produced grain. Thus, exploiting this huge volume of crop residues has great ecological and economic impact (Ivanovs et al., 2018). These LMs are rich in dietary fibers which are designated with low or no feeding value owing to low content of protein and digestible dry matter (Nasehia et al., 2016; Kumar et al., 2024). By applying microbial treatment on LMs, they become digestible and can be used as feed or substrate for SCP production. In the latter, their nutritional value will be significantly enhanced due to an increase in protein content (Ellamar et al., 2019). Filamentous fungi are one of the common microorganisms that can break down lignin with the use of the multi-enzyme complex to convert nondigestible complex polysaccharides into simple carbohydrates (Sharma & Arora, 2015; Patyshakuliyeva et al., 2016). Bakratsas et al. used an edible mushroom (*Pleurotus ostreatus*) to metabolize the main sugars of aspen wood chips hydrolysate as an agro-industrial byproduct, to generate low-cost added value high protein product (Bakratsas et al., 2023). In a recent study, Upcraft et al. presented a novel lignocellulosic solution rich in mycoprotein and emphasized the benefit of an ionic liquid pretreatment in lignocellulosic mycoprotein in combination with cellulose enzyme to produce higher levels of glucose which in turn lead to higher mycoprotein production (Upcraft et al., 2021).

Soybean Hull

Being an agricultural byproduct of soybean oil and meal, soybean hull is regarded as an abundant byproduct. It takes 8–10% of the total soybean mass, which has approximately 350 million tons harvested in the world annually. Although it can be used as a fiber-rich source for ruminants, due to its cellulose and hemicellulose content, untreated soybean hull has no nutritional value for monogastric animals. Nevertheless, applying bioprocessing and fermentation on this byproduct had an enhancing effect on simple sugar and protein content (Wongputtisin et al., 2014).

Starch and Sugar Processing Waste

Similar to dairy wastewater, liquid wastes from starch and sugar production have high COD and BOD values; therefore, these wastewaters can cause environmental damage if discharged untreated. For this to be prevented, these residues can be used in favor of the production of added-value products (Liu et al., 2014; Ivanovs et al., 2018). Numerous kinds of research have been performed on the bioconversion of potato starch processing residues into value-added products and high-protein animal feeds (Liu et al., 2014). Similar studies have been conducted on sugarcane bagasse and beet pulp to produce high-protein SCP (Ivanovs et al., 2018). All the above-mentioned residues, similar to cereal straws, have a high content of cellulose and hemicellulose and hence require treatment before cultivation for SCP production. In this case, bioconversion can be conducted in two phases: In the first stage cellulolytic MO is used for pretreatment and preparation of the substrate for the second phase which is SCP-generating MO (Ivanovs et al., 2018; Liu et al., 2014). Furthermore, the pea-industry byproduct can be another potential starch-rich source for SCP production. It has about 10.3 C: N ratio, about 55% w/w of starch, and more than 18% protein (dry weight basis). According to research, the pea processing industry byproduct was an efficient substrate for the cultivation of filamentous fungi in order to obtain a protein concentrate product. It was also estimated that around 680 kg of fungal biomass ascribing to about 38% of additional protein could be generated for each 1 metric ton of pea-processing byproduct (Souza Filho et al., 2018).

Fruit Waste (Fiber-Rich)

Fruit processing residues are commonly discarded due to the transportation cost of using them as cattle feed (Ivanovs et al., 2018). Fruit processing waste with high levels of dietary fiber usually are byproduct of juice and essential oil production industries. Considering that about 25% of the harvested fruits are used in industrial processing, the global annual amount of pomace and juice pulp produced from apples and citrus fruits is estimated at around 15 million tons (FAO, 2020). Due to the large amount of generated waste and the limited use (owing to its low digestibility and protein), studies have been conducted to enhance the nutritional value of fiber-rich fruit waste and use it as the carbon source for the production of SCP (Aggelopoulos et al., 2014).

Poultry Waste

Disposal of poultry waste with 5–7% nitrogen compounds has similar disadvantages for the environment, as well as its poor nutritional value to further use as feed/fertilizer considering its pathogen contamination (Jalasutram et al., 2013). Thus, as an environmental and economic solution, it can be used as the base substrate to produce SCP. Again, due to its high fiber content, relevant pretreatment is required to degrade polysaccharides and suppress potential pathogens (Jalasutram et al., 2013). Among various microorganisms, *Rhodotorula*, *Candida,* and *Saccharomyces* spp. were reported suitable to produce high-nitrogen in SCP (Ivanovs et al., 2018; Jalasutram et al., 2013).

Spent Grains

Hemicellulose, lignin, and proteins consist of most of the spent grains compounds along with minerals and vitamins. Hence, it could be a suitable substrate for the SCP production. Pretreatment, in order to hydrolyze complex carbohydrates, ameliorates its simple carbohydrate content, such as glucose and xylose. Moreover, their composition varies depending on the target final product, for which they were used (Duarte et al., 2008). As such, an analytical measurement of the concentrations of simple sugars, polysaccharides, and proteins is necessary prior to cultivation and choosing relevant pretreatments and microorganisms to obtain optimum results (Ivanovs et al., 2018).

Crustacean Waste

As FAO reported, 9–11 out of 13.9 million tons of crustacean products are being disposed of as processing waste (FAO, 2020). Some of them are being directly disposed of in the ocean, and some are being burned or landfilled (Ivanovs et al., 2018). Chitin as a main constituent of these residues is considered a structural polysaccharide with glucose being its monosaccharide unit. Therefore, similar to other complex carbohydrates, chitin can be pretreated to degrade its structure to simpler carbohydrates to be accessible by microorganisms to produce SCP. As such, hydrolyzed crustacean waste can account for a suitable waste source to produce high-protein products such as SCP (Ivanovs et al., 2018).

7.3.4 *Protein or Lipid-Rich Sources*

By using protein-rich waste products in the production of SCP, it is possible to obtain very high protein concentrations in the final biomass (Table 7.1). In order to break down the fibrous protein compounds in the waste, they need to be hydrolyzed using enzymes from proteolytic MO. The hydrolysis of proteins usually complicates and raises the cost of SCP production. If it is possible to hydrolyze the waste products by providing associative fermentations (Ivanovs et al., 2018), then protein-rich waste products can become one of the most suitable waste products for the production of SCP due to their high protein yields.

Wastewaters (Protein Rich)

Protein-rich wastewaters mainly have high COD and BOD and need complex and costly waste treatments to protect the environment from pathogen and pollution threats (Kam et al., 2012). Stick water is a liquid byproduct from fish feed production (Kam et al., 2012). The use of stick water in the production of SCP was first described relatively in recent years. In general, stick water, as a substrate, can provide high protein concentrations in microbial biomass. Stick water has high concentrations of protein, phosphorus, and calcium(Kam et al., 2012). Another source of protein-rich wastewater is waste liquor from the glutamic acid factory. It has been reported that the L-glutamate industry generates a huge amount of wastewater with high-protein solid particles (around 20%). Its use in animal nutrition is restricted to prevent consequent metabolic disorders (Ivanovs et al., 2018). Hence, exploiting glutamic acid-rich wastewater in the production of SCP is highly recommended. In the industrial processing of coffee, in addition, wastewater from coffee processing industries has been used as the substrate to generate nutritious extract with sugars, proteins, and minerals (Pillaca-Pullo et al., 2023). In a recent study, *Candida sorboxylosa* was used to incorporate reducing sugars such as glucose, mannose, and fructose from the above-mentioned substrate and generate SCP with about 38% yield. In another study, a similar process was conducted on potato peel wastewater from potato peels enriched with 5% glycerol, and using *Candida utilis* led to the production of SCP with a yield value of 30 g dw (dry weight)/L and protein efficiency of 12.2 g/L (Kurcz et al., 2018; Salazar-López et al., 2022). They recommend using this yeast in both food and feed industries since biomass and metabolites of *Candida utilis* have been regarded as safe by the FDA (Food and Drug Administration, USA). Therefore, after conducting meticulous analysis in vivo, wastewater from different factories could be bio-converted into SCP (Salazar-López et al., 2022).

Soybean Meal

Similar to soybean hull, soybean meal is a byproduct of the soybean oil industry with 50% protein and about 35% carbohydrate content. As earlier discussed, soybean hull which is a polysaccharide-rich soy processing byproduct can be utilized beside soybean meal to adjust its protein concentration (Wongputtisin et al., 2014). Various reports confirmed that the fermentation of soybean meal can ameliorate its nutritional quality, enhance feed conversion ratios, and suppress its nondesirable compounds in animal feeds (Wongputtisin et al., 2014).

Slaughterhouse Waste

Slaughterhouse waste mainly consists of horns, nails, feathers, and hair waste. Thus, fibrous protein-rich wastes can undergo an enzymatic bioconversion by proteolytic microorganisms and ultimately harvest amino acids or proteins (Table 7.1).

Combined Agricultural Wastes

As discussed above, different agro-food wastes provide various nutrients as substrates for SCP generation. Thus, sometimes depending on the available resources, integrating different waste categories is highly effective in harvesting better yields. A wide range of studies has been conducted to confirm this suggestion. For instance, a high-yield SCP was generated by employing agricultural waste feedstock combined with molasses (rich in simple carbohydrates), orange and potato pulps (rich in fiber), and brewers' spent grains (rich in protein) (Aggelopoulos et al., 2014). In another research, *Pleurotus albidus* mycoproteins were produced by using brewer-spent grain which led to 22.6 g/100 g (Stoffel et al., 2019). Moreover, orange and lemon pulp, juice, and peel combined with corn stover effluent were utilized as substrates for *Rhodococcus opacus* to produce value-added SCP for feed (Mahan et al., 2018).

7.4 Municipal and Industrial Waste

Municipal and industrial waste resources applicable for SCP production reviewed in this chapter are summarized in Table 7.2.

Table 7.2 Municipal and industrial waste used as substrate for the production of single-cell protein (SCP) and biomass yield

No.	Substrates	Microorganisms	Yield	References
1	Methane	*Methylomonas* sp., *Bacteria isolated* from soil	0.06 mg/g	Yazdian et al. (2005)
2	Methanol	*P. pastoris*	5 g/L biomass ~0.2 g/g	Wegner (1990)
3	Acetic acid	*Bacteria isolated* from soil	Banana peel 58.62%, pomegranate rind 54.28% apple waste 50.86%, mango waste 39.98%, sweet orange peel 26.26%	Bewersdorff and Dostálek (1971)
4	Formic acid	*Bacteria isolated from soil*	12.43 g/L 12.68 g/L 9.62 g/L 11.98 g/L 10.30 g/L	Dionisi et al. (2008)
5	Waste gases	*M. capsulatus* *R. capsulatus*	Suppl. with glucose: 63.5% crude protein Suppl. with N: 32.5%. Not suppl. 27%	Dürre and Eikmanns (2015)
6	Glycerol	*A. baylyi* *A. limancinum* *S. limacinum*	0.69 g/L Max protein content:44.6%	Liu et al. (2014)
7	Raw glycerol from biodiesel production	*Y. lipolytica*	12.3 g/L from raw glycerol, 0.28–0.56 g/g	Santala et al. (2011)
8	Gas oil	BP yeast	14.8 g/L ~0.74 g/g 13.77 g/L	Shacklady (1970)
9	Biogas slurry (derived from chicken manure) Bottles	*P. kudriavzevii*	6.8 39.4	Zhang et al. (2021)
10	Biogas slurry (derived from chicken manure) Batch, flasks	*P. jadinii* *G. candidum* *C. tropicalis* *S. cerevisiae* *N. rattus*	12.6 35.9	Zhou et al. (2022), Koukoumaki et al. (2024)
11	Inulin, crude oil, glycerol waste hydrocarbons	*Y. lipolytica*	48–54% protein content	Cui et al. (2011)
12	n-paraffins	BP yeast	*N. intermedia*: 16 g/L, 0.17 g/g 56% w/w protein, *A. oryzae*: 19 g/L, 0.2 g/g TS 48% crude protein, *Rhizopus* sp.: 15 g/L (55%), *F. venenatum*: 14 g/L (56%), *M. purpureus*: 12 g/L	Shacklady (1970)

(continued)

Table 7.2 (continued)

No.	Substrates	Microorganisms	Yield	References
13	Waste paper	*S. acidophilum*	Orange 13.3 g/L, plantain 6.7 g/L, banana 6.7 g/L unsupplemented: 16.0, 13.3, 6.7 g/L, supplemented: 32 g/L	Ivarson and Morita (1982)
14	Sulfite waste liquor	*P. variotii* *F. venenatum,* *C. utilis*	Yeast dry weight: *C. utilis* contained 26% protein	Alriksson et al. (2014)
15	Lignin residues	*C. sitophila*	With ammonium sulfate 15.3, 13.7, 15.7, Mixed culture of *K. lactis* and *S. cerevisiae* 22.38	Rodríguez et al. (1997)
16	Latex rubber sheet wastewater	*R. palustris*	0.743 g/g, total viable and non-viable cells 37.0 g/L	Kornochalert et al. (2014)
17	Petrochemical wastewater	*Haloarcula* sp. *IRU1*	76	Taran and Asadi (2014)
18	Effluents of biogas plants	*R. capsulata*	Cell mass yield: 0.65 g/g	Vrati (1984)
19	Saline sewage effluents	*C. salina*	21.89 g/L	Wong and Chan (1980)
20	Synthetic brewery wastewater	*H. pluvialis*	27×10^5 cells/mL biomass ~65% protein content	Yap et al. (2022)
21	*Food processing wastes(tofu)*	*Chlorella* sp.	52.3% 42.5×10^6 cells/mL	Putri et al. (2018)
22	Wastewater effluents	*Micractinium,* *Scenedesmus,* *Oocystis,* *Franceia,* *Euglena*	Biomass dry conc. 50 g/L sweet whey, 65 g/L sour whey conc.	Moraine et al. (1979)
23	Sewage sludge and discarded effluent	*Methylomonas* and *Methylophilus*	11.54 g/g	Zha et al. (2021)
24	Thin stillage (an ethanol production facility based on wheat)	*N. intermedia,* *Rhizopus* sp., *A. oryzae,* *F. venenatum,* *M. purpureus*	16 g/L, 0.17 g/g, 56% 19 g/L, 0.2 g/g TS 15 g/L, 55% 14 g/L, 56%, 12 g/L	Ferreira et al. (2014), Yap et al. (2022)

Abbreviations: *A.(baylyi): Acinetobacter, A.* (*limancinum*): *Aurantiochytrium, A.: Aspergillus, BP* yeast*: British Petroleum protein yeast, C.: Candida, C.(sitophila): Chrysonilia, C. (salina): Chlorella, F.: Fusarium, G.: Geotrichum, H.: Haematococcus, K.: Kluyveromyces, M..(capsulatus): Methylococcus, Methylococcus, M. (purpureus): Monascus, N. (rattus): Nectaromyces, N. (intermedia): Neurospora, P.: Pichia, P. (variotii), Paecilomyces, R.(capsulatus): Rhodobacter, R. (capsulate): Rhodopseudomonas, S.(limacinum): Schizochytrium, S.(acidophilum): Scytalidium, S.:Saccharomyces, Y.: Yarrowia*

7.4.1 Carbon Compounds

C1-C3 Compounds

Methane, methanol, acetic acid, and formic acid are C1-C3 carbon compounds that are abundant in the environment due to natural biochemical processes in plants and animals (Mehta, 1973; Spalvins et al., 2018). They are gradually collected during industrial processes in huge factories and plants where a wide range of waste resources are handled, and therefore, these wasted compounds can be utilized for the benefit of the economy and environment. Despite its disadvantages namely poor solubility and high flammability, methane has a long history of use for SCP production. Hence, natural gas has been regarded as one of the major carbon sources for generating SCP due to its high methane content (85–90%). It is reported that annually about 140 billion m^3 of natural gas is flared and wasted in the world. This leads to huge amounts of CO_2 and methane emissions that have a health adverse impact (Elvidge et al., 2018; Mehta, 1973). The use of natural gas in SCP production has some advantages, such as its low toxicity, selectivity, and volatility, and an end-product with high yield (Yazdian et al., 2005). Another carbon source suitable for SCP production is acetic acid. It has an extensive application in the production of various industries in products such as polymers, glues, fibers, and fabrics, as well as in food production as an acidity regulator. Nevertheless, huge amounts of acetic acid are being wasted from these industries and released into wastewater leading to environmental issues (Gangadwala et al., 2008). Although the use of acetic acid in the cultivation of microorganisms is hardly discussed, some reports demonstrate a rapid growth rate of microbial biomass when they access acetic acid as a carbon source (Spalvins et al., 2018). Further researches are required to investigate the quality of SCP production by using acetic acid-rich wastewaters. Formic acid is widely produced as a byproduct of generating other chemicals and is released into industrial wastewater. Thus, it can be consumed as a carbon source by microorganisms to produce SCP (Mehta, 1973).

Waste Gases

Gases such as CO (carbon monoxide), CO_2 (carbon dioxide), and CH_4 (methane) are released from steel and oil refining industries, as well as coal, natural/shale gas, and other industries (Schuchmann & Müller, 2014; Spalvins et al., 2018). Appropriate exploitation of waste gases suppresses greenhouse gas emissions. There are several researches on producing value-added chemicals from waste gases, such as acetic acid (Schuchmann & Müller, 2014; Dürre & Eikmanns, 2015), fuel (Khan et al., 2014), and high-protein biomass/SCP (Dürre & Eikmanns, 2015). They have the potential to be utilized as substrates for the cultivation of photosynthetic microorganisms, which is discussed later in this chapter.

Glycerol, Petroleum Residues, and Other Carbon Compounds

By increasing biodiesel manufacturing in recent years, glycerol production—as a byproduct of biodiesel—has increased. According to studies, each ton of biodiesel production leads to generating 100 kg of crude glycerol (Johnson & Taconi, 2007). As a result, annual glycerol production from biodiesel manufacturing exceeds two million tons. Due to the impurities of crude glycerol, low market price, and high transportation cost, biodiesel manufacturers have no much options but to sell this byproduct as a fuel or as a feed supplement, with low added value (Ito et al., 2005; Johnson & Taconi, 2007). However, novel biotechnological methods such as their use in SCP production can enhance their waste use in the biodiesel industry (Li et al., 2015; Patil & Gogate, 2015; Spalvins et al., 2018). Fuel oil, n-paraffin, and other residues from petroleum distillation can be utilized as carbon sources for SCP-generating microorganisms (Shacklady, 1970; Spalvins et al., 2018). Moreover, other carbon sources such as methanol, hexane, heptane, heptanol, nonanol, necanol, propionic acid, caprylic acid, and capric acid also can be exploited in SCP production (Bewersdorff & Dostálek, 1971; Spalvins et al., 2018).

7.4.2 Polymer-Rich Sources

Polymer compounds such as cellulose, lignin, hemicellulose, and latex waste, are produced from wood and cotton manufacture, paper and fuel processing, latex, and other industrial productions (Klemm et al., 2005). Although polymer waste and, in particular, lignocellulosic waste are the most abundant industrial waste, rigorous mechanical, chemical, or biological pretreatment processes are necessary for them to be performed solely or in combination prior to the next stage with exploiting SCP-producing microorganisms to generate the final product. In other words, utilization of polymer compounds especially lignocellulosic waste has a direct positive impact the SCP production.

Cellulosic Waste and Waste Papers

Being a major component of agriculture, wood and municipal waste, cellulose, and cellulose-rich byproducts are produced by wood processing factories as well as paper industries (Ivarson & Morita, 1982). Among the most used pretreatment techniques, microbial cellulolysis by cellulolytic MO, steam thermolysis of cellulose as the mechanical pretreatment (Ivarson & Morita, 1982), chemical hydrolysis by using acids and alkali or biochemical pretreatment methods using enzymes are frequently being used (Ivarson & Morita, 1982; Spalvins et al., 2018). Usually, SCP-producing microorganisms can be exploited in the last stage after the waste pretreatment phase, by employing one or several of the abovementioned pretreatments to generate ready-to-ferment waste mixture that can be further utilized by SCP microorganisms. Nowadays, unfortunately, the amount of waste paper is very high and as predicted, will have a rising trend in the future (Bajpai, 2014). Several pretreatments should be applied for the waste paper, considering its main component, that is, cellulose. Hydrolyzed paper waste consists of 60–70% simple sugar, mostly glucose, and the other 30–40% constitute of lignin and other residues (Ivarson & Morita, 1982). Similar to other wastes requiring pretreatment technology before using SCP-producing MO, using waste paper as the substrate has a high total cost considering the microbial contamination during the process (Bajpai, 2014). It has been suggested to exploit extremophilic such as *Scytalidium acidophilum* as an acid-tolerant fungus and thus suppress the costs by providing selective growth conditions (Ivarson & Morita, 1982; Spalvins et al., 2018).

Sulfite Waste Liquor, Lignin, and Latex Waste

As a byproduct of the sulfite process in wood pulp production from wood chips, spent sulfite liquor can be utilized in the cultivation of microorganisms. It has been used to generate ethanol, vanillin, SCP, and other products (Hocking, 1997; Alriksson et al., 2014; Spalvins et al., 2018). Recently, employing the sulfite process has decreased due to using an alternative technique namely the Kraft process, which allows the recycling majority of used chemical compounds in the pulping process (Monte et al., 2009). Theoretically, in the case of the presence of pulping mills and consequently spent sulfite liquor, it can be used as SCP substrate due to its local abundance and low price. Its fermentation leads to a decrease in BOD which means it helps to reduce environmental damage (Spalvins et al., 2018). As the major waste product of the paper industry, lignin is burnt as a fuel, and instead, it can be utilized as a carbon source in microbial fermentation (Ferrari et al., 2021). However, similar to other lignocellulosic waste products, chemical and biological pretreatments are necessary. These treatments lead to more accessible polysaccharides for the microbiological fermentation process (Ferrari et al., 2021). Nowadays latex is mainly derived from plants and is used to produce latex rubber. Latex rubber production generates wastewater that is rich in ammonia, formic acid, sodium metabisulfite, and sodium sulfite and thus is very hazardous to the environment (Kornochalert et al., 2014). Hence, this wastewater should be treated before releasing into local waters. Reports showed that a high yield of SCP can be obtained by the utilization of latex rubber sheet wastewater as substrate as well as a significant reduction of COD values (Kornochalert et al., 2014).

7.4.3 Sources for Photosynthetic MO

In order to conduct wastewater treatments, microorganisms such as algae, zooplankton, and bacteria are cultivated in special ponds (Mohseni et al., 2021). Consequently, they consume and bind to the inorganic nitrogen and phosphorus in the wastewater residue. These sewage-grown MO can be utilized as animal feed owing to their high-protein content (Mohseni et al., 2021). It is quite an efficient strategy to perform treatment on sewage effluent by using microbial co-cultures with organic material oxidizing bacteria, releasing CO_2, and algae which assimilates generated CO_2 and releases O_2 during daytime by photosynthesis. It has been reported that high SCP concentrations have been obtained from such wastewaters as saline and freshwater domestic sewage effluents and effluents from biogas plants (Mohseni et al., 2021; Spalvins et al., 2018).

7.5 Conclusions and Future Trends

A synergistic combination of bioconversion methods has the potential to converge food/protein security and waste recovery towards a bioresource-circular protein future. This recovery of food or feed-grade protein from contaminant-free organic waste requires considering the global regional characteristics, waste resource abundance, and composition along with waste recovery infrastructure. In this chapter, agro-industrial, municipal, and industrial wastes that are safe substrates for the production of SCP were classified and discussed. There are advantages and disadvantages in using each category of waste; nevertheless, those with polymer-rich constituents require rigorous pretreatments before being used for SCP production and thus have higher total production costs. In the case of industrial wastewater, although there is sufficient fundamental infrastructure to use them as substrate for SCP production, several concerns exist that discourage using this category of waste: the potential presence of toxic elements and heavy metals bound to the biomass, as well as inefficient waste and biomass separation methods. To choose the most suitable waste source for SCP production over the vast range of options, it is crucial to consider the following key points: (a) local availability of waste source; (b) cost of the required pretreatments; (c) cost of transport; (d) SCP yield and quality (in terms of biomass and protein), (e) efficient maintenance of cultivation condition, (f) efficient SCP extraction solutions; (g) efficiency of C and N utilization. Based on the abovementioned key points, various substrates can be extensively compared and the economic benefits of resource recovery can be precisely analyzed.

References

Aggelopoulos, T., Katsieris, K., Bekatorou, A., Pandey, A., Banat, I. M., & Koutinas, A. A. (2014). Solid state fermentation of food waste mixtures for single cell protein, aroma volatiles and fat production. *Food Chemistry, 145*, 710–716.

Ahmadi, A. R., Ghoorchian, H., Hajihosaini, R., & Khanifar, J. (2010). Determination of the amount of protein and amino acids extracted from the microbial protein (SCP) of lignocellulosic wastes. *Pakistan Journal of Biological Sciences, 13*(8), 355–361.

Alexander, P., Brown, C., Arneth, A., Finnigan, J., Moran, D., & Rounsevell, M. D. A. (2017). Losses, inefficiencies and waste in the global food system. *Agricultural Systems, 153*, 190–200.

Alriksson, B., Hörnberg, A., Gudnason, A., Knobloch, S., Arnason, J. T., & Johannsson, R. (2014). Fish feed from wood. *Cellulose Chemistry and Technology, 48*, 843–848.

Amran, M. A., Palaniveloo, K., Fauzi, R., Satar, N. M., Mohidin, T. B. M., Mohan, G., et al. (2021). Value-added metabolites from agricultural waste and application of green extraction techniques. *Sustainability, 13*, 11432.

Anvari, M., & Khayati, G. (2011). Submerged yeast fermentation of cheese whey for protein production and nutritional profile analysis. *Advance Journal of Food Science and Technology, 3*(2), 122–126.

Bajić, B., Vučurović, D., Vasić, Đ., Jevtić-Mučibabić, R., & Dodić, S. (2022). Biotechnological production of sustainable microbial proteins from agro-industrial residues and by-products. *Food, 12*(1).

Bajpai, P. (2014). Process steps in recycled fibre processing. In P. Bajpai (Ed.), *Recycling and deinking of recovered paper* (pp. 55–83). Elsevier.

Bakratsas, G., Polydera, A., Nilson, O., Chatzikonstantinou, A. V., Xiros, C., Katapodis, P., et al. (2023). Mycoprotein production by submerged fermentation of the edible mushroom *Pleurotus ostreatus* in a batch stirred tank bioreactor using agro-industrial hydrolysate. *Food, 12*(12), 2295.

Bewersdorff, M., & Dostálek, M. (1971). The use of methane for production of bacterial protein. *Biotechnology and Bioengineering, 13*(1), 49–62.

Chanda, S., & Chakrabarti, S. (1996). Plant origin liquid waste: A resource for single cell protein production by yeast. *Bioresource Technology, 57*(1), 51–54.

Cui, W., Wang, Q., Zhang, F., Zhang, S., Chi, Z., & Madzak, C. (2011). Direct conversion of inulin into single cell protein by the engineered *Yarrowia lipolytica* carrying inulinase gene. *Process Biochemistry, 46*(7), 1442–1448.

Dionisi, D., Majone, M., Bellani, A., Viggi, C. C., & Beccari, M. (2008). Role of biomass adaptation in the removal of formic acid in sequencing batch reactors. *Water Science and Technology, 58*(2), 303–307.

Duarte, L. C., Carvalheiro, F., Lopes, S., Neves, I., & Gírio, F. M. (2008). Yeast biomass production in brewery's spent grains hemicellulosic hydrolyzate. *Applied Biochemistry and Biotechnology, 148*(1–3), 119–129.

Dürre, P., & Eikmanns, B. J. (2015). C1-carbon sources for chemical and fuel production by microbial gas fermentation. *Current Opinion in Biotechnology, 35*, 63–72.

Ellamar, J. B., Valdez, M. T. S. J., Inalvez, A. E., Maniquez, M. C., Rodríguez, J. M., & Tarozza, D. B. (2019). Bioconversion of lignocellulosic agricultural by-products by microorganisms into high mycoprotein feeds. *Philippine Journal of Veterinary and Animal Science, 45*, 75–86.

Elvidge, C. D., Bazilian, M. D., Zhizhin, M., Ghosh, T., Baugh, K., & Hsu, F. (2018). The potential role of natural gas flaring in meeting greenhouse gas mitigation targets. *Energy Strategy Reviews, 20*, 156–162.

Food and Agricultural Organization of United Nations (FAO) (2020). https://www.fao.org/faostat/en/#data/QCL. Accessed 21.11.2023

Ferrari, R., Gautier, V., & Silar, P. (2021). Lignin degradation by ascomycetes. In M. Morel-Rouhier & R. Sormani (Eds.), *Advances in botanical research* (Vol. 99, pp. 77–113). Academic Press.

Ferreira, J. A., Lennartsson, P. R., & Taherzadeh, M. J. (2014). Production of ethanol and biomass from thin stillage using food-grade zygomycetes and ascomycetes filamentous fungi. *Energies, 7*(6), 3872–3885.

Ferreira, J. A., Mahboubi, A., Lennartsson, P. R., & Taherzadeh, M. J. (2016). Waste biorefineries using filamentous ascomycetes fungi: Present status and future prospects. *Bioresource Technology, 215*, 334–345.

Gangadwala, J., Radulescu, G., Kienle, A., Steyer, F., & Sundmacher, K. (2008). New processes for recovery of acetic acid from waste water. *Clean Technologies and Environmental Policy, 10*(3), 245–254.

Gao, Y., Li, D., & Liu, Y. (2012). Production of single cell protein from soy molasses using Candida tropicalis. *Annales de Microbiologie, 62*(3), 1165–1172.

Hashem, M., Hesham, A., Alamri, S. A., & Alrumman, S. A. (2014). Production of single-cell protein from wasted date fruits by *Hanseniaspora uvarum* KKUY-0084 and *Zygosaccharomyces rouxii* KKUY-0157. *Annales de Microbiologie, 64*(4), 1505–1511.

Hashempour-Baltork, F., Hosseini, S. M., Assarehzadegan, M. A., Khosravi-Darani, K., & Hosseini, H. (2020a). Safety assays and nutritional values of mycoprotein produced by *Fusarium venenatum* IR372C from date waste as substrate. *Journal of the Science of Food and Agriculture, 100*(12), 4433–4441.

Hashempour-Baltork, F., Khosravi-Darani, K., Hosseini, H., Farshi, P., & Reihani, S. F. S. (2020b). Mycoproteins as safe meat substitutes. *Journal of Cleaner Production, 253*, 119958.

Hocking, M. B. (1997). Vanillin: Synthetic flavoring from spent sulfite liquor. *Journal of Chemical Education, 74*(9), 1055.

Hosseini, S. M., Khosravi-Darani, K., Mohammadifar, M. A., & Nikoopour, H. (2011). Response surface methodology for mycoprotein production by *Fusarium venenatum* ATCC 20334. *The International Open Access Journal of Bioprocessing and Biotechniques Executive Editors Bioprocessing and Biotechniques, 01*, 2–6.

Ito, T., Nakashimada, Y., Senba, K., Matsui, T., & Nishio, N. (2005). Hydrogen and ethanol production from glycerol-containing wastes discharged after biodiesel manufacturing process. *Journal of Bioscience and Bioengineering, 100*(3), 260–265.

Ivanovs, K., Spalvins, K., & Blumberga, D. (2018). Single cell protein production from waste biomass: Review of various agricultural by-products. *Energy Procedia, 147*(2018), 409–418.

Ivarson, K. C., & Morita, H. (1982). Single-cell protein production by the acid-tolerant fungus Scytalidium acidophilum from acid hydrolysates of waste paper. *Applied and Environmental Microbiology, 43*(3), 643–647.

Jalasutram, V., Kataram, S., Gandu, B., & Anupoju, G. R. (2013). Single cell protein production from digested and undigested poultry litter by *Candida utilis*: Optimization of process parameters using response surface methodology. *Clean Technologies and Environmental Policy, 15*(2), 265–273.

Johnson, D. T., & Taconi, K. A. (2007). The glycerin glut: Options for the value-added conversion of crude glycerol resulting from biodiesel production. *Environmental Progress, 26*(4), 338–348.

Kam, A., Abedian Kenari, A., & Younesi, H. (2012). Production of single cell protein in stickwater by *Lactobacillus acido*philus and *Aspergillus Niger*. *Journal of Aquatic Food Product Technology, 21*(3), 403–417.

Kaza, S., Yao, L., Bhada-Tata, P., & Van Woerden, F. (2018). *What a waste 2.0: A global snapshot of solid waste management to 2050*. The World Bank.

Khan, M., Khan, S. S., Ahmed, Z., & Tanveer, A. (2010). Production of single cell protein from *Saccharomyces cerevisiae* by utilizing fruit wastes. *Nanobiotechnica Universale, 1*(2), 127–132.

Khan, N. E., Myers, J. A., Tuerk, A. L., & Curtis, W. R. (2014). A process economic assessment of hydrocarbon biofuels production using chemoautotrophic organisms. *Bioresource Technology, 172*, 201–211.

Khan, M. K. I., Asif, M., Razzaq, Z., Nazir, A., & Maan, A. (2022). Sustainable food industrial waste management through single cell protein production and characterization of protein enriched bread. *Food Bioscience, 46*, 101406.

Klemm, D., Heublein, B., Fink, H.-P., & Bohn, A. (2005). Cellulose: Fascinating biopolymer and sustainable raw material. *Angewandte Chemie International Edition, 44*(22), 3358–3393.

Kornochalert, N., Kantachote, D., Chaiprapat, S., & Techkarnjanaruk, S. (2014). Use of *Rhodopseudomonas p*alustris P1 stimulated growth by fermented pineapple extract to treat latex rubber sheet wastewater to obtain single cell protein. *Annales de Microbiologie, 64*(3), 1021–1032.

Koukoumaki, D. I., Tsouko, E., Papanikolaou, S., Ioannou, Z., Diamantopoulou, P., & Sarris, D. (2024). Recent advances in the production of single cell protein from renewable resources and applications. *Carbon Resources Conversion, 7*(2), 100195.

Kumar, R., Raj, T., Næss, G., Sørensen, M., & Dhawan, V. (2024). Opportunities and challenges in single-cell protein production using lignocellulosic material. *Biofuels, Bioproducts and Biorefining, 18*(1), 310–321.

Kurbanoglu, E. B., & Algur, O. F. (2002). Single-cell protein production from ram horn hydrolysate by bacteria. *Bioresource Technology, 85*(2), 125–129.

Kurcz, A., Błażejak, S., Kot, A. M., Bzducha-Wróbel, A., & Kieliszek, M. (2018). Application of industrial wastes for the production of microbial single-cell protein by fodder yeast *Candida utilis*. *Waste and Biomass Valorization, 9*(1), 57–64.

Lee, J. Z., Logan, A., Terry, S., & Spear, J. R. (2015). Microbial response to single-cell protein production and brewery wastewater treatment. *Microbial Biotechnology, 8*(1), 65–76.

Li, J., Liu, R., Chang, G., Li, X., Chang, M., Liu, Y., et al. (2015). A strategy for the highly efficient production of docosahexaenoic acid by *Aurantiochytrium limacinum* SR21 using glucose and glycerol as the mixed carbon sources. *Bioresource Technology, 177*, 51–57.

Liu, B., Song, J., Li, Y., Niu, J., Wang, Z., & Yang, Q. (2013). Towards industrially feasible treatment of potato starch processing waste by mixed cultures. *Applied Biochemistry and Biotechnology, 171*(4), 1001–1010.

Liu, B., Li, Y., Song, J., Zhang, L., Dong, J., & Yang, Q. (2014). Production of single-cell protein with two-step fermentation for treatment of potato starch processing waste. *Cellulose, 21*(5), 3637–3645.

Mahan, K. M., Le, R. K., Wells, T. J., Anderson, S., Yuan, J. S., Stoklosa, R. J., et al. (2018). Production of single cell protein from agro-waste using Rhodococcus opacus. *Journal of Industrial Microbiology and Biotechnology, 45*(9), 795–801.

Mehta, R. J. (1973). Studies on methanol-oxidizing bacteria. *Antonie Van Leeuwenhoek, 39*(1), 295–302.

Mohseni, A., Fan, L., & Roddick, F. A. (2021). Impact of microalgae species and solution salinity on algal treatment of wastewater reverse osmosis concentrate. *Chemosphere, 285*, 131487.

Molfetta, M., Morais, E. G., Barreira, L., Bruno, G. L., Porcelli, F., Dugat-Bony, E., et al. (2022). Protein sources alternative to meat: State of the art and involvement of fermentation. *Food, 11*(14).

Monte, M. C., Fuente, E., Blanco, A., & Negro, C. (2009). Waste management from pulp and paper production in the European Union. *Waste Management, 29*(1), 293–308.

Moraine, R., Shelef, G., Meydan, A., & Levi, A. (1979). Algal single cell protein from wastewater treatment and renovation process. *Biotechnology and Bioengineering, 21*(7), 1191–1207.

Nasehia, M., Torbatinejad, N. M., Zerehdaran, S., & Safaie, A. R. (2016). Effect of solid-state fermentation by oyster mushroom (*Pleurotus florida*) on nutritive value of some agro by-products. *Journal of Applied Animal Research, 45*, 221–226.

Patil, K. P., & Gogate, P. R. (2015). Improved synthesis of docosahexaenoic acid (DHA) using *Schizochytrium limacinum* SR21 and sustainable media. *Chemical Engineering Journal, 268*, 187–196.

Patyshakuliyeva, A., Arentshorst, M., Allijn, I. E., Ram, A. F., de Vries, R. P., & Gelber, I. B. (2016). Improving cellulase production by *Aspergillus Niger* using adaptive evolution. *Biotechnology Letters, 38*(6), 969–974.

Piercy, E., Verstraete, W., Ellis, P. R., Banks, M., Rockström, J., Smith, P., et al. (2023). A sustainable waste-to-protein system to maximise waste resource utilisation for developing food- and feed-grade protein solutions. *Green Chemistry, 25*(3), 808–832. https://doi.org/10.1039/D2GC03095K

Pillaca-Pullo, O. S., Lopes, A. M., Rodriguez-Portilla, L., & Estela-Escalante, W. (2023). Optimizing medium composition with wastewater from *Coffea arabica* processing to produce single-cell protein using *Candida sorboxylosa*. *Journal of Chemical Technology and Biotechnology, 98*(1), 106–116.

Pires, J. F., Ferreira, G. M. R., Reis, K. C., Schwan, R. F., & Silva, C. F. (2016). Mixed yeasts inocula for simultaneous production of SCP and treatment of vinasse to reduce soil and fresh water pollution. *Journal of Environmental Management, 182*, 455–463.
Poore, J., & Nemecek, T. (2018). Reducing food's environmental impacts through producers and consumers. *Science, 360*(6392), 987–992.
Putri, D., Ulhidayati, A., Musthofa, I. A., & Wardani, A. K. (2018). Single cell protein production of Chlorella sp. using food processing waste as a cultivation medium. *IOP Conference Series: Earth and Environmental Science, 131*(1), 012052.
Rajoka, M. I., Ahmed, S., Hashmi, A., & Athar, M. (2012). Production of microbial biomass protein from mixed substrates by sequential culture fermentation of *Candida* utilis and *Brevibacterium lactofermentum. Annals of Microbiology, 62*(3), 1173–1179.
Rao, M. B., Varma, A. J., & Deshmukh, S. S. (2010). Production of single cell protein, essential amino acids, and xylanase by *Penicillium janthinellum. BioResources, 5*(4), 2470–2477.
Raziq, A., Lateef, M., Ullah, A., Ullah, H., & Khan, M. W. (2020). 02. Single cell protein (SCP) production and potential substrates: A comprehensive review. *Pure and Applied Biology, 9*(3), 1743.
Reihani, S. F. S., & Khosravi-Darani, K. (2019a). Influencing factors on single-cell protein production by submerged fermentation: A review. *Electronic Journal of Biotechnology, 37*, 34–40.
Reihani, S. F. S., & Khosravi-Darani, K. (2019b). Mycoprotein production from date waste using *Fusarium venenatum* in a submerged culture. *Applied Food Biotechnology, 5*(4), 243–352.
Reihani, S. F. S., Alemzadeh, I., & Vossoughi, M. (2023). A submerged production of glucoamylase from selected agro-waste by using *Aspergillus Niger. Iranian Journal of Chemistry and Chemical Engineering*. https://www.ijcce.ac.ir/article_709087.html
Rhishipal, R., & Philip, R. (1998). Selection of marine yeasts for the generation of single cell protein from prawn-shell waste. *Bioresource Technology, 65*(3), 255–256.
Rischer, H., Szilvay, G. R., & Oksman-Caldentey, K.-M. (2020). Cellular agriculture — industrial biotechnology for food and materials. *Current Opinion in Biotechnology, 61*, 128–134.
Rodríguez, J., Ferraz, A., Nogueira, R. F., Ferrer, I., Esposito, E., & Durán, N. (1997). Lignin biodegradation by the ascomycete *Chrysonilia sitophila. Applied Biochemistry and Biotechnology, 62*(2–3), 233–242.
Sadh, P. K., Duhan, S., & Duhan, J. S. (2018). Agro-industrial wastes and their utilization using solid state fermentation: A review. *Bioresources and Bioprocessing, 5*(1), 1.
Saejung, C., & Salasook, P. (2020). Recycling of sugar industry wastewater for single-cell protein production with supplemental carotenoids. *Environmental Technology, 41*(1), 59–70.
Salazar-López, N. J., Barco-Mendoza, G. A., Zuñiga-Martínez, B. S., Domínguez-Avila, J. A., Robles-Sánchez, R. M., Ochoa, M. A. V., et al. (2022). Single-cell protein production as a strategy to reincorporate food waste and agro by-products back into the processing chain. *Bioengineering (Basel), 9*(11), 623. https://doi.org/10.3390/bioengineering9110623
Santala, S., Efimova, E., Kivinen, V., Larjo, A., Aho, T., Karp, M., et al. (2011). Improved triacylglycerol production in *Acinetobacter baylyi* ADP1 by metabolic engineering. *Microbial Cell Factories, 10*(1), 36.
Schuchmann, K., & Müller, V. (2014). Autotrophy at the thermodynamic limit of life: A model for energy conservation in acetogenic bacteria. *Nature Reviews Microbiology, 12*(12), 809–821.
Shacklady, C. A. (1970). Single cell proteins from hydrocarbons. *Outlook on Agriculture, 6*(3), 102–107.
Sharif, M., Zafar, M. H., Aqib, A. I., Saeed, M., Farag, M. R., & Alagawany, M. (2021). Single cell protein: Sources, mechanism of production, nutritional value and its uses in aquaculture nutrition. *Aquaculture, 531*, 735885.
Sharma, R. K., & Arora, D. S. (2015). Fungal degradation of lignocellulosic residues: An aspect of improved nutritive quality. *Critical Reviews in Microbiology, 41*(1), 52–60.

Şişman, T., Gür, Ö., Doğan, N., Özdal, M., Algur, Ö. F., & Ergon, T. (2013). Single-cell protein as an alternative food for zebrafish, Danio rerio: A toxicological assessment. *Toxicology and Industrial Health, 29*(9), 792–799.

Souza Filho, P. F., Nair, R. B., Andersson, D., Lennartsson, P. R., & Taherzadeh, M. J. (2018). Vegan-mycoprotein concentrate from pea-processing industry byproduct using edible filamentous fungi. *Fungal Biology and Biotechnology, 5*(1). https://doi.org/10.1186/s40694-018-0050-9

Spalvins, K., Zihare, L., & Blumberga, D. (2018). Single cell protein production from waste biomass: Comparison of various industrial by-products. *Energy Procedia, 147*, 409–418.

Stoffel, F., Santana, W., Gregolon, J. N., Kist, T. B. L., Fontana, R. C., & Camassola, M. (2019). Production of edible mycoprotein using agroindustrial wastes: Influence on nutritional, chemical and biological properties. *Innovative Food Science and Emerging Technologies, 58*, 102227.

Suman, G., Nupur, M., Anuradha, S., & Pradeep, B. (2017). Characterization of dairy waste and its utilisation as substrate for production of single cell protein. *Journal of Biotechnology and Biochemistry, 3*, 73–78.

Taran, M., & Asadi, N. (2014). A novel approach for environmentally friendly production of single cell protein from petrochemical wastewater using a halophilic mo in different conditions. *Petroleum Science and Technology, 32*(5), 625–630.

Tropea, A., Ferracane, A., Albergamo, A., Potortì, A. G., Lo Turco, V., & Di Bella, G. (2022). Single cell protein production through multi food-waste substrate fermentation. *Fermentation, 8*, 91.

UNEP, U. N. E. P. (2021). *Food waste index report*. United Nations Environment Programme.

Upcraft, T., Tu, W.-C., Johnson, R., Finnigan, T., Van Hung, N., Hallett, J., et al. (2021). Protein from renewable resources: Mycoprotein production from agricultural residues. *Green Chemistry, 23*(14), 5150–5165. https://doi.org/10.1039/D1GC01021B

Uščа, M., & Tisenkopfs, T. (2023). The resilience of short food supply chains during the COVID-19 pandemic: A case study of a direct purchasing network [Original Research]. *Frontiers in Sustainable Food Systems, 7*(7), 1146446. https://doi.org/10.3389/fsufs.2023.1146446

Valentino, M. J. G., Ganado, L. S., & Undan, J. R. (2016). Single cell protein potential of endophytic fungi associated with bamboo using rice bran as substrate. *Advances in Applied Science Research, 7*(3), 68–72.

Voutilainen, E., Pihlajaniemi, V., & Parviainen, T. (2021). Economic comparison of food protein production with single-cell organisms from lignocellulose side-streams. *Bioresource Technology Reports, 14*, 100683.

Vrati, S. (1984). Single cell protein production by photosynthetic bacteria grown on the clarified effluents of biogas plant. *Applied Microbiology and Biotechnology, 19*(3), 199–202.

Wegner, G. H. (1990). Emerging applications of the methylotrophic yeasts. *FEMS Microbiology Reviews, 7*(3–4), 279–283.

Wen, C., Zhang, J., Duan, Y., Zhang, H., & Ma, H. (2019). A mini-review on brewer's spent grain protein: Isolation, physicochemical properties, application of protein, and functional properties of hydrolysates. *Journal of Food Science, 84*(12), 3330–3340.

Wong, P. K., & Chan, K. (1980). Algal single cell protein production from sewage effluent with high salinity. *Experientia, 36*(9), 1065–1066.

Wongputtisin, P., Khanongnuch, C., Khongbantad, W., Niamsup, P., & Lumyong, S. (2012). Screening and selection of *Bacillus* spp. for fermented corticate soybean meal production. *Journal of Applied Microbiology, 113*(4), 798–806.

Wongputtisin, P., Khanongnuch, C., Kongbuntad, W., Niamsup, P., Lumyong, S., & Sarkar, P. K. (2014). Use of *Bacillus subtilis* isolates from Tua-nao towards nutritional improvement of soya bean hull for monogastric feed application. *Letters in Applied Microbiology, 59*(3), 328–333.

Yadav, J. S., Bezawada, J., Elharche, S., Yan, S., Tyagi, R. D., & Surampalli, R. Y. (2014a). Simultaneous single-cell protein production and COD removal with characterization of residual protein and intermediate metabolites during whey fermentation by *K. marxianus*. *Bioprocess and Biosystems Engineering, 37*(6), 1017–1029.

Yadav, S., Dubey, A. K., Anand, G., Kumar, R., & Yadav, D. (2014b). Purification and biochemical characterization of an alkaline pectin lyase from Fusarium decemcellulare MTCC 2079 suitable for *Crotalaria juncea* fiber retting. *Journal of Basic Microbiology, 54*(Suppl 1), S161–S169.

Yap, S. M., Lan, J. C., Kee, P. E., Ng, H. S., & Yim, H. S. (2022). Enhancement of protein production using synthetic brewery wastewater by *Haematococcus pluvialis*. *Journal of Biotechnology, 350*, 1–10.

Yazdian, F., Hajizadeh, S., Shojaosadati, S. A., Khalilzadeh, R., Jahanshahi, M., & Nosrati, M. (2005). Production of single cell protein from natural gas: Parameter optimization and RNA evaluation. *Iranian Journal of Biotechnology, 3*(4), 235–242.

Zha, X., Tsapekos, P., Zhu, X., Khoshnevisan, B., Lu, X., & Angelidaki, I. (2021). Bioconversion of wastewater to single cell protein by methanotrophic bacteria. *Bioresource Technology, 320*, 124351.

Zhang, L., Zhou, P., Chen, Y. C., Cao, Q., Liu, X. F., & Li, D. (2021). The production of single cell protein from biogas slurry with high ammonia-nitrogen content by screened Nectaromyces rattus. *Poultry Science, 100*(9), 101334.

Zhao, G., Zhang, W., & Zhang, G. (2010). Production of single cell protein using waste capsicum powder produced during capsanthin extraction. *Letters in Applied Microbiology, 50*(2), 187–191.

Zhou, P., Zhang, L., Ding, H., Gao, X., Chen, Y., & Li, D. (2022). Optimization of culture conditions of screened *Galactomyces candidum* for the production of single cell protein from biogas slurry. *Electronic Journal of Biotechnology, 55*, 47–54.

Zhu, Z., Wu, Y., Hu, W., Zheng, X., & Chen, Y. (2022). Valorization of food waste fermentation liquid into single cell protein by photosynthetic bacteria via stimulating carbon metabolic pathway and environmental behaviour. *Bioresource Technology, 361*, 127704.

Chapter 8
Agricultural Waste for Biofertilizers and Soil Amendments

Varsha Thadiyan, Manpreet Kaur, and Reena Gupta

Abbreviations

AD	Anaerobic digestion
CEC	Cation exchanging capacity
GHG	Greenhouse gases
N_2O	Nitrous oxide
PGPR	Plant growth promoting bacteria

8.1 Introduction

Agricultural waste refers to the materials that are no longer utilized in the primary agricultural processes and are left over as surplus or byproducts (Donner et al., 2021). Agriculture is the backbone of many developing countries and one of the largest contributors to the resource sector. Millions of people practice agriculture as their occupation. With the increase in the population, there is a rise in demand for food and food products, so practising modern agriculture can be the only way to meet this demand. These agricultural practices leave a lot of waste residues behind during pre- and post-harvesting; it encompasses various forms of waste which include crop residues, such as stalks, leaves, and husks that remain after harvesting. It also comprises animal manure and other byproducts from livestock farming; additionally, the food waste resulting from agricultural processing and post-harvest activities is considered a form of agricultural waste.

V. Thadiyan · M. Kaur · R. Gupta (✉)
Department of Biotechnology, Himachal Pradesh University, Shimla, Himachal Pradesh, India

J. Arora et al. (eds.), *Transforming Agriculture Residues for Sustainable Development*, Waste as a Resource, https://doi.org/10.1007/978-3-031-61133-9_8

8.2 Definition and Background of Agricultural Waste

The term "Agricultural waste" implies byproducts of various agricultural activities and encompasses a wide range of materials generated during farming processes which includes crop residues, such as stalks, leaves, husks, and stems (Ray, 2022). It also holds potential as a valuable resource that can be recycled or transformed into useful products. If proper disposal procedures are not followed, the discharge of such residues into the environment may lead to environmental pollution and adverse effects on the health of humans or animals. Agricultural waste generation is an inherent part of agricultural practices worldwide (Ramawat et al., 2023). As the global population continues to grow, the demand for food and fiber increases, leading to an increase in agricultural activities. Consequently, the volume of agricultural waste is escalating, posing significant challenges in terms of waste management, environmental impact, and resource utilization. Improper disposal of waste can lead to pollution of water bodies, emissions of greenhouse gases (GHG), soil degradation, and other negative consequences (Arora et al., 2023; Singh & Singh, 2017). Therefore, it is necessary to find sustainable solutions for the administration of agricultural waste to mitigate these environmental impacts and promote sustainable agriculture (Joshi & Arora, 2023a, b). By adopting innovative and responsible approaches to agricultural waste management, one can contribute to the development of a more sustainable and resilient agricultural sector.

8.3 Importance of Agricultural Waste Management

Agricultural waste management is of paramount importance for several reasons. Proper management of agricultural waste helps to prevent environmental pollution. Agricultural activities produce various waste materials including crop residues and animal manure, which can be harmful to the ecosystem if released into the air, water, and soil without being managed effectively. Contaminants from agricultural waste can disrupt ecosystems by harming wildlife and degrading water quality. By implementing appropriate waste management practices, the negative environmental impacts associated with agricultural waste can be minimized. Agricultural waste contains valuable nutrients and major components are organic that can be recycled and reused. Managing agricultural waste plays a crucial role in maintaining soil health and fertility. Properly managed waste materials, such as compost and manure, can improve soil structure, enhance its water-holding capacity, and promote nutrient cycling.

Improper disposal of waste materials can result in the release of greenhouse gases, such as methane and nitrous oxide (N_2O), which contribute to global warming (Joshi & Arora, 2023a, b). However, through practices, biogas can be produced from agricultural waste through processes such as anaerobic digestion (AD) and composting and biochar, respectively. These processes help in capturing and utilizing methane, which is a potent GHG and also contributes to carbon sequestration. By managing agricultural waste effectively, greenhouse gas emissions can be reduced and the impacts of climate change can also be mitigated.

8.3.1 Objectives of Biofertilizers and Soil Amendment

Biofertilizers and soil amendments play crucial roles in sustainable agricultural practices by enhancing soil fertility and promoting plant growth. The main objectives of using biofertilizers and soil amendments include:

Improving Soil Fertility
One of the primary objectives of biofertilizers and soil amendments is to enhance the soil's structure, texture, nutrient supply, water-holding capacity, and the proliferation of beneficial soil microorganisms. They are also less expensive, renewable, and pollution-free ways to enrich the soil. They contain microbial diversity that is capable of atmospheric nitrogen fixation and solubilizing and mobilizing micro- and macro-nutrients (such as phosphorus (P), zinc (Zn), and potassium (K)) and making them available for plants in simple forms for uptake. The application of biofertilizers, or fertilizers containing live microorganisms, is one of the management techniques that can help to maintain or increase the amount of organic matter and improve soil fertility (Zhang et al., 2019; Pandey et al., 2021).

Sustainable Agriculture
Sustainable agriculture refers to the practice of producing food, fiber, or other plant-based and animal-based products in a way that takes into account the long-term health of the environment, economic viability, and social well-being. Biofertilizers and soil amendments are environmentally friendly alternatives to synthetic chemical fertilizers excessive use of which leads to soil degradation, nitrogen leaching, soil compaction, reduction in soil organic matter, and loss of soil carbon (Pandey et al., 2021).

Enhancing Plant Growth and Yield
Since biofertilizers include live or dormant microorganisms that aid in crop plant nutrient uptake through their interactions in the rhizosphere after being applied through the soil, they are probably better known as bioinoculants. The microorganisms present in biofertilizers secrete growth-promoting substances that stimulate plant growth and development. They also improve root development and nutrient uptake, leading to increased crop yields.

Soil Structure and Aeration
Some soil amendments, like organic matter and compost, can improve soil structure by enhancing its water-holding capacity, drainage, and aeration. This creates a better environment for root growth and prevents soil erosion. Organic matter causes soil to clump and form soil aggregates, which improves soil structure. Better soil structure and permeability improve the ability of soil to take up and hold water.

Disease Suppression
Certain biofertilizers and soil amendments contain beneficial microorganisms that can suppress the growth of harmful pathogens that cause diseases in plants. Organic fertilizers are rich in biocontrol microorganisms (i.e., bioorganic fertilizers) and have been shown to improve resistance against plant pathogens and act as natural antagonists against plant diseases which help in maintaining plant health. Microbial isolates like *Bacillus* and *Trichoderma* sp. are added to biofertilizers, which are chosen based on their ability to inhibit Fusarium wilt disease in lab settings (Wang et al., 2019; Yao et al., 2023).

Reducing Environmental Pollution
By reducing the use of chemical fertilizers, the application of biofertilizers and soil amendments such as cow dung helps minimize the risk of nutrient leaching into groundwater as well as the emission of GHG, thus reducing environmental pollution (Behera & Ray, 2021). Biofertilizers are also helpful for the remediation of polluted soils. They are environment friendly and do not cause pollution, unlike chemical fertilizers which often run off into the water bodies leading to their eutrophication. They also help to enhance the soil quality by providing nutrients and a natural environment in the rhizosphere. Microbial inoculants in biofertilizers and soil amendments aid in reducing the use of chemical fertilizers (Singh et al., 2022).

Restoring Degraded Soils
Restoring the quality of soil for agricultural practices and crop production by application of appropriate soil management and conservation techniques is important for all nations, primarily for those who are at risk concerning food security (Imran & Al-Tawaha, 2020). Soil amendments like organic matter and compostor biochar can help to restore degraded and nutrient-deficient soils. Organic amendments improve the physical and chemical nature of soils such as sorption capacity, mobilization of nutrients by mineralization, short-term immobilization into soil organisms, long-term fixation into stable humic substances, and supply of organic substances (Imran, 2018).

Promoting Biological Diversity
The use of biofertilizers and organic amendments encourages beneficial microbial diversity in the soil, which contributes to the overall health and resilience of the agroecosystem. Biofertilizers contain living cells of different types of beneficial microbes like bacteria, fungi, protozoa, and algae. Some common microorganisms in agriculture include *Rhizobium*, *Mycorrhizae*, *Azospirillum, Pseudomonas,* and *Trichoderma Streptomyces* species.

8.4 Types of Agricultural Wastes

Agricultural waste refers to organic and inorganic materials generated during agricultural activities. These materials are byproducts of crop cultivation, livestock rearing, and agro-industrial processes (Fig. 8.1).

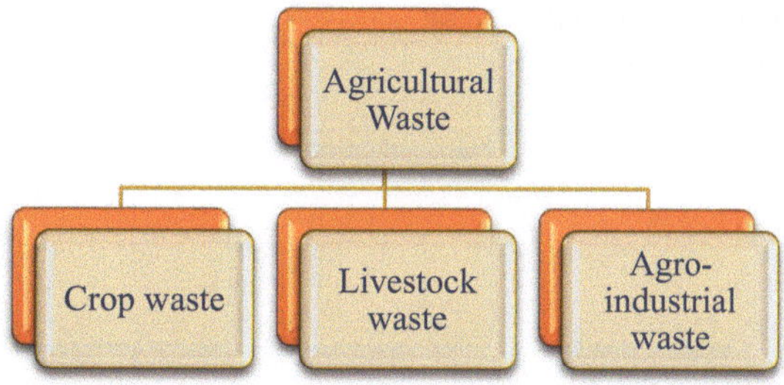

Fig. 8.1 Types of agricultural waste

Crop residues are the materials left on cultivated land after the crop has been harvested such as stems, leaves, husks, and shells. Crop residues are generated from crops like rice straw, wheat straw, corn stover, and sugarcane bagasse (Reicosky et al., 2002). Livestock waste includes the excreta of livestock animals like cattle, poultry, pigs, and sheep. It is a mixture of animal excrement, urine, and materials used in the handling and storage of animals, including waste feed, wash waters, rice hulls, straw, sawdust, and other materials related to animal production (Sims & Maguire, 2005).

Agro-waste includes crop residues, animal manure, and other organic byproducts. Agro-industries produce a variety of byproducts and waste materials as a result of processing agricultural raw materials into finished products. For example, fruit and vegetable peels, bagasse, seed husks, and bran are common byproducts in agro-processing (Ahuja et al., 2020; Ray, 2022, 2024). These byproducts and waste materials can have both potential economic value and environmental impacts, depending on how they are managed. The harnessing of this nutrient-rich content of agro-waste for biofertilizer production offers a sustainable solution to enhance soil fertility, improve crop yields, and reduce reliance on chemical fertilizers (Kumar et al., 2022).

8.4.1 Crop Waste: Utilization Potential

Crop residues are considered as waste after harvesting but they have several valuable utilization potentials. Several strategies have been developed to produce biochemicals from biological wastes (Fig. 8.2).

Chemicals like carbohydrates, minerals, proteins, and other nutritional compounds have been isolated from various crop residues (Jena & Singh, 2022). Soil amendments can correct deficiencies, enhance nutrient availability, improve water retention, and create a more favorable environment for plants (Lim et al., 2016). They are commonly used in gardening, agriculture, and landscaping to optimize soil conditions. They are rich in organic and inorganic compounds and can be classified into organic amendments, inorganic amendments, and nutrient-specific amendments. Organic amendments are the materials that are derived from plants and animals and provide valuable organic matter to the soil and such amendments

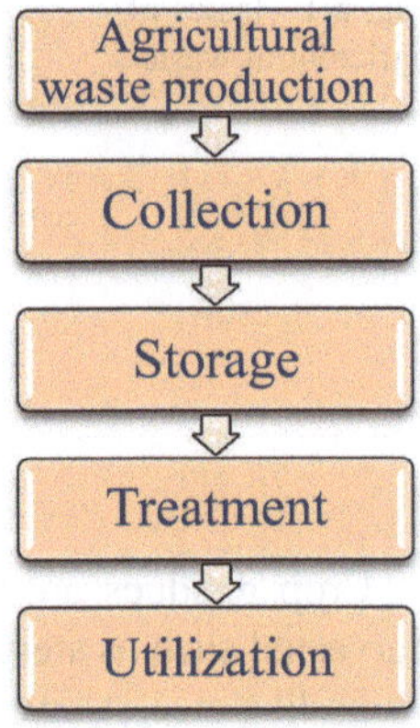

Fig. 8.2 Agriculture waste management

include compost and manure. Inorganic amendments are mineral-based materials and nutrient-specific amendments are used to supply specific nutrients to the soil and enhance plant growth. These include bone meal, blood meal, and rock phosphate.

8.4.2 Animal Manure: A Valuable Source of Nutrients

Natural fertilizer made of animal dung gives plants vital elements like K, P, and N (Behera & Ray, 2021). It has been used as a natural fertilizer for centuries due to its nutrient-rich composition and ability to improve soil fertility. It is an organic material derived from the waste of animals like cattle, poultry, horses, and pigs. When properly managed and applied, it can provide several benefits to soil and plant health (Swain et al., 2007, 2008; Swain & Ray, 2009; Ray & Swain, 2011). It contains essential nutrients like N, P, and K, along with secondary nutrients and micronutrients. Animal manure contains organic matter that, when incorporated into the soil, helps in the binding of soil particles together into aggregates. This results in improved soil structure, which allows for better water infiltration, root penetration, and air movement within the soil (Lin et al., 2010). These microorganisms help break down organic matter and make nutrients more available to plants. The organic matter in manure acts like a sponge, increasing the soil's ability to hold onto water (Mahajan et al., 2023a). This is particularly beneficial in both sandy soils that drain quickly and clay soils that can hold too much water (Mahajan et al., 2022). The presence of organic matter in manure prevents soil compaction and promotes aeration. Well-aerated soil allows plant roots to access oxygen, which is vital for their growth and overall health. The organic matter in manure serves as a food source for beneficial soil microorganisms (Mahajan et al., 2023b, c). This stimulates microbial activity, which further enhances nutrient cycling and the breakdown of organic matter, contributing to soil health. It can be utilized as a renewable energy source through a process called anaerobic digestion (Malandrino et al., 2011).

8.5 Sustainable Agricultural Waste Management Practices

Sustainable agriculture waste management practices are essential to minimize environmental impacts, promote resource efficiency, and support long-term agricultural productivity (Fig. 8.3).

8.5.1 Composting

Recycling organic matter back into the soil to maintain soil fertility has been done for a long time through the practice of composting municipal solid waste and agricultural waste (Anon, 1993; Maynard, 1994; Golueke & Diaz, 1996). Demand for eco-friendly waste treatment technology has given birth to the recent rise in interest in composting (Bujang & Lopez-Real, 1993; Lau et al., 1993; Liao et al., 1993; Maynard, 1994; Georgacakis et al., 1996). Temperature, oxygen supply (also known as aeration), moisture content, pH, C:N ratio, particle size, and degree of compaction are a few variables that might affect how effectively compost is made (Lau et al., 1993; Stentiford, 1996). Cattle manure, chicken litter, fisheries mortalities, and even municipal sewage sludge have been disposed of using composting. Compost can be used as a soil amendment, enhancing soil fertility, structure, and microbial activity.

8.5.2 Anaerobic Digestion (AD)

AD is a process that breaks down organic waste, particularly animal manure in the absence of oxygen to produce biogas and digestate (Karagiannidis & Perkoulidis, 2009). AD appears to be a promising method for handling organic waste, even though there are numerous approaches. The gases that are released during the

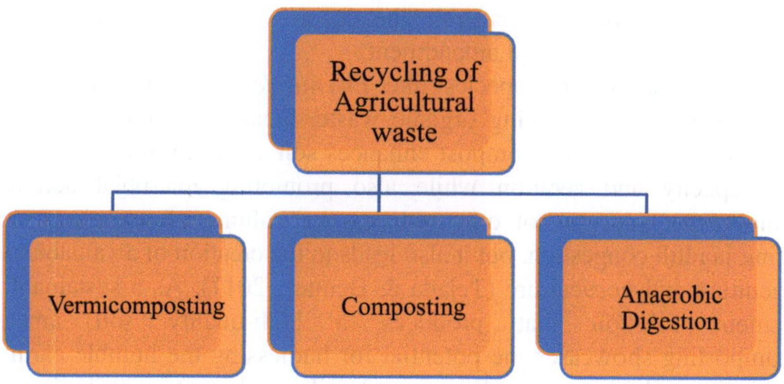

Fig. 8.3 Sustainable waste management practices

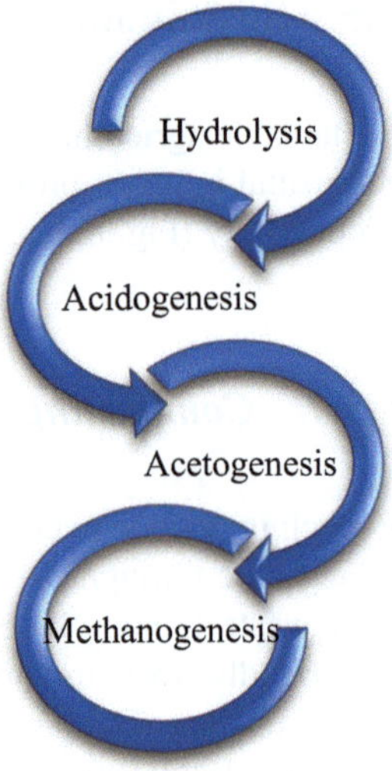

Fig. 8.4 Steps involved in anaerobic digestion

anaerobic digestion of organic waste in landfills are carbon dioxide and methane, and they have the potential to pollute the environment (Mazid & Khan, 2015; McWeeney, 1994; Meegoda et al., 2018). The digestate can be used as fertilizer and biogas as a renewable energy source. The four phases of anaerobic digestion are hydrolysis, acidogenesis, acetogenesis, and methanogenesis; each stage is followed by the next (Fig. 8.4). The anaerobic digestion process depends on interactions between the various bacteria (Verma, 2002).

8.5.3 Vermicomposting: Harnessing the Power of Worms

Vermicomposting is a composting process that involves the use of earthworms to break down organic waste materials into nutrient-rich compost. This process harnesses the power of worms, particularly species like *Eisenia fetida* (red wigglers) and *Lumbricus rubellus* to accelerate the decomposition of organic matter and transform it into a valuable soil amendment.

Earthworms, particularly species like *Eisenia fetida*, play a pivotal role in vermicomposting by consuming organic waste, breaking it down, and excreting nutrient-rich castings. Vermicompost enhances soil structure, improving its water-holding capacity and aeration while also promoting microbial activity. The vermicomposting process not only reduces the volume of organic waste, thus alleviating landfill congestion, but it also leads to the creation of a valuable resource for agriculture and horticulture (Tejada & Benítez, 2017). As a sustainable waste management solution that produces a high-quality soil amendment, vermicomposting showcases the potential for harnessing the humble earthworm's power to address both waste reduction and soil enrichment, contributing to more sustainable agricultural and environmental practices (Muthiniet al., 2020).

8.6 Biofertilizers: Harnessing Nature's Power

Biofertilizers are natural fertilizers that harness the power of beneficial microorganisms to improve the uptake of nutrients and plant growth in agricultural systems. Unlike chemical fertilizers which contain synthetic nutrients, biofertilizers are derived from living organisms, such as bacteria, fungi, and algae (Singh et al., 2016). These microorganisms work symbiotically with plants, promoting their growth, improving soil fertility, and providing various other benefits. The various types of biofertilizers that help the plant to grow at different levels of its growth are shown in Fig. 8.5.

The N_2-fixing bacteria convert atmospheric nitrogen into a plant-available form, enhance the nitrogen content in the soil, and stimulate plant growth, for example., *Rhizobium*, *Azotobacter,* and *Azospirillum* (Phat et al., 2019). Phosphorus-solubilizing biofertilizers contain microorganisms that break down insoluble forms of phosphorus present in the soil into soluble forms (Swain et al., 2012). Potassium-mobilizing biofertilizers contain microorganisms that facilitate the mobilization of potassium in the soil making it more accessible to plants and supporting their overall health and vigor. Plant growth-promoting rhizobacteria (PGPR) are biofertilizers that comprise a diverse group of beneficial bacteria that enhance plant growth through various mechanisms such as producing growth-promoting substances, suppressing plant pathogens, and improving nutrient availability. Mycorrhizal biofertilizers are fungi that form a mutualistic association with plant roots, extending their reach into the soil and helping plants access nutrients like phosphorus and water more efficiently.

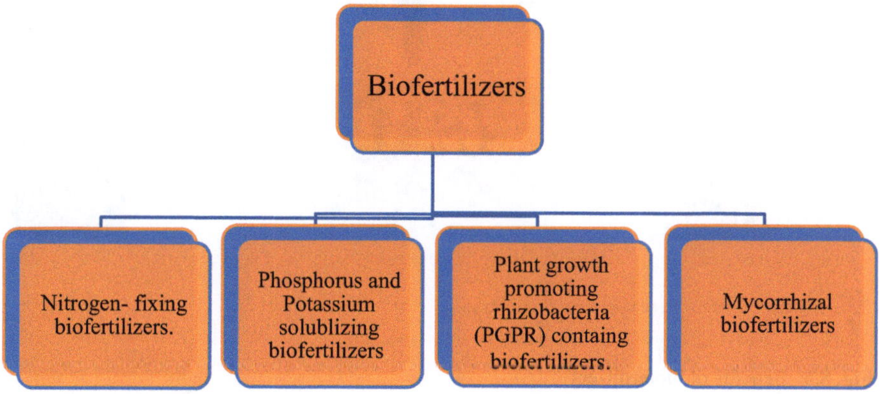

Fig. 8.5 Types of biofertilizers

8.6.1 Benefits of Biofertilizers for Sustainable Agriculture

Biofertilizers offer numerous benefits for sustainable agriculture, promoting environmentally friendly and resource-efficient farming practices (Toledo et al., 2018). Figure 8.6 shows benefits of biofertilizer. Biofertilizers reduce the reliance on chemical fertilizers, which can lead to soil and water pollution when overused. They help to enhance soil structure, water retention, and nutrient availability leading to healthier and more fertile soils over time. As a result, crops grown with biofertilizers often exhibit improved growth rates, increased productivity, and better yield quality (Truckell et al., 2019). Some biofertilizers particularly those containing PGPR have biocontrol properties. This makes crops more resilient to drought conditions and reduces the overall water demand for irrigation (Verma et al., 2011). By reducing the use of chemical inputs, biofertilizers support biodiversity in the soil and surrounding environment. Beneficial microorganisms enhance the diversity and activity of soil microbes, fostering a more balanced and resilient ecosystem.

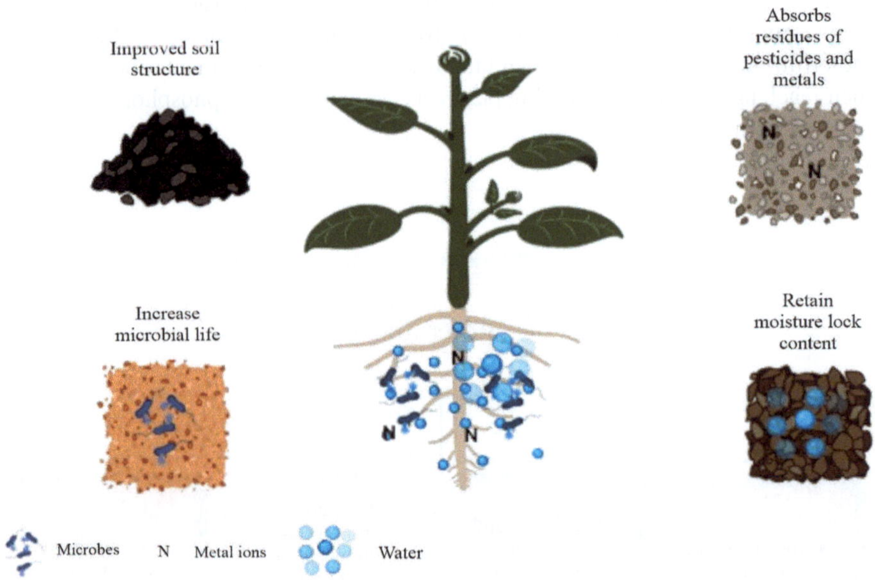

Fig. 8.6 Benefits of biofertilizers

8.6.2 Properties of Biofertilizers

Nutrient Content (e.g., N, P, K)

The nutrient content of biofertilizers varies depending on the type of biofertilizer and the specific microorganisms it contains. Different biofertilizers offer various nutrients to plants, with some specializing in N, P, and K fixation and solubilization. N_2-fixing bacteria, such as species of *Rhizobium, Azotobacter,* and *Azospirillum* are the main components of N_2-fixing biofertilizers. They convert atmospheric N_2 into plant-available forms such as ammonia (NH_3) and nitrate (NO_3^-). P-solubilizing biofertilizers contain microorganisms, mainly P-solubilizing bacteria and fungi, such as species of *Bacillus* and *Pseudomonas*. They solubilize insoluble forms of P present in the soil converting them into plant-available forms like soluble phosphate ($H_2PO_4^-$and HPO_4^{2-}) (Swain et al., 2012). Potassium-mobilizing biofertilizers contain microorganisms that facilitate the mobilization of potassium in the soil making it more available to plants. It is important to note that biofertilizers may not contain all three major nutrients (nitrogen, phosphorus, and potassium) simultaneously. Different formulations may focus on one specific nutrient or a combination of nutrients, depending on the crop's requirements and the soil's nutrient status.

Organic Matter Content

The organic matter content of biofertilizers refers to the presence of organic materials typically derived from the growth media or carrier materials used during their production. Organic matter in biofertilizers can have several benefits for soil health and plant growth. The organic matter in biofertilizers helps to improve soil structure. It enhances the ability of soil to retain moisture, increases aeration, and reduces soil compaction, promoting better root growth and nutrient uptake. Organic matter serves as a slow-release source of nutrients for plants. As the organic matter decomposes, it releases essential nutrients such as N, P, and micronutrients into the soil, providing a steady supply of nutrients to plants over time (Zainudin et al., 2022). Organic matter supports a diverse and active microbial population in the soil. Incorporating organic matter into the soil through biofertilizers helps to sequester carbon thereby reducing the release of greenhouse gases into the atmosphere and contributing to climate change mitigation efforts.

pH and Salinity

It is essential to consider the pH of the biofertilizer and the pH of the soil when applying biofertilizers to ensure compatibility and maximum benefits for the plants. The pH of a biofertilizer can impact several factors. The pH requirements of different crops vary and using a biofertilizer with a pH closer to the optimal range for the specific crop can enhance its effectiveness in promoting plant growth and nutrient uptake. The salinity of a biofertilizer refers to the concentration of salts present in the product. High salinity levels in biofertilizers can have adverse effects on plants and soil, affecting water availability and nutrient uptake. High salinity levels in the biofertilizer can create osmotic stress on plants, reducing water uptake and negatively affecting their growth. If the biofertilizer contains a high concentration of salts and is applied to soils already prone to salinization, it can exacerbate soil salinity issues (Zangaroo et al., 2009). Extremely high salinity levels can harm the viability of beneficial microorganisms in the biofertilizer, reducing their effectiveness in promoting plant growth. To ensure the success of biofertilizer application, it is crucial to use products with appropriate salinity levels and consider the salt content of both the biofertilizer and the target soil. Before applying biofertilizers to crops or soils, it is advisable to conduct soil testing and consult with agricultural experts to determine the best application rates and timings to maximize the benefits of the biofertilizer while minimizing any potential adverse effects.

Microbial Content

The microbial content of biofertilizers refers to the presence and concentration of beneficial microorganisms that provide specific functions to enhance plant growth and soil fertility. These microorganisms play a crucial role in biofertilizer's effectiveness and can be broadly categorized into different groups based on their functions. The main groups include N_2-fixing bacteria, P-solubilizing microorganisms, K-mobilizing microorganisms, PGPR, and mycorrhizal fungi.

8.7 Soil Amendments: Enriching Soil Health

Soil amendments play a critical role in enhancing soil health and fertility, promoting optimal plant growth, and supporting sustainable agricultural practices. It is important to note that the selection of soil amendments should be based on soil testing and analysis to identify specific soil deficiencies and pH imbalances. Soil amendments, especially organic matter like compost and manure, help improve soil structure by increasing soil aggregation. This results in better soil porosity, aeration, and water infiltration allowing plant roots to access water, nutrients, and oxygen more effectively. Organic amendments such as compost and green manure provide a wide range of essential nutrients to plants. Inorganic amendments like rock phosphate and bone meal release nutrients slowly ensuring a continuous supply for plant uptake. Amendments can also help retain nutrients in the root zone, reducing nutrient leaching and loss. Lime is a common soil amendment used to adjust soil pH levels. In acidic soils, lime raises the pH, making essential nutrients more available to plants (Ray, 1985). An abundant and dynamic microbial community is promoted in the soil by organic amendments. These microorganisms contribute to nutrient cycling, decomposition of organic matter, and disease suppression enhancing overall soil health.

Organic amendments improve water retention in the soil, reduce water evaporation, improve drought resistance in plants, and contribute to carbon sequestration in the soil playing a role in lowering CO_2 emissions to mitigate climate change levels in the atmosphere.

8.7.1 Types of Soil Amendments and Their Functions

There are various types of soil amendments, both organic and inorganic, each serving specific functions to improve soil health and support plant growth (Fig. 8.7).

Compost, manure, and cover crops are examples of organic matter that enhances soil structure, water retention, and nutrient-holding capacity (Behera & Ray, 2021). It enhances microbial activity, promotes nutrient cycling, and supports overall soil health. Inorganic amendments provide P, an essential nutrient for plant growth, and help to replenish P-deficient soils and support root development. Nitrogen sources supply plants with nitrogen, a vital element for leaf and stem development, promoting overall plant growth and improving crop yield. Potassium sources provide K, which plays a key role in root development, enzyme activation, and overall plant

Fig. 8.7 Types of soil amendments

health. Vermicompost is a nutrient-rich organic amendment produced by earthworms. It improves soil structure, enhances nutrient availability, and stimulates plant growth.

8.7.2 Enhancing Soil Structure and Water Holding Capacity

Soil amendments can enhance soil structure and water-holding capacity through various mechanisms and contributions. Organic matter such as compost, well-rotted manure, or organic mulches are primary soil amendments for enhancing soil structure and water-holding capacity. Some soil amendments like organic matter and clay minerals can increase the CEC (cation exchange capacity) of soil. CEC refers to the soil's ability to retain and exchange essential plant nutrients, including calcium, magnesium, potassium, and ammonium ions. Soil amendments like organic matter and hydrogel crystals have a high water-holding capacity. They can absorb and retain water, releasing it slowly to plant roots as needed. This helps prevent water runoff and allows plants to access moisture during dry periods. Amendments like gypsum can help to reduce soil compaction, particularly in clay soils. Gypsum improves soil structure by loosening clay particles, making it easier for roots to penetrate and water to infiltrate. Perlite and vermiculite are lightweight and provide aeration to the soil. They create air pockets that improve root respiration and prevent soil from becoming too compact. They are rich in organic matter such as compost and provide a habitat for beneficial soil microorganisms. These microorganisms play a crucial role in maintaining soil structure by breaking down organic materials and producing substances that bind soil particles.

8.7.3 Promoting Nutrient Availability and Microbial Activity

Soil amendments can promote nutrient availability and microbial activity by creating a more hospitable environment for beneficial soil organisms and by altering the chemical properties of the soil. Soil amendments like lime (to raise pH) or sulfur (to lower pH) are used to adjust soil pH levels. Proper pH levels are crucial for nutrient availability because they influence chemical reactions that release or bind nutrients in the soil. Certain soil amendments, such as bone meal (rich in P) or wood ash (rich in K), can be used to directly supply specific nutrients that may be deficient in the soil. They contain beneficial microorganisms or microbial inoculants. These can include mycorrhizal fungi, rhizobia for legume crops, and other soil biota. When introduced into the soil, these microorganisms form symbiotic relationships with plant roots enhancing nutrient uptake and overall plant health. By improving soil structure and adding organic matter, soil amendments encourage a diverse microbial community. Different types of microbes play specific roles in nutrient cycling, such as N_2-fixing bacteria, decomposers, and nutrient-mineralizing microorganisms.

Beneficial microbes in the amendments outcompete and inhibit the growth of disease-causing organisms. In some cases, soil amendments like gypsum can help mitigate soil toxicities such as excess soluble salts by improving soil structure and leaching out harmful ions.

8.8 Applications of Biofertilizers and Soil Amendments Derived from Agricultural Waste

Biofertilizers and soil amendments derived from agricultural waste offer sustainable and eco-friendly solutions for improving soil fertility and crop productivity. Biofertilizers containing nitrogen-fixing bacteria such as *Rhizobium* or *Bradyrhizobium* species are applied to legume crops like soybeans, peanuts, and beans (Silva et al., 2017). These bacteria form symbiotic relationships with legume roots, converting atmospheric nitrogen into a plant-usable form, thereby reducing the need for synthetic nitrogen fertilizers. Certain biofertilizers contain phosphate-solubilizing bacteria, such as *Bacillus* and *Pseudomonas* species (Kiener & Leisinger, 1983; Ray & Swain, 2011). These bacteria release P from insoluble compounds in the soil, making it more accessible to plants (Swain et al., 2012). This reduces the reliance on chemical phosphorus fertilizers.

Mycorrhizal fungi are beneficial soil organisms that form symbiotic relationships with plant roots. They improve nutrient and water uptake for a wide range of crops. Mycorrhizal inoculants derived from agricultural waste materials can be used to establish or enhance mycorrhizal associations in the soil. Agricultural waste materials, including crop residues, straw, and plant trimmings, can be composted to create nutrient-rich organic matter (Kabaloev et al., 2018). This compost can serve as an excellent soil amendment, improving soil structure, enhancing water retention, and supplying a wide range of essential nutrients to plants. Rice bran and husk, byproducts of rice milling, can be used as organic soil amendments to enhance soil fertility due to their nutrient content, including organic matter, P and K.

8.8.1 Improving Soil Fertility and Structure

Biofertilizers, which include beneficial microorganisms like N_2-fixing bacteria and mycorrhizal fungi, play a crucial role in boosting soil fertility. These microorganisms establish symbiotic relationships with plants, facilitating nutrient uptake, particularly N and P, while also enhancing the organic matter content in the soil. Organic matter-rich amendments, such as compost, crop residues, and manure, bolster soil structure by increasing its water-holding capacity and nutrient-holding capacity. These materials act as a source of slow-release nutrients, gradually feeding plants and enhancing microbial activity in the soil. The combined application of biofertilizers and soil amendments is a holistic approach to soil fertility and structure improvement

(Bamdad et al., 2022). Biofertilizers provide targeted nutrient enhancements, while organic-rich soil amendments offer sustained nutrient release and improved water retention. These practices, in turn, increase crop yields and quality, reduce the need for synthetic inputs, and support long-term soil health (Lim & Matu, 2015). Moreover, the use of agricultural waste as a source of biofertilizers and amendments contributes to waste reduction, converting a disposal problem into a valuable resource for sustainable agriculture.

8.8.2 Reducing Soil Erosion and Runoff

Biofertilizers and soil amendments derived from agricultural waste play a crucial role in reducing soil erosion and runoff. Soil amendments like organic matter from compost and biochar enhance soil structure by increasing its ability to hold water and resist erosion (Doostikhah et al., 2020). Well-structured soils have better aggregation, reducing the risk of soil particles being washed away by rainwater or carried away by wind. They improve soil porosity and water-holding capacity, allowing rainwater to infiltrate more effectively into the soil. This reduces surface runoff, as water is absorbed by the soil rather than flowing over its surface. Biofertilizers and soil amendments, such as mycorrhizal fungi and beneficial bacteria, promote root growth and development. Stronger and deeper root systems help anchor the soil in place, preventing erosion. Leaving crop residues on the field surface or using them as mulch helps protect the soil from the impact of rainfall, reducing erosion (Bomfim et al., 2021). These residues create a physical barrier that shields the soil from raindrop impact and slows down water flow.

Cover crops, which can be considered a form of soil amendment, are planted specifically to cover the soil surface during periods when the main crop is not growing. They help reduce erosion by protecting the soil from wind and water erosion and improving soil structure. The use of biofertilizers can reduce the need for chemical fertilizers, which can contribute to soil acidification and degradation. Healthier, more balanced soils are less susceptible to erosion (Hamidpour et al., 2017). Incorporating agricultural waste and organic matter into the soil can reduce the need for intensive tillage, which can disrupt soil structure and increase erosion risk. Overall, the use of biofertilizers and soil amendments encourages sustainable land management practices that prioritize soil conservation. This, in turn, reduces the likelihood of soil erosion and runoff events.

8.8.3 Reducing the Need for Synthetic Fertilizers and Pesticides

Biofertilizers and soil amendments derived from agricultural waste play a crucial role in reducing the need for synthetic fertilizers and pesticides. These sustainable agricultural practices offer several environmental and economic benefits.

Biofertilizers containing N_2-fixing bacteria (e.g., *Rhizobium)* and mycorrhizal fungi enhance nutrient availability in the soil (Glisic et al., 2009). These microorganisms help in fixing atmospheric nitrogen, making it accessible to plants, which reduces the reliance on synthetic nitrogen fertilizers. This increases soil organic matter content, improves soil structure, and enhances water retention, nutrient-holding capacity, and microbial activity (Gujer & Zehnder, 1983). Healthy soils require fewer synthetic inputs and are more resilient to pests and diseases. Some biofertilizers and soil amendments contain beneficial microorganisms (e.g., *Bacillus thuringiensis, B. subtilis*) that can suppress harmful pests and diseases (Swain & Ray, 2009; Nedunchezhiyan et al., 2011; Ray & Swain, 2011), reduce the need for chemical pesticides, and minimize environmental contamination. Synthetic fertilizers and pesticides can have adverse effects on the environment, including water pollution, soil degradation, and harm to non-target organisms. Using biofertilizers and organic amendments minimizes these negative impacts, contributing to sustainable agriculture, and switching to biofertilizers and soil amendments derived from agricultural waste can reduce the cost of purchasing synthetic inputs. They contribute to the long-term sustainability of agriculture by improving soil health and reducing dependence on finite resources like synthetic fertilizers. This is especially important in the face of global challenges like climate change and resource depletion. Organic farming practices often require the use of natural fertilizers and soil amendments instead of synthetic chemicals (Arsenov et al., 2001).

8.8.4 Reducing GHG Emissions

Biofertilizers and soil amendments derived from agricultural waste can play a significant role in reducing GHG emissions in agriculture. The production and application of synthetic fertilizers involve energy-intensive processes, including the extraction and processing of raw materials (Bao et al., 2021). Biofertilizers and soil amendments derived from agricultural waste often have lower energy requirements for production and transportation, reducing the carbon footprint associated with their use. Synthetic nitrogen fertilizers are a major source of nitrous oxide (N_2O) emissions, a potent greenhouse gas. Biofertilizers, such as N_2-fixing bacteria and mycorrhizal fungi, enhance nitrogen use efficiency by making atmospheric nitrogen available to plants. This reduces the need for synthetic nitrogen fertilizers and consequently decreases nitrous oxide emissions. Some biofertilizers and soil amendments contain microorganisms that can reduce methane emissions from paddy rice fields. Biofertilizers and organic soil amendments are often less prone to leaching, which can help reduce the movement of nutrients into water bodies and minimize the associated emissions of greenhouse gases. Incorporating agricultural waste-based soil amendments into crop rotations can support more sustainable and diversified agricultural systems (Barzee et al., 2019). This can lead to reduced greenhouse gas emissions by optimizing land use and reducing the need for synthetic inputs.

8.8.5 Enhancing Food Security and Rural Livelihoods

The use of biofertilizers and soil amendments derived from agricultural waste can have a positive impact on enhancing food security and rural livelihoods in several ways. They provide essential nutrients to crops, improving soil fertility and overall crop productivity (Chauhan et al., 2023a). This increased yield can contribute to higher food production, addressing food security challenges by providing more food for local communities. Using biofertilizers and agricultural waste-derived soil amendments can reduce the cost of purchasing synthetic fertilizers and soil conditioners. This can lower production costs for farmers, making agriculture more financially viable, and potentially leading to increased income for rural communities. They promote sustainable farming practices that focus on soil health and long-term productivity. These practices help maintain soil fertility over time, reducing the risk of soil degradation and the need for costly inputs in the future. The adoption of biofertilizers and organic soil amendments can reduce the environmental impact of agriculture (Chauhan et al., 2023b). Fewer synthetic chemicals mean less pollution of water bodies, reduced soil erosion, and lower greenhouse gas emissions, contributing to a healthier environment for rural communities. Some agricultural waste-derived products, such as compost, can be produced locally and sold to neighboring farms, providing an additional income source for rural communities. This diversification of income can enhance the economic stability of rural households. They are often well-suited for small-scale farmers who may not have access to extensive resources or capital (Chauhan et al., 2021). These sustainable practices enable smallholders to improve their yields and income, contributing to food security at the local level.

8.9 Conclusion

Agricultural waste has significant potential for promoting sustainable agriculture. Sustainable agriculture aims to meet the current and future demands for food, while also ensuring the long-term health of ecosystems, minimizing environmental impact, and improving the livelihoods of farmers. To harness the potential of agricultural waste for sustainable agriculture, it is crucial to develop policies and practices that encourage responsible waste management, promote the use of waste-derived products, and provide education and support to farmers in adopting sustainable practices. By doing so, agricultural waste can be a valuable resource for achieving sustainable and environmentally friendly farming. Research and implementation of agriculture waste as biofertilizer and soil amendments hold great promise for sustainable agriculture.

Acknowledgments The financial support from the Department of Biotechnology, Ministry of Science and Technology, Government of India to Himachal Pradesh University, Shimla is thankfully acknowledged. Council of Scientific and Industrial Research (CSIR) is thankfully acknowledged for providing financial assistance to Ms. Manpreet Kaur in the form of SRF (File No. 09/237 (0170)/2018-EMR-1).

References

Ahuja, I., Dauksas, E., Remme, J. F., Richardsen, R., & Løes, A. K. (2020). Fish and fish waste-based fertilizers in organic farming with status in Norway: A review. *Waste Management, 115*, 95–112.

Anon. (1993). Forging a link between composting and sustainable agriculture. *Compost Science and Utilization, 1*, 48–51.

Arora, J., Ramawat, K. G., & Mérillon, J. M. (2023). Disposal of agricultural waste and its effects on the environment, production of useful metabolites and energy: Potential and challenges. In K. Ramawat, J. M. Mérillon, & J. Arora (Eds.), *Agricultural waste: Environmental impact, useful metabolites and energy production. Sustainable development and biodiversity* (Vol. 31, pp. 3–20). Springer. https://doi.org/10.1007/978-981-19-8774-8_1

Arsenov, D. V., Krasnitsky, V. M., & Kuzmichev, A. V. (2001). New technologies for the distillery industry and feed production. *Production of Alcohol and Alcoholic Beverages, 4*, 24–25.

Bamdad, H., Papari, S., Lazarovits, G., & Berruti, F. (2022). Soil amendments for sustainable agriculture: Microbial organic fertilizers. *Soil Use and Management, 38*, 94–120.

Bao, J., Zhuo, C., Zhang, D., Li, Y., Hu, F., Li, H., & He, H. (2021). Potential applicability of a *cyanobacterium* as a biofertilizer and biopesticide in rice fields. *Plant and Soil, 463*, 97–112.

Barzee, T. J., Edalati, A., El-Mashad, H., Wang, D., Scow, K., & Zhang, R. (2019). Digestate biofertilizers support similar or higher tomato yields and quality than mineral fertilizers in a subsurface drip fertigation system. *Frontiers in Sustainable Food Systems, 3*, 58.

Behera, S. S., & Ray, R. C. (2021). Bioprospecting of cow dung microflora for sustainable agricultural, biotechnological, and environmental applications. *Current Research in Microbial. Science, 2*, 1000.

Bomfim, C. A., Coelho, L. G. F., do Vale, H. M. M., de Carvalho Mendes, I., Megías, M., Ollero, F. J., & dos Reis Junior, F. B. (2021). Brief history of biofertilizers in Brazil: from conventional approaches to new biotechnological solutions. *Brazilian Journal of Microbiology, 52*, 2215–2232.

Bujang, K. B., & Lopez-Real, J. M. (1993). Composting for the treatment of cattle wastes. *Compost Science and Utilization, 1*, 38–40.

Chauhan, V., Dhiman, V., & Kanwar, S. S. (2021). Combination of classical and statistical approaches to enhance the fermentation conditions and increase the yield of Lipopeptide (s) by Pseudomonas sp. OXDC12: Its partial purification and determining antifungal property. *Turkish Journal of Biology, 45*, 695–710.

Chauhan, V., Dhiman, V. K., Mahajan, G., Pandey, A., & Kanwar, S. S. (2023a). Synthesis and characterization of silver nanoparticles developed using a novel lipopeptide (s) biosurfactant and evaluating its antimicrobial and cytotoxic efficacy. *Process Biochemistry, 124*, 51–62.

Chauhan, V., Mazumdar, S., Pandey, A., & Kanwar, S. S. (2023b). *Pseudomonas* lipopeptide: An excellent biomedical agent. *MedComm–Biomaterials and Applications, 2*, 27. https://doi.org/10.1002/mba2.27

Donner, M., Verniquet, A., Broeze, J., Kayser, K., & De Vries, H. (2021). Critical success and risk factors for circular business models valorizing agricultural waste and by-products. *Resources, Conservation and Recycling, 165*, 105–236.

Doostikhah, N., Panahpour, E., Nadian, H., & Gholami, A. J. P. B. (2020). Tomato *(Lycopersiconesculentum L.)* nutrient and lead uptake affected by zeolite and DTPA in a lead-polluted soil. *Plant Biology, 22*, 317–322.
Georgacakis, D., Tsavdaris, A., Bakouli, J., & Symeonidis, S. (1996). Composting solid swine manure and lignite mixtures with selected plant residues. *Bioresource Technology, 56*, 195–200.
Glisic, I. P., Milosevic, T. M., Glisic, I. S., & Milosevic, N. T. (2009). The effect of natural zeolites and organic fertilizers on the characteristics of degraded soils and yield of crops grown in western Serbia. *Land Degradation and Development, 20*, 33–40.
Golueke, C. G., & Diaz, L. F. (1996). Historical review of composting and its role in municipal waste management. In *The science of composting* (pp. 3–14). Springer.
Gujer, W., & Zehnder, A. J. (1983). Conversion processes in anaerobic digestion. *Water Science and Technology, 15*, 127–167.
Hamidpour, M., Akbari, L., & Shirani, H. (2017). Effects of co-application of zeolites and vermicompost on speciation and phytoavailability of cadmium, lead and zinc in a contaminated soil. *Communications in Soil Science and Plant Analysis, 48*, 262–273.
Imran. (2018). Physiological and morphological traits of agronomic crops influenced by climate change. *Modern Concepts and Developments in Agronomy, 1*, 1–4.
Imran, A., & Al-Tawaha, A. R. (2020). The productivity of subsequent wheat enhanced with residual carbon sources and phosphorus under improved irrigation system. *Communications in Soil Science and Plant Analysis, 51*, 1–9.
Jena, S., & Singh, R. (2022). Agricultural crop waste materials–A potential reservoir of molecules. *Environmental Research, 206*, 112–284.
Joshi, A., & Arora, J. (2023a). Recent trends and future prospects of nanotechnology for agro-waste valorization into biofuels. In *Agricultural waste: Environmental impact, useful metabolites and energy production* (pp. 559–582). Springer Nature.
Joshi, A., & Arora, J. (2023b). Nanotechnology: Applications in biofuel production. In P. Moodley, G. B. Kana, & R. C. Ray (Eds.), *Advances in lignocellulosic biofuel production systems* (pp. 305–328).
Kabaloev, T. X., Dzanagov, S. C., Bekuzarova, S. A., Subbotin, I. M., & Esenov, I. K. (2018). Use of distillery waste as a fertilizer. *Journal of Pharmaceutical Sciences and Research, 10*, 2314–2316.
Karagiannidis, A., & Perkoulidis, G. (2009). A multi-criteria ranking of different technologies for the anaerobic digestion for energy recovery of the organic fraction of municipal solid wastes. *Bioresource Technology, 100*, 2355–2360.
Kiener, A., & Leisinger, T. (1983). Oxygen sensitivity of methanogenic bacteria. *Systematic and Applied Microbiology, 4*, 305–312.
Kumar, V., Mitra, D., Rani, A., Suyal, D. C., Gautam, B. P. S., Jain, L., & Soni, R. (2022). Bio-inoculants for biodegradation and bioconversion of agro-waste: Status and prospects. In *Bioremediation of environmental pollutants: Emerging trends and strategies* (pp. 351–367).
Lau, A. K., Liao, P. H., & Lo, K. V. (1993). Evaluation of swine waste composting in vertical reactors. *Journal of Environ Science and Health, 28*, 761–777.
Liao, P. H., Vizcarra, A. T., Chen, A., & Lo, K. V. (1993). Composting separated solid swine manure. *Journal of Environmental Science and Health, 28*, 1889–1901.
Lim, S. F., & Matu, S. U. (2015). Utilization of agro-wastes to produce biofertilizer. *International Journal of Energy and Environmental Engineering, 6*, 31–35.
Lim, S. L., Lee, L. H., & Wu, T. Y. (2016). Sustainability of using composting and vermicomposting technologies for organic solid waste biotransformation: Recent overview, greenhouse gases emissions and economic analysis. *Journal of Cleaner Production, 111*, 262–278.
Lin, L., Yan, R., Liu, Y., & Jiang, W. (2010). In-depth investigation of enzymatic hydrolysis of biomass wastes based on three major components: Cellulose, hemicellulose and lignin. *Bioresource Technology, 101*, 8217–8223.

Mahajan, G., Kaur, M., & Gupta, R. (2022). Green functionalized nanomaterials: Fundamentals and future opportunities. In *Green functionalized nanomaterials for environmental applications* (pp. 21–41). Elsevier.

Mahajan, G., Chauhan, V., Sharma, V., & Gupta, R. (2023a). Combining in vitro and in silico studies to unravel the antifungal potential of chitinase from a marine bacterial isolate *Cellulosimicrobium cellulans* RS7. *Process Biochemistry, 134*, 88–100.

Mahajan, G., Sharma, V., & Gupta, R. (2023b). Chitinase: A potent biocatalyst and its diverse applications. *Biocatalysis and Biotransformation, 1*, 25. https://doi.org/10.1080/10242422.2023.2218524

Mahajan, G., Sharma, V., Ray, R. C., & Gupta, R. (2023c). Lactic acid bacteria as potential sources of enzymes: From genes to industrial applications. In *Lactic acid bacteria as cell factories* (pp. 183–197). Woodhead Publishing.

Malandrino, M., Abollino, O., Buoso, S., Giacomino, A., La Gioia, C., & Mentasti, E. (2011). Accumulation of heavy metals from contaminated soil to plants and evaluation of soil remediation by vermiculite. *Chemosphere, 82*, 169–178.

Maynard, A. A. (1994). Seventy years of research on waste composting and utilization at the Connecticut Agricultural Experimental Station. *Compost Science and Utilization, 2*, 13–21.

Mazid, M., & Khan, T. A. (2015). Future of bio-fertilizers in Indian agriculture: An overview. *International Journal of Agricultural and Food Research, 3*. https://doi.org/10.24102/IJAFR.V3I3.132

McWeeney, L. J. (1994). *Archaeological settlement patterns and vegetation dynamics in southern New England in the Late Quaternary*. Yale University.

Meegoda, J. N., Li, B., Patel, K., & Wang, L. B. (2018). A review of the processes, parameters, and optimization of anaerobic digestion. *International Journal of Environmental Research and Public Health, 15*, 22–24.

Muthini, M., Awino, R., Kirui, K. C., Koech, K., Jalloh, A. A., & Njeru, E. M. (2020). Optimizing Rhizobium-legume symbiosis in smallholder agroecosystems. *Sustainable Agriculture Reviews: Legume Agriculture and Biotechnology, 1*, 159–177.

Nedunchezhiyan, M., Jata, S. K., Ray, R. C., & Misra, R. S. (2011). Management of mealybug (*Rhizoecus amorphophalli*) in elephant foot yam (*Amorphophallus paeoniifolius*). *Experimental Agriculture, 47*, 717–728.

Pandey, V., Papnai, N., Singh, D., & Prasad, R. (2021). Role of bio-fertilizer for improving soil health. *Vigyan Varta an International E-Magazine for Science Enthusiasts, 2*(9).

Phat, T. D., Phuong, T. V., & Diep, C. N. (2019). Effect of compost, NPK and plant promoting rhizobacteria (PGPR) on growth and yield of three vegetables cultivated on arenosols. *International Journal of Environmental and Agriculture Research, 5*, 27–34.

Ramawat, K. G., Mérillon, J. M., & Arora, J. (Eds.). (2023). *Agricultural waste: Environmental impact, useful metabolites and energy production*. Springer.

Ray, R. C. (1985). Effect of liming on dehydrogenase activity of flooded and non-flooded acid sulfate soils. *Fertilizer research, 6*, 193–197.

Ray, R. C. (2022). *Fruits and vegetable wastes: Valorization into bioproducts and platform chemicals* (p. 448). Springer.

Ray, R. C. (2024). *Roots, tubers and bulb crop wastes; management by biorefinery approaches* (p. 378). Springer.

Ray, R. C., & Swain, M. R. (2011). *Agricultural and biotechnological applications of Bacillus subtilis* (p. 205). LAMBERT Academic Publishing GmbH and Co.

Reicosky, D. C., Evans, S. D., Cambardella, C. A., Allmaras, R. R., Wilts, A. R., & Huggins, D. R. (2002). Continuous corn with moldboard tillage: Residue and fertility effects on soil carbon. *Journal of Soil and Water Conservation, 57*, 277–284.

Silva, L. R., Bento, C., Gonçalves, A. C., Flores-Félix, J. D., Ramírez-Bahena, M. H., Peix, A., & Velázquez, E. (2017). Legume bioactive compounds: Influence of rhizobial inoculation. *AIMS Microbiology, 3*, 267–278.

Sims, J. T., & Maguire, R. O. (2005). Manure management. In *Encyclopedia of soils in the environment* (pp. 402–410).

Singh, R. L., & Singh, P. K. (2017). Global environmental problems. In *Principles and applications of environmental biotechnology for a sustainable future* (pp. 13–41).

Singh, M., Dotaniya, M., Mishra, A., Dotaniya, C., Regar, K., & Lata, M. (2016). Role of biofertilizers in conservation agriculture. In *Conservation agriculture: an approach to combat climate change in Indian Himalaya* (pp. 113–134).

Singh, P., Behera, H. T., Mishra, S., & Ray, L. (2022). Biofertilization of biogas digestates: An insight on nutrient management, soil microbial diversity and greenhouse gas emission. In *New and future developments in microbial biotechnology and bioengineering* (pp. 199–215). Elsevier.

Stentiford, E. I. (1996). Composting control: principles and practice. In *The science of composting* (pp. 49–59).

Swain, M. R., & Ray, R. C. (2009). Bio-control and other beneficial activities of *Bacillus subtilis* isolated from cow dung microflora. *Microbiological Research, 164*, 121–130.

Swain, M. R., Naskar, S. K., & Ray, R. C. (2007). Indole-3 –acetic acid production and effect on sprouting of yam (*Dioscorea rotundata* L.) ministers by *Bacillus subtilis* isolated from culturable cow dung microflora. *Polish Journal of Microbiology, 56*(2), 103–110.

Swain, M. R., Ray, R. C., & Nautiyal, C. S. (2008). Biocontrol efficacy of *Bacillus subtilis* strains isolated from cow dung against post-harvest yam (*Dioscorea rotundata* L.) pathogens. *Current Microbiology, 55*, 407–411.

Swain, M. R., Laxminarayana, K., & Ray, R. C. (2012). Phosphorus solubilization by a thermotolerant *Bacillus subtilis* strain isolated from cow dung micro-flora. *Agricultural Research, 1*(3), 273–279.

Tejada, M., & Benítez, C. (2017). Flazasulfuron behavior in a soil amended with different organic wastes. *Applied Soil Ecology, 117*, 81–87.

Toledo, M., Siles, J. A., Gutiérrez, M. C., & Martín, M. A. (2018). Monitoring of the composting process of different agroindustrial waste: Influence of the operational variables on the odorous impact. *Waste Management, 76*, 266–274.

Truckell, I. G., Shah, S. H., Baillie, I. C., Hallett, S. H., & Sakrabani, R. (2019). Soil and transport factors in potential distribution systems for biofertilizers derived from palm oil mill residues in Malaysia. *Computers and Electronics in Agriculture, 166*, 105005.

Verma, S. (2002). *Anaerobic digestion of biodegradable organics in municipal solid wastes.* Columbia University.

Verma, R., Chourasia, S. K., & Jha, M. N. (2011). Population dynamics and identification of efficient strains of Azospirillum in maize ecosystems of Bihar, India. *3 Biotech, 1*, 247–253.

Wang, J., Liu, Z., Xia, J., & Chen, Y. (2019). Effect of microbial inoculation on physicochemical properties and bacterial community structure of citrus peel composting. *Bioresource Technology, 291*, 121843.

Yao, X., Guo, H., Zhang, K., Zhao, M., Ruan, J., & Chen, J. (2023). Trichoderma and its role in biological control of plant fungal and nematode disease. *Frontiers in Microbiology, 14*, 1160551.

Zainudin, M. H. M., Zulkarnain, A., Azmi, A. S., Muniandy, S., Sakai, K., Shirai, Y., & Hassan, M. A. (2022). Enhancement of agro-industrial waste composting process *via* the microbial inoculation: A brief review. *Agronomy, 12*, 198.

Zangaro, W., Nogueira, M. A., & Andrade, G. (2009). Arbuscular mycorrhizal fungi used as biofertilizers in revegetation programs. In *Advances in fungal biotechnology* (pp. 351–378).

Zhang, J., Jiang, X., Miao, Q., Yu, B., Xu, L., & Cui, Z. (2019). Combining mineral amendments improves wheat yield and soil properties in a coastal saline area. *Agronomy, 9*, 48.

Chapter 9
Conversion of Agricultural Residues into High-Value Animal Feed

Thameridus B. Marak, Ajita Tiwari, and Aniruddha Roy

Abbreviations

MT	Million tons
SmF	Submerged fermentation
SSF	Solid-state fermentation
TBtu	Trillion British thermal units
TVS	Total volatile solid

9.1 Introduction

The agricultural sector in the last few decades has taken a sharp rise. Due to the continuous growth in human population, an increase in agricultural production has become a necessity of an hour. Other factors such as advancements in technology toward green revolution and expansion of soil for agricultural production are responsible for higher agricultural production (FAO, 2017). Each step of agricultural production, processing, and consumption generates quantities of agricultural wastes, depending on the type of agricultural produce or product, processing techniques, and purpose of use. The agricultural sector is one of the main sectors generating the largest quantities of waste, which, if not treated or managed appropriately, may result in a threat to global health and food security (Bracco et al., 2018). Agricultural wastes are the residues from the production and processing of agricultural and livestock products. They are also known as the non-product outputs of production

T. B. Marak · A. Tiwari (✉)
Department of Agricultural Engineering, Assam University, Silchar, Assam, India

A. Roy
ICAR Research Complex for NEH Region, Umiam, Meghalaya, India

J. Arora et al. (eds.), *Transforming Agriculture Residues for Sustainable Development*, Waste as a Resource, https://doi.org/10.1007/978-3-031-61133-9_9

and processing of agricultural and livestock products that may contain substances that can be beneficial but whose economic values are less than the cost of their production. Their composition varies with the system and type of agricultural activities and can be in different forms such as solids, liquids, or slurries. Agricultural waste otherwise called agro-waste is comprised of animal waste, food processing waste, crop waste, and hazardous and toxic waste. Agricultural waste also occurs due to post-harvest losses. Approximately 20% of agro-products are damaged due to poor post-harvesting facilities and 10% are eaten by rodents (Lakshmi et al., 2017). In developed countries, agricultural waste generally contributes a notable proportion of the total waste matter. With increased agricultural production, there is an increase in quantities of agricultural crop residues, livestock waste, and agro-industrial by-products. This raises a concern that if developing countries continue to intensify farming systems, there is likely to be a significant increase in agricultural waste globally.

Accelerated agricultural activities and food industries have resulted in the generation of millions of tons of agricultural waste (Ytrestøyl et al., 2015). It is estimated that about 998 million tons of agro-waste or residues per year are being produced and are likely to have a significant increase in the coming years (Agamuthu, 2009). Figure 9.1 depicts the estimation of agricultural waste generated globally based on the main producer countries. In the year 2020, the total agricultural waste generated from different activities of the main producer countries was estimated to be 60 million tons of wheat bran, 150 million tons of soy pulp, 45 million tons of rice bran, and 200 million tons of palm kernel cake (Nath et al., 2023). According to the literature, about 4532 trillion British Thermal units (TBtu) of biomass were generated by the United States alone (Dey et al., 2021). Canada is reported to generate 36.1 MTs of waste per year, which accounts for 10 MTs more per person than the United States. Out of 1.33 billion MTs of waste generated in Canada every year, 1.12

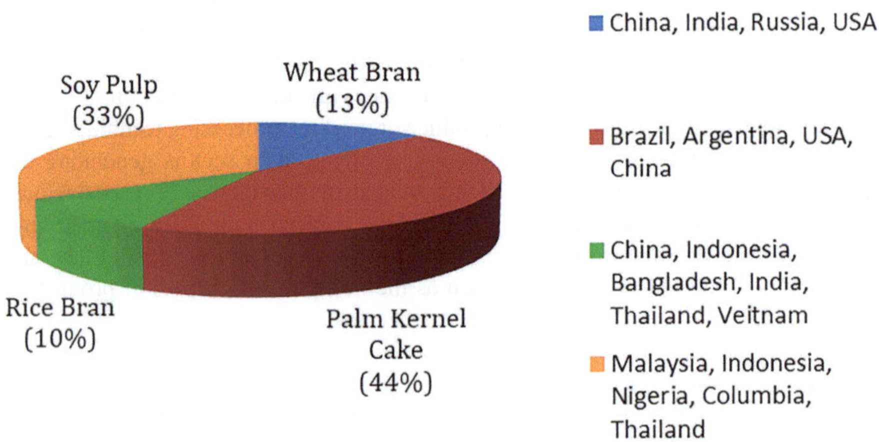

Fig. 9.1 Global agricultural waste estimation among the main producer countries

billion MTs come from agro-industries (Weisz et al., 2020). These wastes impair the economic and environmental sustainability of the ecosystem. A lack of effective disposal of this waste can have an adverse effect on the environment contributing to all forms of pollution. Generally, these wastes are disposed of by burning, dumping, random piling, or landfilling, which results in environmental pollution becoming hazardous to human health and therefore raising a huge concern (Bölük & Mert, 2014). This situation raises the need for developing an integrated and innovative technology for the management of agro-waste and to improve sustainability footprints. Utilization of agro-industry waste through recycling for resource recovery and manufacturing of value-added products such as animal feed caters great potential. Research emphasizing proper management and processing techniques for the conversion of agro-waste into high-value animal feed is the need of the hour. This chapter discusses and summarizes some aspects of animal feed conversion techniques.

9.2 Nutritional Value of Agro-Waste

Agro-waste consists of residuals from agro-operations like processing and manufacturing of fruits, vegetables, roots, tubers, and dairy products (Pattanaik et al., 2019; Ray, 2022; Chakraborty et al., 2024). There are four main types of agricultural waste:

- Livestock manure.
- Agricultural sector manure.
- Fruit and vegetable manure.
- Crop residues (Saini et al., 2015).

These residues offer a rich source of nutrients including carbohydrates, proteins, fibers, vitamins, and minerals that are essential for animal nutrition. Table 9.1 shows the proximate composition and ultimate constituents of several agro-wastes that can be utilized as animal feed. Between 40% and 85% of the total solids in the agro-wastes are made up of carbohydrate compounds. Agro-waste contains good sources of protein ranging from 3.48% to 15.4% and lipids ranging from 0.7% to 5.9%, making it a viable source of nutrients for animal feeds. These residues are also rich in macronutrients such as nitrogen, potassium, and phosphorous (Table 9.2). Agricultural residue contains lignocellulose material ranging from 80% to 85%. Many crop residues are made up of 30–50% cellulose and 20–38% hemicellulose content (Table 9.3). In contrast, the lignin concentration of crops ranges from 7% to 21%. Based on the varied sources, agro-processing wastes comprise 22–44% cellulose, 16–34% hemicelluloses, and 6–25% lignin. Rice bran waste contains just 5% lignin and has been demonstrated to be a viable feedstock (Sun & Cheng, 2002: Ray, 2020).

Table 9.1 Proximate and ultimate composition of agro-residues

	Proximate composition (Wt%)						Ultimate composition (Wt%)					References
Residue	Moisture	Carbohydrate	Protein	Lipid	Ash	Total volatile solid (TVS)	C	H	O	N	S	
Rice straw	4.2–6.0	23.58	14.6–15.4	5.9	8.2–16.0	71.6–92.8	34.0–41.5	4.6–6.7	32.8–41.2	0.2–0.8	4.2–6.0	Kumar et al. (2018)
Oat straw	5.38	–	5.34	1.66	5.9	74.48	48.8	6.0	44.6	0.5	0.08	Nath et al. (2022), Mani et al. (2011)
Corn straw	5.3–7.4	7.9	3.6–8.7	0.7–1.3	4.2–6.3	86.5–96.8	35.2–45.6	5.4–6.3	43.4–45.7	0.3–0.8	0.1–0.3	Morgan et al. (2015)
Barley straw	–	–	3.48	5.34	5.3–9.8	76.2	49.4	6.2	43.6	0.7	0.13	Serrano et al. (2011)
Wheat straw	4.4–8.4	–	3.62	1.91	7.12.8	74.4–92.7	41.7–46.7	5.1–6.3	34.1–51.4	0.4–0.5	0.1–0.3	Katileviciute et al. (2019), Bridgeman et al. (2008)
Rice bran	7.70	50.00	12.00	17.00	7.0–18.8	64.30	38.69	5.40	0.67	0.21	–	Khir et al. (2019), Choi et al. (2011), Ji-Lu (2007)
Wheat bran	–	12.0	16.0	5.0	6.0	–	–	–	–	–	–	Kari et al. (2022)
Coffee husk	–	–	7–8	–	2.4	78.50	47.50	6.40	–	–	43.7	Murugesan et al. (2021), Yin (2011)
Sugarcane bagasse	16.07	–	–	–	3.20–4.34	79–83.66	45.48	5.96	45.21	0.15	–	Nath et al. (2022), Tsai et al. (2006)
Oil palm empty fruit bunch	67	–	–	–	3.02–7.54	71.2–79.7	45.0–48.8	6.40–7.33	47.3–40.2	0.25	0.68–1.06	Nath et al. (2022), Omar et al. (2011)
Orange peel	–	–	–	–	–	–	16.23	17.1	60.28	0.76	2.99	Salem et al. (2019), Joshi et al. (2015)

Table 9.2 Macronutrients of different agro-wastes

Agricultural waste	Nitrogen (%)	Calcium (%)	Potassium (%)	Phosphorus (%)	References
Jackfruit waste	–	0.8	–	0.1	Murugesan et al. (2021)
Coffee husk	1.27	0.51	2.46	0.25	Murugesan et al. (2021)
Rice husk	–	0.56	1.28	0.38	Murugesan et al. (2021), Jayant et al. (2020)
Corn cob	1.38	–	0.46	0.09	Nath et al. (2022)
Peach leaves	0.00	–	2.45	0.14	Nath et al. (2022)
Sugarcane bagasse	0.73	0.71	0.33	0.30	Nath et al. (2022)
Corn leaves	1.30	–	1.48	0.21	Ahmad et al. (2021)
Corn stem	0.48	–	1.23	0.09	Ahmad et al. (2021)
Corn grain	2.16	–	0.43	0.35	Ahmad et al. (2021)
Corn silage	1.35	–	0.90	0.16	Ahmad et al. (2021)

Table 9.3 Lignocellulosic composition of different agro-wastes

Feedstocks	Lignocellulosic composition (Wt%)			References
	Cellulose	Hemicellulose	Lignin	
Rice straw	30.3–52.3	19.8–31.6	7.2–12.8	Kumar et al. (2018)
Oat straw	37.6–44.8	23.3–33.4	12.9–21.8	Mani et al. (2011)
Corn straw	31.3–49.4	21.1–26.2	3.1–8.8	Morgan et al. (2015)
Wheat straw	32.9–44.5	37.8–33.2	8.5–22.3	Wieser et al. (2020), Bridgeman et al. (2008)
Barley straw	29.2–48.6	35.8–29.7	6.7–21.7	Serrano et al. (2011)
Sweet sorghum bagasse	44.6	27.1	20.7	Kim and Day (2011)
Sugarcane bagasse	43.6–45.8	31.3–33.5	18.1–22.9	Tinh et al. (2021), Tsai et al. (2006)
Soya stalks	34.5	24.8	19.8	Hiloidhari et al. (2020)
Coffee husk	24.5–43	7–29.7	9–23.7	Sinha et al. (2021)
Barley straw	33.8	21.9	13.8	Serrano et al. (2011)
Apple pomace	21.22	14.75	18.50	Yahya et al. (2015)
Oil palm empty fruit bunch	23.6–63.0	21.5–32.0	29.2–36.6	Aziz et al. (2020), Yahya et al. (2015)
Olive de-oiled cake	2.0	18.3	50.0	Fazio et al. (2022), Tsonis et al. (1989)
Pineapple peel	18.11	–	1.37	Kyawt et al. (2020)

9.3 Overview of Conversion/Valorization

Agro-waste being used as animal feed should be economical, rich in nutrients, and free of toxins or other substances that may be harmful to animals. Most of the time, the by-products of agro-industries are being used directly without any modification or sometimes after fermentation. Conversion of by-products to animal feed has numerous benefits. It allows the agro-industry to cut off unnecessary costs by reducing waste discharges, reducing waste-management costs, and also preventing environmental pollution (Ajila et al., 2012). The recent global pandemic has led to an increase in commercial animal feed prices (corn and soybean meal) which has resulted in huge feed cost inflation accounting for an average of 60–70% of the total cost of animal production. Agro-industrial waste or residues are major valuable biomass and can contribute as a potential solution to the problems of increasing the cost of animal feeds. Agro-residues often offer comparatively less expensive sources of nutrients to traditional feeds and also support an acceptable performance in animal growth. However, conversion of agro-residues into animal feed is a challenging task concerning various components of animal industries. Improper conversion techniques may result in serious feed deficits leading to low animal productivity, and the inability of the components of the animal industries to meet national targets, especially that of animal proteins. In addition, untreated agro-waste may sometimes have an adverse effect on animal health (Rajeh et al., 2021).

A range of processing processes are used to increase the nutritional content, digestibility, feeding efficiency, removal of toxins, pathogen sanitation, removal of non-edible components, feasibility for long-term storage, transportability, and marketability of agro-waste (Bhatia et al., 2023). Processing is a crucial stage in any such conversion process to improve feed quality, stable feed in the animal diet, and decrease loss during feeding (Peng et al., 2023). Commonly used processing techniques include dehydration and/or drying, pelleting/extrusion, fermentation, and silage production that can be utilized to convert agro-waste into animal feed. These processing technologies can either be combined or used independently.

9.3.1 *Dehydration or Drying*

Drying or dehydration is one of the most commonly used methods of processing. Agro-waste can be converted into animal feed through various drying techniques, such as solar-drying, spray-drying, freeze-drying, and hot air-drying (Ray, 2022). Agro-waste is dried to a less than 20% moisture percentage which results in a longer shelf life and a smaller bulk volume. This makes the management process easier and less expensive to store and transport. They are also simpler to incorporate into diet regimens (Rajeh et al., 2021). Spoiled and small-sized bananas, banana peels, apple pomace (the residue remaining after the juice has been squeezed out of the fruit), and vegetables such as fresh potatoes and different types of waste and all other types of animal feed that can be produced using drying techniques (Choi et al., 2016). Many researchers across the globe have developed several solar dryers for the valorization of animal feed having an evaporation rate ranging from 3.85% to 19.62% (Bulgakov et al., 2020; Rahmani et al., 2022). Spray-drying is one of the simplest ways to extend the shelf life of liquid extracts and improve the organoleptic features of products by converting extracts into stable dry powder (Nath et al., 2023). With spray-drying, liquid feeds (extracts) are atomized into the drying chamber, and then the resulting droplets pass through a hot-air (or sometimes nitrogen) stream to evaporate (Filková & Mujumdar, 2020). The evaporation of water takes place at a faster pace when the droplets themselves are smaller since this increases the surface-to-mass ratio. Because of the speed and intensity with which this procedure is carried out, there is very little heat damage caused to sensitive material. Cutting-edge household garbage dryer involves a novel approach for the separation and dehydration of food and agro-waste separated at the source. This drying method resulted in a large mass reduction, approximately 70% w/w, due to the removal of moisture content throughout the drying process, while the energy requirements remained economically acceptable (Sotiropoulos et al., 2015). Freeze-drying techniques involve a process of sublimation where both freezing, in which almost all of the solvent is transformed into a solid that is frozen, and drying, in which the mixture almost completely gets rid of all of the solvent (frozen or unfrozen), take place (Assegehegn et al., 2019). Many studies have shown that freeze-drying is an effective method for minimizing agro- and food waste (Wheeler et al., 2007). Microwave-assisted drying techniques have great benefits in the processing of high volatile content food wastes due to three main factors: its speed, low energy usage, and good product quality (Karam et al., 2016). The dehydrated biomass contains low moisture, preventing biological decomposition, limiting odor emissions, and so minimizing the frequency with which residential trash must be collected. Additionally, dry biomass is considerably easier to handle than wet waste. Moreover, the waste can be used in environmentally friendly and alternative methods to create high-value goods such as compost, bioethanol, biogas, animal feed, and thermal energy (Yaashikaa et al., 2020).

9.3.2 Pelleting/Extrusion

Pelleted feeds have been defined as agglomerated feeds formed by extruding individual ingredients or mixtures by compacting and forcing through die openings by any mechanical process. The combination of moisture, heat, and pressure on feed ingredients results in a degree of gelatinization preventing the nutrient loss in animal feed and improving its conversion process. Pelleting prevents the segregation of ingredients in a mixing, handling, or feeding process. By feeding a pelleted feed, the animal is more apt to receive a mixed ration than one that has separated through these processes. Pellet feed has increased bulk density which enhances the storage capabilities of most bulk facilities. A better flow and handling characteristic of pellets is one of the least mentioned advantages but probably the most important when it comes to animal handlers. The process of producing feed pellets involves a plastic molding operation of the extrusion technique. Feed ingredients are made up of various compounds such as proteins, acids, sugars, fibers, and minerals. The process involves the application of controlled compression to the conditioned feed ingredients that results in the formation of dense mass and shaped to conform through a die against which they are pressed.

9.3.3 Silage

Since the 1960s, there have been major changes in agriculture, including a greater emphasis on employing science to produce crops, increased mechanization, and larger livestock farms (Stanton et al., 2018). Improved animal nutrition and genetics led to a rise in cattle productivity per head. Most European countries had an increase in silage dry matter (DM) production between 1975 and 2000 (Wilkinson et al., 2018). This was partially caused by both an increase in the size of animal farms and a decline in the output of hay at that time. On the other hand, hay production rose significantly in the United States over the same period (Maclaren et al., 2020). To manage seasonal feed shortages, farmers can use commercial feeds like grains, seasonally excess grass other fodder crops, or a combination of these. Prior to the 1960s, fodder was mostly kept in hay storage, which meant that it was usually harvested at a mature growth stage and subject to weather fluctuations during the long periods between cutting and harvest (Wilkinson et al., 2018). The gathering and preservation of fermented materials, such as moist grass and fermented whole-crop cereals and crop by-products, for use as animal feed are referred to as silage. Production of silage requires understanding the physical, chemical, and biological components that impact the entire conservation process. Silage is an essential component of livestock diets for two reasons: (i) it preserves digestible nutrients for high-producing animals, helping to maintain optimal rumen function and reducing the risk of diseases such as displaced abomasa and rumen acidosis; and (ii) it can be used as supplemental feed during dry spells and winters when pasture growth isn't keeping up with animal needs (Wilkanowska & Kokoszyński, 2015; Shinde & Mahanta, 2020).

9.3.4 Fermentation

Another feed conversion technique is fermentation procedures/ techniques which can be either solid state fermentation (SSF) or submerged fermentation (SmF). The SSF technique is gaining popularity in producing several bulk chemicals and enzymes from agro-waste or residues (Robinson & Nigam, 2003; Behera & Ray, 2016). It is defined as any fermentation process carried out on a solid material in the

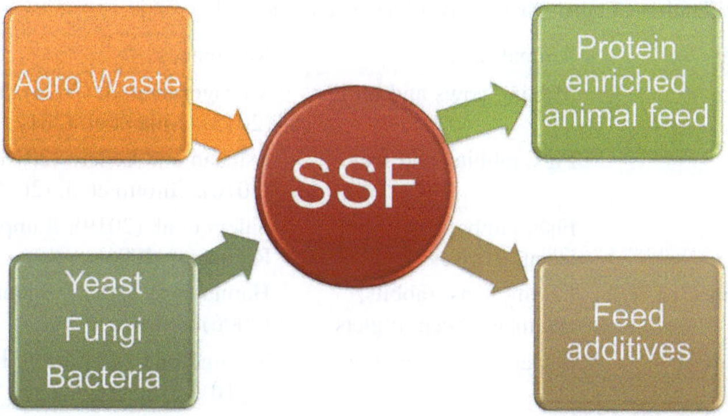

Fig. 9.2 Solid State Fermentation (SSF) in the conversion of agro-waste to animal feed

absence of free-flowing liquid (Pandey, 1992; Behera et al., 2023). SSF technique is known to produce a more stable product, with less energy consumption, in smaller fermenters and smaller volumes of polluting effluents, as compared to the SmF systems (Ajila et al., 2012). A wide range of solid materials are employed as supports in the SSF technique and they are usually classified into two categories, viz., noninert and inert materials (Krishna, 2005). In the first category, a divided and humidified solid (e.g., cereal grain, flour, bran, sawdust) behaves as a support and nutrient source, whereas in the second category, a nutritionally inert solid (e.g., synthetic foam) acts as a support medium which is soaked in a nutrient solution. Using this technique, various useful additives for animal feed can be obtained from agro-waste (Robinson & Nigam, 2003), as shown in Fig. 9.2.

9.4 Utilization of Valorized Residue for Animal Feed

High-quality feed and bedding derived from valorized residues contribute to improved animal health, growth, and productivity. Enhanced digestibility and nutrient absorption translate to higher yields and profitability for livestock producers. In addition, valorization creates new revenue streams for farmers and agribusinesses. By adding value to agricultural residues, it enhances profitability and reduces dependency on conventional feed and bedding sources. Table 9.4 shows the potential utilization of various agro-waste as animal feed.

Converting agricultural residues or food waste into animal feed is an age-old practice. Thirty percent of what is being fed to animals across the world either comes from food waste or by-products of agro-production activities (Westendorf et al., 1998). Food and agro-industry by-products and agricultural wastes used as animal feed can be categorized into three main groups, viz., plant origin, animal origin, and fermentation industry. Plant origin includes organic crop residues such as processing by products, pomace, straw, peels, husks, stones, stover, and factory vegetable oil

Table 9.4 Utilization of agro-waste or residues as animal feed

Type of waste	Animal feed	References
Apple pomace	Milking cows and broiler	Murugesan et al. (2021), Manju et al. (2015), Ajila et al. (2012)
Banana peels	Pigs, rabbits, and fish	Esteban and Ladero (2018), Giri et al. (2016), Girotto et al. (2015)
Citrus pulp	Fish, lambs, cows, and lactating ewes	Salem et al. (2019), Bampidis and Robinson (2006)
Carrots residues, papaya residues	Laying hens, rabbits, Rahmani sheep, piglets	Hamid et al. (2022), Stienfeldt et al. (2006)
Potato residues	Lactating cows and pigs	Murugesan et al. (2021), Halliday (2010)
Mango byproducts, pineapple byproducts	Fish	Sukri et al. (2021), Mahadevaswamy and Venkataraman (1990)
Apple byproduct	Poultry	Murugesan et al. (2021), Manju et al. (2015), Zafar et al. (2005)
Tomato waste	Poultry	Murugesan et al. (2021), Lira et al. (2010)
Juice waste	Poultry	Rizal et al. (2010)
Corn liquor waste	Fish	Rahmatullah and Rahardja (2020), Chovatiya et al. (2011)
Brewery waste, molasses	Chicken, cow, hamster, shrimps	Tinh et al. (2021), Dhiman et al. (2003)
Wheat bran, rye bran	Pig, fish	Liñan-Vidriales et al. (2021)
Olive byproduct	Sheep and goat	Fazio et al. (2022), Molina-Alcaide and Yáñez-Ruiz (2008)
Soyhull, soy pulp, corn husk	Cow, buffalo, catfish	Kari et al. (2022), Ipharraguerre and Clark (2003)
Rice bran, wheat bran	Aqua feed	Liñan-Vidriales et al. (2021), Wieser et al. (2020)
Sweet potato leaves	Animal feed	Chakraborty et al.(2024)

and oleochemical residues that are rich in carbohydrates, lipids, sugars, and inorganic substances. The byproducts of the oil industry, such as solvent-extracted cakes from soybean and oil-yielding rape, sunflower, flax, and products formed during the refining of plant oils, lecithin, and fatty acids, can also be used for the production of animal feed (Ajila et al., 2012). Plant byproducts such as husk and pods of common pulses with leaves and tender stems can be used as nutritious cattle feed and are also good sources of digestible protein (Ranjhan, 1993; Ray, 2024). Pineapple waste is a by-product from pineapple production and cannery plants consisting of the crown, core, peel, leaves, and waste from flesh trimming. The transportation and treatment of such organic crop wastes are expensive and thus can be used as animal feed (Verma, 1997). Animal-origin feed includes waste from meat and poultry production farms that has long been and will continue to be utilized as animal feed. Animal feed often contains a combination of blood, feathers, and bones that have been processed into a meal. Leftover cuts of meat deemed unfit for human consumption are either marketed to third-party rendering operations or transferred to such facilities directly for processing into animal and pet meals. Fermentation industry residues mainly consist of grain, molasses, potato distillers' soluble, brewer's and wine yeast, bacteria, and fungi biomass. Molasses, a byproduct of the sugarcane or sugarbeets processing industry, is one of the major sources of fermentable carbohydrates in animal feed (Cheeke, 1991). Sugarcane tops are palatable and are mostly used as cattle feed along with a little supplement of concentrate mixture or leguminous feeds (Verma, 1997). Bagasse is another good source of cellulose and has been used as roughage for animals across the world (Bhatti & Khan, 1996).

9.5 Challenges or Limitations Involved in the Utilization of Agro-Waste to Animal Feed

Although valorization has become a significant strategy for recycling biomass waste into value-added products (especially lignocellulosic biomass and food wastes), there are certain difficulties and restrictions associated with the potential chemical, biochemical, and thermochemical technologies used in the process. These include high transportation costs and difficulties collecting wastes due to their wide distribution; short storage periods of collected wastes, particularly food wastes, which do not support their constant supply to industry; high costs of various pre-treatment methods involved in food wastes, which depend on the types of substrate; the use of chemicals (strong acids and bases) for biomass waste hydrolysis, which is environmentally unfriendly; and the conversion of wastes into useful resources (Yan et al., 2020). Several characteristics that determine the quality of animal feed include its protein concentration and its biological value, digestibility, level of energy, fats and carbohydrates, the quantitative and qualitative composition of amino acids, vitamin, and mineral content, and the amount of fiber and the presence of hazardous or toxic substances. Despite having huge potential, the animal feed from agro-residues or by-products of agro-industries is underutilized or less utilized. The raw materials obtained from agro-residues are scattered and seasonal and in some cases, the quality is low or not up to the mark. Due to its complex characteristics, processing or treatment of agro-by-products often becomes difficult, costly, and problematic. Lack of capital resources, insufficient managerial skills, and lack of channelized market of the final products prevent the better utilization of agro-processing residues or by-products. Other factors include nutritional and technological aspects such as variability in nutritional level and quantity, presence of naturally occurring antinutritional and toxic factors, presence of pathogenic microorganisms, and lack of proper research and development efforts (Ravindran & Blair, 1992). It's interesting to note that most research on the upscale conversion of biomass wastes is still in the laboratory, which accounts for the dearth of knowledge on this topic (Ray et al., 2024). The diverse protocols of biomass waste valorization remain difficult to execute in practice due to their high cost and unsustainable waste supply. The integration of current technologies may result in cost savings and may be an environmentally friendly approach that significantly improves the economic and environmental values of recycling and utilizing biomass waste in the circular bioeconomy, despite the obstacles and limitations associated with biomass waste valorization.

9.6 Conclusion and Future Perspectives

Agro-waste or residues are often composted or discarded in landfills and rivers which result in environmental damage. These residues or waste can be recycled instead for animal feed or further processed to extract or manufacture value-added items addressing the waste management and food security challenges. Agro-waste mainly consists of residues from agricultural production and byproducts of agro-industries, which are valuable biomass and can be utilized as potential solutions to meet several issues related to animal nutrition. The agro-waste or agro-industrial byproducts can be converted into animal feed in three categories, viz., energy-rich feeds, protein supplements, and byproducts from various refining industries. Strategies for conversion technologies must be developed considering better utilization of agro-waste into valuable animal feeds delivering products that are safe not only for animal feed use but also from the perspective of human feeding. This will ultimately lead to the production of animal feed with reduced food costs and lower prices contributing to self-sufficiency in nutrients from locally available waste. Proper utilization of agro-waste for animal feed will serve as a sound basis for continued technological development and management of animal farming.

References

Agamuthu, P. (2009). Challenges and opportunities in Agro-waste management: An Asian perspective. In *Inaugural meeting of First Regional 3R Forum in Asia*, 11–12 November, Tokyo, Japan.

Ahmad, F., Tauqir, N., Faraz, A., Asghar, I., Wadood, F., Tahir, M. & Mujahid, M. (2021). Performance of Lactating Sahiwal Cows Fed Corn Stovers Ensiled with Molasses, Urea and Lime Solution. *Iran. J. Appl. Anim. Sci.*, 11, 59–66.

Ajila, C. M., Brar, S. K., Verma, M., Tyagi, R. D., Godbout, S., & Valéro, J. R. (2012). Bio-processing of agro-byproducts to animal feed. *Critical Reviews in Biotechnology, 32*(4), 382–400.

Assegehegn, G., Brito-de la Fuente, E., Franco, J. M., & Gallegos, C. (2019). The importance of understanding the freezing step and its impact on freeze-drying process performance. *Journal of Pharmaceutical Sciences, 108*, 1378–1395.

Aziz, M. M. A., Kassim, K. A., ElSergany, M., Anuar, S., Jorat, M. E., Yaa-cob, H., Ahsan, A., & Imteaz, M. A. (2020). Recent advances on palm oil mill effluent (POME) pretreatment and

anaerobic reactor for sustainable biogas production. *Renewable and Sustainable Energy Reviews, 119*, 109603.
Bampidis, V. A., & Robinson, P. H. (2006). Citrus by-products as ruminant feeds: A review. *Animal Feed Science and Technology, 128*, 175–217.
Behera, S. S., & Ray, R. C. (2016). Solid state fermentation for production of microbial cellulases: Recent advances and improvement strategies. *International Journal of Biological Macromolecules (Elsevier), 86*, 656–669.
Behera, S. S., Kerketta, A., & Ray, R. C. (2023). Solid-state fermentation for the production of microbial cellulases. In G. Brahmachari (Ed.), *Biotechnology of microbial enzymes* (2nd ed., pp. 59–82). Academic Press.
Bhatia, L., Jha, H., Sarkar, T., & Sarangi, P. K. (2023). Food waste utilization for reducing carbon footprints towards sustainable and cleaner environment: A review. *International Journal of Environmental Research and Public Health, 20*, 2318.
Bhatti, M. B., & Khan, S. (1996). *Fodder production in Pakistan.* FAO, PARC.
Bölük, G., & Mert, M. (2014). Fossil and renewable energy consumption, GHGs (greenhouse gases) and economic growth: Evidence from a panel of EU (European Union) countries. *Energy, 74*, 439–446.
Bracco, S., Calicioglu, O., Juan, M. G. S., & Flammini, A. (2018). Assessing the contribution of bioeconomy to the total economy: A review of national frameworks. *Sustainability., 10*, 1698.
Bridgeman, T., Jones, J., Shield, I., & Williams, P. (2008). Torrefaction of reed canary grass, wheat straw and willow to enhance solid fuel qualities and combustion properties. *Fuel, 87*(6), 844–856.
Bulgakov, V., Sevostianov, I., Kaletnik, G., Babyn, I., Ivanovs, S., Holovach, I., & Ihnatiev, Y. (2020). Theoretical studies of the vibration process of the dryer for waste of food. *Rural Sustainable Resources., 44*, 32–45.
Chakraborty, I., Chattopadhyay, A., Maity, P., et al. (2024). Biovalorization of sweet potato bagasse into food additives, feeds and fuels. In R. C. Ray (Ed.), *Roots, tubers and bulb crop wastes: Management by biorefinery approaches* (pp. 133–147). Springer.
Cheeke, P. R. (1991). Cereal milling by-products. In *Applied animal nutrition: Feeds and feeding* (pp. 53–57). Macmillan.
Choi, Y. S., Choi, J. H., Han, D. J., Kim, H.-Y., Lee, M. A., Kim, H.-W., Jeong, J. Y., & Kim, C. J. (2011). Effects of rice bran fiber on heat-induced gel prepared with pork salt-soluble meat proteins in model system. *Meat Science, 88*, 59–66.
Choi, W. M., Lam, C. L., Mo, W. Y., & Wong, M. H. (2016). The use of food wastes as feed ingredients for culturing grass carp (*Ctenopharyngodon idellus*) in Hong Kong. *Environmental Science and Pollution Research, 23*(8), 7178–7185.
Chovatiya, S. G., Bhatt, S. S., & Shah, A. R. (2011). Evaluation of corn steep liquor as a supplementary feed for *Labeo rohita* (Ham.) fingerlings. *Aquaculture International, 19*, 1–12.
Dey, T., Bhattacharjee, T., Nag, P., Ghati, A., & Kuila, A. (2021). Valorization of agro-waste into value added products for sustainable development. *Bioresource Technology Reports, 16*, 100834.
Dhiman, T. R., Bingham, H. R., & Radloff, H. D. (2003). Production response of lactating cows fed dried versus wet brewers' grain in diets with similar dry matter content. *Journal of Dairy Science, 86*, 2914–2921.
Esteban, J., & Ladero, M. (2018). Food waste as a source of value-added chemicals and materials: A biorefinery perspective. *International Journal of Food Science and Technology, 53*, 1095–1108.
Fazio, F., Habib, S. S., Naz, S., Filiciotto, F., Cicero, N., Rehman, H. U., Saddozai, S., Rind, K. H., Rind, N. A., & Shar, A. H. (2022). Effect of fortified feed with olive leaves extracts on the hematological and biochemical parameters of *Oreochromis niloticus* (Nile tilapia). *Natural Product Research, 36*, 1575–1580.
Filková, I., & Mujumdar, A. S. (2020). Industrial spray drying systems. In *Handbook of industrial drying* (Vol. 2020, pp. 263–307). CRC Press.

Food and Agriculture Organization of the United Nations (FAO). (2017). *Strategic work of FAO for sustainable food and agriculture*. Available from: http://www.fao.org/3/a-i6488e.pdf

Giri, S. S., Jun, J. W., Sukumaran, V., & Park, S. C. (2016). Dietary administration of banana (*Musa acuminata*) peel flour affects the growth, antioxidant status, cytokine responses, and disease susceptibility of rohu, *Labeo rohita. Journal of Immunology Research*, 4086591.

Girotto, F., Alibardi, L., & Cossu, R. (2015). Food waste generation and industrial uses: A review. *Waste Management, 45*, 32–41.

Halliday, L. (2010). Feeding Cull Potatoes to Beef Cattle. Department of Agriculture. Prince Edward Island, Canada. http://www.gov.pe.ca/photos/original/af_fact_cull.pdf

Hamid, N. K. A., Somdare, P. O., Md Harashid, K. A., Othman, N. A., Kari, Z. A., Wei, L. S., & Dawood, M. A. O. (2022). Effect of papaya (*Carica papaya*) leaf extract as dietary growth promoter supplement in red hybrid tilapia (*Oreochromis mossambicus* × *Oreochromis niloticus*) diet. *Saudi Journal of Biological Sciences, 29*, 3911–3917.

Hiloidhari, M., Bhuyan, N., Gogoi, N., Seth, D., Garg, A., Singh, A., et al. (2020). Agroindustry wastes: Biofuels and biomaterials feedstocks for sustainable rural development. In *Refining biomass residues for sustainable energy and bioproducts* (pp. 357–388). Elsevier.

Ipharraguerre, I. R., & Clark, J. H. (2003). Soyhulls as an alternative feed for lactating dairy cows: Review. *Journal of Dairy Science, 86*, 1052–1073.

Jayant, M., Sahu, N. P., Deo, A. D., Subodh, G., Garg, C. K., & Valappil, R. K. (2020). Nutritional evaluation of fermented sweet potato leaf meal as a replacer of deoiled rice bran in the diet of *Labeo rohita* fingerlings. *Journal of Experimental Zoology, 23*, 61–70.

Ji-Lu, Z. (2007). Bio-oil from fast pyrolysis of rice husk: Yields and related properties and improvement of the pyrolysis system. *Journal of Analytical and Applied Pyrolysis, 80*(1), 30–35.

Joshi, S., Waghmare, J., Sonawane, K., & Waghmare, S. (2015). Bio-ethanol and bio-butanol production from orange peel waste. *Biofuels, 6*(1–2), 55–61.

Karam, M. C., Petit, J., Zimmer, D., Djantou, E. B., & Scher, J. (2016). Effects of drying and grinding in production of fruit and vegetable powders: A review. *Journal of Food Engineering, 188*, 32–49.

Kari, Z. A., Kabir, M. A., Dawood, M. A. O., Razab, M. K. A. A., Ariff, N. S. N. A., Sarkar, T., Pati, S., Edinur, H. A., Mat, K., & Ismail, T. A. (2022). Effect of fish meal substitution with fermented soy pulp on growth performance, digestive enzyme, amino acid profile, and immune-related gene expression of African catfish (*Clarias gariepinus*). *Aquaculture, 546*, 737418.

Katileviciute, A., Plakys, G., Budreviciute, A., Onder, K., Damiati, S., & Kodzius, R. (2019). A sight to wheat bran: High value-added products. *Biomolecules, 9*, 887.

Khir, R., Venkitasamy, C., & Pan, Z. (2019). Infrared heating for improved drying efficiency, food safety, and quality of rice. In J. Jia, D. Liu, & H. Ma (Eds.), *Advances in food processing technology* (pp. 231–251). Springer.

Kim, M. & Day, D. (2011). Composition of sugar cane, energy cane, and sweet sorghum suitable for ethanol production at Louisiana sugar mills. *Journal of industrial microbiology & biotechnology,* 38, 803-807. 10.1007/s10295-010-0812-8.

Krishna, C. (2005). Solid-state fermentation systems-an overview. *Critical Reviews in Biotechnology, 25*, 1–30.

Kumar, S., Paritosh, K., Pareek, N., Chawade, A., & Vivekanand, V. (2018). De-construction of major Indian cereal crop residues through chemical pretreatment for improved biogas production: An overview. *Renewable and Sustainable Energy Reviews, 90*, 160–170.

Kyawt, Y. Y., San Win, K., San Mu, K., Aung, A., & Aung, M. (2020). Feeding pineapple waste silage as roughage source improved the nutrient intakes, energy status and growth performances of growing Myanmar local cattle. *Journal of Advanced Veterinary and Animal Research, 7*(3), 436.

Lakshmi, M. V., Goutami, N., & Kumari, A. H. (2017). Agricultural waste concept, generation, utilization and management. *International Journal of Multidisciplinary Advanced Research Trends, 1*(3), 1–4.

Liñan-Vidriales, M. A., Peña-Rodríguez, A., Tovar-Ramírez, D., Elizondo-González, R., Barajas-Sandoval, D. R., Ponce-Gracía, E. I., Rodríguez-Jaramillo, C., Balcázar, J. L., & Quiroz-Guzmán, E. (2021). Effect of rice bran fermented with *Bacillus* and *Lysinibacillus* species on dynamic microbial activity of Pacific white shrimp (*Penaeus vannamei*). *Aquaculture, 531*, 735958.

Lira, R. C., Rabello, C. B., Ludke, M. C. M. M., Ferreira, P. V., Lana, G. R. Q., & Lana, S. R. V. (2010). Productive performance of broiler chickens fed tomato waste. *Revista Brasileira de Zootecnia, 39*, 1074–1081.

MacLaren, C., Storkey, J., Menegat, A., Metcalfe, H., & Dehnen-Schmutz, K. (2020). An ecological future for weed science to sustain crop production and the environment. A review. *Agronomy for Sustainable Development, 40*, 24.

Mahadevaswamy, M., & Venkataraman, L. V. (1990). Integrated utilization of fruit processing wastes for biogas and fish production. *Biological Wastes, 32*, 243–251.

Mani, T., Murugan, P., & Mahinpey, N. (2011). Pyrolysis of oat straw and the comparison of the product yield to wheat and flax straw pyrolysis. *Energy and Fuels, 25*(7), 2803–2807.

Manju, W. M. W., Bakshi, M. P., & Makkar, H. P. (2015). Waste to worth: Fruit wastes and by-products as animal feed. *CABI Reviews*, 1–26.

Molina, A. E., & Yáñez-Ruiz, D. R. (2008). Potential use of olive by-products in ruminant feeding: A review. *Animal Feed Science and Technology, 147*, 247–264.

Morgan, T. J., George, A., Boulamanti, A. K., Álvarez, P., Adanouj, I., Dean, C., et al. (2015). Quantitative X-ray fluorescence analysis of biomass (switchgrass, corn stover, eucalyptus, beech, and pine wood) with a typical commercial multi-element method on a WD-XRF spectrometer. *Energy and Fuels, 29*(3), 1669–1685.

Murugesan, K., Srinivasan, K. R., Paramasivam, K., Selvam, A., & Wong, J. (2021). Conversion of food waste to animal feeds. In *Current developments in biotechnology and bioengineering* (pp. 305–324). Elsevier.

Nath, P. C., Saikia, D., Debnath, S., Tiwari, A., & Nayak, P. K. (2022). Recent advances in valorization of agro-waste: A step towards sustainable development. *Indian Journal of Hill Farming, 35*(2), 73–91.

Nath, P. C., Ojha, A., Debnath, S., Sharma, M., Nayak, P. K., Sridhar, K., & Inbaraj, B. S. (2023). Valorization of food waste as animal feed: A step towards sustainable food waste management and circular bioeconomy. *Animals, 13*(8), 1366.

Omar, R., Idris, A., Yunus, R., Khalid, K., & Isma, M. A. (2011). Characterization of empty fruit bunch for microwave-assisted pyrolysis. *Fuel, 90*(4), 1536–1544.

Pandey, A. (1992). Recent process developments in solid-state fermentation. *Process Biochemistry, 27*, 109–117.

Pattanaik, L., Pattnaik, F., Saxena, D. K., & Naik, S. N. (2019). Biofuels from agricultural wastes. In *Second and third generation of feedstocks* (pp. 103–142). Elsevier.

Peng, X., Jiang, Y., Chen, Z., Osman, A. I., Farghali, M., Rooney, D. W., & Yap, P.-S. (2023). Recycling municipal, agricultural and industrial waste into energy, fertilizers, food and construction materials, and economic feasibility: A review. *Environmental Chemistry Letters, 21*, 765–801.

Rahmani, M., Azadbakht, M., Dastar, B., & Esmaeilzadeh, E. (2022). Production of animal feed from food waste or corn? Analyses of energy and exergy. *Bioresource Technology Reports, 20*, 101213.

Rahmatullah, H., & Rahardja, B. (2020). Different addition of molasses on feed conversion ratio and water quality in catfish (*Clarias* sp.) rearing with the bio floc-aquaponic system. *IOP Conference Series: Earth and Environmental Sciences, 441*, 012122.

Rajeh, C., Saoud, I. P., Kharroubi, S., Naalbandian, S., & Abiad, M. G. (2021). Food loss and food waste recovery as animal feed: A systematic review. *Journal of Material Cycles and Waste Management, 23*(1), 1–17.

Ranjhan, S. K. (1993). Agro-industrial by-products as a component of livestock rations. In *Animal nutrition in the tropics* (pp. 222–258). Vikas Publishing House PVT Ltd..

Ravindran, V., & Blair, R. (1992). Feed resources for poultry production in Asia and the Pacific. II. Plant protein sources. *World Poultry Sci J., 48*, 205–231.

Ray, R. C. (2020). *Sustainable biofuels: Challenges and opportunities* (p. 502). Academic Press.

Ray, R. C. (2022). *Fruits and vegetable wastes: Valorization into bioproducts and platform chemicals* (p. 448). Springer.

Ray, R. C. (2024). *Roots, tubers and bulb crop wastes; management by biorefinery approaches* (p. 374). Springer.

Ray, R. C., Noe, A.-R., & Sooch, B. S. (2024). Prospective for biorefineries development from roots, tubers and bulb crop wastes and by-products: Value addition and circular economy. In R. C. Ray (Ed.), *Roots, tubers and bulb crop wastes: Management by biorefinery approaches* (pp. 351–274). Springer.

Rizal, Y., Mahata, M. E., Andriani, M., & Wu, G. (2010). Utilization of juice wastes as corn replacement in the broiler diet. *International Journal of Poultry Science, 9*, 886–889.

Robinson, T., & Nigam, P. (2003). Bioreactor design for protein enrichment of agricultural residues by solid state fermentation. *Biochemical Engineering Journal, 13*, 197–203.

Saini, J. K., Saini, R., & Tewari, L. (2015). Lignocellulosic agriculture wastes as biomass feedstocks for second-generation bioethanol production: Concepts and recent developments. *Biotech, 5*(4), 337–353.

Salem, M.E., Abdel-Ghany, H.M., Sallam, A.E., El-Feky, M.M.M. & Almisherfi, H.M. (2019). Effects of dietary orange peel on growth performance, antioxidant activity, intestinal microbiota and liver histology of gilthead sea bream (Sparus aurata) larvae. *Aquatic Nutrition, 25*, 1087–1097.

Serrano, C., Monedero, E., Lapuerta, M., & Portero, H. (2011). Effect of moisture content, particle size and pine addition on quality parameters of barley straw pellets. *Fuel Processing Technology, 92*(3), 699–706.

Shinde, A., & Mahanta, S. K. (2020). Nutrition of small ruminants on grazing lands in dry zones of India. *Range Management and Agroforestry, 41*, 1–14.

Sinha, A. K., Rakesh, S., Mitra, B., Roy, N., Sahoo, S., Saha, B., et al. (2021). Agricultural waste management policies and programme for environment and nutritional security. In *Input use efficiency for food and environmental security* (pp. 627–664). Springer.

Sotiropoulos, A., Malamis, D., & Loizidou, M. J. (2015). Dehydration of domestic food waste at source as an alternative approach for food waste management. *Waste and Biomass Valorization, 6*, 167–176.

Stanton, R., Morrissey, C. A., & Clark, R. G. J. (2018). Analysis of trends and agricultural drivers of farmland bird declines in North America: A review. *Ecosystem Environment, 254*, 244–254.

Steinfeld, H., Gerber, P., Wassenaar, T. D., Castel, V., Rosales, M., Rosales, M., & de Haan, C. (2006). *Livestock's long shadow: Environmental issues and options*. Food and Agriculture Organization.

Sukri, S. A. M., Andu, Y., Harith, Z. T., Sarijan, S., Pauzi, M. N. F., Wei, L. S., Dawood, M. A. O., & Kari, Z. A. (2021). Effect of feeding pineapple waste on growth performance, texture quality and flesh colour of nile tilapia (*Oreochromis niloticus*) fingerlings. *Saudi Journal of Biological Sciences, 29*, 2514–2519.

Sun, Y., & Cheng, J. (2002). Hydrolysis of lignocellulosic materials for ethanol production: A review. *Bioresource Technology, 83*(1), 1–11.

Tinh, T. H., Koppenol, T., Hai, T. N., Verreth, J. A., & Verdegem, M. C. (2021). Effects of carbohydrate sources on a biofloc nursery system for whiteleg shrimp (*Litopenaeus vannamei*). *Aquaculture, 531*, 735795.

Tsai, W., Lee, M., & Chang, D. Y. (2006). Fast pyrolysis of rice straw, sugarcane bagasse and coconut shell in an induction-heating reactor. *Journal of Analytical and Applied Pyrolysis, 76*(1–2), 230–237.

Tsonis, S. P., Tsola, V. P., & Grigoropoulos, S. G. (1989). Systematic characterization and chemical treatment of olive oil mill wastewater. *Toxicological and Environmental Chemistry, 20*(1), 437–457.

Verma, D.N. (1997). *A text book of animal nutrition* (1st ed.). R. 814. New Rajinder Nagar Publications, New Delhi, India.

Weisz, U., Pichler, P.-P., Jaccard, I. S., Haas, W., Matej, S., Bachner, F., et al. (2020). Carbon emission trends and sustainability options in Austrian health care. *Resources, Conservation and Recycling, 160*, 104862.

Westendorf, M., Dong, Z., & Schoknecht, P. (1998). Recycled cafeteria food waste as a feed for swine: Nutrient content digestibility, growth and meat quality. *Journal of Animal Science, 76*(12), 2976–2983.

Wheeler, R. R., Hadley, N. M., Dahl, R. W., Williams, T. W., Zavala, D. B., Akse, J. R., & Fisher, J. W. (2007). Microwave enhanced freeze drying of solid waste. *SAE Transactions, 116*, 510–537.

Wieser, H., Koehler, P., & Scherf, K. (2020). Wheat-based raw materials. In *Wheat—An exceptional crop* (pp. 103–131). Woodhead Publishing.

Wilkanowska, A., & Kokoszyński, D. (2015). Effect of diet and physical activity of farm animals on their health and reproductive performance. In *Handbook of fertility* (Vol. 2015, pp. 159–171). Elsevier.

Wilkinson, J., Rinne, M. J. G., & Science, F. (2018). Highlights of progress in silage conservation and future perspectives. *Grass Forage Science, 73*, 40–52.

Yaashikaa, P., Kumar, P. S., Saravanan, A., Varjani, S., & Ramamurthy, R. (2020). Bioconversion of municipal solid waste into bio-based products: A review on valorisation and sustainable approach for circular bioeconomy. *Science of the Total Environment, 748*, 141312.

Yahya, M. A., Al-Qodah, Z., & Ngah, C. Z. (2015). Agricultural bio-waste materials as potential sustainable precursors used for activated carbon production: A review. *Renewable and Sustainable Energy Reviews, 46*, 218–235.

Yan, G., Feng, Y., Long, S., Zeng, X., Sun, Y., Tang, X., & Lin, L. (2020). Challenges with biomass waste valorization. In C. S. K. Lin, G. Kaur, C. Li, & X. Yang (Eds.), *Waste valorization: Waste streams in a circular economy* (pp. 183–202). Wiley.

Yin, C.-Y. (2011). Prediction of higher heating values of biomass from proximate and ultimate analyses. *Fuel, 90*(3), 1128–1132.

Ytrestøyl, T., Aas, T. S., & Åsgård, T. (2015). Utilisation of feed resources in production of Atlantic salmon (Salmo salar) in Norway. *Aquaculture, 448*, 365–374.

Zafar, F., Idrees, M., & Ahmed, Z. (2005). Use of apple by-products in poultry rations of broiler chicks in Karachi. *Pakistan Journal of Physiology, 1*, 1–2.

Chapter 10
Sustainable Packaging Solutions from Agri-food Waste: An Overview

Daniel Castro-Criado, Mercedes Jiménez-Rosado, Victor M. Perez-Puyana, and Alberto Romero

Abbreviations

FAO	Food and Agriculture Organization of the United Nations
SDGs	Sustainable Development Goals
PHA	Polyhydroxyalkanoates
PLA	Polylactic acid

10.1 Introduction

Agricultural waste refers to the undesired materials or byproducts produced by agricultural processes. These materials generate a great problem due to the negative consequences that they have on the environment, economy, and society (Capanoglu et al., 2022). The rapid increase in global population and food consumption patterns are the main causes of agri-food loss and waste (Yukesh Kannah et al., 2020). It is worth mentioning the difference between loss and waste in this area. Agri-food loss refers to discards produced during production, whether due to quantity or quality. On the other hand, agri-food waste is due to discards made by food suppliers or consumers. Whether they are losses or waste, these materials have a high organic content that is discarded, decomposing, and frequently contaminating soil and water (Kavitha et al., 2020; Ray, 2022). The global food system is supported by an extensive network of agricultural producers, processors, distributors, wholesalers,

D. Castro-Criado · V. M. Perez-Puyana · A. Romero (✉)
Departamento de Ingeniería Química, Universidad de Sevilla, Sevilla, Spain
e-mail: alromero@us.es

M. Jiménez-Rosado
Departamento de Química y Física Aplicadas, Universidad de León, León, Spain

J. Arora et al. (eds.), *Transforming Agriculture Residues for Sustainable Development*, Waste as a Resource, https://doi.org/10.1007/978-3-031-61133-9_10

retailers, and consumers. Food loss and waste occur at every level (Bancal & Ray, 2022). The food system also needs a lot of inputs, including energy, pesticides, fertilizers, and irrigation water, all of which can have negative effects on the environment (such as soil erosion, biodiversity loss, water pollution). In this sense, agri-food waste causes resource waste and undesired environmental effects (Conrad & Blackstone, 2021).

Agri-food loss in the food system worsens the environmental effects of food production (Food and Agriculture Organization of the United Nations, 2019). Initial estimates made by FAO indicate that around 14% of the world's food is lost from post-harvest up to (but not including) the retail level. According to the Food Waste Index Report, 931 million tons of agri-food waste were produced in 2019, with 61% of it coming from households, 26% from food services, and 13% from retail (United Nations Environment Programme, 2021). Around 14% of the world's food (valued at $400 billion) is lost every year, according to FAO's estimations which could (Food and Agriculture Organization of the United Nations, 2019). At the same time, 32.6 million people cannot afford a quality meal every second day (European Commission, 2023b). In this sense, different strategies for change towards more sustainable systems are being implemented. Among them, the most well-known is the Sustainable Development Goals (SDGs) of the 2030 Agenda (Carrillo, 2022). Thus, the European Union (EU) and EU countries are committed to meeting the SDG 12.3 target to halve per capita waste at the retail and consumer level by 2030 and reduce agri-food losses along the food production and supply chains (Carrillo, 2022). However, it is important to carefully investigate the numerous and context-dependent mechanisms through which reduction in food losses and waste has an impact on food security and nutrition. The impact relies on how and where food losses or waste are decreased, as well as the location of nutritionally vulnerable communities. It is not a guarantee that reducing food loss or waste will increase food security and nutrition; in some circumstances, its effects might even be detrimental (Food and Agriculture Organization of the United Nations, 2019). The Commission will work to intensify efforts to stop agri-food loss and waste throughout the EU as demanded by the new Farm to Fork Strategy. This strategy, established by the Commission as a component of the European Green Deal, outlines several initiatives to make it possible to move to a sustainable EU food system that ensures food security and offers access to healthy diets sourced from a healthy environment.

These initiatives are also observed globally. In this sense, the FAO established the International Day of Awareness about Food Loss and Waste (29 September) intending to increase social awareness about this problem. In addition, programs such as the "United Nations Environment Programme" have been established. However, more collective action is needed to scale up efforts to reduce food loss and waste while limiting greenhouse gas emissions (Food and Agriculture Organization of the United Nations, 2019).

However, even if these strategies were effective, there is still a large amount of agri-food waste that cannot be consumed and continues to create a great environmental impact. Examples of these are non-edible crop discards (leaves, stems, roots, etc.) and those discarded by the consumer (skins, bones, shells, etc.). Agriculture has intensified over the past 50 years, and the use of fertilizers and pesticides has resulted in an imbalance between productivity and environmental conditions and the need for more organic material in the soil. Under these conditions, agricultural wastes could be utilized by clean technologies or returned as an organic substrate to the soil as valuable materials, with the least amount of risk to ecosystems and the environment. In addition, these materials could be disposed of in landfills, used as animal food, or used as fuel for combustion. However, agri-food wastes must be treated because of the possible threats they provide to the environment due to their microbial degradation. Agri-food wastes typically contain complex structures and compositions made up of polysaccharides, proteins, sugars, and polyphenolic components, among other things. These characteristics might make them suitable for use as renewable natural resources. Agri-food wastes are also accessible, affordable, sustainable, and environmentally beneficial. These wastes are assessed as eco-materials based on these factors (Matei et al., 2021). In this sense, many researchers are betting on another alternative for recovering this waste (Notarnicola et al., 2017). In this way, the initial chemical composition of agri-food wastes plays a significant role in both the valorization process and application. However, agri-food waste mostly has biopolymers in its composition. This is why their use as raw materials for the production of bio-based materials (bioplastics) has a great impact (Visco et al., 2022). For the reasons stated above, this chapter will be focused on agri-food waste used directly as raw material to form biopolymers and bioplastics and their application of bioplastics in various sectors.

10.2 Bioplastics and Biopolymers

Bioplastics are materials that include two aspects: materials made from bio-based matter (biopolymers) or biodegradable materials (Ibrahim et al., 2021). The use of agri-food waste as raw material for the production of bioplastics is a potential alternative since they could have both characteristics: being bio-based and biodegradable (Jõgi & Bhat, 2020). For this reason, there are many researchers who study the processing of this type of materials to replace other less environmentally friendly ones such as conventional plastics. In this sense, Bugatti et al. (2019) used tomato peels to produce food packaging (Bugatti et al., 2019). Tan et al. (2020) propose kefiran as a raw material for biomedical products (Tan et al., 2020). Khoukhi et al. (2021) process thermal insulation materials based on rice (Khoukhi et al., 2021), among many other examples.

Bioplastics are very versatile and can be used in many industrial sectors. However, its use as packaging is the most promising since it allows for a completely disposable container that allows us to be in tune with the circular economy (Versino et al., 2023). In this way, 60% of bioplastic research is focused on packaging (Jagannath & Rao, 2022). For this reason, this work aims to show the different sectors where agri-food waste is used as raw material for packaging different components, giving examples of the most studied agri-food waste. This study aims to create a guide to potential uses of agri-food waste to improve its understanding and application at an industrial level.

10.2.1 Bioplastic Processing

As mentioned previously, the definition of bioplastic has two connotations: bio-based and biodegradable. This has generated great confusion at a social level since people think that all bioplastics are biodegradable, continuing to carelessly discard them in landfills when the majority of bioplastics on the market do not have the quality of biodegradable (Dilkes-Hoffman et al., 2019). It is for this reason that the term biodegradable is becoming increasingly important in this type of materials to avoid the great environmental impact they entail.

In the last decade, a large amount of research has been carried out on the development of bio-based and biodegradable materials with improved, such as improved barrier and mechanical properties as well as low diffusion properties compared to conventional packaging materials (Rameshkumar et al., 2020). Among the processing methods, waste fermentation is the most studied and used (Tsang et al., 2019).

The fermentation of waste to obtain bioplastics consists of using the large source of carbon that agri-food waste presents so that different bacteria are capable of producing volatile fatty acids, precursors to the formation of bioplastics such as PHA (polyhydroxyalkanoates) and PLA (polylactic acid)(Chan et al., 2021). This process is complex because the ideal conditions must be present (pH, temperature, humidity, amount of nutrients, among others) so that the bacteria can ferment the waste optimally. The monitoring of these parameters, as well as the variability of agri-food waste, has made this process have a high production cost, which is today too high to compete with petroleum-derived polymers (Gironi & Piemonte, 2011; Nozari et al., 2023).

For these reasons, it is being considered more and more to produce bioplastics directly from the biopolymers found in agri-food waste rather than using intermediate steps like fermentation (Chantapet et al., 2013; Czigány et al., 2007). One of its benefits is how easily bioplastics can be produced using the same processing techniques as conventional plastics. Thus, many industries are forced to rely on government subsidies to adapt their industries to the fabrication of bioplastics, which lowers the investment made for the development of these systems (Pirzada et al., 2020). The most used processing methods are compression molding, extrusion, and injection, although in most cases, it is necessary to incorporate additional steps to improve the process (Azeredo et al., 2015; Bourny et al., 2017; Cuadri et al., 2016; Gamero et al., 2019; Perez-Puyana et al., 2022; Pereira et al., 2015) They all agree on the fundamental stages: the breaking of intermolecular bonds that stabilize the biopolymers to give them mobility; rearrangement and orientation of the mobile polymer chains in the desired shape; and, finally, the formation of new bonds and intermolecular interactions that allow the stabilization of the three-dimensional matrix created. All this is thanks to the action of temperature and pressure in the thermomechanical process. Although there are also other well-studied processing methods such as electrospinning, casting, or 3D printing, their poor assimilation to processes already used industrially makes them economically unprofitable, being used in more exclusive sectors where the cost of the final product is less important, such as cosmetics or biomaterials (Guilbert & Gontard, 2005; Nida et al., 2021; Tsang et al., 2019).

10.2.2 Bioplastics in Food Packaging

Packaging is a fundamental part of the food industry due to the benefits it can generate when consuming, preserving, transporting, and storing food, in addition to being an advertising medium for the brand (Marsh & Bugusu, 2007). All this has led to an excessive use of packaging in the food sector (Hudson et al., 2016). In this way, you can see containers wrapped in more packaging (i.e., sweets) and the packaging of foods that already have their natural protection (i.e., fruits and vegetables). The most used food packaging is film and polypropylene or polyethylene bags due to the great versatility they present (Ebnesajjad & Andrew, 2012). However, it is also worth mentioning polyethylene terephthalate, polyvinyl chloride, and polystyrene (Porta, 2019). All this excessive packaging has a large environmental impact since they are single-use plastics that usually end up in landfills or the environment (Hudson et al., 2016; Ncube et al., 2020). Consumers are increasingly aware of this impact, choosing more sustainable packaging during their purchases. However, packaging is still necessary in this industry for the preservation of many foods. For this reason, the use of bioplastics in this sector is widely studied (Shlush & Davidovich-Pinhas, 2022).

The food industry is especially sensitive due to the nature of its product, which has to meet hygiene and quality requirements to be suitable for human consumption (Trienekens & Zuurbier, 2008). That is why finding a replacement for the current packaging is not an easy task. Among the requirements that food packaging must have these must provide physical protection; they must protect the food against vibrations, changes in pressure, temperature, shocks, and bacteria. In addition, they must be a protective barrier against the penetration of oxygen, water vapor, and dust to increase the stability of the food for a longer period of time (increase its best before or expiration date). On the other hand, it must allow the inclusion of information about the product necessary for its sale (Lee et al., 2008).

The valorization of agri-food waste for the development of bioplastics that serve as food packaging is a sustainable solution that can generate high-added value to this waste (Cristofoli et al., 2023; Otoni et al., 2021). Furthermore, it can be highly welcomed by producers and consumers thanks to the new perception of the circular economy. On the other hand, the use of these wastes is advantageous because they are renewable, non-toxic, capable of being recycled, and low cost. Table 10.1 shows the most used agri-food waste for the production of food packaging. As can be seen, fruit and vegetable waste are the most abundant due to the high biopolymeric content (proteins and polysaccharides) they present.

Table 10.1 Most used agri-food waste as raw materials for food packaging and its specific application

Biopolymer	Applications	References
Wheat	Active/smart food packaging Films	Chavoshizadeh et al. (2020) and Rezaei et al. (2020)
Pea protein	Generic Active food packaging	Perez-Puyana et al. (2022)
Rice	Generic Films	Elhussieny et al. (2020), Gupta et al. (2019), and Sain (2020)
Sugarcane bagasse	Generic Films	Gupta et al. (2019) and Nida et al. (2021)
Banana	Active food packaging Films	Mostafa (2021) and Othman et al. (2020)
Fish skin	Active food packaging	Lionetto and Esposito Corcione (2021)
Cellulose and derivatives	Generic Films	Liu et al. (2021)
Cocoa	Films	Azmin et al. (2020, 2022)
Corn	Fiber	Chong et al. (2021)
Cassava	Generic	Ray and Swain (2011), Dasumiati et al. (2019), and Nee and Othman (2022)
Avocado	Active food packaging	Merino et al. (2021)
Soy	Generic	Dey et al. (2023)

Advances in this sector have meant that the benefits of waste are used to generate added value in food packaging. In this sense, new strategies are being developed to use active compounds and pigments from agri-food waste so that packaging is active and/or intelligent (Valencia et al., 2021). Active packaging allows the quality and safety of food to be maintained for a longer period of time, increasing its useful life and reducing the amount of food waste (Yildirim et al., 2018). Thus, they allow the organoleptic properties of foods to be maintained, humidity or permeability to carbon dioxide to be controlled, or ethylene to be eliminated. In this sense, both fish skin and pea packaging have been used for their antimicrobial properties that can reduce the putrefaction of the food they contain (Lionetto & Esposito Corcione, 2021; Perez-Puyana et al., 2022). On the other hand, smart packaging allows end users to know the status of the product by monitoring the growth of microbes, chemical changes, or gas production (Amin et al., 2022). For example, in the packaging developed by Chavoshizadeh et al. (2020), the color changes (from green to yellow) when the oil oxidizes, indicating the expiration time of the oil (Chavoshizadeh et al., 2020). Also noteworthy is the production of edible films from this agri-food waste, which allows packaging waste to be reduced to zero (Mileti et al., 2023; Yildirim-Yalcin et al., 2022).

10.2.3 Bioplastics in Personal Care and Hygiene

The personal care and hygiene industry is highly competitive, needing to constantly innovate in the performance of existing products (Smith et al., 2013). This is due to the high number of companies dedicated to this sector and the ever-increasing consumer demand. All this has led to the use of attractive or voluminous packaging to attract the buyer's interest. However, this creates a significant problem due to the excessive use of unsustainable packaging, with the environmental impact that it entails (Montes et al., 2022).

People are becoming more and more aware of these negative effects, looking for products that are committed to the environment (Kotler et al., 2019). This is why many companies in this sector are using packaging as a sales strategy. Thus, as is happening in the food sector, the hygiene and personal care industry is adapting to offer more sustainable packaging alternatives (Boz et al., 2020; Tang, 2011). However, these containers must meet certain format, material, and sustainability criteria that allow them to maintain the qualities of the products they retain while being easy and comfortable for the consumer to use and transport, as well as attractive to the public (King, 2011). The main actors have committed to using only recyclable, reusable, or compostable plastic packaging by 2025(Strangl et al., 2018). This is why the most used packaging at the moment is high-density polyethylene due to its flexibility and lightness, which allows easy transport and controlled dosage of the product, which reduces waste (Strangl et al., 2018). In addition, it can be 100% recyclable.

However, this material is not the most sustainable, so many of these companies are looking for other alternatives, such as biodegradable materials. In this sense, L'Occitane is investigating the use of tubes made from recycled oils (De et al., 2023) and L'Oreal is developing bioplastic bottles from different waste (Nikolova, 2022). Many researchers are evaluating the use of different agri-food waste as raw materials for packaging in this sector. The most used agri-food waste is shown in Table 10.2. It should be noted that most of the agri-food waste used in this type of product has bioactive components or nutrients that can increase the benefits of the products. In this sense, grapevines can provide vitamin E, dragon fruit provides nutrients that protect against solar radiation, pomegranate has antioxidant potential, and animal skin contains collagen that helps regenerate the skin. Thus, the use of these agri-food waste as raw materials for personal care and hygienic packaging allows these nutrients to end up in the products, improving them.

Table 10.2 Most used agri-food waste as raw materials for personal care and hygiene packaging and its specific application

Biopolymer	Applications	References
Grapevine	Cosmetics	Maamoun (2022)
Dragon fruit	Generic	Taharuddin et al. (2023)
Pomegranate	Cosmetic	Ko et al. (2021)
Animal skin	Generic	Matinong et al. (2022)

10.2.4 Bioplastics in the Pharmaceutical Industry

Packaging is essential in the pharmaceutical sector as it allows medications to be protected so that they do not cause adverse effects when taken (Nasa, 2014). In addition, it also plays an essential role in its administration since its design and functionality can affect dosage and ease of use for the patient (Hanson & Haddad, 2022). This importance means that it is regulated by different regulations (in Europe, Directive 2001/83/EC).

Despite the importance of pharmaceutical packaging, there are some challenges to solve, such as its environmental impact, since its majority of components are plastic and aluminum, which are difficult to recycle (Raju et al., 2016). For this reason, the pharmaceutical industry is working on the development of more sustainable packaging materials.

Pharmaceutical packaging can be compared to food packaging in terms of requirements: impermeability, oxygen barrier, opaque, among others. In this sense, many of the wastes listed in Table 10.1 could be used in these packaging. However, due to the high regulatory restrictions for medicine packaging and the fact that current packaging is usually recycled at SIGRE points (at least 68%), there is no defined trend toward the use of agri-food waste for this application (de Oliveira Souza et al., 2021).

Nevertheless, the use of biodegradable materials has generated a new concept of controlled-release drug packaging in the pharmaceutical industry. Controlled-release systems are inert matrices (package) that allow the release of the drug at a specific speed and/or location to achieve adequate bioavailability values and guarantee a correct pharmacotherapeutic effect on the patient (Huang & Brazel, 2001). Among the characteristics to consider for this matrix are its biodegradability, biocompatibility, and zero toxicity. In this sense, different natural materials have been tested as a matrix that contains the drug to be released (Tong et al., 2020). Zhang et al. (2021) used microcapsules of chitosan-based film for controlled drug release. Fu et al. (2019) administered intracellular proteins with hyaluronic acid-controlled systems. Finally, many researchers have used collagen/gelatin scaffolds for drug delivery into wounds to improve wound healing (Gaspar-Pintiliescu et al., 2019).

10.2.5 Bioplastics in Agriculture

The increase in the consumption of fruits, vegetables, and cereals has generated a new way of perceiving agriculture, opting for intensive crops to satisfy demand (Hazell & Wood, 2008). This intensive agriculture, along with the increased presence of pests and changes in weather conditions (due to climate change), has led to an increase in the use of plastic devices in this sector. These include packaging, which in this sector is divided into two aspects: on the one hand, there is the packaging intended to carry the plants or seeds (i.e., pots or seedbeds) along with the coatings to protect the crop from weeds and insects. As well as more sophisticated packaging that aims to supply water and fertilizers in a controlled manner (similar to that obtained in the pharmaceutical industry). However, the use of these plastics has an extra cost for the products since it is not only necessary to pay for their purchase, but it is also necessary to pay for their removal once their work is finished to avoid soil contamination (Karan et al., 2019). In this way, the use of bioplastics based on agri-food waste can solve these problems due to their biodegradability without the need to collect them after use. Furthermore, their raw materials can be obtained from cheaper sources, making the cost of final bioplastics cheaper than conventional plastics (Palani & Karthikeyan, 2020).

In this way, bioplastics based on agri-food waste have attracted a lot of attention in recent years in both scientific and industrial research. Agri-food waste (such as chitosan, starch, or cellulose) derived from plants, microorganisms, and animals are abundant in nature, biodegradable, and hence eco-friendly, and can be used to develop devices for agriculture. These materials can be used as slow-release packaging to distribute agrochemicals (fertilizers/nutrients) or pesticides to crop plants to improve yields. Similarly, they can be used as direct product coating to increase product shelf life, boost seed germination, or protect against infections and pests (Wypij et al., 2023).

Agriculture needs to improve its ability to increase food production while using a smaller land area. Because of this, the use of fertilizers and other agrochemicals increases the quantity of agricultural products, but with a negative impact on the environment. For example, up to 70% of soil nitrogen can be lost through volatilization (Avila-Quezada et al., 2022). Furthermore, some micronutrients might not be available to plants. This makes the availability of novel fertilizers based on nanotechnology crucial (Materu et al., 2021).

Moreover, there is a growing interest in novel alternatives to traditional pots made of non-renewable oil-based raw materials, such as polystyrene. In this sense, biodegradable pots are being manufactured from agri-food waste, thus favoring the development of plants due to the characteristics of these containers, avoiding the excessive generation of plastic waste, and allowing the valorization of these agri-food waste.

10.2.6 Biopolymers as Carriers of Nutrients

As an alternative to bulk synthetic fertilizers, biopolymers obtained from agri-food waste, like alginate, cellulose, or chitosan, might be used as matrices (packaging) to encapsulate nutrients and avoid dissolution and oxidation due to their high biodegradability, low toxicity, and additional nutrient contribution during their decomposition (Mishra et al., 2018). In this sense, the matrix based on agri-food waste has a large nutritional load (especially N and C) that allows for stimulating the growth of plants during their biodegradation, in addition to being able to supply another extra nutrient that is incorporated into the matrix and allows biofortification of the crop

(Jiménez-Rosado et al., 2023). Additionally, while experiencing a significant volume increase, these materials can absorb and retain water from the environment while preserving their structural integrity (Jiménez-Rosado et al., 2018).

These materials have demonstrated exceptional efficacy in the treatment of disorders caused by infections and nutrient deficiencies. Thus, they can enhance conventional methods of applying agrochemicals to fields to decrease biotic stress and collateral damage, such as the loss of soil fertility, or to boost the effectiveness against crop diseases, pests, and nutrient depletion (Machado et al., 2022). Nevertheless, the processing of these matrices is very important since their structure can affect the way nutrients are released or how they can be assimilated by plants (Bandyopadhyay et al., 2014; Chen et al., 2020; Sipponen et al., 2017). Thus, a lot of nutrient packaging systems have recently been investigated, along with the mechanism by which they interact with active ingredients. These mechanisms include (1) encapsulation, emulsion, and entrapment; (2) application of nanoscale delivery agents to crop fields; and (3) nutrient release from nanoscale delivery agents (Dutta et al., 2022).

Among all the possibilities, chitosan is the biopolymer that is most widely used in agriculture because of its non-toxic origin. The adsorption and gradual release of the desired active ingredient can also be programmed using this matrix, which is simple to modify (Pandey et al., 2018). For example, the water-soluble insecticide acetamiprid's quick release is delayed when it is enclosed in a chitosan alginate nanocarrier, which can also be used to nanoencapsulated other substances to lessen their toxicity (Kumar et al., 2015a, b). Two naturally occurring substances, chitosan and hydroxyapatite, were employed to supply micro- and macronutrients to the soil. Wood stems' natural cellulose and other biopolymer pores can be employed to store nanofertilizers for the release of nitrogen and other elements under various soil samples (Kottegoda et al., 2017).

Some of the biopolymers used as packaging for sustained release in agriculture are shown in Table 10.3.

Table 10.3 Most used agri-food waste as raw materials for agriculture and its specific application

Biopolymer	Applications	References
Alginate	Pesticide delivery Growth promoter Sustained irrigation	Sharma et al. (2014) and Yin et al. (2020)
Cellulose	Pesticide delivery Fertilizer delivery	Kumar et al. (2015a, b) and Olad et al. (2018)
Chitosan	Pesticide delivery Fertilizer/nutrient delivery Growth promoter	Choudhary et al. (2019), Saberi Riseh et al. (2023), Saharan et al. (2016), and Sharma et al. (2019)
Lignin	Pesticide delivery	Machado et al. (2020)
Polypeptides	Pesticide/nutrient delivery Gene delivery	Chen et al. (2015), Chuah et al. (2016), and Xin et al. (2018)
Soy	Nutrient delivery Plant simulation	Jiménez-Rosado et al. (2022, 2023)

10.2.7 Biodegradable Pots

Commonly used pots are made of polyethylene, which is not biodegradable, so they must be removed before the plant is transplanted into the soil. Plastic pots are widely used, resulting in an excess of waste that cannot be recycled. This fact shows a drawback of using plastic pots in nursery and greenhouse operations that cannot only be attributed to human action but also to the material (Rizarelli et al., 2020). From an agronomical perspective, the usage of plastic pots affects the development of plants (McKay, 1997). In their search for water and nutrients, roots are forced by the walls of the pot to grow lower and encircle one another. As a result, a few long rots become intertwined, causing them to choke (Fuentes et al., 2021). Because the root constriction prevents nutrients and water from being absorbed, the plant is placed in a high-stress state. Additionally, plants only receive oxygen from the top of the plastic pots because of the high oxygen barrier, which prevents root oxygenation. As a result, since oxygen helps growing media maintain the temperature, inadequate intake helps to suffocate the soil and eradicate the beneficial bacteria for plant growth. This unfavorable factor impacts the entire plant, making it more susceptible to heat, drought, and disease, impacting aboveground production (Markham et al., 2011). Moreover, after plastic pots are retired, plants are subjected to a variety of challenges, including extreme temperatures, mechanical damage, and alterations in the growing environment, which might lead to a "transplant shock" (Koeser et al., 2009; McKay, 1997).

Due to the aforementioned problems, there is an increasing interest in the usage of biodegradable pots. These containers are made using materials that come from renewable resources or byproducts of agriculture, animal husbandry, etc. (Schettini et al., 2013). Depending on the biodegradation rate, which is mostly established by the raw materials, they may be classified as plantable or compostable (Nambuthiri et al., 2015). On the one hand, plantable pots can be directly planted in the soil together with the plant. These containers are made to endure handling, watering, and short-term production and shipping circumstances. They are designed to decompose quickly after being planted, allowing plant roots to spread through the pot and into the soil. In these containers, root disruption and transplant shock might not occur. They can be made from byproducts such as rice hull, bamboo fiber, and coconut coir fiber, among others. On the other hand, plants in compostable pots must be taken out during installation, and the containers are composted separately. The majority of bioplastics, hard rice hulls, peat, and thick-walled paper or wood fiber containers designed for long-term production come into this category (Fuentes et al., 2021; Nambuthiri et al., 2015).

These pots are biocomposites with a continuous phase made up of biopolymers from accessible and renewable sources, like polysaccharides, and a solid phase made up of natural fillers and fibers dispersed through a polymeric matrix made from waste products from the agri-food and textile processing industries. The structure created by combining the phases results in a system with increased structural, mechanical, and physico-chemical performance (Schettini et al., 2013).

Polysaccharides, available in large quantities, such as starch, cellulose, or alginate, among others, are renewable, biocompatible, and biodegradable polymers that can be employed as binders for pot applications (Kapanen et al., 2008). Additionally, polysaccharides have a strong affinity for water molecules due to the presence of polar groups on macromolecular chains and as a result of their inherent physicochemical characteristics. They promote the formation of hydrogels, which are three-dimensional water-stable networks that are crosslinked by ionic, covalent, thermos-reversible, or pH-reversible mechanisms (Coviello et al., 2007; Russo et al., 2007).

The natural fibers work as a reinforcement in the manufacture of biodegradable pots, enhancing the stiffness and strength of the final biocomposite structures. However, when compared to the most common competitive man-made reinforcing fibers, natural fibers perform mechanically worse. The characteristics of natural fibers depend on the plant source, plant age, separation method, moisture content, etc., whereas conventional fibers like glass, carbon, and aramid can be created with a specific range of characteristics. However, due to their low density, natural fibers can match the strength and stiffness of glass fibers in terms of their specific qualities. Additionally, because polysaccharides and natural fibers, which primarily consist of cellulose, share some chemical similarities, the reinforced polymers' tensile strength was improved. This result has been attributed to the strong attachment between the fiber and matrix, which is promoted by their shared high hydrophobicity (Schettini et al., 2013).

10.3 Conclusions and Future Perspectives

In conclusion, many agri-food wastes have been evaluated for their application as packaging for different applications. All this marks the high potential of these wastes as raw materials for the production of packaging materials. However, more research is still necessary so that these products can compete with the materials conventionally used for packaging.

The management of agri-food waste is currently a major challenge due to the increase in world population. Valorization of these wastes into valuable products will benefit the environment. Food agri-waste from many different sources can be extracted and purified to produce packaging components (Varghese et al., 2023). However, current laboratory-scale processing and production techniques for various bioplastics from agri-food waste show a low yield. Therefore, the future of these agri-food-based materials will depend on their industrial scalability and the improvement of their properties (especially mechanical) to make them competitive in the current market. In this sense, investigations into wastes derived from crop and processing residues, as well as byproducts from animals, have been considerable. The mechanical barrier and thermal properties of composite materials are significantly influenced by extraction techniques, processing setups, and material blending. While feather keratin has been developed as a new material, cellulose, and eggshell powder have been effectively employed as fillers or reinforcements in rigid packagings to improve their mechanical properties (Varghese et al., 2023). Therefore, the future of these materials depends on their ability to reinforce them, and in this way, they can match the mechanical characteristics and stability of current packaging.

Acknowledgments This work is a part of an I + D + i PID2021-124294OB-C21 project which is sponsored by MCIN/AEI/10.13039/501100011033/ and "ERDF A way of making Europe". The authors thank their financial support. In addition, the authors would also acknowledge for the postdoctoral contract of Víctor M. Pérez Puyana from the "Contratación de Personal Investigador Doctor" supported by the European Social Fund and Junta de Andalucía (PAIDI DOCTOR—Convocatoria 2019-2020).

References

Amin, U., Khan, M. K. I., Maan, A. A., Nazir, A., Riaz, S., Khan, M. U., Sultan, M., Munekata, P. E. S., & Lorenzo, J. M. (2022). Biodegradable active, intelligent, and smart packaging materials for food applications. *Food Packaging and Shelf Life, 33*, 100903. https://doi.org/10.1016/j.fpsl.2022.100903

Avila-Quezada, G. D., Ingle, A. P., Golińska, P., & Rai, M. (2022). Strategic applications of nano-fertilizers for sustainable agriculture: Benefits and bottlenecks. *Nanotechnology Reviews, 11*(1), 2123–2140. https://doi.org/10.1515/ntrev-2022-0126

Azeredo, H. M. C., Kontou-Vrettou, C., Moates, G. K., Wellner, N., Cross, K., Pereira, P. H. F., & Waldron, K. W. (2015). Wheat straw hemicellulose films as affected by citric acid. Food Hydrocolloids, 50, 1–6. https://doi.org/10.1016/j.foodhyd.2015.04.005

Azmin, S. N. H. M., Hayat, N. A. binti M., & Nor, M. S. M. (2020). Development and characterization of food packaging bioplastic film from cocoa pod husk cellulose incorporated with sugarcane bagasse fibre. Journal of Bioresources and Bioproducts, 5(4), 248–255. https://doi.org/10.1016/j.jobab.2020.10.003

Bancal, V., & Ray, R. C. (2022). Overview of food loss and wastes in fruits and vegetables: From issues to resources. In R. C. Ray (Ed.), *Fruits and vegetable wastes: Valorization to bioproducts and platform chemicals* (pp. 3–32). Springer.

Bandyopadhyay, S., Ghosh, K., & Varadachari, C. (2014). Multimicronutrient slow-release fertilizer of zinc, iron, manganese, and copper. *International Journal of Chemical Engineering, 2014*, 1–7. https://doi.org/10.1155/2014/327153

Bourny, V., Perez-Puyana, V., Felix, M., Romero, A., & Guerrero, A. (2017). Evaluation of the injection moulding conditions in soy/nanoclay based composites. European Polymer Journal, 95, 539–546. https://doi.org/10.1016/j.eurpolymj.2017.08.036

Boz, Z., Korhonen, V., & Koelsch Sand, C. (2020). Consumer considerations for the implementation of sustainable packaging: A review. *Sustainability, 12*(6), 2192. https://doi.org/10.3390/su12062192

Bugatti, V., Brachi, P., Viscusi, G., & Gorrasi, G. (2019). Valorization of tomato processing residues through the production of active bio-composites for packaging applications. *Frontiers in Materials, 6*. https://doi.org/10.3389/fmats.2019.00034

Capanoglu, E., Nemli, E., & Tomas-Barberan, F. (2022). Novel approaches in the valorization of agricultural wastes and their applications. *Journal of Agricultural and Food Chemistry, 70*(23), 6787–6804. https://doi.org/10.1021/acs.jafc.1c07104

Carrillo, M. (2022). Measuring progress towards sustainability in the European Union within the 2030 agenda framework. *Mathematics, 10*(12), 2095. https://doi.org/10.3390/math10122095

Chan, J. X., Wong, J. F., Hassan, A., & Zakaria, Z. (2021). Bioplastics from agricultural waste. In Biopolymers and Biocomposites from Agro-Waste for Packaging Applications (pp. 141–169). Elsevier. https://doi.org/10.1016/B978-0-12-819953-4.00005-7
Chantapet, P., Kunanopparat, T., Menut, P., & Siriwattanayotin, S. (2013). Extrusion Processing of Wheat Gluten Bioplastic: Effect of the Addition of Kraft Lignin. Journal of Polymers and the Environment, 21(3), 864–873. https://doi.org/10.1007/s10924-012-0557-8
Chavoshizadeh, S., Pirsa, S., & Mohtarami, F. (2020). Sesame oil oxidation control by active and smart packaging system using wheat gluten/chlorophyll film to increase shelf life and detecting expiration date. *European Journal of Lipid Science and Technology, 122*(3). https://doi.org/10.1002/ejlt.201900385
Chen, M., Jensen, S. P., Hill, M. R., Moore, G., He, Z., & Sumerlin, B. S. (2015). Synthesis of amphiphilic polysuccinimide star copolymers for responsive delivery in plants. *Chemical Communications, 51*(47), 9694–9697. https://doi.org/10.1039/C5CC02726H
Chen, J., Fan, X., Zhang, L., Chen, X., Sun, S., & Sun, R. (2020). Research progress in lignin-based slow/controlled release fertilizer. *ChemSusChem, 13*(17), 4356–4366. https://doi.org/10.1002/cssc.202000455
Chong, T. Y., Law, M. C., & Chan, Y. S. (2021). The Potentials of Corn Waste Lignocellulosic Fibre as an Improved Reinforced Bioplastic Composites. Journal of Polymers and the Environment, 29(2), 363–381. https://doi.org/10.1007/s10924-020-01888-4
Choudhary, R. C., Kumaraswamy, R. V., Kumari, S., Sharma, S. S., Pal, A., Raliya, R., Biswas, P., & Saharan, V. (2019). Zinc encapsulated chitosan nanoparticle to promote maize crop yield. *International Journal of Biological Macromolecules, 127*, 126–135. https://doi.org/10.1016/j.ijbiomac.2018.12.274
Chuah, J.-A., Horii, Y., & Numata, K. (2016). Peptide-derived method to transport genes and proteins across cellular and organellar barriers in plants. *Journal of Visualized Experiments, 118*. https://doi.org/10.3791/54972
Conrad, Z., & Blackstone, N. T. (2021). Identifying the links between consumer food waste, nutrition, and environmental sustainability: A narrative review. *Nutrition Reviews, 79*(3), 301–314. https://doi.org/10.1093/nutrit/nuaa035
Coviello, T., Matricardi, P., Marianecci, C., & Alhaique, F. (2007). Polysaccharide hydrogels for modified release formulations. *Journal of Controlled Release, 119*(1), 5–24. https://doi.org/10.1016/j.jconrel.2007.01.004
Cristofoli, N. L., Lima, A. R., Tchonkouang, R. D. N., Quintino, A. C., & Vieira, M. C. (2023). Advances in the food packaging production from Agri-food waste and by-products: Market trends for a sustainable development. *Sustainability, 15*(7), 6153. https://doi.org/10.3390/su15076153
Cuadri, A. A., Bengoechea, C., Romero, A., & Guerrero, A. (2016). A natural-based polymeric hydrogel based on functionalized soy protein. European Polymer Journal, 85(Complete), 164–174. https://doi.org/10.1016/j.eurpolymj.2016.10.026
Czigány, T., Romhány, G., & Kovács, J. G. (2007). Starch for Injection Molding Purposes. In Handbook of Engineering Biopolymers. https://doi.org/10.3139/9783446442504.003
Dasumiati, Saridewi, N., & Malik, M. (2019). Food packaging development of bioplastic from basic waste of cassava peel (manihot uttilisima) and shrimp shell. IOP Conference Series: Materials Science and Engineering, 602(1), 012053. https://doi.org/10.1088/1757-899X/602/1/012053
Dilkes-Hoffman, L., Ashworth, P., Laycock, B., Pratt, S., & Lant, P. (2019). Public attitudes towards bioplastics – knowledge, perception and end-of-life management. Resources, Conservation and Recycling, 151, 104479. https://doi.org/10.1016/j.resconrec.2019.104479
de Oliveira Souza, H., dos Santos Costa, R., Quadra, G. R., & dos Santos Fernandez, M. A. (2021). Pharmaceutical pollution and sustainable development goals: Going the right way? *Sustainable Chemistry and Pharmacy, 21*, 100428. https://doi.org/10.1016/j.scp.2021.100428

De, S. K., Kawdia, P., Gupta, D., & Pragya, N. (2023). Packaging plastic waste management in the cosmetic industry. *Management of Environmental Quality: An International Journal, 34*(3), 820–842. https://doi.org/10.1108/MEQ-03-2022-0053

Dey, A., Rahman, M. M., Gupta, A., Yodo, N., & Lee, C. W. (2023). A Performance Study on 3D-Printed Bioplastic Pots from Soybean By-Products. Sustainability, 15(13), 10535. https://doi.org/10.3390/su151310535

Dutta, S., Pal, S., Panwar, P., Sharma, R. K., & Bhutia, P. L. (2022). Biopolymeric nanocarriers for nutrient delivery and crop biofortification. *ACS Omega, 7*(30), 25909–25920. https://doi.org/10.1021/acsomega.2c02494

Ebnesajjad, S., & Andrew, W. (2012). *Plastic films in food packaging: Materials, technology and applications* (p. 384p). William Andrew Publishing.

Elhussieny, A., Faisal, M., D'Angelo, G., Aboulkhair, N. T., Everitt, N. M., & Fahim, I. S. (2020). Valorisation of shrimp and rice straw waste into food packaging applications. *Ain Shams Engineering Journal, 11*(4), 1219–1226. https://doi.org/10.1016/j.asej.2020.01.008

European Commission. (2023b). *Food waste*. https://food.ec.europa.eu/safety/food-waste_en

Food and Agriculture Organization of the United Nations. (2019). *The state of food security and nutrition in the world: Safeguarding against economic slowdowns and downturns*. Food and Agriculture Organization of the United Nations.

Fu, Y., Jang, M.-S., Wu, T., Lee, J. H., Li, Y., Lee, D. S., & Yang, H. Y. (2019). Multifunctional hyaluronic acid-mediated quantum dots for targeted intracellular protein delivery and real-time fluorescence imaging. *Carbohydrate Polymers, 224*, 115174. https://doi.org/10.1016/j.carbpol.2019.115174

Fuentes, R. A., Berthe, J. A., Barbosa, S. E., & Castillo, L. A. (2021). Development of biodegradable pots from different agroindustrial wastes and byproducts. *Sustainable Materials and Technologies, 30*, e00338. https://doi.org/10.1016/j.susmat.2021.e00338

Gamero, S., Jiménez-Rosado, M., Romero, A., Bengoechea, C., & Guerrero, A. (2019). Reinforcement of Soy Protein-Based Bioplastics Through Addition of Lignocellulose and Injection Molding Processing Conditions. Journal of Polymers and the Environment. https://doi.org/10.1007/s10924-019-01430-1

Gaspar-Pintiliescu, A., Stanciuc, A.-M., & Craciunescu, O. (2019). Natural composite dressings based on collagen, gelatin and plant bioactive compounds for wound healing: A review. International Journal of Biological Macromolecules, 138, 854–865. https://doi.org/10.1016/j.ijbiomac.2019.07.155

Gironi, F., & Piemonte, V. (2011). Bioplastics and Petroleum-based Plastics: Strengths and Weaknesses. Energy Sources, Part A: Re, 1949–1959. https://doi.org/10.1080/15567030903436830

Guilbert, S., & Gontard, N. (2005). Agro-polymers for edible and biodegradable films: Review of agricultural polymeric materials, physical and mechanical characteristics. In Innovations in Food Packaging (pp. 263–276). Elsevier. https://doi.org/10.1016/B978-012311632-1/50048-6

Gupta, H., Kumar, H., Kumar, M., Gehlaut, A. K., Gaur, A., Sachan, S., & Park, J.-W. (2019). Synthesis of biodegradable films obtained from rice husk and sugarcane bagasse to be used as food packaging material. *Environmental Engineering Research, 25*(4), 506–514. https://doi.org/10.4491/eer.2019.191

Hanson, A., & Haddad, L. M. (2022). Nursing rights of medication administration. In *StatPearls*. StatPearls Publishing.

Hazell, P., & Wood, S. (2008). Drivers of change in global agriculture. Philosophical Transactions of the Royal Society B: Biological Sciences. https://doi.org/10.1098/rstb.2007.2166

Huang, X., & Brazel, C. S. (2001). On the importance and mechanisms of burst release in matrix-controlled drug delivery systems. *Journal of Controlled Release, 73*(2–3), 121–136. https://doi.org/10.1016/S0168-3659(01)00248-6

Hudson, A., Macdonald, M., & Blake, K. (2016). Packing and problematic feeding behaviors in CHARGE syndrome: A qualitative analysis. *International Journal of Pediatric Otorhinolaryngology, 82*, 107–115. https://doi.org/10.1016/j.ijporl.2016.01.009

Ibrahim, N. I., Shahar, F. S., Sultan, M. T. H., Shah, A. U. M., Safri, S. N. A., & Mat Yazik, M. H. (2021). Overview of bioplastic introduction and its applications in product packaging. *Coatings, 11*(11), 1423. https://doi.org/10.3390/coatings11111423

Jagannath, A., & Rao, P. J. (2022). Agri and Food Waste Valorization Through the Production of Biochemicals and Packaging Materials. In Biotechnology for Zero Waste (pp. 521–541). Wiley. https://doi.org/10.1002/9783527832064.ch34

Jiménez-Rosado, M., Perez-Puyana, V., Cordobés, F., Romero, A., & Guerrero, A. (2018). Development of soy protein-based matrices containing zinc as micronutrient for horticulture. *Industrial Crops and Products, 121*, 345–351. https://doi.org/10.1016/j.indcrop.2018.05.039

Jiménez-Rosado, M., Perez-Puyana, V., Guerrero, A., & Romero, A. (2022). Micronutrient-controlled-release protein-based systems for horticulture: Micro vs. nanoparticles. Industrial Crops and Products, 185, 115128. https://doi.org/10.1016/j.indcrop.2022.115128

Jiménez-Rosado, M., Di Foggia, M., Rosignoli, S., Guerrero, A., Rombolà, A. D., & Romero, A. (2023). Effect of zinc and protein content in different barley cultivars: use of controlled release matrices. Renewable Agriculture and Food Systems, 38, e34. https://doi.org/10.1017/S1742170523000297

Jõgi, K., & Bhat, R. (2020). Valorization of food processing wastes and by-products for bioplastic production. *Sustainable Chemistry and Pharmacy, 18*, 100326. https://doi.org/10.1016/j.scp.2020.100326

Kapanen, A., Schettini, E., Vox, G., & Itävaara, M. (2008). Performance and environmental impact of biodegradable films in agriculture: A field study on protected cultivation. *Journal of Polymers and the Environment, 16*(2), 109–122. https://doi.org/10.1007/s10924-008-0091-x

Karan, H., Funk, C., Grabert, M., Oey, M., & Hankamer, B. (2019). Green Bioplastics as Part of a Circular Bioeconomy. Trends in Plant Science, 24(3), 237–249. https://doi.org/10.1016/j.tplants.2018.11.010

Kavitha, S., Yukesh Kannah, R., Kumar, G., Gunasekaran, M., & Rajesh Banu, J. (2020). Introduction: Sources and characterization of food waste and food industry wastes. In *Food waste to valuable resources* (pp. 1–13). Elsevier. https://doi.org/10.1016/B978-0-12-818353-3.00001-8

Khoukhi, M., Dar Saleh, A., Hassan, A., & Abdelbaqi, S. (2021). Thermal characterization of a new bio-based insulation material containing puffed rice. *Energies, 14*(18), 5700. https://doi.org/10.3390/en14185700

King, T. P. (2011). *Sustainable packaging in the healthcare industry* [Imperial College London]. https://doi.org/10.25560/6813

Ko, K., Dadmohammadi, Y., & Abbaspourrad, A. (2021). Nutritional and bioactive components of pomegranate waste used in food and cosmetic applications: A review. *Food, 10*(3), 657. https://doi.org/10.3390/foods10030657

Koeser, A. K., Stewart, J. R., Bollero, G. A., Bullock, D. G., & Struve, D. K. (2009). Impacts of handling and transport on the growth and survival of balled-and-burlapped trees. *HortScience, 44*(1), 53–58. https://doi.org/10.21273/HORTSCI.44.1.53

Kotler, P., Kartajaya, H., & Setiawan, I. (2019). Marketing 3.0: From products to customers to the human spirit. In *Marketing wisdom. Management for professionals* (pp. 139–156). Springer. https://doi.org/10.1007/978-981-10-7724-1_10

Kottegoda, N., Sandaruwan, C., Priyadarshana, G., Siriwardhana, A., Rathnayake, U. A., Berugoda Arachchige, D. M., Kumarasinghe, A. R., Dahanayake, D., Karunaratne, V., & Amaratunga, G. A. J. (2017). Urea-hydroxyapatite nanohybrids for slow release of nitrogen. *ACS Nano, 11*(2), 1214–1221. https://doi.org/10.1021/acsnano.6b07781

Kumar, S., Bhanjana, G., Sharma, A., Sarita, Sidhu, M. C., & Dilbaghi, N. (2015a). Herbicide loaded carboxymethyl cellulose nanocapsules as potential carrier in agrinanotechnology. *Science of Advanced Materials, 7*(6), 1143–1148. https://doi.org/10.1166/sam.2015.2243

Kumar, S., Chauhan, N., Gopal, M., Kumar, R., & Dilbaghi, N. (2015b). Development and evaluation of alginate–chitosan nanocapsules for controlled release of acetamiprid. *International Journal of Biological Macromolecules, 81*, 631–637. https://doi.org/10.1016/j.ijbiomac.2015.08.062

Lee, D. S., Yam, K. L., & Piergiovanni, L. (2008). *Food packaging science and technology* (p. 656p). CRC Press.

Lionetto, F., & Esposito Corcione, C. (2021). Recent applications of biopolymers derived from fish industry waste in food packaging. *Polymers, 13*(14), 2337. https://doi.org/10.3390/polym13142337

Liu, Y., Ahmed, S., Sameen, D. E., Wang, Y., Lu, R., Dai, J., Li, S., & Qin, W. (2021). A review of cellulose and its derivatives in biopolymer-based for food packaging application. Trends in Food Science & Technology, 112, 532–546. https://doi.org/10.1016/j.tifs.2021.04.016

Maamoun, M. A. I. (2022). An insight into the brilliant benefits of grape waste. In *Mediterranean fruits bio-wastes* (pp. 433–465). Springer. https://doi.org/10.1007/978-3-030-84436-3_18

Machado, T. O., Beckers, S. J., Fischer, J., Müller, B., Sayer, C., de Araújo, P. H. H., Landfester, K., & Wurm, F. R. (2020). Bio-based lignin nanocarriers loaded with fungicides as a versatile platform for drug delivery in plants. *Biomacromolecules, 21*(7), 2755–2763. https://doi.org/10.1021/acs.biomac.0c00487

Machado, T. O., Grabow, J., Sayer, C., de Araújo, P. H. H., Ehrenhard, M. L., & Wurm, F. R. (2022). Biopolymer-based nanocarriers for sustained release of agrochemicals: A review on materials and social science perspectives for a sustainable future of agri- and horticulture. *Advances in Colloid and Interface Science, 303*, 102645. https://doi.org/10.1016/j.cis.2022.102645

Markham, J. W., Bremer, D. J., Boyer, C. R., & Schroeder, K. R. (2011). Effect of container color on substrate temperatures and growth of red maple and redbud. *HortScience, 46*(5), 721–726. https://doi.org/10.21273/HORTSCI.46.5.721

Marsh, K., & Bugusu, B. (2007). Food packaging? Roles, materials, and environmental issues. *Journal of Food Science, 72*(3), R39–R55. https://doi.org/10.1111/j.1750-3841.2007.00301.x

Matei E., Râpă M., Predescu A.M., Țurcanu A.A., Vidu R., Predescu C., Bobirica C., Bobirica L. & Orbeci C. (2021) Valorization of Agri-Food Wastes as Sustainable Eco-Materials for Waste-water Treatment: Current State and New Perspectives. *Materials,* 14(16), 4581. https://doi.org/10.3390/ma14164581

Materu, S. F., Heise, S., & Urban, B. (2021). Seasonal and spatial detection of pesticide residues under various weather conditions of agricultural areas of the Kilombero Valley, Ramsar site, Tanzania. *Frontiers in Environmental Science, 9*. https://doi.org/10.3389/fenvs.2021.599814

Matinong, A. M. E., Chisti, Y., Pickering, K. L., & Haverkamp, R. G. (2022). Collagen extraction from animal skin. *Biology, 11*(6), 905. https://doi.org/10.3390/biology11060905

McKay, H. M. (1997). A review of the effect of stresses between lifting and planting on nursery stock quality and performance. *New Forests, 13*, 369–399. https://doi.org/10.1023/A:1006563130976

Merino, D., Bertolacci, L., Paul, U. C., Simonutti, R., & Athanassiou, A. (2021). Avocado Peels and Seeds: Processing Strategies for the Development of Highly Antioxidant Bioplastic Films. ACS Applied Materials & Interfaces, 13(32), 38688–38699. https://doi.org/10.1021/acsami.1c09433

Mileti, O., Baldino, N., Filice, F., Lupi, F. R., Sinicropi, M. S., & Gabriele, D. (2023). Formulation Study on Edible Film from Waste Grape and Red Cabbage. Foods, 12(14), 2804. https://doi.org/10.3390/foods12142804

Mishra, R. K., Ha, S. K., Verma, K., & Tiwari, S. K. (2018). Recent progress in selected bio-nanomaterials and their engineering applications: An overview. *Journal of Science: Advanced Materials and Devices, 3*(3), 263–288. https://doi.org/10.1016/j.jsamd.2018.05.003

Montes, R. M. A., Delapaz, G. V., Oliquino, J. J. D. R., Pascasio, H. K. A., & Lugay, C. I. J. R. P. (2022). A study on the impact of green cosmetics, personal care products, and their packaging on consumers' purchasing behavior in Luzon, Philippines. In *Proceedings of the 7th North American International Conference on Industrial Engineering and Operations Management*.

Mostafa, H. S. (2021). Banana plant as a source of valuable antimicrobial compounds and its current applications in the food sector. *Journal of Food Science, 86*(9), 3778–3797. https://doi.org/10.1111/1750-3841.15854

Nambuthiri, S., Fulcher, A., Koeser, A. K., Geneve, R., & Niu, G. (2015). Moving toward sustainability with alternative containers for greenhouse and nursery crop production: A review and research update. *HortTechnology, 25*(1), 8–16. https://doi.org/10.21273/HORTTECH.25.1.8

Nasa, P. (2014). A review on pharmaceutical packaging material. *World Journal of Pharmaceutical Research, 3*(5), 344–368.

Ncube, L. K., Ude, A. U., Ogunmuyiwa, E. N., Zulkifli, R., & Beas, I. N. (2020). Environmental impact of food packaging materials: A review of contemporary development from conventional plastics to polylactic acid based materials. *Materials, 13*(21), 4994. https://doi.org/10.3390/ma13214994

Nee, F. C., & Othman, S. A. (2022). Preparation and Characterization of Irradiated Bioplastic from Cassava Peel – A Review. Journal of Physics: Conference Series, 2169(1), 012041. https://doi.org/10.1088/1742-6596/2169/1/012041

Nida, S., Moses, J. A., & Anandharamakrishnan, C. (2021). 3D printed food package casings from sugarcane bagasse: A waste valorization study. *Biomass Conversion and Biorefinery*. https://doi.org/10.1007/s13399-021-01982-0

Nikolova, A. (2022). Sustainability in the beauty and personal care industry: Trends, best practices, and opportunities. *SSRN Electronic Journal*. https://doi.org/10.2139/ssrn.4190097

Notarnicola, B., Sala, S., Anton, A., McLaren, S. J., Saouter, E., & Sonesson, U. (2017). The role of life cycle assessment in supporting sustainable agri-food systems: A review of the challenges. *Journal of Cleaner Production, 140*, 399–409. https://doi.org/10.1016/j.jclepro.2016.06.071

Nozari, M., Mohammadi, A., Fouladvand, S., & Ansari Shiri, M. (2023). Novel Sludge Processing Methods Focusing on Advantages and Disadvantages, Energy Production, and Bioplastics: A Systematic Review. Jundishapur Journal of Health Sciences, 15(2). https://doi.org/10.5812/jjhs-133092

Olad, A., Zebhi, H., Salari, D., Mirmohseni, A., & Reyhani Tabar, A. (2018). Slow-release NPK fertilizer encapsulated by carboxymethyl cellulose-based nanocomposite with the function of water retention in soil. *Materials Science and Engineering: C, 90*, 333–340. https://doi.org/10.1016/j.msec.2018.04.083

Othman, S. H., Tarmiti, N. A., Shapi'i, R. A., Zahiruddin, S. M. M., Tawakkal, I. S. M. A., & Basha, R. K. (2020). Starch/banana pseudostem biocomposite films for potential food packaging applications. *BioResources, 15*(2), 3984.

Otoni, C. G., Azeredo, H. M. C., Mattos, B. D., Beaumont, M., Correa, D. S., & Rojas, O. J. (2021). The food–materials nexus: Next generation bioplastics and advanced materials from agri-food residues. *Advanced Materials, 33*(43), e2102520. https://doi.org/10.1002/adma.202102520

Palani, R., & Karthikeyan, M. (2020). Bioplastics from Agricultural Wastes. SSRN Electronic Journal. https://doi.org/10.2139/ssrn.3564039

Pandey, P., Verma, M. K., & De, N. (2018). Chitosan in agricultural context: A review. *Bulletin of Environment, Pharmacology and Life Sciences, 7*(4), 87–96.

Pereira, E. I., da Cruz, C. C. T., Solomon, A., Le, A., Cavigelli, M. A., & Ribeiro, C. (2015). Novel Slow-Release Nanocomposite Nitrogen Fertilizers: The Impact of Polymers on Nanocomposite Properties and Function. Industrial & Engineering Chemistry Research, 54(14), 3717–3725. https://doi.org/10.1021/acs.iecr.5b00176

Perez-Puyana, V., Cuartero, P., Jiménez-Rosado, M., Martínez, I., & Romero, A. (2022). Physical crosslinking of pea protein-based bioplastics: Effect of heat and UV treatments. *Food Packaging and Shelf Life, 32*, 100836. https://doi.org/10.1016/j.fpsl.2022.100836

Pirzada, T., de Farias, B., Mathew, R., Guenther, R., Byrd, M., Sit, T., Pal, L., Opperman, C., & Khan, S. (2020). Recent advances in biodegradable matrices for active ingredient release in crop protection: Towards attaining sustainability in agriculture. Current Opinion in Colloid & Interface Science, 48, 121–136. https://doi.org/10.1016/j.cocis.2020.05.002

Porta, R. (2019). The Plastics Sunset and the Bio-Plastics Sunrise. Coatings, 9(8), 526. https://doi.org/10.3390/coatings9080526

Raju, G., Sarkar, P., Singla, E., Singh, H., & Sharma, R. K. (2016). Comparison of environmental sustainability of pharmaceutical packaging. *Perspectives in Science, 8*, 683–685. https://doi.org/10.1016/j.pisc.2016.06.058

Ray, R. C. (2022). *Fruits and vegetable wastes: Valorization into bioproducts and platform chemicals* (p. 448p). Springer.

Ray, R. C., & Swain, M. R. (2011). Bio-ethanol, bioplastics and other fermented industrial products from cassava starch and flour. In C. M. Pace (Ed.), *Cassava: Farming, uses and economic impacts* (pp. 1–32). Nova Science Publishers.

Rameshkumar, S., Shaiju, P., O'Connor, K. E., & Ramesh Babu, R. (2020). Bio-based and biodegradable polymers - State-of-the-art, challenges and emerging trends. Current Opinion in Green and Sustainable Chemistry, 21, 75–81. https://doi.org/10.1016/j.cogsc.2019.12.005

Rezaei, M., Pirsa, S., & Chavoshizadeh, S. (2020). Photocatalytic/antimicrobial active film based on wheat gluten/ZnO nanoparticles. *Journal of Inorganic and Organometallic Polymers and Materials, 30*(7), 2654–2665. https://doi.org/10.1007/s10904-019-01407-6

Rizarelli, P., Innocenti, F. D., Valenti, G., & Rapisarda, M. (2020). Biodegradation of green polymer composites: Laboratory procedures and standard test methods. *Materials Research Foundations, 68*, 1–44. https://doi.org/10.21741/9781644900659-1

Russo, R., Malinconico, M., & Santagata, G. (2007). Effect of cross-linking with calcium ions on the physical properties of alginate films. *Biomacromolecules, 8*(10), 3193–3197. https://doi.org/10.1021/bm700565h

Saberi Riseh, R., Vatankhah, M., Hassanisaadi, M., & Kennedy, J. F. (2023). Chitosan-based nanocomposites as coatings and packaging materials for the postharvest improvement of agricultural product: A review. Carbohydrate Polymers, 309, 120666. https://doi.org/10.1016/j.carbpol.2023.120666

Saharan, V., Kumaraswamy, R. V., Choudhary, R. C., Kumari, S., Pal, A., Raliya, R., & Biswas, P. (2016). Cu-chitosan nanoparticle mediated sustainable approach to enhance seedling growth in maize by mobilizing reserved food. *Journal of Agricultural and Food Chemistry, 64*(31), 6148–6155. https://doi.org/10.1021/acs.jafc.6b02239

Sain, M. (2020). Production of bioplastics and sustainable packaging materials from rice straw to eradicate stubble burning: A mini-review. *Environment Conservation Journal, 21*(3), 1–5. https://doi.org/10.36953/ECJ.2020.21301

Schettini, E., Santagata, G., Malinconico, M., Immirzi, B., ScarasciaMugnozza, G., & Vox, G. (2013). Recycled wastes of tomato and hemp fibres for biodegradable pots: Physico-chemical characterization and field performance. *Resources, Conservation and Recycling, 70*, 9–19. https://doi.org/10.1016/j.resconrec.2012.11.002

Sharma, R., Bajpai, J., Bajpai, A. K., Acharya, S., Shrivastava, R. B., & Shukla, S. K. (2014). Designing slow water-releasing alginate nano reservoirs for sustained irrigation in scanty rainfall areas. *Carbohydrate Polymers, 102*, 513–520. https://doi.org/10.1016/j.carbpol.2013.11.059

Sharma, A., Sood, K., Kaur, J., & Khatri, M. (2019). Agrochemical loaded biocompatible chitosan nanoparticles for insect pest management. *Biocatalysis and Agricultural Biotechnology, 18*, 101079. https://doi.org/10.1016/j.bcab.2019.101079

Shlush, E., & Davidovich-Pinhas, M. (2022). Bioplastics for food packaging. Trends in Food Science & Technology, 125, 66–80. https://doi.org/10.1016/j.tifs.2022.04.026

Sipponen, M. H., Rojas, O. J., Pihlajaniemi, V., Lintinen, K., & andÖsterberg, M. (2017). Calcium chelation of lignin from pulping spent liquor for water-resistant slow-release urea fertilizer systems. *ACS Sustainable Chemistry and Engineering, 5*(1), 1054–1061. https://doi.org/10.1021/acssuschemeng.6b02348

Smith, M., Saunders, R., Stuckhardt, L., and McGinnis, J. M. (2013). Best care at lower cost: The path to continuously learning health care in America. Institute of Medicine of the National Academies.

Strangl, M., Schlummer, M., Maeurer, A., & Buettner, A. (2018). Comparison of the odorant composition of post-consumer high-density polyethylene waste with corresponding recycled and virgin pellets by combined instrumental and sensory analysis. *Journal of Cleaner Production, 181*, 599–607. https://doi.org/10.1016/j.jclepro.2018.01.137

Taharuddin, N. H., Jumaidin, R., Mansor, M. R., Hazrati, K. Z., Tarique, J., Asyraf, M. R. M., & andRazman, M. R. (2023). Unlocking the potential of lignocellulosic biomass dragon fruit (hylocereuspolyrhizus) in bioplastics, biocomposites and various commercial applications. *Polymers, 15*(12), 2654. https://doi.org/10.3390/polym15122654

Tan, K.-X., Chamundeswari, V. N., & Loo, S. C. J. (2020). Prospects of kefiran as a food-derived biopolymer for agri-food and biomedical applications. *RSC Advances, 10*(42), 25339–25351. https://doi.org/10.1039/D0RA02810J

Tang, S. L. P. (2011). Innovative and sustainable packaging strategies for hygiene products. In *Textiles for hygiene and infection control* (pp. 48–67). Elsevier. https://doi.org/10.1533/9780857093707.1.48

Tong, X., Pan, W., Su, T., Zhang, M., Dong, W., & Qi, X. (2020). Recent advances in natural polymer-based drug delivery systems. *Reactive and Functional Polymers, 148*, 104501. https://doi.org/10.1016/j.reactfunctpolym.2020.104501

Trienekens, J., & Zuurbier, P. (2008). Quality and safety standards in the food industry, developments and challenges. *International Journal of Production Economics, 113*(1), 107–122. https://doi.org/10.1016/j.ijpe.2007.02.050

Tsang, Y. F., Kumar, V., Samadar, P., Yang, Y., Lee, J., Ok, Y. S., Song, H., Kim, K. H., Kwon, E. E., & Jeon, Y. J. (2019). Production of bioplastic through food waste valorization. Environment International, 127, 625–644. https://doi.org/10.1016/j.envint.2019.03.076

United Nations Environment Programme. (2021). *Food waste index report 2021*. United Nations Environment Programme.

Valencia, G. A., de Andrade, C. J., Ienczak, J. L., Monteiro, A. R., & Gutiérrez, T. J. (2021). Valorization of Agri-food wastes. In *Bio-valorization of waste: Trends and perspectives* (pp. 111–132). Springer. https://doi.org/10.1007/978-981-15-9696-4_5

Varghese, S. A., Pulikkalparambil, H., Promhuad, K., Srisa, A., Laorenza, Y., Jarupan, L., Nampitch, T., Chonhenchob, V., & Harnkarnsujarit, N. (2023). Renovation of Agro-Waste for Sustainable Food Packaging: A Review. Polymers, 15(3), 648. https://doi.org/10.3390/polym15030648

Versino, F., Ortega, F., Monroy, Y., Rivero, S., López, O. V., & García, M. A. (2023). Sustainable and bio-based food packaging: A review on past and current design innovations. *Food, 12*(5), 1057. https://doi.org/10.3390/foods12051057

Visco, A., Scolaro, C., Facchin, M., Brahimi, S., Belhamdi, H., Gatto, V., & Beghetto, V. (2022). Agri-food wastes for bioplastics: European prospective on possible applications in their second life for a circular economy. *Polymers, 14*(13), 2752. https://doi.org/10.3390/polym14132752

Wypij, M., Trzcińska-Wencel, J., Golińska, P., Avila-Quezada, G. D., Ingle, A. P., & Rai, M. (2023). The strategic applications of natural polymer nanocomposites in food packaging and agriculture: Chances, challenges, and consumers' perception. *Frontiers in Chemistry, 10*. https://doi.org/10.3389/fchem.2022.1106230

Xin, X., He, Z., Hill, M. R., Niedz, R. P., Jiang, X., & Sumerlin, B. S. (2018). Efficiency of biodegradable and pH-responsive polysuccinimide nanoparticles (PSI-NPs) as smart nanodelivery systems in grapefruit: *In vitro* cellular investigation. *Macromolecular Bioscience, 18*(7), e1800159. https://doi.org/10.1002/mabi.201800159

Yildirim, S., Röcker, B., Pettersen, M. K., Nilsen-Nygaard, J., Ayhan, Z., Rutkaite, R., Radusin, T., Suminska, P., Marcos, B., & Coma, V. (2018). Active packaging applications for food. *Comprehensive Reviews in Food Science and Food Safety, 17*(1), 165–199. https://doi.org/10.1111/1541-4337.12322

Yildirim-Yalcin, M., Tornuk, F., & Toker, O. S. (2022). Recent advances in the improvement of carboxymethyl cellulose-based edible films. Trends in Food Science & Technology, 129, 179–193. https://doi.org/10.1016/j.tifs.2022.09.022

Yin, Y., Yang, M., Xi, J., Cai, W., Yi, Y., He, G., Dai, Y., Zhou, T., & Jiang, M. (2020). A sodium alginate-based nano-pesticide delivery system for enhanced in vitro photostability and insecticidal efficacy of phloxine B. *Carbohydrate Polymers, 247*, 116677. https://doi.org/10.1016/j.carbpol.2020.116677

Yukesh Kannah, R., Merrylin, J., Poornima Devi, T., Kavitha, S., Sivashanmugam, P., Kumar, G., & Rajesh Banu, J. (2020). Food waste valorization: Biofuels and value added product recovery. *Bioresource Technology Reports, 11*, 100524. https://doi.org/10.1016/j.biteb.2020.100524

Zhang, L., Yu, D., Regenstein, J. M., Xia, W., & Dong, J. (2021). A comprehensive review of natural bioactive films with controlled release characteristics and their applications in foods and pharmaceuticals. *Trends in Food Science and Technology, 112*, 690–707. https://doi.org/10.1016/j.tifs.2021.03.053

Chapter 11
Sustainable Packaging Solution from Agriculture Waste: Production of Bioplastic, Biocomposite, Biopolymer

Sakshi Gupta, Akashdeep Kaur, and Gargi Ghoshal

Abbreviations

AL	Alkaline lignin
CA	Cellulose acetate
CH	Chitosan
CMC	Carboxymethyl cellulose
CN	Cellulose nitrate
CNC	Cellulose nanocrystals
FAO	Food and Agriculture Organization
GPE	Garlic peel extract
LDP	Low-density polyethylene
LSS	Lignin sulfonate
PBS	Polybutylene succinate
PE	Polyethylene
PHAs	Polyhydroxylalkanoates
PLA	Polylactic acid
PVA	Polyvinyl alcohol
SB	Starch-based
SPI	Soy protein isolate

S. Gupta · G. Ghoshal (✉)
Dr. S. S. Bhatnagar University Institute of Chemical Engineering & Technology, Panjab University, Chandigarh, India

A. Kaur
Energy Research Centre, Panjab University, Chandigarh, India

J. Arora et al. (eds.), *Transforming Agriculture Residues for Sustainable Development*, Waste as a Resource, https://doi.org/10.1007/978-3-031-61133-9_11

11.1 Introduction

Due to increasing environmental concerns, synthetic plastics continue to be replaced by bioplastics. The use of synthetic plastics has several disadvantages, including resistance to degradation, toxicity after combustion, and accumulation in the environment (Bayram et al., 2021). Food loss and waste caused by the worthless depletion of food resources pose challenges to resilient and sustainable food chain ecosystems. According to the Food and Agriculture Organization of the United Nations, food waste is a reduction in the quality or quantity of food produced for consumption. The Food and Agriculture Organization (FAO) estimates that the food packaging market has an economic impact of $15.4 billion on the packaging sector. The smart and active packaging solutions market is expected to surpass the $26.7 billion smart packaging market by 2024. Losses occur at various stages, such as during food production, handling, storage, and processing. Food waste is generated during distribution and consumption in the food supply. Improving environmental resilience through new environmental legislation is forcing the industry to seek more environmentally friendly solutions, especially in food packaging. (Jayakumar et al., 2023). To minimize food loss, the 3Rs, i.e., reuse, reduce, and recycle can be used. Advances in technology have led to the development of various biodegradable packaging materials. These biodegradable materials are expensive and need to be replaced with cheaper materials, (Acharjee et al., 2023). Increasing population demands lead to product shortages and require cost-effective energy production from agricultural residues. The disposal of agricultural waste and other solid waste is a major problem today, as is food waste and loss. Agricultural waste is the residue produced during the production and processing of various foods (Varghese et al., 2023). Agricultural wastes contain a variety of low-cost resources and byproducts that can be used as bioplastics, biocomposites, antimicrobials, and other antioxidants with different properties. These include bioactive compounds and biopolymers (Merino et al., 2022). Vegetables, fruits, and plant wastes were successfully processed to separate natural biopolymers such as pectin, starch, alginate, lignin, hemicellulose, and cellulose. Bioplastics can serve the purpose of sustainable food packaging as they offer many benefits such as being non-toxic, biodegradable, biocompatible with other materials, and useful for extending the shelf life (Kadzińska et al., 2019). Bioplastics account for approximately 1% of the world's total plastic production of 370 million tons. The annual growth rate is 30% until 2025 (Leal Filho et al., 2021). In 2018, global packaging demand was approximately USD 974 billion, and demand was highest in Asia (40%), followed by North America (21%). In short, the main purpose of food packaging is to extend the shelf life of food. A variety of packaging materials are used to protect the food inside, maintain quality and safety until consumption, and prevent side effects. (Bayram et al., 2021). These bioplastics require modern technology to overcome their associated drawbacks. Extraction and purification of these substances help promote sustainable development. Various plant parts such as stems, leaves, seed pods, and straws have been used for manufacturing packaging materials (Varghese

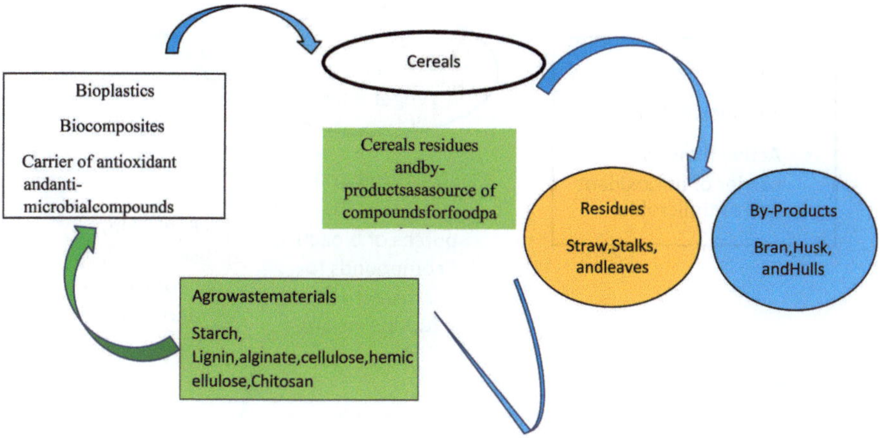

Fig. 11.1 Application of cereal-waste products in food packaging

et al., 2023). Dairy products, cereal, and sugarcane products produce residue that can be used to reduce waste generation. After harvesting grains such as wheat, rice, maize, and barley, residues remain in the form of stalks and stems (stumps) (Ali et al., 2019). Figure 11.1 shows the use of cereal residues and byproducts as raw materials for food packaging materials. These residues are used as animal feed or left on the ground to prevent erosion. However, these approaches are less efficient and can promote the spread of unwanted pests. Incineration of residues causes significant air pollution and nutrient loss, with negative effects on the environment and human health. Therefore, the best option is to use these residues as a significant reserve of fibrous biomass that can be used economically to produce paper and bioplastics. (Bilo et al., 2018). Some agricultural wastes containing lignocellulose can be degraded by fungi using a filamentous network called mycelium, which primarily contains chitin, glucans, proteins, and lipids. Using agricultural waste as raw material for bioplastic production solves the disposal problem, minimizes environmental pollution, and reduces the use of petroleum-based synthetic polymers. These biomaterials have different properties. Mechanical properties are important for processing and packaging applications. The mechanical properties of these materials can be improved by adding lignocellulosic nanofibers, cellulose crystals, and fibers resulting in rigid, stiffer, and harder bio-packaging materials. The shell that covers the body of an organism is composed of calcium carbonate, protein, polysaccharides, chitin, etc. Calcium carbonate is often used in material development. The addition of these materials to films improves the properties of packaging materials (Deshwal et al., 2021).

Food waste is rich in health-promoting bioactive compounds such as polyphenols, proteins, lipids, vitamins, and dietary fibers, which can improve the nutritional, functional, and technical properties of foods. Over the past decade, numerous studies have been conducted to examine how it can be improved. This waste can be used to

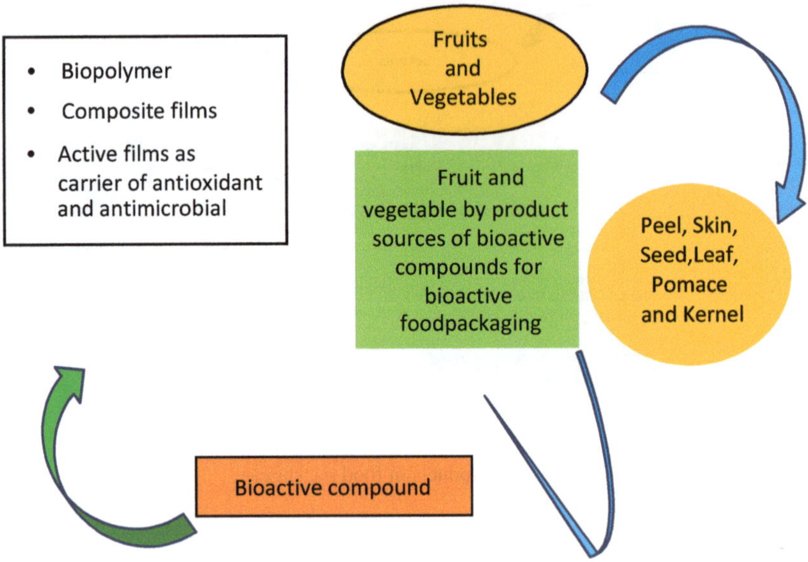

Fig. 11.2 Application of fruits and vegetables waste products in food packaging

produce food raw materials, functional foods, dietary supplements, pharmaceuticals, and cosmetics. (Ng et al., 2020). Plastics are widely used as food packaging materials due to their low cost, ease of processing, ability to manufacture in various shapes, convenience, and desirable physicochemical properties. Figure 11.2 shows the use of fruit and vegetable waste as byproducts in food packaging.

11.2 Agricultural Waste Sources

11.2.1 Cereal Crop Residues and Cereal Processing Byproducts

Fifty percent of the world's population depends on grains for food. Grains include wheat, rice, barley, millet, and rye for food supply. Cereals are milled to make various products. The grinding process produces bran, skin, and other byproducts used in the production of food packaging materials. Various residues and byproducts are produced (Deshwal et al., 2021). Cereal crop residues such as straw, leaf, and stalk residues are used as animal feed or left on the ground to prevent erosion. However, this can lead to the appearance of unwanted pests. Incineration of residue causes air pollution. The best way to use them is to generate renewable energy in the form of biofuels, metabolites, and biofilms. (Kolawole & Iyiola, 2023). A variety of treatments are used to utilize agricultural waste as packaging material. These treatments help to change the nature of agricultural waste. The first physical process is done to reduce the size, such as crushing, grinding, sieving, sifting, and milling. Heat treatment sterilizes raw materials and prevents the development of microorganisms. It also improves substrate utilization by promoting growth and increasing reaction rates. Chemical treatments such as adhesives and plasticizers can help change the properties of agricultural waste and improve its adhesive and assimilability properties. Alkaline treatment is performed to remove impurities. Biological treatments involve the addition of microbes to remove lignin and break down the waxy epidermis (Deshwal et al., 2021).

Rice

Rice is the third most widely produced crop. The main rice-producing countries are China, India, and Indonesia. These countries account for 60% of the total world's production. Two million hectares of rice cultivation generate 1.5 million tons of residue (Deshwal et al., 2021). Rice straw is a solid waste as it is insoluble in organic solvents and has low thermo-plastic properties. The introduction of acetyl groups by modification of rice straw using acetic anhydride in the presence of mild acids contributed to the development of thermoplastic films (Lu et al., 2021). The composition of the rice straw includes cellulose (32–47%), hemicellulose (19–27%), and lignin (5–24%). These are used in the production of fuel ethanol. One kg of harvested crop produces 1–1.5 kg of straw. Rice straw is mainly used in the production of cellulose-based bioplastic films. Rice husk is used as reinforcing material in polymer composites composed of polypropylene, polyethylene, and polylactic acid as a matrix. Rice husks and milling byproducts were incorporated into starch films in the form of cellulose nanofibers. These ensure better mechanical properties and optical and barrier properties. Rice bran is used for film formation (Bilo et al., 2018). Rice husk is mainly used for fuel production, building materials, bioplastics, and fertilizer production. Producing bioplastics from rice husk provides an alternative to synthetic plastics and allows rice husk waste to be used for non-energy production. Three different bioplastics are produced from rice husk: carboxy methyl cellulose (CMC), cellulose acetate (CA), and cellulose nitrate (CN). The first step involves the extraction of cellulose. Cellulose goes through several processes and reactions to produce the final bioplastic. To obtain cellulose, rice husks are pretreated by washing and drying. After removing impurities, it is crushed and sieved to obtain fine particles. The powder is alkaline treated and heated to remove the silica content and some lignin and hemicellulose. The amorphous portion is removed by adding a certain concentration of acid and heating. This helps neutralize the alkali left after the alkali treatment. This is called acid hydrolysis. This step is followed by bleaching to remove the dye. Add bleach to the colored cellulose and heat the mass. White cellulose is formed. The different processes used to produce bioplastics such as CMC, CA, and CN are etherification, acetyl at ion, and nitration (Yeboah et al., 2022).

Wheat

China is the largest producer of wheat, followed by India, the United States, and Russia. Together, they contribute to 45% of the global wheat production (Kogan, 2023). Six-hundred and fifty million tons of wheat straw are produced every year (Cao et al., 2017). These wheat straws can be used as filler in packaging materials and help improve the suitability of food packaging. Pretreatments such as drying and cutting are performed to make the straw suitable for mixing with pellets and developing biocomposites. Other byproducts from wheat plants are bran, hulls, and husk (Bascón-Villegas et al., 2022). Straws can be used to make environmentally friendly plastics used as disposable bottles and packaging materials. Five percent of straw production can be used for biodegradable packaging. Lignin, a component of wheat straw, can be used to make biodegradable plastics (Rai & Choure, 2023). Bran is a waste product generated during grinding. The composition of wheat bran, such as starch and protein, influences its properties. Combining wheat and corn husks increases tensile and impact strength. Citric acid contained in the hemicellulose of wheat straw and wheat byproducts contributes to the softening effect of food packaging and improves the properties of food packaging. Wheat gluten can be used as a binding agent with fiber. The properties of starch and gluten-based biocomposites were improved by reinforcing them with fibers. These play an important role in food packaging. These starch-based biocomposites produce bio-plastic sheets (Mohammed et al., 2021). Table 11.1 shows the utilization of grain waste using different methods.

Barley

The fourth largest grain is barley, with 50 million cultivated areas and 150 million tons of grain produced worldwide. Barley hulls and byproducts of the brewing and milling industries are used as animal feed. The ingredient in barley straw is cellulose,

Table 11.1 Utilization of cereal Agro-waste product using different techniques

Cereal agro-waste product	Techniques	Observation	References
Chitosan-based film with the addition of nano rice straw	Casting and wet mixing	Food packaging bags	Elhussieny et al. (2020)
Wheat straw hemicellulose film with cellulose nanocrystals	Solution casting	Enhanced in water vapor barrier property Application as a coating on foods	Pereira et al. (2017)
Rice straw mixed in cellulose acetate films	Solution casting	Enhanced mechanical properties	Hassan et al. (2019)
Starch film mixed with starch-based nanocellulose as reinforcement	Ultrasonication and Alkaline digestion	Water-resistant packaging film	Asem et al. (2018)

which is used as a packaging material. Barley straw and barley husk are used as leavening agents. Barley husk is composed of 30–35% cellulose, 30–33% hemicellulose, and 17–25% lignin. Various types of antioxidants found in barley husks are used in frozen swordfish along with low-density polyethylene for food packaging. This helped prevent lipid damage and increase the oxidative stability of the fish. Therefore, these byproducts can help improve the properties of packaging films such as optical, thermal, and mechanical properties (Nandiyanto et al., 2021). Cellulose nanocrystals are formed from barley husk and barley straw. These are used as fillers in chitosan-based polyvinyl alcohol (PVA) nanocomposite films with better functional properties. Better tensile strength was observed in barley hull reinforced composites (Perumal et al., 2018). Barley straw is covered with a crystalline wax layer and is used in a wide range of applications as a barrier and protective coating. Bio-based products are made from barley husks using alkaline and acid extraction to yield cellulose and hemicellulose. Natural hemicellulose has short chains and low molecular weight, so it does not form durable shapes. Structural changes are required to improve strength. Reducing the hydrophilicity of hemicellulose requires various techniques such as enzymatic treatment, esterification, etherification, oxidation, coupling, and cross-linking. Arabinoxylan was obtained by alkaline extraction from barley hulls to develop oxygen barrier films or coatings. Cellulose fibers and starch form more resistant films with good mechanical and thermal properties. Cellulose nanocomposites made from barley husk and straw are used as fillers in chitosan-based PVA nanocomposite films. Barley hull reinforced composites showed better tensile strength and elongation at break than softwood composites (Bhardwaj et al., 2020).

Corn

Corn is made up of agricultural residues such as cobs, stalks, husks, and leaves that remain after harvesting, called corn stalks. These byproducts are used in animal feed, ethanol production, and packaging materials. Corn cobs are the largest agricultural waste, accounting for 20% of crop residue. Corn cobs are used to make paper because they are composed of 45% cellulose, 35% hemicellulose, and 15% lignin. Corn cobs help improve the properties of packaging materials. Corn husks are the outer covering of corn and are used as packaging materials. Corn husks consist of high cellulose content (31–39%), low lignin content (2–14%), and ash content of 4–5%. Corn husks can be used as packaging materials in pulp form or nanofiber form. Chitosan and corn cob-based films with improved tensile strength and thermal stability are produced (Bhardwaj et al., 2020). Corn husk is mainly composed of hemicelluloses and lignin. Corn husks are a byproduct of corn harvesting and processing. Corn husks are used to obtain nanocellulose for the reinforcement of composite materials and as an adsorbent. Composite materials reinforced with corn husk have advantages such as heat resistance, flame retardancy, sound absorption, and mechanical properties. Cellulose fibers were obtained from the husk and stalk of corn plants. Biomass serves as a substrate for producing biomaterials to replace non-biodegradable materials. Therefore, corn husks help reduce environmental pollution and waste generation. Various biocomposites such as flame-retardant composites, polyester composites, polypropylene composites, epoxy composites, polylactic acid composites, polyethylene composites, and starch composites are prepared from corn husk (Ratna et al., 2022). Corn starch and chitosan films were prepared in 9 wt% clove oil and the structure of the films was characterized. The produced film has a uniform shape and smooth surface. This film can be used as active packaging and can replace packaging films made from natural starch (Hasan et al., 2023).

Oat

Oat husk is composed of 30–35% cellulose and hemicellulose and 2–10% lignin. Isolation of cellulose nanofibrils from oat hulls by oxidation and high-pressure homogenization is useful for the development of nanopackaging materials from agricultural wastes. Incorporating oat hulls into biodegradable polymers increases tensile and flexural strength. Cellulose nanofibers from oat hulls are removed by high-pressure homogenization and oxidation. This will help in the development of nanopackaging materials from agricultural waste. The addition of 5% nanocellulose from oat hulls helps in enhancing the tensile properties, water vapor barrier, and opacity of packaging films (Deshwal et al., 2021). Oat bran is a byproduct obtained when oats or oat flakes are ground and then the ground material is fractionated into fractions of up to 50% of the starting material. Oat bran contains beta-glucan and total fiber content. These are used for obtaining β-glucan and manufacturing β-glucan products. Oat bran defatted with supercritical carbon dioxide had increased β-glucan enrichment (more than 40%). In contrast, the defatted bran fraction was enriched up to 25%. Oat bran also contains arabinoxylans, oligosaccharides, tools, and phenolic compounds (Dapčević-Hadnađev et al., 2018). The use of natural starches in industry helps solve problems such as synergism, thermal instability, and pH sensitivity. This study was conducted to investigate the interaction between non-starch jicama polysaccharides and oat starch. The microstructure of the composite system consisting of oat starch, non-starch storage, distribution, processing, packaging, retail, and household consumption. The main plant forms producing food waste are roots, tubers, and oilseeds, accounting for 25%. Fruits and vegetables are the second most important source at 21%, and cereals and pulses are the third at 14% (FAO, 2019). The production of wine, juice, sugar, and other industrial products contributes significantly to the production of byproducts. Although such residues are usually left behind or discarded, they are an excellent source of bioactive substances such as polyphenols, vitamins, and minerals. A detailed evaluation of recent developments in synthetic or bio-based films enriched with byproduct components is presented, focusing on the potential use of byproduct extracts in food packaging materials (Dilucia et al., 2020).

Fruit and Vegetable Byproducts

Post-harvest processing and storage of fruits and vegetables generate large amounts of processing waste. This waste is primarily the result of inefficient treatment and storage processes, which inevitably generate many by-products (Sharma et al., 2021). On the other hand, these agricultural and processing waste resources have important biological resource value due to their high carbohydrate and fiber contents and the presence of numerous bioactive chemicals (Majerska et al., 2019). Therefore, byproducts from the processing of fruits and vegetables, which are rich sources of beneficial and nutritional compounds, can be used to enhance artificial materials by adding biologically active ingredients and additional nutrients to foods. They can be used in a variety of ways. They can also be used to produce biodegradable biopolymers (Sani et al., 2022). Fruit and vegetable waste is rich in bioactive substances such as hydroxyl cinnamates, phloretin, quercetin, catechins, procyanidins, epicatechin, and chlorogenic acid from apple pomace (Li et al., 2020a, b). The asparagus peel contains rutin, dietary fiber, peroxidases, phenols, flavonoids, hydroxyl cinnamic acids, and saponins (Cruz-Tirado et al., 2019). Blueberry pomace contains anthocyanins, cinnamic acid derivatives, and flavanols (Avram et al., 2017). Citrus peel contains neohesperidin, naringin, and hesperidin (Singh et al., 2020). Byproducts of the sugar industry consist of sugarcane bagasse and sugar beet pulp. Sugarcane bagasse is used as fuel or animal feed in developed countries. They are used for a variety of biotechnological and non-biotechnological applications. The main component of bagasse is hemicellulose, which accounts for 33.5% and is an important component in the formation of biodegradable packaging. The chemical composition of bagasse is 52.42% cellulose, 21.69% lignin, 73.92% holocellulose, 45.3% α-cellulose, and 2.73% extractable dry weight. Bagasse has various advantages such as low cost, abundant availability, low density, and properties, making it an ideal raw material for food packaging applications. Hemicellulose from sugarcane bagasse was obtained by alkaline treatment and used to develop an environmentally friendly film with good water permeability properties. Sugarcane bagasse fibers reinforced with cellulose nanocrystals were used as packaging material (Bhardwaj et al., 2020). Another byproduct of sugar mills is sugar beet pulp, which is composed of 20% cellulose, 25% hemicelluloses, and 25–30% pectin. These are used as animal feed, fuel sources, and platform chemicals. Film formation involves various processes such as cryogenic grinding, homogenization, alkaline extraction, and bleaching. Using pressure heating technology, sugar beet pulp can be used for packaging (Saeidy et al., 2023). Numerous studies have shown how these substances can be used as an innovative and cost-effective natural source of fiber, pectin, antioxidants, enzymes, antibacterial, antifungal, food additives, and essential oils (Sani et al., 2022). Figure 11.3 shows the classification of fruit and vegetable waste derived from natural colors. New methods are developed to produce biocomposites from plant waste. This is the thermo-mechanical treatment of various residues in the presence of small amounts of water and represents a state-of-the-art advance for the complete conversion of fruit and vegetable biomass into

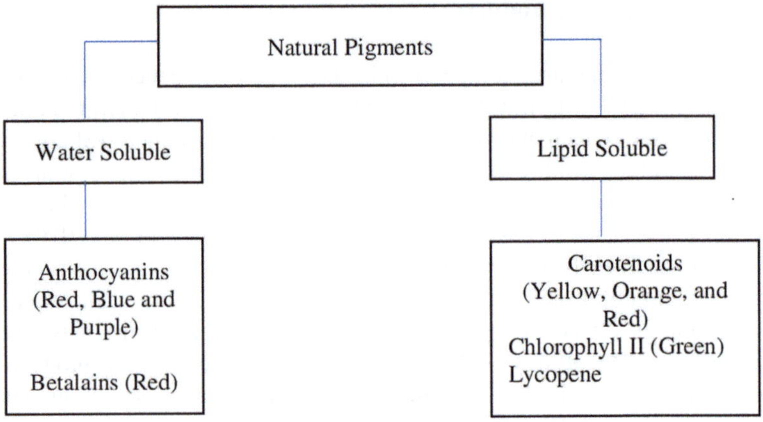

Fig. 11.3 Classification of fruit and vegetable waste-derived natural colors

biocomposites. Molded parts are processed at various temperatures, pressures, and times. The results showed that the morphological, optical, barrier and antioxidant properties of the biocomposites were improved (Merino & Athanassiou, 2023).

Techniques for Obtaining Value-Added Components from Fruit and Vegetable Waste

Processing of byproducts poses serious environmental, economic, and nutrient loss problems, and the processing of fruits and vegetables produces byproducts such as seeds, peels, leaves, gums, and pulp. These materials are valuable because they contain nutrients such as fiber, pectin, starch, carotenoids, polyphenols, vitamins, and enzymes. As a result, the use of these bioactive molecules as a natural complement to synthetic antioxidants and antimicrobials has become widely used in various food, pharmaceutical, and healthcare industries. Traditional extraction techniques (soaking, boiling, osmosis) and new techniques (supercritical fluids, microwaves, dielectric barrier discharge plasma, high hydrostatic pressure, and ultrasound) have been used to extract bioactive valuables from fruit and vegetable byproducts (Rifna et al., 2023). Traditional extraction techniques have many limitations, such as long processing times, high energy, and the need for solvents, making them effective for industrial applications. Table 11.2 shows the extraction techniques for bioactive compounds. However, combining traditional approaches with modern technology has revealed great potential for the processing of fruit and vegetable waste (Mohd Basri et al., 2021).

Table 11.2 Techniques for extracting bioactive components from fruit and vegetable byproducts

Results	Extraction method	Functional components	References
Carotene and tocopherol extraction from pumpkin	Supercritical carbon dioxide	β-carotene and α-tocopherol	Wang et al. (2017)
Recovery of bioactive chemicals from avocado peel waste	Pressurized liquid extraction	Phenolic compounds	Figueroa et al. (2018)
Recovery of carotenoids from carrot peel	Supercritical fluid	Carotenoids	De Andrade Lima et al. (2018)
Pectin extraction from pomelo peel	Acidic normal heat, ultrasonic and microwave energy, and subcritical water dynamics extraction	Pectin	Liew et al. (2019)
Purification of antioxidant phenolics from discarded peels of jackfruit (*Artocarpushetero phyllus* Lam.)	Microwave and ultrasonic	Gallic acid, catechin, and chlorogenic acid	Jiang et al. (2019)
Extraction of phenolic compounds and antioxidant properties from banana peel	Microwave and water	Phenolics and antioxidants	Vu et al. (2019)
Integrated biomaterial made from the peel of the *Opuntia ficus-indica*	Microwave	Pectin and betanin	Ciriminna et al. (2019)
Effects of plasma discharge barrier dielectric excitation Frequency of the enzyme activity, antioxidant capability, and phenolic content of apples and apple juice	Dielectric barrier discharge plasma	Phenolics and polyphenol enzyme inactivation	Farias et al. (2020)
Pectin and polyphenols are recovered from discarded tomato peel	High hydrostatic pressure	Polyphenols and pectin	Grassino et al. (2020)
Bioactive chemicals are recovered from pomegranate (*Punica granatum* L.) peel	Pressurized liquid	Bioactive components	García et al. (2021)

11.3 Biopolymer Synthesis from Agro-wastes

11.3.1 Starch

Starch is the most promising polysaccharide and is composed of amylose (20–25%) and amylopectin (75–80%). Amylose molecules are made up of 200–20,000 glucose units. In amylopectin, α (1→4)-linked glucosyl units are connected by α (1→6) branches. Starch occurs as a stored carbohydrate in rice, potatoes, corn, and cereal grains. Starch from fruit waste has unique properties that have the potential to expand the industrial use of starch. Fruits that represent new starch sources include kiwi fruit, pineapple, mango, litchi, tamarind, longan, loquat, annatto, jackfruit, avocado, apple, and banana (Kringel et al., 2020). These are available in granular form. This starch plays an important role as a filler, coating, and film. They have several advantages as they are readily available, inexpensive, and have good mechanical properties. Starch forms strong hydrogen bonds, resulting in ordered crystalline domains. Starch exhibits gelatinizing properties that allow it to be converted into thermoplastic starch. Although starch exhibits thermoplastic behavior, it is not commonly used because it is sensitive to moisture and has poor mechanical properties (Maraveas, 2020; Bayram et al., 2021). It is used with the addition of fillers and reinforcing materials. Bioplastic films are manufactured using various techniques such as mixing, solution casting, and chemical synthesis. Intracellular polymers such as starch can be obtained by acid-base treatment or solvent-assisted extraction. In wet extraction, various extractants are used to dilute the starch in the liquid phase. Starch is obtained through an extraction process using ultrasound. For alkaline leaching of starch, the material was treated with an alkaline solution at 50°C for 30 minutes and washed with D.H2O to obtain starch. Starch can also be obtained by acid-base or acid treatment. The isolated starch was used to make bioplastics. A fully biodegradable bio-based starch polymer was formed from corn or potato starch using additives. This resin is suitable for producing starch-based bulk packaging materials. Rice bran is composed of 70% starch, and the starch content forms a biodegradable film (Gadhave et al., 2018). Wheat bran is mostly composed of starch, which together with sorbitol has good mechanical properties. After starch is extracted, it must be converted into a thermoplastic starch. This occurs through enzymatic saccharification and microbial fermentation. These are further modified with plasticizers to market them hydrophilic. Cassava starch is modified with distilled water, vinegar, and glycerol to create a bioplastic film. The liquid glucose obtained after enzymatic treatment was further fermented by bacteria to obtain lactic acid, succinic acid, PHA, and 1, 3-propanediol. Condensation polymerization has been used to produce bioplastics (Coppola et al., 2021). Extracting starch from used avocado seeds makes an accounting rate of return of 75% and a break-even analysis of 82%. Since the extraction process uses few chemicals and the resulting waste can be processed into high-value commodities, the technology is environmentally sustainable and can improve company profitability (Tesfaye et al., 2021). Starch is a cost-effective, environmentally friendly, and biodegradable material that is used as an alternative to traditional non-biodegradable plastics.

11.3.2 Cellulose

Cellulose is composed of repeating D-glucose units linked together by β-1, 4 glycosidic bonds. Cellulose is biodegradable, recyclable, biocompatible, and durable. They are used as coating materials, laminates, films, and nanocomposites

(Wang et al., 2021). Cellulose is highly crystalline and insoluble in water, so it is processed into cellophane films. Cellophane can be coated to increase the moisture barrier in meat, cheese, and confectionery products (Bayram et al., 2021). Various biopolymers can be manufactured from monomers obtained from cellulose. Cellulose-based forms such as nanocellulose fibers and cellulose derivatives are used to form biocomposites. These endow the film with many multi-functional properties (Phiri et al., 2023). Lignin acts as a physical barrier to the extraction of cellulose and hemicellulose. Therefore, pretreatment is required to hydrolyze cellulose and hemicellulose using enzymatic and chemical methods. Pretreatment methods include physical, chemical, biological, and physicochemical processes. Alkaline hydrogen peroxide, steam explosion, and organic solvents are used to extract lignocellulosic compounds. Other methods include ammonia fiber blasting, hot-water extraction under pressure, co-solvent enhanced lignocellulose fractionation, ultrasound, microwave, and deep solvent. Using lingo cellulosic materials to produce bioplastics reduces costs and helps to solve environmental problems (Mujtaba et al., 2023). Several studies have been conducted to extract cellulose from grains such as wheat straw, rice husks, corn stalks, and sugarcane bagasse. Extraction methods include physical, chemical, biological, and biochemical methods. Cellulose nanocrystals were formed from sugarcane straw through sulfuric acid hydrolysis (Lu et al., 2021). Cellulose was removed from rice meal and converted to carboxy methyl cellulose for use in packaging. Cellulose fibers were obtained from rice straw by pulping methods using potassium hydroxide, soda, and sodium sulfate method. These fibers were used to make sheets of paper from soda ash pulp. Citric acid contained in hemicelluloses obtained from wheat straw enhances moisture barrier properties. Alkaline treatment of wheat straw nanofibers doubled the cellulose content through hydrolysis of hemicellulose and lignin. Cellulose is found in barley straws and husks used in packaging films. Sugarcane bagasse, which is used to develop biodegradable packaging films and coatings, contains 33.5% hemicellulose. Hemicellulose is obtained from sugarcane bagasse through alkali treatment, alkali-peroxide treatment, and acid hydrolysis. Films can be formed from these using extrusion techniques. Hemicellulose from sugarcane bagasse is used as a plasticizer in cellulose acetate films. Nanocellulose from sugarcane bagasse was used, which is renewable, strong, and environmentally friendly. Its application techniques include homogenization, microfluidization, and mechanical grinding (Deshwal et al., 2021). Combinations of two polysaccharides, such as chitosan and cellulose, are receiving increasing attention. Composite materials made of cellulose and chitosan (and their derivatives) are of great importance. On the other hand, different structures (mainly cellulose) can be coated with thin films (mainly chitosan), nanofibers, or nanoparticles. This wide range of structures offers diverse properties for a variety of high-tech applications such as packaging, medical devices, and advanced sorbents and filters (Strnad & Zemljič, 2023).

11.3.3 Pectin

Pectin is composed of heterogeneous polysaccharides and has α-(1,4)-D-galacturonic acid in its structure. Pectin is commonly found in agricultural waste. These can be modified by a demethylation process in the presence of calcium ions. Due to their low moisture resistance, these films are typically blended with materials such as chitosan and pectin/poly blends (Bayram et al., 2021). The main component of pectin is homo galacturonan. Microwave-assisted extraction, alkaline extraction, ultrasound-assisted extraction, and ultrasound-assisted microwave extraction are used to extract pectin. Pectin is obtained through a multi-step process. Acid treatment or alkaline leaching are the two most used methods for pectin extraction. Citric acid and hydrochloric acid are suitable for acid treatment. Alkaline leaching uses Na2SO4 and NaHSO4 for extraction (Choudhary et al., 2022). Casting is a versatile method for producing pectin-based films. A pectin solution is mixed with a plasticizer. These are dried under controlled conditions of temperature and humidity conditions. Antimicrobial and antioxidant compounds can be added to improve its properties (Mellinas et al., 2020). Citrus peels and apple pomace, which are byproducts of the juice industry, are rich sources of pectin and contain large amounts of pectin. Table 11.3 shows different techniques for extracting pectin from fruit and vegetable waste. Peel (e.g., citrus, pomelo, mango, banana, dragon fruit, grapefruit, jackfruit, and berry fruit), pomace (e.g., apples, sugar beets), rinds (watermelon), and husks (cocoa pods, fava beans) are examples of industrial byproducts. Depending on the source, the byproduct pectin concentration ranges from 5 to 40% (Kumar et al., 2023).

11.3.4 Chitosan

Chitosan has β-(1→4) linked *D*-glucosamine units and *N*-acetyl-*D*-glucosamine units. Chitosan is a pseudo-natural biopolymer derived from chitin. They are biocompatible, biodegradable, and used as cross-linking agents to improve the properties of cellulosic films. These are used as fillers and copolymers to improve the physio-chemical properties and durability of the bio-plastic films. Plastics incorporated into chitosan improve its barrier and anti-bacterial properties. Chitosan/corn cob-based composite films were prepared by solvent casting. Electron beam irradiation at a lower dose (10 kGy) improved the thermal stability and tensile strength of biocomposite films (Bhardwaj et al., 2020). Chitosan has disadvantages when used as a bio-plastic film because it has low solubility and therefore does not interact with other compounds used as a film. It exhibits excellent properties when combined with other polymers. Chitosan films are formed by dissolving chitosan in organic acids. The formed agar and chitosan films give stable films with good properties. Mixing the same concentration, agar, and chitosan decreased the water vapor permeability (Contessa et al., 2023). Composite films of chitosan and polyethylene glycol loaded

Table 11.3 Techniques of extraction of pectin from fruits and vegetables waste

Source	Extraction process	References
Citrus peel	Chemical extraction and microwave-assisted extraction	Kute et al. (2020)
Apple pomace	Chemical extraction and microwave-assisted extraction	Dranca et al. (2021)
Sugar beet	Microwave-assisted extraction	Abou-Elseoud et al. (2021)
Watermelon rinds	Chemical extraction	Manivannan et al. (2020)
Cocoa pod husk	Supercritical water extraction and chemical extraction	Muñoz-Almagro et al. (2019)
Jackfruit peel	Ultrasonic microwave-assisted extraction and chemical extraction	Xu et al. (2018)
Pomelo peel	Chemical extraction, ultrasonic microwave-assisted extraction, and supercritical water extraction	Liew et al. (2018)
Fig skin	Chemical extraction, ultrasonic microwave-assisted extraction, microwave-assisted extraction, and hot water extraction	Gharibzahedi et al. (2019)
Pomegranate peel	Chemical extraction	Pereira et al. (2016)
Sugarcane bagasse nanocellulose-based hydrogel	Induced gelation and ionic cross-linking for intelligent packaging of food	Lu et al. (2020)
Sugarcane bagasse cellulose film incorporated with cashew nut extract	Interfacial assembly for antioxidant and antimicrobial film	Lee et al. (2020)

with date palm fruit waste extract were fabricated (Zidan et al., 2023). Garlic peel extract (GPE) and chitosan (CH) were combined to create a sustainable multifunctional composite film for food packaging. This result suggests that GPE/CH film scan replaces petrochemical plastics in various commercial applications such as food packaging materials (Chaudhary et al., 2021).

11.3.5 Lignin

Lignin is the most well-known polymer substrate. It makes up 40% of the plant cell walls. It is produced from three monomers: *p-coumaric* alcohol, coniferyl alcohol, and sinapyl alcohol. Together they form a three-dimensional network. lignin is used in bio-packaging materials. Every year, large amounts of plant and fruit-based agricultural waste are generated around the world, placing a huge burden on the environment and sustainable development. Plant and fruit wastes are rich in water and valuable organic matter such as cellulose, minerals, carbohydrates, proteins, organic acids, lipids, and soluble sugars. Considering the compositional characteristics and circular economy of vegetable and fruit-based waste, many new conversion technologies have been developed for the treatment of vegetable and fruit-based wastes (hydrothermal gasification, ultrasound-assisted extraction, and synthesis of bioplastics, etc.) (Liu, 2023). Alkaline lignin (AL) and lignin sulfonate (LSS) were added to enzyme-modified soy protein isolate (SPI) biopolymer films at different concentrations to improve the physical and functional properties of the films (Zadeh et al., 2018). A variety of non-edible sources have been investigated as lingo cellulosic feedstocks, including corn straw, corn cob, rice husks, rice straw, wheat straw, and sugarcane bagasse. Biodegradable plastics can be obtained from these raw materials, potentially reducing the waste problems. Pretreatments such as alkaline hydrogen peroxide, steam explosion, and organic solvents are used to extract lingo cellulosic compounds. New methods such as ammonia fiber blasting have been developed for lignin extraction (Mujtaba et al., 2023). Pea hulls are renewable, readily available, abundant, and highly valuable agricultural waste. The purpose of the study was to isolate cellulose nanocrystals (CNC) from pea hulls and assess their ability to strengthen carboxy methyl cellulose (CMC) films (Li et al., 2020a, b).

11.4 Agro-waste Biocomposites

Biocomposites are the process of combining materials from natural resources to create new composite materials. Biocomposites can be formed naturally, either as reinforcements or as matrices, or as containing both. Traditional methods of producing biocomposites include compression molding, injection, and extraction. Other methods are continuously being developed to improve on traditional methods. One of these methods is additive manufacturing. These help form lightweight composite structures. Agro-industrial waste is converted into various biocomposite products by first extracting fibers, proteins, carbohydrates, organic acids, and oils from the waste followed by fermentation and enzymatic treatment. Examples of biodegradable materials include biocomposites made from natural fibers and fillers, and biodegradable polymers. Starch can be obtained through an extraction process. On the other hand, polybutylene succinate (PBS), polyethylene (PE), and polylactic acid (PLA) are produced by fermentation, chemical treatment, and polymerization of their respective monomers.

11.4.1 Poly Hydroxyl Alkanoates

(PHAs) are produced by bacterial fermentation (Kringel et al., 2020). Etherification is a method that can be used to produce carboxy methyl cellulose (CMC). Some biopolymers are made from modified starches, with PLA commonly used for food packaging, and PHA for injection molding and water-resistant packaging (Nayak &

Bhushan, 2019). The choice of manufacturing method is determined by many factors such as manufacturing cost, geometry, final design, raw material properties, size, and process limitations. Plant fibers can be divided into cottonseed fiber, rice husk fiber, and coir. Cellulose nanocrystals were obtained from potato peel waste and used as reinforcement in thermoplastic starch composites. A decrease in water vapor permeability was observed (Olad et al., 2020). Sweet potato residues are used to produce biocomposites for food packaging applications. A portion of it was added to poly (3-hydroxy butyrate-*co*-3-hydroxy valerate). Amounts range from 5 to 40% by weight at various temperatures up to 200 °C for 5 min, or 180 °C for 6 min. Good mixing of residues in the biopolymer matrix was achieved through interphase hydrogen interactions (Vannini et al., 2021). Rice straw properties such as fiber length, fiber extraction method, and conditions influence the formation of biocomposites. The strength of green composites based on polyacetic acid and rice straw increased with increasing fiber content. Rice straw and other polycopolymers were used to increase tensile strength. Therefore, the remaining rice straw residue can be used for packaging applications (Deshwal et al., 2021). Agricultural wastes such as manure, weeds, and sawdust are converted into energy and fertilizer, benefiting both the environment and the economy. The epoxy-based composite material is made from corn cob waste, creating an environmentally friendly waste disposal solution. They are applied in packaging materials and household products (Phiri et al., 2023). Additionally, fibers from fruit and vegetable seeds, pomace, and peel/skin have been used as fillers in the film matrix to improve the properties of biopolymers (Sohany et al., 2021).

11.5 Applications in Food Packaging

11.5.1 Antimicrobial and Antioxidant Properties

Antioxidants suppress free radical reactions and prevent cell damage. Antimicrobial compounds inactivate pathogenic and harmful microorganisms from food. Bioactive compounds found in agricultural waste provide antibacterial and antioxidant properties to bioplastics. Chitosan polymers, organic acids, and plant extracts are the products of choice. Compounds found in essential oils help prevent the growth of microorganisms such as. *Escherichia coli, Salmonella enterica*, and *Listeria monocytogenes*. Lignocellulosic biomass provides antioxidant properties to bioplastic films. Thermoplastic corn starch composites were prepared from rice and coffee husks through melt mixing and compression molding (Merino et al., 2022). A chitosan matrix made with the addition of purple sweet potato extract provides antioxidant properties to the bioplastic. The addition of these extracts improves the properties of bioplastic packaging materials by imparting antibacterial and antioxidant properties (Contessa et al., 2023). Antioxidants were obtained from barley hulls by acid hydrolysis. Frozen swordfish wrapped in low-density polyethylene (LDPE) material. The LDPE (low-density polyethylene) film was sprayed with an antioxidant that prevents lipid damage and improves oxidative stability in frozen swordfish. (Deshwal et al., 2021). It has been reported that a pea starch coating aqueous solution was combined with grape seed extract to analyze the antibacterial activity of the extract containing *L. monocytogenes*. After 4 days of storage, active packaging was able to significantly slow down the growth of the bacterial strain *Brochothrix thermosphacta* (Bayram et al., 2021). Green tea extract was added to a pectin-based coating solution to create a package with both antioxidant and antibacterial properties. Although the pectin coating reduced the number of bacteria in the samples, the antimicrobial effect of the included green tea extract tended to enhance this effect while significantly reducing lipid oxidation (Kang et al., 2007).

11.5.2 Mechanical and Barrier Properties

Various cereal residues provide mechanical and barrier properties to bioplastic films. Rice straw cellulose nanofibers were extracted by acid and alkali treatments, enzymes, and high-pressure homogenization, which resulted in improved tensile strength, high surface area-to-volume ratio, and mechanical properties. Strain and applied force are two parameters that influence the physical and mechanical properties of biopolymers and are influenced by polymer type, molecular weight, crystallinity, chemical composition, and polymer shape. On the other hand, the combination of films with different polymers can change their physical–mechanical and functional properties. For example, the properties of composite films can be improved and modified by combining polysaccharides (starch, alginate, cellulose, chitosan) and proteins (milk protein, soy protein, collagen, gelatin, zein, gluten). The addition of other substances such as emulsifiers or surfactants (improves emulsion stability), plasticizers (improves flexibility), and reinforcing agents (improves physical/mechanical properties) improves the physical properties of biodegradable films (Sani et al., 2022). The addition of rice straw composite nanofibers to PLA provides an environmentally friendly alternative in the development of biocomposites with improved mechanical, barrier, and optical properties that can be used in packaging films. Using 5–10% rice bran as a filler also improved elongation and biodegradability. The polyurethane manufacturing process produces intermediate polymerization products that can be added to wheat straw or LDPE composites to improve their mechanical properties. The citric acid present in hemicellulose derived from wheat-based byproducts improves moisture resistance and tensile properties and contributes to flexible packaging. The addition of starch and pectin affects the properties of extruded wheat bran leaves. High starch content improves tensile strength, water vapor barrier properties, and elongation at break. Corn cobs have been incorporated into biocomposites to improve interfacial bonding and thus mechanical properties (Deshwal et al., 2021). Figure 11.4 shows the use of fruit and vegetable byproducts in food packaging.

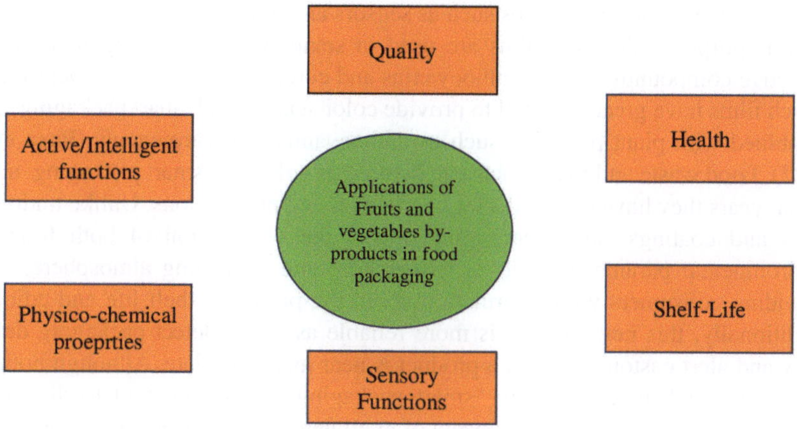

Fig. 11.4 Application of fruits and vegetable byproducts in food packaging

11.5.3 Active Packaging

A combination of edible packaging and active packaging is recommended. The bioactive compounds contained help increase the value of the product. The use of edible films and coatings made from food waste is receiving more attention, although the development of edible and active packaging materials is also encouraged. These coatings add value to objects by containing bioactive substances made from materials that would normally be thrown away (de Moraes Crizel et al., 2018). These bioactive compounds provide antimicrobial and antioxidant properties and help the emitter generate CO_2 to maintain a stable atmosphere within the package. Antimicrobial packaging prevents the growth of *L. monocytogenes,* which can survive and grow in the refrigerator at 2–4 °C. The addition of green tea extract to pectin-based coating solutions prevented microbial growth and provided antioxidant properties. This reduces lipid oxidation (Bayram et al., 2021). Edible films can be used as coatings and release agents in new packaging systems by modifying their features such as improving mechanical properties, reducing permeability, improving barrier properties, creating active antimicrobial surfaces, and increasing heat resistance (Alizadeh Sani et al., 2022).

11.5.4 Intelligent Packaging

Intelligent packaging measures changes that occur in food and communicates that information to consumers. It is useful for observing environmental conditions such as food environmental conditions and temperature. Color changes indicate food freshness or spoilage of food. Intelligent packaging helps identify gas molecules and pH values. Various systems such as sensors and indicators have been developed for this purpose. Chitosan films are used in smart packaging along with various bioactive compounds such as anthocyanins and curcumin derived from purple corn. Starch films have great potential to provide color sensors and smart packaging when combined with plant pigments such as anthocyanins and carotenoids (Pirsa et al., 2022). Food waste and byproducts can be used as colors for smart packaging, and in recent years they have received a lot of attention and publications. Unlike traditional films and coatings, smart packaging allows the observation of both food and environmental parameters such as temperature and packaging atmosphere, while providing consumers with information about the product's shelf life and edibility. Additionally, this new method is more reliable as it can detect packaging defects early and alert customers when a product is near its expiry date. Although both can be summarized under the term "smart packaging," it is important to distinguish between these benefits and the concept of intelligent packaging and active packaging technique. While active packaging serves a specific purpose for food and protects it from influencing factors like oxidative processes, intelligent packaging measures these changes and allows consumers to use them without affecting the food (Sohail et al., 2018). The focus of smart packaging is on detecting pH levels and gas molecules (H_2, O_2, NO_2, and CO_2) that are less visible to the consumer, especially when the package remains closed (Rawdkuen & Kaewprachu, 2019). Table 11.4 shows the fruit and vegetable processing byproducts being used in food packaging.

11.6 Future Prospects and Challenges

Large-scale production of environmentally sustainable plastics will not only help reduce the production of synthetic plastics but also consume waste from the food industry to produce bioplastics, biocomposites made from cellulose, pectin, and chitosan. This helps in waste management in agriculture. In biofilms, substrate composition, production processes, and other parameters can be modified to improve various properties such as thermal and mechanical properties. Filler particles can be arranged into specific shapes and layered structures to maximize properties. The production of bioplastics faces challenges in several ways. These include not only strength but also feasibility, and durability. Another major challenge is the availability of raw materials and desirable properties for bioplastics. Fruit and vegetable waste not only provides a variety of valuable and adaptable polymers but is also a rich source of bioactive substances that can be used to produce sustainable and environmentally friendly packaging materials. However, to develop a commercially viable packaging system based on food waste, future research will need to address costs, industrial feasibility, high productivity, low moisture, mechanical resistance, regulatory issues, customer concerns, and customer acceptance. Since they are made from natural resources, the use of food waste and byproducts in the production of edible films does not pose a threat to consumers or food safety and is considered an environmentally friendly material. Optimizing food-packaging interactions based on

Table 11.4 Fruit and vegetable processing by-products being used in food packaging

Byproducts	Film/coating matrics	Results	References
Licorice root extract	Chitosan	Lipid oxidation and microbial growth were inhibited by citric acid or licorice extract. Fish fillets with coatings containing citric acid or licorice extract have better quality and have a longer shelf life	Qiu et al. (2014)
Lemon peel essential oil and oregano leaves	Chitosan	Less mesophilic, lactic acid, entero bacteriaceae, mold, and yeast bacteria represent the extended shelf life of grilled breast meat when stored in the refrigerator	Sharafati Chaleshtori and Sharafati Chaleshtori (2017)
Apple peel	Apple peel	Lipid oxidation and microbial development were prevented by treatments. Beef flesh quality can be preserved using an edible covering made from apple peel	Shin et al. (2017)
Grape seed extract	Chitosan	The physicochemical characteristics of active films should be improved. Extend the shelf life of salmon in the refrigerator	Alves et al. (2018)
Pomegranate peel extract	Zein	The tensile strength, elongation, total phenolic content, and antioxidant activity of the films were improved by pomegranate peel extract. Integrating fruit waste reduces fat, protein oxidation, and cheese spoilage	Mushtaq et al. (2018)
Grape seed extract	Chitosan/ polycapro lactone/ nanocellulose	Improved physicochemical properties lead to a reduction in thiobarbituric acid. The quality of chicken breast meat has improved	Sogut and Seydim (2019)
Pineapple peel	Alginate	Increased antibacterial and antioxidant activities. Using pineapple skin can effectively preserve the color of red beef meat. Antioxidants in pineapple peels prevented lipid oxidation in beef samples	Lourenco et al. (2019)
Sesame pulp extract	Starch	The total amount of living bacteria, total nitrogen, and Thio-barbituric acid all decreased. When temperatures are extremely low, beef's shelf life is extended	Guo et al. (2021)
Kiwi fruit (Actinidia chinensis) peel extract	Watermelon rind pectin	Watermelon peel pectin films showed higher tensile strength and positive modulus The films' turbidity, elongation, and permeability to the water vapor increased together with the concentration of kiwi peel extract	Han and Song (2021)

(continued)

Table 11.4 (continued)

Byproducts	Film/coating matrics	Results	References
Grape seed extract	Pectin	The mechanical resistance, antimicrobial, and antioxidant properties raise the shelf life of peanuts and prevent acidity	Priyadarshi et al. (2022)

food waste or byproducts requires a forward-looking perspective based on legal considerations and continued technology input. The particles must be very thin and wide. However, this is a difficult task as it requires sophisticated machinery such as thin-line dispensers and high-pressure homogenizers. Attempts to reach nanoscale particles often lead to large size ranges and discrepancies. Governments should encourage biocomposite innovation through the implementation of policies and programs.

11.7 Conclusion

In recent years, there has been an increase in interest in the use of waste and byproducts from the agricultural industry and food production. The formation of sustainable biopolymer-based composites from agro-waste is a promising field that helps in addressing environmental issues and economic challenges with time. There has been an increase in interest in creating edible and biodegradable films and coatings to replace non-biodegradable plastic packaging materials. As an alternative to plastics, biopolymer-based packaging films or coatings utilizing carbohydrates, proteins, and lipids are becoming more popular. The potential uses of fruit and vegetable byproducts as biopolymers, biocomposites, active or intelligent packaging, edible films, and coatings, as well as other products, have all been highlighted in this chapter. Promising characteristics include biodegradability, renewability, lower production costs, reduced waste, and reduced environmental impact in bio-based safe, value-added, green materials. Recent advancements in electrospinning/electrospray technology, along with innovations in nanofibrillated cellulose concentration and the fabrication of bio-based films. Cellulose reinforcement is contributing to the demand for biofilms in packaging. Future research should therefore concentrate on the treatment of additives to enhance the mechanical, thermal, and barrier capabilities of associated biomaterials.

References

Abou-Elseoud, W. S., Hassan, E. A., & Hassan, M. L. (2021). Extraction of pectin from sugar beet pulp by enzymatic and ultrasound-assisted treatments. *Carbohydrate Polymer Technologies and Applications, 2*, 100042.

Acharjee, S. A., Bharali, P., Gogoi, B., Sorhie, V., Walling, B., & Alemtoshi. (2023). PHA-based bioplastic: A potential alternative to address microplastic pollution. *Water, Air, and Soil Pollution, 234*(1), 21.

Ali, M., Saleem, M., Khan, Z., & Watson, I. A. (2019). The use of crop residues for biofuel production. In *Biomass, biopolymer-based materials, and bioenergy* (pp. 369–395). Woodhead Publishing.

Alizadeh Sani, M., Tavassoli, M., Salim, S. A., Azizi-lalabadi, M., & McClements, D. J. (2022). Development of green halochromic smart and active packaging materials: TiO2 nanoparticle- and anthocyanin-loaded gelatin/κ-carrageenan films. *Food Hydrocolloids, 124*, 107324.

Alves, V. L., Rico, B. P., Cruz, R. M., Vicente, A. A., Khmelinskii, I., & Vieira, M. C. (2018). Preparation and characterization of a chitosan film with grape seed extract-carvacrol microcapsules and its effect on the shelf-life of refrigerated Salmon (*Salmo salar*). *LWT Journal, 89*, 525–534.

Asem, M., Nawawi, W. M. F. W., & Jimat, D. N. (2018, June). Evaluation of water absorption of polyvinyl alcohol-starch biocomposite reinforced with sugarcane bagasse nanofibre: Optimization using two-level factorial design. In *IOP conference series: Materials science and engineering* (Vol. 368, p. 012005). IOP Publishing.

Avram, A. M., Morin, P., Brownmiller, C., Howard, L. R., Sengupta, A., & Wickramasinghe, S. R. (2017). Concentrations of polyphenols from blueberry pomace extract using nanofiltration. *Food and Bioproducts Processing, 106*, 91–101.

Bascón-Villegas, I., Pereira, M., Espinosa, E., Sánchez-Gutiérrez, M., Rodríguez, A., & Pérez-Rodríguez, F. (2022). A new eco-friendly packaging system incorporating lignocellulose nanofibres from agri-food residues applied to fresh-cut lettuce. *Journal of Cleaner Production, 372*, 133597.

Bayram, B., Ozkan, G., Kostka, T., Capanoglu, E., & Esatbeyoglu, T. (2021). Valorization and application of fruit and vegetable wastes and by-products for food packaging materials. *Molecules, 26*(13), 4031.

Bhardwaj, A., Alam, T., Sharma, V., Alam, M. S., Hamid, H., & Deshwal, G. K. (2020). Lignocellulosic agricultural biomass as a biodegradable and eco-friendly alternative for polymer-based food packaging. *Journal of Packaging Technology and Research, 4*, 205–216.

Bilo, F., Pandini, S., Sartore, L., Depero, L. E., Gargiulo, G., Bonassi, A., et al. (2018). A sustainable bioplastic obtained from rice straw. *Journal of Cleaner Production, 200*, 357–368.

Cao, C., Xu, L., He, Y., Guo, L., Jin, H., & Huo, Z. (2017). High-efficiency gasification of wheat straw black liquor in supercritical water at high temperatures for hydrogen production. *Energy and Fuels, 31*(4), 3970–3978.

Chaudhary, B. U., Lingayat, S., Banerjee, A. N., & Kale, R. D. (2021). Development of multifunctional food packaging films based on waste garlic peel extract and chitosan. *International Journal of Biological Macromolecules, 192*, 479–490.

Choudhary, P., Pathak, A., Kumar, P., & Sharma, N. (2022). Commercial production of bioplastic from organic waste–derived biopolymers viz-a-viz waste treatment: A mini-review. *Biomass Conversion and Biorefinery*, 1–11.

Ciriminna, R., Fidalgo, A., Avellone, G., Danzì, C., Timpanaro, G., Locatelli, M., et al. (2019). Integral extraction of *Opuntia ficus-indica* peel bioproducts via microwave-assisted hydrodiffusion and hydrodistillation. *ACS Sustainable Chemistry and Engineering, 7*(8), 7884–7891.

Contessa, C. R., Rosa, G. S. D., Moraes, C. C., & Burkert, J. F. D. M. (2023). Agar-agar and chitosan as precursors in the synthesis of functional film for foods: A review. *Macromolecules, 3*(2), 275–289.

Coppola, G., Gaudio, M. T., Lopresto, C. G., Calabro, V., Curcio, S., & Chakraborty, S. (2021). Bioplastic from renewable biomass: A facile solution for a greener environment. *Earth Systems and Environment, 5*, 231–251.

Cruz-Tirado, J. P., Vejarano, R., Tapia-Blácido, D. R., Angelats-Silva, L. M., & Siche, R. (2019). The addition of sugarcane bagasse and asparagus peel enhances the properties of sweet potato starch foams. *Packaging Technology and Science, 32*(5), 227–237.

Dapčević-Hadnađev, T., Hadnađev, M., & Pojić, M. (2018). The healthy components of cereal by-products and their functional properties. In *Sustainable recovery and reutilization of cereal processing by-products* (pp. 27–61). Woodhead Publishing.

de Andrade Lima, M., Charalampopoulos, D., & Chatzifragkou, A. (2018). Optimisation and modelling of supercritical CO_2 extraction process of carotenoids from carrot peels. *The Journal of Supercritical Fluids, 133*, 94–102.

de Moraes Crizel, T., de Oliveira Rios, A., Alves, V. D., Bandarra, N., Moldão-Martins, M., & Flôres, S. H. (2018). Active food packaging prepared with chitosan and olive pomace. *Food Hydrocolloids, 74*, 139–150.

Deshwal, G. K., Alam, T., Panjagari, N. R., & Bhardwaj, A. (2021). Utilization of cereal crop residues, cereal milling, sugarcane and dairy processing by-products for sustainable packaging solutions. *Journal of Polymers and the Environment, 29*, 2046–2061.

Dilucia, F., Lacivita, V., Conte, A., & Del Nobile, M. A. (2020). Sustainable use of fruit and vegetable by-products to enhance food packaging performance. *Foods, 9*(7), 857.

Dranca, F., Talón, E., Vargas, M., & Oroian, M. (2021). Microwave vs. conventional extraction of pectin from *Malus domestica* 'Fălticeni'pomace and its potential use in hydrocolloid-based films. *Food Hydrocolloids, 121*, 107,026.

Elhussieny, A., Faisal, M., D'Angelo, G., Aboulkhair, N. T., Everitt, N. M., & Fahim, I. S. (2020). Valorisation of shrimp and rice straw waste into food packaging applications. *Ain Shams Engineering Journal, 11*(4), 1219–1226.

FAO. (2019). *The State of Food and Agriculture 2019*. Moving Forward on Food Loss and Waste Reduction., Licence: CC BY-NC-SA 3.0 IGO, Rome.

Farias, T. R., Rodrigues, S., & Fernandes, F. A. (2020). Effect of dielectric barrier discharge plasma excitation frequency on the enzymatic activity, antioxidant capacity and phenolic content of apple cubes and apple juice. *Food Research International, 136*, 109617.

Figueroa, J. G., Borrás-Linares, I., Lozano-Sánchez, J., Quirantes-Piné, R., & Segura-Carretero, A. (2018). Optimization of drying process and pressurized liquid extraction for recovery of bioactive compounds from avocado peel by-product. *Electrophoresis, 39*(15), 1908–1916.

Gadhave, R. V., Das, A., Mahanwar, P. A., & Gadekar, P. T. (2018, May). Starch based bio-plastics: The future of sustainable packaging. *Open Journal of Polymer Chemistry, 8*(2). https://www.scirp.org/journal/journalarticles.aspx?journalid=791

García, P., Fredes, C., Cea, I., Lozano-Sánchez, J., Leyva-Jiménez, F. J., Robert, P., et al. (2021). Recovery of bioactive compounds from pomegranate (*Punica granatum* L.) peel using pressurized liquid extraction. *Foods, 10*(2), 203.

Gharibzahedi, S. M. T., Smith, B., & Guo, Y. (2019). Pectin extraction from common fig skin by different methods: The physicochemical, rheological, functional, and structural evaluations. *International Journal of Biological Macromolecules, 136*, 275–283.

Grassino, A. N., Ostojić, J., Miletić, V., Djaković, S., Bosiljkov, T., Zorić, Z., et al. (2020). Application of high hydrostatic pressure and ultrasound-assisted extractions as a novel approach for pectin and polyphenols recovery from tomato peel waste. *Innovative Food Science and Emerging Technologies, 64*, 102424.

Guo, Z., Ge, X., Gou, Q., Yang, L., Han, M., Han, G., et al. (2021). Changes in chilled beef packaged in starch film containing sea buckthorn pomace extract and quality changes in the film during super-chilled storage. *Meat Science, 182*, 108620.

Han, H. S., & Song, K. B. (2021). Antioxidant properties of watermelon (*Citrullus lanatus*) rind Pectin films containing kiwifruit (Actnidia chinensis) peel extract and their application as chicken thigh packaging. *Food Packaging and Shelf Life, 28*(1), 10,063. https://doi.org/10.1016/J.FPSL.2021.100636

Hasan, M., Khaldun, I., Zatya, I., Rusman, R., & Nasir, M. (2023). Facile fabrication and characterization of an economical active packaging film based on corn starch–chitosan biocomposites incorporated with clove oil. *Journal of Food Measurement and Characterization, 17*(1), 306–316.

Hassan, M., Berglund, L., Abou-Zeid, R., Hassan, E., Abou-Elseoud, W., & Oksman, K. (2019). Nanocomposite film based on cellulose acetate and lignin-rich rice straw nanofibers. *Materials, 12*(4), 595.

Jayakumar, A., Radoor, S., Siengchin, S., Shin, G. H., & Kim, J. T. (2023). Recent progress of bioplastics in their properties, standards, certifications and regulations: A review. *Science of the Total Environment, 878*, 163156.

Jiang, Z., Shi, R., Chen, H., & Wang, Y. (2019). Ultrasonic microwave-assisted extraction coupled with macroporous resin chromatography for the purification of antioxidant phenolics from waste jackfruit (*Artocarpus heterophyllus* Lam.) peels. *Journal of Food Science and Technology, 56*, 3877–3886.

Kadzińska, J., Janowicz, M., Kalisz, S., Bryś, J., & Lenart, A. (2019). An overview of fruit and vegetable edible packaging materials. *Packaging Technology and Science, 32*(10), 483–495.

Kang, H. J., Jo, C., Kwon, J. H., Kim, J. H., Chung, H. J., & Byun, M. W. (2007). Effect of a pectin-based edible coating containing green tea powder on the quality of irradiated pork patty. *Food Control, 18*(5), 430–435.

Kogan, F. (2023). Global warming crop yield and food security. In *Remote sensing land surface changes: The 1981–2020 intensive global warming* (pp. 217–275). Springer.

Kolawole, A. S., & Iyiola, A. O. (2023). Environmental pollution: Threats, impact on biodiversity, and protection strategies. In *Sustainable utilization and conservation of Africa's biological resources and environment* (pp. 377–409). Springer Nature.

Kringel, D. H., Dias, A. R. G., Zavareze, E. D. R., & Gandra, E. A. (2020). Fruit wastes as promising sources of starch: Extraction, properties, and applications. *Starch-Stärke, 72*(3-4), 1900200.

Kumar, A., Ramakanth, D., Akhila, K., & Gaikwad, K. K. (2023). Influence of halloysite nanotubes/microfibrillated cellulose on pine leaves waste based ethylene scavenging composite paper for food packaging applications. *Applied Clay Science, 231*, 106726.

Kute, A. B., Mohapatra, D., Kotwaliwale, N., Giri, S. K., & Sawant, B. P. (2020). Characterization of pectin extracted from orange peel powder using microwave-assisted and acid extraction methods. *Agricultural Research, 9*, 241–248.

Leal Filho, W., Salvia, A. L., Bonoli, A., Saari, U. A., Voronova, V., Klõga, M., ... Barbir, J. (2021). An assessment of attitudes towards plastics and bioplastics in Europe. *Science of the Total Environment, 755*, 142732.

Lee, J. J. L., Cui, X., Chai, K. F., Zhao, G., & Chen, W. N. (2020). Interfacial assembly of a cashew nut (*Anacardium occidentale*) testa extract onto a cellulose-based film from sugarcane bagasse to produce an active packaging film with a pH-triggered release mechanism. *Food and Bioprocess Technology, 13*, 501–510.

Li, H., Shi, H., He, Y., Fei, X., & Peng, L. (2020a). Preparation and characterization of carboxymethyl cellulose-based composite films reinforced by cellulose nanocrystals derived from pea hull waste for food packaging applications. *International Journal of Biological Macromolecules, 164*, 4104–4112.

Li, W., Yang, R., Ying, D., Yu, J., Sanguansri, L., & Augustin, M. A. (2020b). Analysis of polyphenols in apple pomace: A comparative study of different extraction and hydrolysis procedures. *Industrial Crops and Products, 147*, 112250.

Liew, S. Q., Ngoh, G. C., Yusoff, R., & Teoh, W. H. (2018). Acid and deep eutectic solvent (DES) extraction of pectin from pomelo (*Citrus grandis* (L.) Osbeck) peels. *Biocatalysis and agricultural biotechnology, 13*, 1–11.

Liew, S. Q., Teoh, W. H., Yusoff, R., & Ngoh, G. C. (2019). Comparisons of process intensifying methods in the extraction of pectin from pomelo peel. *Chemical Engineering and Processing-Process Intensification, 143*, 107586.

Liu, Z. (2023). A review on the emerging conversion technology of cellulose, starch, lignin, protein and other organics from vegetable-fruit-based waste. *International Journal of Biological Macromolecules, 124804*.

Lourenço, S. C., Moldão-Martins, M., & Alves, V. D. (2019). Antioxidants of natural plant origins: From sources to food industry applications. *Molecules, 24*, 4132.

Lu, P., Yang, Y., Liu, R., Liu, X., Ma, J., Wu, M., & Wang, S. (2020). Preparation of sugarcane bagasse nanocellulose hydrogel as a colourimetric freshness indicator for intelligent food packaging. *Carbohydrate Polymers, 249*, 116831.

Lu, Y., He, Q., Fan, G., Cheng, Q., & Song, G. (2021). Extraction and modification of hemicellulose from lignocellulosic biomass: A review. *Green Processing and Synthesis, 10*(1), 779–804.

Majerska, J., Michalska, A., & Figiel, A. (2019). A review of new directions in managing fruit and vegetable processing by-products. *Trends in Food Science and Technology, 88*, 207–219.

Manivannan, A., Lee, E. S., Han, K., Lee, H. E., & Kim, D. S. (2020). Versatile nutraceutical potentials of watermelon—A modest fruit loaded with pharmaceutically valuable phytochemicals. *Molecules, 25*(22), 5258.

Maraveas, C. (2020). Production of sustainable and biodegradable polymers from agricultural waste. *Polymers, 12*(5), 1127.

Mellinas, C., Ramos, M., Jiménez, A., & Garrigós, M. C. (2020). Recent trends in the use of pectin from agro-waste residues as a natural-based biopolymer for food packaging applications. *Materials, 13*(3), 673.

Merino, D., & Athanassiou, A. (2023). Thermomechanical plasticization of fruits and vegetables processing byproducts for the preparation of bioplastics. *Advanced Sustainable Systems, 7*(9), 2300179.

Merino, D., Quilez-Molina, A. I., Perotto, G., Bassani, A., Spigno, G., & Athanassiou, A. (2022). A second life for fruit and vegetable waste: a review on bioplastic films and coatings for potential food protection applications. *Green Chemistry, 24*(12), 4703–4727.

Mohammed, A. A., Omran, A. A. B., Hasan, Z., Ilyas, R. A., & Sapuan, S. M. (2021). Wheat biocomposite extraction, structure, properties and characterization: A review. *Polymers, 13*(21), 3624.

Mohd Basri, M. S., Abdul Karim Shah, N. N., Sulaiman, A., Mohamed Amin Tawakkal, I. S., Mohd Nor, M. Z., Ariffin, S. H., et al. (2021). Progress in the valorization of fruit and vegetable wastes: Active packaging, biocomposites, by-products, and innovative technologies used for bioactive compound extraction. *Polymers, 13*(20), 3503.

Mujtaba, M., Fraceto, L., Fazeli, M., Mukherjee, S., Savassa, S. M., de Medeiros, G. A., et al. (2023). Lignocellulosic biomass from agricultural waste to the circular economy: A review with focus on biofuels, biocomposites and bioplastics. *Journal of Cleaner Production, 136815*.

Muñoz-Almagro, N., Valadez-Carmona, L., Mendiola, J. A., Ibáñez, E., & Villamiel, M. (2019). Structural characterisation of pectin obtained from cacao pod husk. Comparison of conventional and subcritical water extraction. *Carbohydrate Polymers, 217*, 69–78.

Mushtaq, M., Gani, A., Gani, A., Punoo, H. A., & Masoodi, F. A. (2018). Use of pomegranate peel extract incorporated zein film with improved properties for prolonged shelf life of fresh Himalayan cheese (Kalari/kradi). *Innovative Food Science and Emerging Technologies, 48*, 25–32.

Nandiyanto, A. B. D., Triawan, F., Fiandini, M., Suryani, I. O., & Sunnardianto, G. K. (2021). influence of the size of turmeric microparticles reinforcing agent on mechanical and biodegradation properties of cornstarch-based bioplastic material: current studies, experimental results, and proposal material crack phenomena during mechanical testing. *Materials Physics and Mechanics, 47*(2).

Nayak, A., & Bhushan, B. (2019). An overview of the recent trends on the waste valorization techniques for food wastes. *Journal of Environmental Management, 233*, 352–370.

Ng, H. S., Kee, P. E., Yim, H. S., Chen, P. T., Wei, Y. H., & Lan, J. C. W. (2020). Recent advances on the sustainable approaches for conversion and reutilization of food wastes to valuable bioproducts. *Bioresource Technology, 302*, 122889.

Olad, A., Doustdar, F., & Gharekhani, H. (2020). Fabrication and characterization of a starch-based superabsorbent hydrogel composite reinforced with cellulose nanocrystals from potato peel waste. *Colloids and Surfaces A: Physicochemical and Engineering Aspects, 601*, 124962.

Pereira, P. H. F., Oliveira, T. Í. S., Rosa, M. F., Cavalcante, F. L., Moates, G. K., Wellner, N., et al. (2016). Pectin extraction from pomegranate peels with citric acid. *International Journal of Biological Macromolecules, 88*, 373–379.

Pereira, P. H., Waldron, K. W., Wilson, D. R., Cunha, A. P., de Brito, E. S., Rodrigues, T. H., et al. (2017). Wheat straw hemicelluloses added with cellulose nanocrystals and citric acid. Effect on film physical properties. *Carbohydrate Polymers, 164*, 317–324.

Perumal, A. B., Sellamuthu, P. S., Nambiar, R. B., & Sadiku, E. R. (2018). Development of polyvinyl alcohol/chitosan bio-nanocomposite films reinforced with cellulose nanocrystals isolated from rice straw. *Applied Surface Science, 449*, 591–602.

Phiri, R., Rangappa, S. M., Siengchin, S., Oladijo, O. P., & Dhakal, H. N. (2023). *Development of sustainable biopolymer-based composites for lightweight applications from agricultural waste biomass: A review. Advanced Industrial and Engineering Polymer Research.*

Pirsa, S., Sani, I. K., & Mirtalebi, S. S. (2022). Nano-biocomposite based color sensors: Investigation of structure, function, and applications in intelligent food packaging. *Food Packaging and Shelf Life, 31*, 100789.

Priyadarshi, R., Riahi, Z., & Rhim, J. W. (2022). Antioxidant pectin/pullulan edible coating incorporated with Vitis vinifera grape seed extract for extending the shelf life of peanuts. *Postharvest Biology and Technology, 183*, 111740.

Qiu, X., Chen, S., Liu, G., & Yang, Q. (2014). Quality enhancement in the Japanese sea bass (*Lateolabrax japonicas*) fillets stored at 4 C by chitosan coating incorporated with citric acid or licorice extract. *Food Chemistry, 162*, 156–160.

Rai, P. K., & Choure, K. (2023). Agriculture waste to bioplastics: A perfect substitution of plastics. In *Value-addition in agri-food industry waste through enzyme technology* (pp. 299–314). Academic.

Ratna, A. S., Ghosh, A., & Mukhopadhyay, S. (2022). Advances and prospects of corn husk as a sustainable material in composites and other technical applications. *Journal of Cleaner Production, 133563*.

Rawdkuen, S., & Kaewprachu, P. (2019). Valorization of food processing by-products as smart food packaging materials and their application. *Food Preservation and Waste Exploitation, 10*. https://doi.org/10.5772/intechopen.86245

Rifna, E. J., Misra, N. N., & Dwivedi, M. (2023). Recent advances in extraction technologies for recovery of bioactive compounds derived from fruit and vegetable waste peels: A review. *Critical Reviews in Food Science and Nutrition, 63*(6), 719–752.

Saeidy, S., Omidi, P., Nasirpour, A., & Keramat, J. (2023). Physicochemical and functional properties of cross-linked and high pressure homogenized sugar beet pectin: A comparative study. *Food Hydrocolloids, 134*, 108041.

Sani, M. A., Tavassoli, M., Salim, S. A., Azizi-lalabadi, M., & McClements, D. J. (2022). Development of green halochromic smart and active packaging materials: TiO_2 nanoparticle- and anthocyanin-loaded gelatin/κ-carrageenan films. *Food Hydrocolloids, 124*, 107324.

Sharafati Chaleshtori, F., & Sharafati Chaleshtori, R. (2017). Antimicrobial activity of chitosan incorporated with lemon and oregano essential oils on broiler breast meat during refrigerated storage. *Nutrition and Food Science, 47*(3), 306–317.

Sharma, M., Usmani, Z., Gupta, V. K., & Bhat, R. (2021). Valorization of fruits and vegetable wastes and by-products to produce natural pigments. *Critical Reviews in Biotechnology, 41*(4), 535–563.

Shin, S. H., Chang, Y., Lacroix, M., & Han, J. (2017). Control of microbial growth and lipid oxidation on beef product using an apple peel-based edible coating treatment. *LWT, 84*, 183–188.

Singh, B., Singh, J. P., Kaur, A., & Singh, N. (2020). Phenolic composition, antioxidant potential and health benefits of citrus peel. *Food Research International, 132*, 109114.

Sogut, E., & Seydim, A. C. (2019). The effects of chitosan-and polycaprolactone-based bilayer films incorporated with grape seed extract and nanocellulose on the quality of chicken breast fillets. *LWT, 101*, 799–805.

Sohail, M., Sun, D. W., & Zhu, Z. (2018). Recent developments in intelligent packaging for enhancing food quality and safety. *Critical Reviews in Food Science and Nutrition, 58*(15), 2650–2662.

Sohany, M., Tawakkal, I. S. M. A., Ariffin, S. H., Shah, N. N. A. K., & Yusof, Y. A. (2021). Characterization of anthocyanin-associated purple sweet potato starch and peel-based pH indicator films. *Foods, 10*(9), 2005.

Strnad, S., & Zemljič, L. F. (2023). Cellulose–chitosan functional biocomposites. *Polymers, 15*(2), 425.

Tesfaye, T., Ayele, M., Ferede, E., Gibril, M., Kong, F., & Sithole, B. (2021). A techno-economic feasibility of a process for extraction of starch from waste avocado seeds. *Clean Technologies and Environmental Policy, 23*, 581–595.

Vannini, M., Marchese, P., Sisti, L., Saccani, A., Mu, T., Sun, H., & Celli, A. (2021). Integrated efforts for the valorization of sweet potato by-products within a circular economy concept: Biocomposites for packaging applications close the loop. *Polymers, 13*(7), 1048.

Varghese, S. A., Pulikkalparambil, H., Promhuad, K., Srisa, A., Laorenza, Y., Jarupan, L., et al. (2023). Renovation of Agro-Waste for sustainable food packaging: A review. *Polymers, 15*(3), 648.

Vu, H. T., Scarlett, C. J., & Vuong, Q. V. (2019). Maximising recovery of phenolic compounds and antioxidant properties from banana peel using microwave-assisted extraction and water. *Journal of Food Science and Technology, 56*, 1360–1370.

Wang, X., Wang, C., Zha, X., Mei, Y., Xia, J., & Jiao, Z. (2017). Supercritical carbon dioxide extraction of β-carotene and α-tocopherol from pumpkin: A Box–Behnken design for extraction variables. *Analytical Methods, 9*(2), 294–303.

Wang, J. X., Cao, Y., Jaquet, B., Gerhard, C., Li, W., Xia, X. D., et al. (2021). Self-compounded nanocomposites: toward multifunctional membranes with superior mechanical, gas/oil barrier, UV-shielding, and photothermal conversion properties. *ACS Applied Materials & Interfaces, 13*, 28668–28678.

Xu, S. Y., Liu, J. P., Huang, X., Du, L. P., Shi, F. L., Dong, R., et al. (2018). Ultrasonic-microwave assisted extraction, characterization and biological activity of pectin from jackfruit peel. *LWT, 90*, 577–582.

Yeboah, W. O., Kwofie, E. M., & Wang, D. (2022). Circular bioeconomy potential of rice husk as a bioplastic resource: Techno-environmental assessment. *Bioresource Technology Reports, 20*, 101248.

Zadeh, E. M., O'Keefe, S. F., & Kim, Y. T. (2018). Utilization of lignin in biopolymeric packaging films. *ACS Omega, 3*(7), 7388–7398.

Zidan, N. S., Alalawy, A. I., Al-Duais, M. A., Alzahrani, S., & Kasem, M. (2023). Modification of edible chitosan/polyethylene glycol films fortified with date palm fruit waste extract as promising antimicrobial food packaging materials for fresh strawberry conservation. *European Polymer Journal, 194*, 112171.

Chapter 12
Conversion of Agriculture Residues for Bioenergy Production

Omojola Awogbemi, Daramy Vandi Von Kallon, and Ramesh C. Ray

Abbreviations

ABE	Acetone-butanol-ethanol
AD	Anaerobic digestion.
AI	Artificial intelligence.
ANFIS	Adaptive neuro-fuzzy inference system
ANNs	Artificial neural networks
CFD	Computational fluid dynamics
CO_2	Carbon dioxide
COD	Chemical oxygen demand
FB	Fossil-based
GHGs	Greenhouse gases
$GtCO_2$	Billion metric tons
ML	Machine learning
MMT	Million metric tons
RSM	Response surface methods
TWh	Tetrawatt-hours

O. Awogbemi (✉) · D. V. Von Kallon
Department of Mechanical and Industrial Engineering Technology, University of Johannesburg, Johannesburg, South Africa

R. C. Ray
Center for Industrial Biotechnology Research, Siksha 'O' Anusandhan (SOA) deemed to be University, Khandagiri, Bhubaneswar, India

Formerly, ICAR-Central Tuber Crops Research Institute (Regional Center), Bhubaneswar, Odisha, India

J. Arora et al. (eds.), *Transforming Agriculture Residues for Sustainable Development*, Waste as a Resource, https://doi.org/10.1007/978-3-031-61133-9_12

12.1 Introduction

The global energy scenario is still largely dominated by nonrenewable energy resources. Despite the mounting environmental challenges, diminishing oil reserves, biting economic impact, and the challenges associated with its exploration, transportation, and storage. The total global energy consumption is from fossil-based (FB) fuel sources, namely oil, coal, and gas. The total primary nonrenewable energy consumed in 1990 rose from 37.68 terawatt-hours (TWh), 25.91 TWh, and 19.48 TWh in 1990 to 47.97 TWh, 41.94 TWH, and 31.39 TWh in 2010 and further to 52. 97 TWH, 48.85 TWh, and 39.41 TWh in 2022 for oil, coal, and gas, respectively (Our World in Data, 2023). Currently, the contribution of oil, coal, and gas to the global primary energy consumption is around 29.61%, 25.07%, and 22.03%, respectively. But, by the current rate of consumption, the percentage of oil, coal, and gas consumption has been predicted to become 26%, 24%, and 24%, respectively (Baruya, 2015; Rahimi et al., 2022). Justifiably, the increase in the consumption of oil, coal, and gas has continued to result in increased emissions of carbon dioxide (CO_2) and other greenhouse gases (GHGs). Available data showed that the annual CO_2 emissions which was 22.76 billion metric tons ($GtCO_2$) in 1990 became 33.36 $GtCO_2$ in 2010 and further increased to 37.5 $GtCO_2$ in 2022 (Our World in Data, 2022). Rising CO_2 emissions exacerbate rising global warming, climate change, and environmental degradation. Figure 12.1 shows the consumption of oil, coal, and gas and the corresponding CO_2 emission from 1990 to 2022. To mitigate the impact of FB fuel consumption, meet the demand for clean energy, and ensure energy security, there must be a paradigm shift to renewable energy resources as sustainable alternatives to FB energy sources.

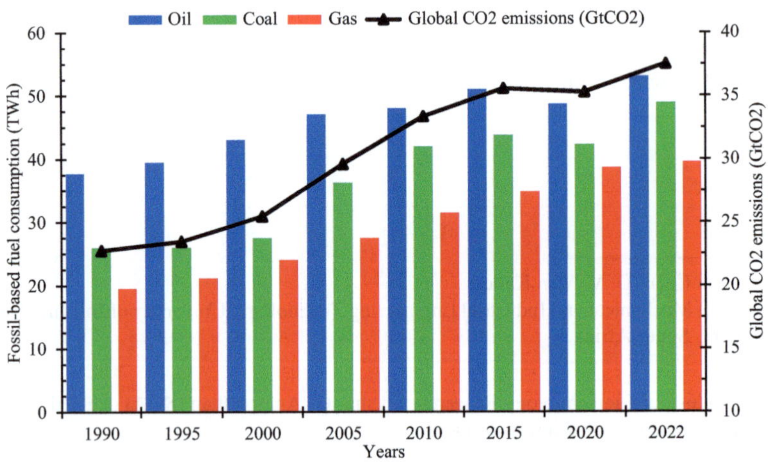

Fig. 12.1 Global fossil fuels consumption and CO_2 emissions 1990–2022

Renewable energy, due to its renewability, degradability, ready accessibility, economic, and ecological benefits, is a viable alternative and has attracted sustained attention among researchers and other stakeholders. Among the renewable energy technologies, bioenergy is a prominent, easily achievable, and eco-friendly option for meeting global energy security crises. Bioenergy is a form of renewable energy derived from feedstock such as lignocellulosic biomass, agricultural waste, crop residues, food waste, household waste, animal dung, algae, energy crops, etc. bioenergy in inexhaustible, biodegradable, and contributes to a reduction in GHGs emissions (Singh et al., 2023). Bioenergy is used for producing vehicle fuels, heat generation, production of electric power, and valuable products. The application of bioenergy provides an alternative to a petroleum-based economy, promotes bioeconomy, and reduces oil importation. The dependency. Despite the benefits and applications, the production of bioenergy has been hampered by high production costs, inadequate infrastructure, and inadequate financial and manpower investments. For example, the cost of feedstock makes up more than 70% of the total production cost of bioenergy (Awogbemi & Kallon, 2022a). The application of agricultural residues as feedstock contributes to efforts at reducing the cost of feedstock, and ultimately, the production cost for bioenergy.

Due to the escalating global population, there has been an increase in food demand and agricultural activities over the past decades. Escalated food production has led to a significant increment in the generation of agricultural residues and other forms of waste from agricultural activities. Agricultural residues are the leftovers after harvesting and processing of crops and animals in the fields, homes, and industries. They include straws, peels, shells, bagasse, dung, feathers, husk, leaves, and woods, and they vary considerably in appearance, species, composition, properties, processing methods, and applications (Awogbemi & Kallon, 2023; Behera & Ray, 2021). Residues from agricultural activities such as leaves, stubbles, straws, husks, and shells are used as feedstuffs for animals, building materials, and as fuels for cooking and heating. However, a huge quantity of residues is either left in the farms to decompose, burn in situ, or disposed of inappropriately thereby resulting in bush burning, soil and water contamination, generation of GHGs, and environmental pollution. To avoid these unpleasant scenarios, waste conversion has become one of the sustainable strategies for agricultural residue management. Proper management of agricultural residues through conversion of agricultural residues to bioenergy is cost-effective, improves sanitation, minimizes the cost of waste disposal, reduces deforestation, and contributes to the circular economy. Precious materials are recovered from agricultural residues during conversion and many useful chemicals and other products are generated from the residues.

Previous research efforts have shown the applicability of the conversion of many agricultural residues to bioenergy and other valuable products. In separate studies, Rahimi et al. (2022), Sarkar et al. (2023), Song et al. (2023), and Abdullah et al. (2021) explained the diverse pathways and technologies for the generation of bioenergy from various agricultural residues for various applications. Specifically, bioethanol, biohydrogen, and biochar, from banana peels, potato peel, and coconut shells by John et al. (2020), Bhurat et al. (2021), and Saeed et al. (2023), respectively. According to these authors, agricultural residues remain a feasible, low-cost, readily available, easily convertible, and eco-friendly feedstock for bioenergy production. Converting agricultural residues to biofuels requires no additional land, ensures effective waste management, supports a circular economy, and contributes to global energy security.

Notwithstanding the successes of the previous studies, ample research opportunities still exist in bioenergy production from the agricultural residues research domain. The main objective of the present study is to review the outcome of the recent research on the generation of bioethanol, biohydrogen, biochar, biojet fuel, biogas, and biobutanol from agricultural residues. The methods of feedstock collection and pretreatment, bioenergy production and utilization, and future directions in the research domain are also discussed. The outcome of this intervention will engender renewed interest in sustainable waste management, agricultural residue conversion, and bioenergy generation from low-cost feedstock for various applications.

12.2 Collection, Pretreatment, and Conversion Techniques for Agricultural Residues

12.2.1 Collection of Agricultural Residues

The conversion of agricultural residues to bioenergy can be divided into three stages, namely, collection, pretreatment, storage, and conversion (Fig. 12.2). Each stage must be handled with utmost care to ensure effective conversion and the sanctity of the product. Since agricultural wastes are generated in different locations, there is a need for logistics for collection, during generation, and almost immediately. This is to prevent the scattering of the materials which might result in material loss, contamination, and decay (Sarkar et al., 2023). Peels, pods, bagasse, straws, husks,

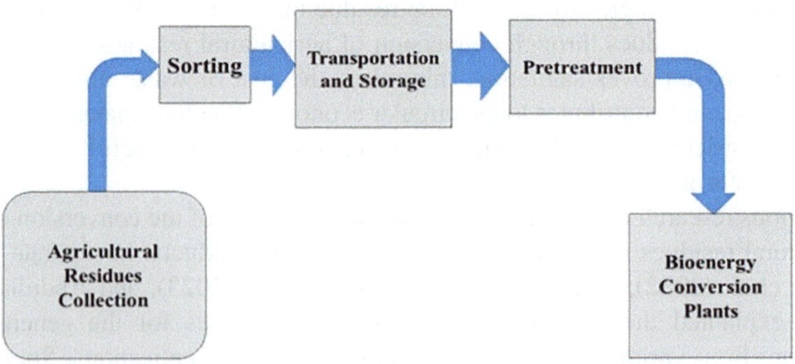

Fig. 12.2 Flowchart of agricultural residues processing

stubbles, cobs, and other crop residues are collected at the farm site during harvesting while feathers, dungs, bones, poultry droppings, etc. are collected from slaughterhouses, poultry farms, animal pens, and other generation sites. Untimely collection of these materials can result in contamination, environmental pollution, insect infestation, unpleasant odor, and becoming habitats for rodents, flies, cockroaches, and other disease-causing agents (Tallentire & Steubing, 2020). The collected materials must thereafter be sorted to remove dirt, stones, and other unwanted substances before being transferred to the conversion plants. The transportation of the sorted waste must be done in a clean sealed and nonreactive container. In cases, where the conversion plants or reactors are within the farms and generating site, the transportation stage with the associated costs and other infrastructure are drastically reduced or eliminated. In some cases, waste collectors and scavengers can be engaged to collect and transport the agricultural residues from the point of collection to the conversion site. Though this method increases this production cost, it nonetheless creates employment, provides a means of livelihood to people, particularly the rural dwellers, and ensures inclusivity.

12.2.2 Pretreatment of Agricultural Residues

In most cases, the collected agricultural residues are wet, recalcitrant, and unsuitable for immediate conversion unless they are treated. This is called the pretreatment stage. Naturally, lignocellulosic biomass is not easily degraded as it possesses some resistance that must be overcome before conversion. The compositional make-up, physicochemical structure, source, type, and state of materials and the conversion techniques to be adopted significantly determine the type of retreatment techniques to be adopted. The pretreatment process enhances the surface area and volume of pores, increases the amorphous region to ease the conversion, enhances the porosity of the porous matrix, and separates the cellulose from lignin, and hemicellulose (Tran et al., 2020; Maache-Rezzoug et al., 2023). But generally, irrespective of the pretreatment techniques, the purpose of pretreatment is to enhance the digestibility, biodegradability, and conversion efficiency of the materials by weakening the bonds between the constituents of the biomass. The benefits of agricultural residue pretreatment include reduced energy consumption, overall cost reduction, easy conversion of biomass to sugars, and a significant decline in the formation of inhibitory products (Awogbemi & Kallon, 2022b; Moodley et al., 2023). Common pretreatment techniques are physical, chemical, biological, physicochemical, and green solvent-based processes. Figure 12.3 shows examples of the five pretreatment techniques for agricultural residues, while Table 12.1 compiles the description, pros, and cons of the major biomass pretreatment techniques.

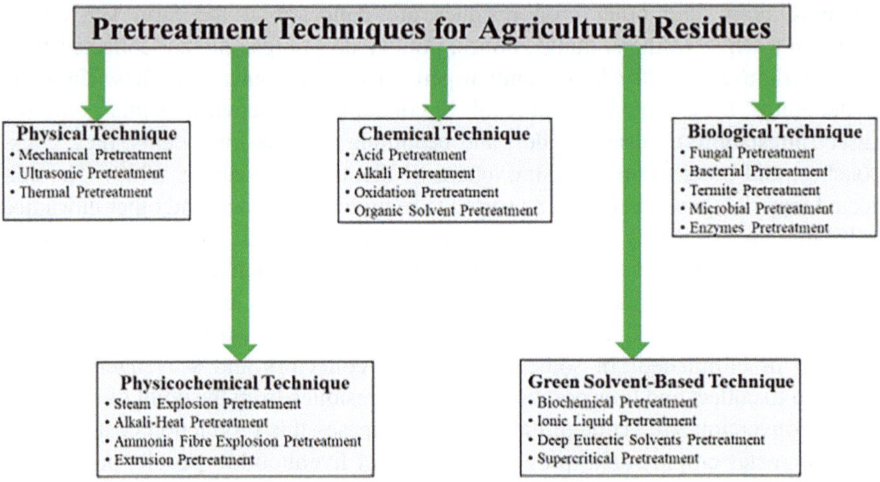

Fig. 12.3 Common pretreatment techniques and their major examples

12.3 Bioenergy Production from Agricultural Residues

Using agricultural residues as feedstock for bioenergy production is one of the sustainable and eco-friendly approaches for utilizing the enormous waste generated in the agricultural sector. This utilization pathway ensures effective waste management, contributes to clean sanitation, enhances resource recovery, and helps maintain carbon balance in the environment (Moodley et al., 2023; Ray & Ramachandran, 2018) Primarily, the conversion of agricultural residues to bioenergy is achieved through chemical, biological, and thermal techniques. Advancements in research have led the scientific community to resort to the combination and amalgamation of any two of these techniques. To this end, biochemical and thermochemical techniques remain the most effective and most commonly used in recent times. This sect. x-rays the outcome of recent studies in the production of bioethanol, biohydrogen, biochar, biojet fuel, biogas, and biobutanol from agricultural residues.

12.3.1 Bioethanol

Bioethanol is a renewable liquid biofuel and a prominent clean energy source produced by fermentation of natural materials. Bioethanol is produced from three main groups of feedstock, namely sugar-containing feedstock, starch-containing feedstocks, and lignocellulosic biomass. Common examples of feedstocks containing sugar and starch for bioethanol production are mainly grains and root crops such as corn, sugarcane, sweet sorghum, cassava, wheat, and potato (Ray &

Table 12.1 Mechanisms, advantages, and disadvantages of agricultural residue pretreatment techniques (Awogbemi & Kallon, 2022b; Khanh Nguyen et al., 2021; Cheah et al., 2020; Awogbemi & Kallon, 2023)

Pretreatment technique	Mechanism involved	Advantages	Disadvantages
Physical	Grinding, milling, extrusion, heating, drying, freezing, Ultrasound, and microwave irradiation	• Low-cost. • Easy and simple to achieve. • Significant size reduction. • Improves digestibility. • Produces no inhibitor. • Enhances the surface area of the feedstock. • Safe and nontoxic. • Minimum waste generation. • Eco-friendly.	• Additional cost. • High energy consumption. • Does not degrade lignin. • Must be combined with other pretreatments.
Chemical	Uses chemical reagents to improve the digestibility and biodegradability of biomass	• Low operating temperature. • High rate of delignification. • Fast and eco-friendly. • Eliminates lignin and hemicellulose. • Reduces biomass crystallinity.	• Management of spent chemicals. • Not eco-friendly. • Additional cost of chemicals and reagents. • Likelihood of poisonous substances and inhibitor. • Uses toxic and corrosive chemicals.
Biological	Applies bacteria, enzymes, or fungi to degrade lignin and enhance biomass conversion	Eco-friendly Low-energy consumption Produces less hazardous byproducts Improved digestibility of cellulose Economic viability	Low efficiency Requires further research Long incubation time Low cellulose recovery Slow and time-consuming Possible microbial instability Unpredictable reaction outcomes

(continued)

Table 12.1 (continued)

Pretreatment technique	Mechanism involved	Advantages	Disadvantages
Physicochemical	Combines physical and chemical pretreatments to improve lignin denigration and biomass digestibility	Safe, effective, and nonpolluting Eco-friendly and cost-effective Simple and easy-to-achieve operating conditions No inhibitor production	High cost of equipment Requires high temperature High energy consumption Not cost-effective The process needs further research
Green solvent-based	Employs innovative and eco-friendly pretreatment technologies	High efficiency Fast and affordable process No production of inhibitors Nonflammable, eco-friendly, and recyclable Highly biodegradable and low toxicity	Can be complex Viscous and expensive process Requires high temperature and pressure Requires special skills and infrastructure

Compiled by the authors

Ramachandran, 2018). The use of these groups of edible feedstock for bioethanol production is costly, escalates the food vs fuel debate, conflicts with the food chain, and exacerbates food security. This has made lignocellulosic biomass including agricultural residue a feasible alternative feedstock for eco-friendly, cost-effective, and sustainable bioethanol production. Ethanol is a major industrial raw material for the production of preservatives, solvents, disinfectants, alcoholic drinks, paints, makers, personal care products, reagents, fuel additives, and numerous pharmaceutical products. The increased utilization of bioethanol as a renewable fuel and a source of bioenergy has escalated its production and utilization. Currently, bioethanol is the most consumed renewable liquid fuel for transport and stationary compression ignition engines, globally (Awogbemi & Kallon, 2022a, b, c). The multifaceted application of bioethanol, low emission, biodegradability, and environmental benefits have led to increased production of bioethanol. The global production of bioethanol which was about 100.2 billion liters in 2016 has been estimated to rise to 134.5 billion liters by 2024, and further to 132 billion liters by 2030 (Awogbemi et al., 2021). To compensate for the expected growth, the global market share of bioethanol which was worth USD 40.8 billion in 2021 has been projected to become USD 124.5 billion by 2030 (GlobeNewsire, 2022).

The production of bioethanol from agricultural residues has enjoyed patronage in recent years. There has also been improved product yield, conversion efficiency, lower energy consumption, and faster production time due to the adaptation of innovative production techniques. One of the innovative strategies for enhanced bioethanol production from agricultural residues is the deployment of integrated feedstock pretreatment. Shukla et al. (2023) and Pant et al. (2022) in their separate studies carried out integrated pretreatment of agricultural residue feedstock for bioethanol production. The integrated approaches involve a combination of two or more pretreatment techniques to drastically reduce the resistance to the degradation of the biomass. The simultaneous multiple pretreatment techniques led to improvement in feedstock digestibility, faster process, reduction in energy consumption, and better product yield. They reported that the combination of extrusion and ultrasonic pretreatment methods excluded the use of any chemical generated less emissions, and led to about 77.7% product yield when rice husk was converted to bioethanol. Similarly, the strategic combined steam explosion and alkalis pretreatment of corn stover resulted in 84.4% lignin removal and about 4.5 times bioethanol yield over a single pretreatment process. The combination of microwave radiation and acid pretreatments resulted in significant improvement in the breakdown of the hemicellulosic recalcitrant structure of wheat and rye stillages and resulted in a 75% bioethanol yield as compared with only microwave radiation pretreatment method (Mikulski & Kłosowski, 2020).

The combination of multiple production methods also contributes to reduced production duration and improved product yield. A recent investigation showed that adopting combined hydrolysis and co-fermentation resulted in between 78.8% and 89.1% increased bioethanol yield from barley straw and *Escherichia coli* SL100 (Díaz et al., 2022). A similar study with the co-fermentation of microwave-assisted pretreated elephant grass with *Saccharomyces cerevisiae* ITB-R89 and *Pichia stipitis* ITB-R58 resulted in about 76.15% sugar mixture, 69.48% fermentation efficiency, and 34.74% better bioethanol yield (Kolo et al., 2020). The co-fermentation of the cotton stalk with *S. cerevisiae* YPH499 and *Pachysolen tannophilus* 32,691 was achieved at a lesser duration and 94.1% bioethanol yield. This was attributed to enhanced disintegration and abrasion in corn stalk structure during pretreatment and co-fermentation processes (Malik et al., 2021). Recently, researchers have explored strategic manipulation of the process parameters to reduce process time, minimize material consumption, control energy consumption, improve conversion efficiency, and increase product yield. Adesina et al. (2023), Chaudhary et al. (2022), and Bakari et al. (2023) optimized the production of bioethanol from sugarcane bagasse, pomegranate peel, and sweet sorghum stalks and reported 32%, 36.1%, and 27.5% improved conversion efficiency, respectively. They attributed these results to the effect of the optimization that predicted optimal conditions for bioethanol production. The advancement in information and communication technologies and modeling tools has contributed immensely to bioethanol production. The application of computational fluid dynamics (CFD), response surface method (RSM), artificial neural networks (ANNs), adaptive neuro-fuzzy inference system (ANFIS), and other intelligent predictive tools has led to the intensification and significant improvements in the production process and outcomes of bioethanol production from agricultural residues (Novia et al., 2023; Taiwo & Musonge, 2023). The combined effect of integrated pretreatment techniques, co-fermentation, and deployment of intelligent predictive, optimization, and modeling tools have greatly advanced bioethanol production.

12.3.2 Biohydrogen

Biohydrogen is a clean, highly combustible renewable fuel, and a major bioenergy produced by the biological and thermochemical conversion of biomass containing cellulose and starch. Owing to its high energy content, improved calorific value, renewable and readily available feedstock, zero GHG emissions, and contribution to energy security, biohydrogen has gained prominence as a sustainable and cost-effective alternative to FB fuels. Notable areas of application of biohydrogen include as fuel for transportation engines, heat, and electricity generation, and industrial raw material, especially in food, beverage, chemical, and pharmaceutical industries (Machineni et al., 2023). Numerous advantages of biohydrogen including its ready availability, low-cost feedstock, environmental friendliness, and multiple application pathways have resulted in increased production and market share. From the 1.8 million metric tons (MMT) produced in 2015, global biohydrogen production has been predicted to net 19 MMT by 2030 and 65 MMT by 2050 (Statista, 2021). To meet up with the anticipated increased production, the market share is expected to increase from the USD 103 billion recorded in 2017 to USD 183 billion by 2023 and further to USD 207 billion by 2026 (PRNewswire, 2018). Major biohydrogen production methods include fermentation, biophotolysis, microbial electrolysis, thermochemical gasification, pyrolysis, solar gasification, and supercritical (Awogbemi & Kallon, 2022c). Table 12.2 shows the process description, advantages, and disadvantages of major biohydrogen production pathways.

One of the strategies for the improvement of biohydrogen production from agricultural residues is the application of nanotechnology. In separate studies, Zhou et al. (2023), Gökçek et al. (2023), and Ren et al. (2023) deployed zinc molybdate, magnetite, and milli-magnetite for the production of biohydrogen from rice straw, apple pulp waste, and potato peel, respectively. The authors reported about 46% to 140% increased production of biohydrogen when compared with the control experiments without the nanomaterials. The deployment of magnetite nanoparticles also significantly influenced the conversion of coconut shells to biohydrogen and caused about 63.99% increased productivity rate compared to nonnanoparticle process (Mohd Jamaludin et al., 2023). Blending two or more agricultural residues as feedstock has been found to significantly reduce reaction time and increase biohydrogen production. Srivastava et al. (2023) experimented with co-substrate fermentation using wheat straw and sugarcane bagasse in the presence of copper oxide-based nanocatalyst. The experiment led to improved bioconversion of the feedstocks and yielded more biohydrogen at reduced reaction time. A similar investigation with the co-fermentation of dairy whey and sugarcane bagasse showed that co-substrate proceeded faster and yielded more biohydrogen when compared with mono-substrate (Mumtha et al., 2023). Earlier, Rodríguez-Valderrama et al. (2020) produced biohydrogen from the fermentation of vegetable wastes and corn stover as co-substrates, while Zhang et al. (2023) fermented a blend of corn straw and duckweed for the generation of biohydrogen. The outcome of these studies demonstrates the positive influence of the use of co-substrates on

Table 12.2 Description, advantages, and disadvantages of major biohydrogen production pathways (Awogbemi & Kallon, 2022a, b, c; Ahmed et al., 2021)

Production method	Process description	Advantages	Disadvantages
Photo-fermentation	Conversion of feedstock to biohydrogen through photosynthetic microorganisms	Can process many substrates Opportunity for bioremediation Efficient conversion High product yield Simplicity	Expensive implementation Time-consuming Requires pretreatment Low conversion rate
Dark fermentation	Fermentation of biomass into biohydrogen in the absence of light	Independent on light Accommodates a variety of feedstocks Bioremediation Simple reactor	Hard separation of products Inefficient COD removal Erratic product yield Thermodynamically unsupportive
Direct biophotolysis	Biohydrogen production through the breakdown of water molecules using light by microalgae	Cost-effective Efficient conversion Requires only water and light	Low photochemical efficiency Highlight can be an inhibitor
Indirect biophotolysis	Biohydrogen production from water by photosynthetic conversion	Affordable Requires water and light only Enhanced energy conversion	Reduced hydrogenate Removal of hydrogenate
Microbial electrolysis	Microbial electrolytic conversion of biomass in biohydrogen	High yield process High removal of COD Eco-friendly process Improved substrate degradation	High capital cost Scalability challenge Loss of product Concerns about product quality
Thermochemical gasification	Partial oxidation of biomass to produce biohydrogen	High conversion rate High superior yield	Expensive equipment Tar elimination
Pyrolysis	Application of heat to degrade biomass into biohydrogen	High product yield Other products are generated	Catalysts made inactive High cost of equipment

(continued)

Table 12.2 (continued)

Production method	Process description	Advantages	Disadvantages
Solar gasification	Thermochemical synthesis of biohydrogen from biomass using solar energy	High product yield Product generation at low temperature No need to grind feedstock Low CO_2 emission	High cost of solar energy infrastructure
Supercritical conversion	Thermochemical biohydrogen using supercritical water as a gasifying agent	Quality product Wet substrate can be converted Effective conversion process	Supercritical process conditions and medium not easily achieved

Compiled by the authors; *COD* chemical oxygen demand, CO_2 carbo dioxide

biohydrogen production. The application of artificial intelligence (AI), machine learning (ML), CFD, ANN, and other innovative technologies in biohydrogen production has contributed to advancing biohydrogen production from agricultural residues. These technologies are applied in smart-driven bioprocess, bioreactor design and modeling, process prediction, techno-economic and scientific simulations, and real-time data acquisition (Khanal et al., 2023; Kumar Sharma et al., 2022).

12.3.3 Biochar

Biochar is a solid, rigid, and carbon-rich product of thermochemical degradation of agricultural residue, food waste, animal waste, and other forms of biomass at a controlled temperature under airless conditions. Biochar is used as solid fuel, catalysts for biofuel production, air, soil, and wastewater treatment, sanitation of human and kitchen wastes, animal feed, or treated and converted to activated carbon (Awogbemi & Kallon, 2023). Biochar is also used for carbon sequestration, soil amendments, compost additives, production of organic fertilizers, immobilization of heavy metals, and other agronomic applications (Sachdeva et al., 2023; Enaime & Lübken, 2021). Apart from the numerous industrial, horticultural, agricultural, and industrial applications of biochar, the growing demand for eco-friendly fuels and advancement in pyrolysis and gasification technologies have escalated the demand and production capacity of biochar, globally. From a production volume of 394.09 kilotons recorded in 2021, the global biochar production volume has been estimated to become 781.09 kilotons by 2028, while the market size which was USD 1320 million in 2021 has been predicted to reach USD 3171 million by 2028 (Global Biochar Market, 2022–2028). In a recent study, Awogbemi and Kallon (2023) listed the techniques for biochar production from agricultural waste to include the traditional burning of biomass in an airless ditch covered by soil, pyrolysis, gasification, torrefaction, hydrothermal carbonization, liquefaction, while biochar modification techniques include electro-modified biochar, magnetic biochar, and plasma.

Biochar is a good source of bioenergy and a feasible renewable energy source. The use of biochar enhances sanitation, does not lead to the emission of GHGs, and contributes to environmental sustainability. Biochar is used to catalyze transesterification, anaerobic digestion (AD), and fermentation reactions for the synthesis of biodiesel, biomethane, and biohydrogen from diverse feedstocks. In a series of recently reported studies, biochar derived from various agricultural residues achieved more than 92% conversion efficiency during the transesterification of various feedstocks to biodiesel. The biochar constructed from agricultural residues was successfully modified and demonstrated good surface conditions and better active sites for the generation of quality biodiesel (Quah et al., 2020, Rajendran et al., 2022). The addition of biochar manufactured from agricultural residues to the slurry in a digester improved the production of biomethane in an AD process. This was demonstrated by Shen et al. (2015), Chiappero et al. (2021), and Waqas et al. (2018) during the production of biomethane from various substrates using biochar constructed from corn stover, rice husk, and plant leaves, respectively. The authors reported that the biochar enhanced biomethane production at a shorter hydraulic retention time. The improved biomethane generation was ascribed to the good surface configuration, high porosity, thermal stability, and alkaline pH of the biochar that energized the AD reactions. Recent studies have shown that biochar derived from agricultural residues promotes simultaneous saccharification and fermentation for better production of biohydrogen. Cornstalk and sugarcane bagasse converted to biochar were used to improve the digestibility of the substrate, quicken bacterial growth, advance enzymatic activities, and boost the electron transfer efficiency of the simultaneous saccharification and fermentation process (Zhao et al., 2021; Bu et al., 2021). The preponderance of this research shows that biochar constructed from agricultural residues stimulates the generation of biofuels from various substrates and contributes to bioenergy.

12.3.4 Biogas

Biogas is a renewable gaseous fuel produced from anaerobic digestion (AD) of organic matter by microorganisms. During the process, raw materials such as food waste, agricultural residues, animal waste, municipal waste, plant materials, wastewater, and sewage are broken down by microorganisms in the absence of oxygen to produce a methane-rich gas called biogas. Typically, raw biogas consists of methane (50–75%), carbon dioxide (25–50%), nitrogen (2–8%), and the trace of hydrogen sulfide, ammonia, hydrogen, and volatile organic compounds. The process of biogas

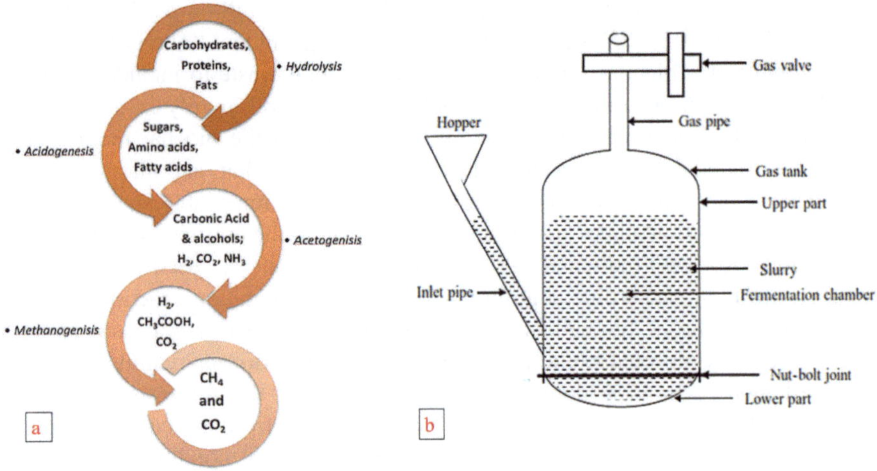

Fig. 12.4 (**a**) Four processes in AD and (**b**) schematic diagram of a digester

production involves hydrolysis, acidogenesis, acetogenesis, and methanogenesis (Fig. 12.4a), which occur in an airtight reactor called a digester (Fig. 12.4b) (Awogbemi & Kallon, 2024). Notable areas of application of biogas include engine fuel, water and space heating, electricity generation, microturbines, and fuel cells. The CO_2 and other impurities in raw biogas can be removed through a series of upgrading technologies to produce a renewable natural gas and methane-rich fuel called biomethane.

Biomethane is a clean pure and clean version of biogas with applications as vehicle fuel and for the generation of heat and electricity. Biomethane can be injected into the gas grid, used as a vehicle fuel, or used for combined heat and electricity generation. The increased utilization avenues of biogas have led to the escalation of the production capacity in recent decades. The global production of biogas rose from 0.29 exajoules (EJ) in 2000 to 1.34 EJ in 2015 and further to 1.46 EJ in 2020 (Statista, 2023). Similarly, the global biogas market size rose from which was USD 24.03 billion in 2020 to USD 58.19 billion in 2023 and has been predicted to become USD 66.05 billion and USD 78.8 billion in 2030. The increased market value of biogas was due to the growing demand and preference for clean energy and low-carbon electricity (Precedence Research, 2022). The conversion of agricultural residue to biogas contributes to sanitation and waste recycling and reduces the contamination of soil, air, and water bodies. As a feasible bioenergy, biogas utilization contributes to the reduction of GHG emissions and promotes circular economy and environmental sustainability.

The production of biogas from agricultural residue has been demonstrated by many researchers in recent years. In a recent study, Yan et al. (2021), Tamburini et al. (2020), Weide et al. (2020), and Liu et al. (2023) demonstrated that the anaerobic conversion of agricultural waste to biogas contributes to increasing renewable energy penetration, reduction in carbon emission, and sustainable ecosystem. In order to escalate the production of biogas from agricultural residue, various pretreatment, enhancement, and process intensification have been adopted. The use of feedstock pretreatment techniques, the addition of enzymes and inoculants, optimization of process parameters, co-digestion, substrate particle size reduction, the use of additives, and improved design of digesters have contributed to biogas yield, product quality, and reduced hydraulic retention time during AD processes (Liu et al., 2021; Wang et al., 2023; Kasinath et al., 2023). The aggregate opinions of these studies show the contribution of agricultural residue to cost-effective biogas and bioenergy production.

12.3.5 Biojet Fuel

The use of fossil-based fuels, such as jet kerosene, to power aviation engines has led to the emission of carbon and other GHGs in recent decades. In 2022, the aviation sector contributed about 2% of the total global energy-related CO_2 emissions (IEA, 2023). The use of renewable aviation fuels, including biojet fuel, is one of the strategies for decarbonizing the aviation sector of the transport industry. Biojet fuel, also known as bio-aviation fuel, is a renewable fuel used to power aircraft. Though the majority of biojet fuel is produced by the hydrotreatment of esters, vegetable oil, waste cooking oils, fats, and greases, the use of agricultural residue is gradually gaining traction due to its ready availability, low-cost, and renewability (Moreno-Gómez et al., 2020). The global biojet fuel production was about 0.29 million liters per year between 2013 and 2015 to about 4.5 million liters in 2016, seven million liters in 2018, and 140 million liters in 2019 (Lim et al., 2023; IRENA, 2021). The global biojet fuel market value, which was USD 186.81 million in 2021, has been predicted to rise by 17.95% between 2022 and 2030 (Market Research Report, 2022). The urgent need to reduce the rising GHG emissions in the aviation industry is the driving factor behind the estimated growth in market value.

The adaptation of crop residues, palm oil mill effluent, forest residues, woody waste, and other forms of agricultural waste as feedstock for biojet fuel production has been demonstrated. For example, Vela-García et al. (2022), Hao et al. (2021), Muanruksa et al. (2021), and Cervi et al. (2021) successfully generated biojet fuel from diverse agricultural residues. The technical and economic analysis of the process shows an effective, easily achievable, and economically viable project. The utilization of biojet fuel for powering aircraft engines significantly reduces carbon and GHG emissions, enhances circular economy, and promotes environmental sustainability, and the product meets ASTM D7566 specifications. Going forward, more innovative investigations are expected on feedstock selection and pretreatment, process optimization, cost reduction production process, product quality improvement, combustion enhancement, engine performance, and emission characteristics. The conversion of agricultural residue to renewable aviation-range fuels through gasification, pyrolysis, Fischer–Tropsch, and alcohol-to-jet processes is feasible, cost-effective, and eco-friendly. However, more collaborative research, advanced technologies, and infusion of innovative technologies are required to ensure sustainability, real-life applications, and commercialization (Emmanouilidou et al., 2023).

12.3.6 Biobutanol

Biobutanol is the production of butanol using biomass, waste materials, and other renewable feedstock. Due to its high energy density, low Reid vapor pressure, generation of fewer emissions, and molecular similarity to gasoline, biobutanol is seen as a sustainable fuel for internal combustion engines without any form of engine modifications. Microbial acetone-butanol-ethanol (ABE) fermentation is the most used method of producing biobutanol from waste biomass and the process has been escalated to the industrial level. The production of biobutanol from waste biomass including food waste, and agricultural residue is seen as a novel process of increasing bioenergy penetration, sustainable energy production, and reduction in environmental pollution (Kapil et al., 2023). The global production of biobutanol increased from 4.39 MMT in 2015 to 5.23 MMT in 2021 and has been estimated to become 6.72 MMT by 2029. Similarly, the market value of biobutanol has been predicted to rise to USD 114.7 million by 2026 from the USD 90 million recorded in 2020. The driving factor for the increased demand for biobutanol is attributed to the increased global population, drift towards renewable fuel for internal combustion engines, greater pursuit of carbon neutrality, and favorable government policy toward bioenergy utilization (Awogbemi & Kallon, 2022a, b, c).

Unlike corn, cassava, wheat, and other edible feedstock commonly used for biobutanol production, the use of agricultural residue as feedstock is cheap and does not affect food security. This has encouraged the scientific communities to continue to explore the production of biobutanol from sugarcane bagasse, rice straw, sweet sorghum bagasse, and other agricultural residue. Researchers across various jurisdictions have generated biobutanol from sweet sorghum bagasse (Khalili & Amiri, 2020), sugarcane molasses (Owuna et al., 2018), oil palm fruit bunch (Md Razali et al., 2018), corn cob (Shanmugam et al., 2019), rice straw (Cheng et al., 2012), etc., for various applications. Despite the benefits of producing biobutanol from diverse agricultural residues, there is a need for targeted studies to improve and optimize the production process, promote economic viability, and ensure consolidation of bioprocessing technologies.

12.4 Implications and Future Research

Inappropriate disposal of the huge agricultural residues generated from diverse agricultural activities exacerbates sanitation, generates offensive odors, and is inimical to environmental sustainability. Adapting the concept of waste to energy, recycling, and waste conversion is a sustainable waste management strategy and a way to increase renewable energy penetration, contributing to energy security. The use of agricultural residues to meet the clean energy demand contributes to employment generation, generates extra income for farmers, and supports a circular economy. Other benefits of the conversion of agricultural residues to bioenergy include reduction in landfill waste, minimization in the cost of waste collection and transportation, serving as an alternative energy source, reduction in the reliance on FB fuels, lowering greenhouse gas emissions, and help in meeting the global energy demand.

Despite the multifarious benefits of the conversion of agricultural residues to bioenergy, there are socioeconomic and environmental issues that must be addressed to ensure the sustainability of the process. The socioeconomic factors might include access to land, income generation, wealth redistribution, job creation, increased access to clean and modern energy, and the likelihood of occupational injury, exposure to hazardous environments, and fatalities. Conversion of agricultural residues to bioenergy may impact soil quality, deforestation, soil and water contamination, and change in biodiversity (Sarkar et al., 2023). However, inappropriate handling of agricultural residues, inadequate personal protective equipment, and obsolete conversion technologies can make the process laborious, and injurious, and cause fatalities. Lack of incentives and appropriate legal and policy frameworks will discourage practitioners, investors, and stakeholders. There are opportunities exist in the agricultural waste conversion to bioenergy including the provision of incentives and subsidies, the introduction of innovative technologies for waste handling band conversion, and sensitization to remove sociocultural biases that impeded waste conversion and utilization (Awogbemi & Kallon, 2024).

Going forward, there is a need to commission multidisciplinary research into waste conversion and utilization to modernize the process. There must be aggressive training and manpower development in the waste conversion sector to discover novel conversion technologies and ensure an investment return. The issue of the sustainability of the process such as techno-economic analysis, life cycle assessment, energy and exergy analysis, and energy cost. Needs further study. Adequate research and development are desirable to mitigate and overcome the technological, socioeconomic, cultural, environmental, legal, policies, and safety barriers in waste conversion and bioenergy production and application.

12.5 Conclusion

This chapter has presented the various social, economic, and environmental benefits of the conversion of huge waste from the agricultural sectors to bioethanol, biohydrogen, biochar, biojet fuel, biogas, and biobutanol. Indiscriminate disposal and inappropriate management of agricultural residues exacerbate sanitation, contaminate the ecosystem, increase landfill waste, increase emissions of GHGs, and impede environmental sustainability. The conversion of agricultural residues reduces the cost of waste disposal, prevents bush burning and deforestation, ensures resource recovery, supports a circular economy, and contributes to global energy security. To overcome the challenges of effective conversion of agricultural residues to bioenergy, the government must be sensitive to the populace on the benefits of waste conversion, incentivize waste conversion to clean renewable energy and establish appropriate legal, institutional, and financial frameworks for proper functioning of the waste conversion and utilization sector. There must be the development of innovative technologies, and the establishment of robust waste conversion protocols to build investors' confidence in the sector. Multidisciplinary research should be commissioned for capacity building in waste conversion and bioenergy production and utilization.

References

Abdullah, A., Ahmed, A., Akhter, P., Razzaq, A., Hussain, M., Hossain, N., Abu Bakar, M. S., Khurram, S., Majeed, K., & Park, Y. K. (2021). Potential for sustainable utilisation of agricultural residues for bioenergy production in Pakistan: An overview. *Journal of Cleaner Production, 287*, 125047.

Adesina, O. A., Taiwo, A. E., Akintola, O., & Igbafe, A. I. (2023). Optimization of process variables for metallic nanoparticle inclusion in bioethanol synthesis of sugar cane bagasse. *Biofuels, 14*, 1–6.

Ahmed, S. F., Rafa, N., Mofijur, M., Badruddin, I. A., Inayat, A., Ali, M. S., Farrok, O., & Yunus Khan, T. (2021). Biohydrogen production from biomass sources: Metabolic pathways and economic analysis. *Frontiers in Energy Research, 9*, 753878.

Awogbemi, O., & Kallon, D. V. V. (2022a). Application of tubular reactor technologies for the acceleration of biodiesel production. *Bioengineering, 9*, 347.

Awogbemi, O., & Kallon, D. V. V. (2022b). Pretreatment techniques for agricultural waste. *Case Studies in Chemical and Environmental Engineering, 6*, 100229.

Awogbemi, O., & Kallon, D. V. V. (2022c). Valorization of agricultural wastes for biofuel applications. *Heliyon, 8*, e11117.

Awogbemi, O., & Kallon, D. V. V. (2023). Application of biochar derived from crops residues for biofuel production. *Fuel Communications, 15*, 100088.

Awogbemi, O., & Kallon, D. V. V. (2024). Recent advances in the application of nanomaterials for improved biodiesel, biogas, biohydrogen, and bioethanol production. *Fuel, 358*, 130261.

Awogbemi, O., Kallon, D. V. V., Onuh, E. I., & Aigbodion, V. S. (2021). An overview of the classification, production and utilization of biofuels for internal combustion engine applications. *Energies, 14*, 5687.

Bakari, H., Djomdi, D., Falama Ruben, Z., Roger, D. D., Cedric, D., Guillaume, P., Pascal, D., Philippe, M., & Gwendoline, C. (2023). Optimization of bioethanol production after enzymatic treatment of sweet sorghum stalks. *Waste and Biomass Valorization, 14*, 1–15.

Baruya, P. (2015). *World forest and agricultural crop residue resources for cofiring*. Available from: http://www.usea.org/sites/default/files/042015

Behera, S. S., & Ray, R. C. (2021). Bioprospecting of cowdung microflora for sustainable agricultural, biotechnological and environmental applications. *Current Research in Microbial Sciences, 2*, 100018.

Bhurat, K. S., Banerjee, T., Pandey, J. K., & Bhurat, S. S. (2021). A lab fermenter level study on anaerobic hydrogen fermentation using potato peel waste: Effect of pH, temperature, and substrate pre-treatment. *Journal of Material Cycles and Waste Management, 23*, 1617–1625.

Bu, J., Wei, H. L., Wang, Y. T., Cheng, J. R., & Zhu, M. J. (2021). Biochar boosts dark fermentative H2 production from sugarcane bagasse by selective enrichment/colonization of functional bacteria and enhancing extracellular electron transfer. *Water Research, 202*, 117440.

Cervi, W. R., Lamparelli, R. A. C., Gallo, B. C., De Oliveira Bordonal, R., Seabra, J. E. A., Junginger, M., & Van Der Hilst, F. (2021). Mapping the environmental and techno-economic potential of biojet fuel production from biomass residues in Brazil. *Biofuels, Bioproducts and Biorefining, 15*, 282–304.

Chaudhary, A., Hussain, A., Ahmad, Q.-U.-A., Manzoor, M., Tahira, S. A., & Karita, S. (2022). Statistical optimization of alkaline treatment of pomegranate peel waste for bioethanol production. *Biomass Conversion and Biorefinery*, 1–13.

Cheah, W. Y., Sankaran, R., Show, P. L., Ibrahim, T., Baizura, T. N., Chew, K. W., Culaba, A., & Chang, J. S. (2020). Pretreatment methods for lignocellulosic biofuels production: Current advances, challenges and future prospects. *Biofuel Research Journal, 7*, 1115–1127.

Cheng, C. L., Che, P. Y., Chen, B. Y., Lee, W. J., Lin, C. Y., & Chang, J. S. (2012). Biobutanol production from agricultural waste by an acclimated mixed bacterial microflora. *Applied Energy, 100*, 3–9.

Chiappero, M., Cillerai, F., Berruti, F., Mašek, O., & Fiore, S. (2021). Addition of different biochars as catalysts during the mesophilic anaerobic digestion of mixed wastewater sludge. *Catalysts, 11*, 1094.

Díaz, M. J., Moya, M., & Castro, E. (2022). Bioethanol production from steam-exploded barley straw by co-fermentation with Escherichia coli SL100. *Agronomy, 12*, 874.

Emmanouilidou, E., Mitkidou, S., Agapiou, A., & Kokkinos, N. C. (2023). Solid waste biomass as a potential feedstock for producing sustainable aviation fuel: A systematic review. *Renewable Energy, 206*, 897–907.

Enaime, G., & Lübken, M. (2021). Agricultural waste-based biochar for agronomic applications. *Applied Sciences, 11*, 8914.

Global Biochar Market. (2022–2028). Available online https://www.tritonmarketresearch.com/reports/biochar-market#report

GlobeNewsire. (2022). *Bioethanol market size*. Available online https://www.globenewswire.com/en/news-release/2022/10/11/2532393/0/en/Bioethanol-Market-Size-to-Worth-Around-USD-124-5-Billion-by-2030.html

Gökçek, Ö. B., Baş, F., Muratçobanoğlu, H., & Demirel, S. (2023). Investigation of the effects of magnetite addition on biohydrogen production from apple pulp waste. *Fuel, 339*, 127475.

Hao, J., Xiao, J., Song, G., & Zhang, Q. (2021). Energy and exergy analysis of bio-jet fuel production from lignocellulosic biomass via aqueous conversion. *Case Studies in Thermal Engineering, 26*, 101006.

IEA. (2023). *Aviation*. Available online https://www.iea.org/energy-system/transport/aviation

IRENA. (2021). *Reaching zero with renewables: Biojet fuels*. Available online www.irena.org/publications

John, I., Yaragarla, P., & Appusamy, A. (2020). Production of bioethanol from Banana Peel using isolated cellulase from *Aspergillus Niger*. In V. Sivasubramanian & S. Subramanian (Eds.), *Global challenges in energy and environment: Select proceedings of ICEE 2018* (pp. 9–18). Springer.

Kapil, S., Vinayak, A., Pal, N., & Sharma, V. (2023). Application of nanotechnology in biobutanol production. In J. G. Segovia-Hernandez, S. Behera, & E. Sanchez-Ramirez (Eds.), *Advances and developments in biobutanol production* (pp. 363–379). Woodhead Publishing.

Kasinath, A., Byliński, H., Artichowicz, W., Remiszewska Skwarek, A., Szopińska, M., Zaborowska, E., Luczkiewicz, A., & Fudala-Ksiazek, S. (2023). Biochemical assays of intensified methane content in biogas from low-temperature processing of waste activated sludge. *Energy, 282*, 128855.

Khalili, F., & Amiri, H. (2020). Integrated processes for production of cellulosic and hemicellulosic biobutanol from sweet sorghum bagasse using autohydrolysis. *Industrial Crops and Products, 145*, 111918.

Khanal, S. K., Tarafdar, A., & You, S. (2023). Artificial intelligence and machine learning for smart bioprocesses. *Bioresource Technology, 375*, 128826.
Khanh Nguyen, V., Kumar Chaudhary, D., Hari Dahal, R., Hoang Trinh, N., Kim, J., Chang, S. W., Hong, Y., Duc La, D., Nguyen, X. C., Hao Ngo, H., Chung, W. J., & Nguyen, D. D. (2021). Review on pretreatment techniques to improve anaerobic digestion of sewage sludge. *Fuel, 285*, 119105.
Kolo, S. M. D., Wahyuningrum, D., & Hertadi, R. (2020). The effects of microwave-assisted pretreatment and cofermentation on bioethanol production from elephant grass. *International Journal of Microbiology, 2020*, 1–11.
Kumar Sharma, A., Kumar Ghodke, P., Goyal, N., Nethaji, S., & Chen, W. H. (2022). Machine learning technology in biohydrogen production from agriculture waste: Recent advances and future perspectives. *Bioresource Technology, 364*, 128076.
Lim, M., Luckert, M. K., & Qiu, F. (2023). Economic opportunities and challenges in biojet production: A literature review and analysis. *Biomass and Bioenergy, 170*, 106727.
Liu, M., Wei, Y., & Leng, X. (2021). Improving biogas production using additives in anaerobic digestion: A review. *Journal of Cleaner Production, 297*, 126666.
Liu, T., Ferrari, G., Pezzuolo, A., Alengebawy, A., Jin, K., Yang, G., Li, Q., & Ai, P. (2023). Evaluation and analysis of biogas potential from agricultural waste in Hubei Province. *China. Agricultural Systems, 205*, 103577.
Maache-Rezzoug, Z., Semhaoui, I., Maugard, T., Zarguili, I., Zhao, J. M. Q., Nouviaire, A., Pontoire, B., & Rezzoug, S. A. (2023). Enhancing enzymatic hydrolysis of industrial hemp hurds (*Cannabis sativa* L.) by combination of soaking in dilute acid and steam pretreatment. *Biomass Conversion and Biorefinery*, 1–15.
Machineni, L., Deepanraj, B., Chew, K. W., & Rao, A. G. (2023). Biohydrogen production from lignocellulosic feedstock: Abiotic and biotic methods. *Renewable and Sustainable Energy Reviews, 182*, 113344.
Malik, K., Salama, E.-S., El-Dalatony, M. M., Jalalah, M., Harraz, F. A., Al-Assiri, M. S., Zheng, Y., Sharma, P., & Li, X. (2021). Co-fermentation of immobilized yeasts boosted bioethanol production from pretreated cotton stalk lignocellulosic biomass: Long-term investigation. *Industrial Crops and Products, 159*, 113122.
Market Research Report. (2022). *Renewable/biojet market analysis*. Available online. https://www.coherentmarketinsights.com/market-insight/renewable-bio-jet-fuel-market-
Md Razali, N. A. A., Ibrahim, M. F., Kamal Bahrin, E., & Abd-Aziz, S. (2018). Optimisation of simultaneous saccharification and fermentation (SSF) for biobutanol production using pretreated oil palm empty fruit bunch. *Molecules, 23*, 1944.
Mikulski, D., & Kłosowski, G. (2020). Microwave-assisted dilute acid pretreatment in bioethanol production from wheat and rye stillages. *Biomass and Bioenergy, 136*, 105528.
Mohd Jamaludin, N. F., Jamali, N. S., Abdullah, L. C., Idrus, S., Engliman, N. S., & Abdul, P. M. (2023). Biohydrogen production with utilisation of magnetite nanoparticles embedded in granular activated carbon from coconut shell. *International Journal of Hydrogen Energy, 48*, 11695–11708.
Moodley, P., Ray, R. C., & Kana, E. B. G. (2023). Advances in Lignocellulosic biofuel production systems. *Elsevier, 436*.
Moreno-Gómez, A. L., Gutiérrez-Antonio, C., Gómez-Castro, F. I., & Hernández, S. (2020). Production of biojet fuel from waste raw materials: A review. In A. Bonilla-Petriciolet & G. P. Rangaiah (Eds.), *Process systems engineering for biofuels development* (pp. 149–171). Wiley.
Muanruksa, P., Winterburn, J., & Kaewkannetra, P. (2021). Biojet fuel production from waste of palm oil mill effluent through enzymatic hydrolysis and decarboxylation. *Catalysts, 11*, 78.
Mumtha, C., Subashri, D., & Mahalingam, P. U. (2023). Enhancing biohydrogen production from mono-substrates and co-substrates using a novel bacterial strains. *Biotech, 13*, 270.
Novia, N., Hasanudin, H., Hermansyah, H., Fudholi, A., & Pareek, V. K. (2023). Recent advances in CFD modeling of bioethanol production processes. *Renewable and Sustainable Energy Reviews, 183*, 113522.

Our Wolrd in Data. (2023). *Energy consumption by source, world.* Availabe online https://ourworldindata.org/grapher/energy-consumption-by-source-and-country?stackMode=absolute&time=1990

Our World in Data. (2022). *Global CO_2 emissions from fossil fuels.* Available online https://ourworldindata.org/co2-emissions

Owuna, G., Makut, M., Ekeleme, I., & Obiekezie, S. (2018). Isolation, identification and production of biobutanol by different clostridium species isolated from soil using waste paper and sugar cane molasses. *South Asian Journal of Research in Microbiology, 2*, 1–9.

Pant, S., Ritika, & Kuila, A. (2022). Pretreatment of lignocellulosic biomass for bioethanol production. In D. Tuli, S. Kasture, & A. Kuila (Eds.), *Advanced biofuel technologies* (pp. 177–194). Elsevier.

Precedence Research. (2022). *Biogas market.* Available Online https://www.precedenceresearch.com/biogas-market

PRNewswire. (2018). *Global hydrogen generation market 2017–2018 & 2026.* Available online https://www.prnewswire.com/news-releases/global-hydrogen-generation-market-2017-2018-2026

Quah, R. V., Tan, Y. H., Mubarak, N. M., Kansedo, J., Khalid, M., Abdullah, E. C., & Abdullah, M. O. (2020). Magnetic biochar derived from waste palm kernel shell for biodiesel production via sulfonation. *Waste Management, 118*, 626–636.

Rahimi, Z., Anand, A., & Gautam, S. (2022). An overview on thermochemical conversion and potential evaluation of biofuels derived from agricultural wastes. *Energy Nexus, 7*, 100125.

Rajendran, N., Kang, D., Han, J., & Gurunathan, B. (2022). Process optimization, economic and environmental analysis of biodiesel production from food waste using a citrus fruit peel biochar catalyst. *Journal of Cleaner Production, 365*, 132712.

Ray, R. C., & Ramachandran, S. (2018). *Bioethanol production from food crops: Sustainable sources, interventions, and challenges* (p. 460p). Academic.

Ren, Y., Tang, S., Hong, F., Jiang, W., Liu, Z., Lu, H., Wang, C., & Si, B. (2023). Effects of milli-magnetite on biohydrogen production from potato peels: Insight of metabolism mechanisms. *Fuel, 348*, 128576.

Rodríguez-Valderrama, S., Escamilla-Alvarado, C., Magnin, J. P., Rivas-García, P., Valdez-Vazquez, I., & Ríos-Leal, E. (2020). Batch biohydrogen production from dilute acid hydrolyzates of fruits-and-vegetables wastes and corn Stover as co-substrates. *Biomass and Bioenergy, 140*, 105666.

Sachdeva, S., Kumar, R., Sahoo, P. K., & Nadda, A. K. (2023). Recent advances in biochar amendments for immobilization of heavy metals in an agricultural ecosystem: A systematic review. *Environmental Pollution, 319*, 120937.

Saeed, M., Mohammad, A., Singh, P., Lal, B., Suliman, M., Alshahrani, M. Y., & Sharma, M. (2023). Coconut waste valorization to produce biochar catalyst and its application in cellulose-degrading enzymes production via SSF. *International Journal of Biological Macromolecules, 240*, 124382.

Sarkar, J. D., Sarkar, A. K., & Mondal, P. (2023). Sustainable manipulation of agricultural residues in bioenergy production: Asian perspective. In A. Rakshit, A. Biswas, D. Sarkar, V. S. Meena, & R. Datta (Eds.), *Handbook of energy management in agriculture* (pp. 1–25). Springer.

Shanmugam, S., Sun, C., Chen, Z., & Wu, Y. R. (2019). Enhanced bioconversion of hemicellulosic biomass by microbial consortium for biobutanol production with bioaugmentation strategy. *Bioresource Technology, 279*, 149–155.

Shen, Y., Linville, J. L., Urgun-Demirtas, M., Schoene, R. P., & Snyder, S. W. (2015). Producing pipeline-quality biomethane via anaerobic digestion of sludge amended with corn Stover biochar with in-situ CO2 removal. *Applied Energy, 158*, 300–309.

Shukla, A., Kumar, D., Girdhar, M., Kumar, A., Goyal, A., Malik, T., & Mohan, A. (2023). Strategies of pretreatment of feedstocks for optimized bioethanol production: Distinct and integrated approaches. *Biotechnology for Biofuels and Bioproducts, 16*, 44.

Singh, A., Prajapati, P., Vyas, S., Gaur, V. K., Sindhu, R., Binod, P., Kumar, V., Singhania, R. R., Awasthi, M. K., & Zhang, Z. (2023). A comprehensive review of feedstocks as sustainable substrates for next-generation biofuels. *Bioenergy Research, 16*, 105–122.

Song, J., Liu, C., Xing, J., Yang, W., & Ren, J. (2023). Linking bioenergy production by agricultural residues to sustainable development goals: Prospects by 2030 in China. *Energy Conversion and Management, 276*, 116568.

Srivastava, N., Singh, R., Ahmad, I., Asiri, M., Tripathi, S. C., Rai, A. K., Mishra, P. K., & Gupta, V. K. (2023). Biologically derived copper oxide-based nanocatalyst using Moringa oleifera leaves and its applications in hydrolytic enzymes and biohydrogen production. *Bioresource Technology, 376*, 128847.

Statista. (2021). *Global hydrogen production outlook by type 2015–2050*. Available online https://www.statista.com/statistics/859104/hydrogen-production-outlook-worldwide-by-type/

Statista. (2023). *Global biogas production 2000–2020*. Available online https://www.statista.com/statistics/481791/biogas-production-worldwide/

Taiwo, A. E., & Musonge, P. (2023). Comparative evaluation of bioethanol fermentation process parameters using RSM, ANN and ANFIS. *Biofuels, Bioproducts and Biorefining, 17*, 961–975.

Tallentire, C. W., & Steubing, B. (2020). The environmental benefits of improving packaging waste collection in Europe. *Waste Management, 103*, 426–436.

Tamburini, E., Gaglio, M., Castaldelli, G., & Fano, E. A. (2020). Biogas from agri-food and agricultural waste can appreciate agro-ecosystem services: The case study of Emilia Romagna region. *Sustainability, 12*, 8392.

Tran, T. T. A., Le, T. K. P., Mai, T. P., & Nguyen, D. Q. (2020). Bioethanol production from lignocellulosic biomass. In Y. Yun (Ed.), *Alcohol fuels-current technologies and future prospect* (pp. 1–8). Rijeka.

Vela-García, N., Bolonio, D., García-Martínez, M.-J., Ortega, M. F., & Canoira, L. (2022). Thermochemical conversion of agricultural waste to biojet fuel. In A. Yousuf & C. Gonzalez-Fernandez (Eds.), *Sustainable alternatives for aviation fuels* (p. 27). Elsevier.

Wang, J., Ma, D., Lou, Y., Ma, J., & Xing, D. (2023). Optimization of biogas production from straw wastes by different pretreatments: Progress, challenges, and prospects. *Science of the Total Environment, 905*, 166992.

Waqas, M., Aburiazaiza, A. S., Miandad, R., Rehan, M., Barakat, M. A., & Nizami, A. S. (2018). Development of biochar as fuel and catalyst in energy recovery technologies. *Journal of Cleaner Production, 188*, 477–488.

Weide, T., Baquero, C. D., Schomaker, M., Brügging, E., & Wetter, C. (2020). Effects of enzyme addition on biogas and methane yields in the batch anaerobic digestion of agricultural waste (silage, straw, and animal manure). *Biomass and Bioenergy, 132*, 105442.

Yan, B., Yan, J., Li, Y., Qin, Y., & Yang, L. (2021). Spatial distribution of biogas potential, utilization ratio and development potential of biogas from agricultural waste in China. *Journal of Cleaner Production, 292*, 126077.

Zhang, X., Zhang, Q., Li, Y., & Zhang, H. (2023). Modeling and optimization of photo-fermentation biohydrogen production from co-substrates based on response surface methodology and artificial neural network integrated genetic algorithm. *Bioresource Technology, 374*, 128789.

Zhao, L., Wu, K. K., Chen, C., Ren, H. Y., Wang, Z.-H., Nan, J., Yang, S. S., Cao, G. L., & Ren, N. Q. (2021). Role of residue cornstalk derived biochar for the enhanced bio-hydrogen production via simultaneous saccharification and fermentation of cornstalk. *Bioresource Technology, 330*, 125006.

Zhou, Y., Wang, X., Huang, X., Deng, H., Hu, Y., & Lu, L. (2023). Tuning the isoelectric point of zinc molybdate nanomaterials to enhance the biohydrogen production of rice straws. *Chemical Engineering Journal, 452*, 139458.

Chapter 13
Utilization of Agricultural Waste for Water and Wastewater Treatment Processes

Steve Carly Zangué Desobgo

Abbreviations

AOP	Advanced oxidation processes
COD	Chemical oxygen demand.
GAC	Granular activated carbon
MF	Microfiltration
NF	Nanofiltration
PAC	Powdered activated carbon
RC	Reverse osmosis
TOC	Total organic carbon
UF	Ultrafiltration

13.1 Introduction

The utilization of agricultural waste for water and wastewater treatment processes has become increasingly important in recent years. This is due to its potential to solve two critical environmental issues: the disposal of agricultural waste and the treatment of water and wastewater (Khunt et al., 2019; Kumar et al., 2023). Agricultural waste includes byproducts from agricultural activities, such as crop residues, animal manure, and food processing waste. If not properly managed, these waste materials can cause environmental pollution and health hazards (Green, 2019; Siddiqua et al., 2022). Water scarcity and pollution are global problems that require innovative and sustainable solutions (He et al., 2021; Mulwa et al., 2021; Shemer

S. C. Z. Desobgo (✉)
Department of Food Engineering and Quality Control, University Institute of Technology of the University of Ngaoundere, Ngaoundere, Cameroon

J. Arora et al. (eds.), *Transforming Agriculture Residues for Sustainable Development*, Waste as a Resource, https://doi.org/10.1007/978-3-031-61133-9_13

et al., 2023). Traditional water and wastewater treatment methods often involve the use of chemicals and energy-intensive processes, which can be expensive and harmful to the environment. Therefore, the utilization of agricultural waste for water and wastewater treatment has emerged as a promising alternative (Harshwardhan & Upadhyay, 2017; Khunt et al., 2019). Various methods and technologies harness the potential of agricultural waste for water and wastewater treatment (Elbeshbishy et al., 2012). These methods not only effectively treat water and wastewater but also utilize agricultural waste as a valuable resource. For example, biofiltration uses microorganisms present in agricultural waste to remove pollutants from water and wastewater. Anaerobic digestion decomposes organic matter in the absence of oxygen, producing biogas and nutrient-rich digestate. Biochar, produced through the pyrolysis of agricultural waste, can effectively adsorb pollutants from water and wastewater and improve soil fertility (Kamali et al., 2021; Mota et al., 2023). The utilization of agricultural waste for water and wastewater treatment offers a sustainable and cost-effective solution to address agricultural waste management and water pollution challenges. By implementing these innovative approaches, there will be a contribution to a more sustainable and circular economy while protecting our water resources and the environment.

This chapter aims to explore the utilization of agricultural waste for water and wastewater treatment processes, including its types, properties, challenges, and opportunities. It will also discuss various methods and technologies using agricultural waste for water treatment and present case studies. Additionally, it will highlight the advantages, limitations, and future perspectives of utilizing agricultural waste for water treatment.

13.2 Agricultural Waste as a Resource

13.2.1 Types of Agricultural Waste

There are various agricultural waste types suitable for water and wastewater treatment (Chang & Li, 2019; Koul et al., 2022; Rao & Rathod, 2019). Crop residues like corn stalks, rice straw, wheat straw, and sugarcane bagasse can be used as a filter medium or a carbon source for biological treatment (Pandey & Soccol, 2000). Manure from animals, including pigs, cows, and poultry, offers valuable nutrients and organic matter (Bicudo & Goyal, 2003). Fruit and vegetable peels, husks, and shells are examples of food processing waste that can supply nutrients and carbon (Esparza et al., 2020). Then, waste from food manufacturing and animal dung can both be used in biological treatment systems. Aquaculture waste, such as fish scales, fish waste, and shrimp shells, can serve as a carbon and nutrient source in biological treatment processes(Mirabella et al., 2014). Certain bioenergy crops like switchgrass and miscanthus offer their residues, such as leaves and stalks, for water and wastewater treatment. Biogas digestate, a byproduct rich in nutrients and organic matter from biogas production, can be used as a fertilizer or carbon source in wastewater treatment (Al Seadi et al., 2013). Proper management of agricultural runoff is crucial to prevent water pollution and safeguard water resources. The choice of agricultural waste depends on local practices and available resources.

13.2.2 Properties and Composition of Agricultural Waste

Agricultural waste refers to the byproducts and residues generated during agricultural activities, including crop production, livestock farming, and food processing. It is predominantly composed of organic matter such as plant materials, animal manure, and food residues, which provide valuable nutrients and can be used for composting and energy production (Adeyi, 2010; Titiloye et al., 2013). The moisture content of agricultural waste varies widely, with high moisture content leading to issues such as microbial growth and increased decomposition rates. Essential nutrients like nitrogen, phosphorus, and potassium can be found in significant amounts in agricultural waste, derived from plant residues, animal manure, and fertilizers. The carbon-to-nitrogen (C/N) ratio is an important parameter that determines the rate of decomposition and nutrient availability in agricultural waste (Adeyi, 2010). Different types of waste have varying C/N ratios, with lower ratios indicating a higher nitrogen content and faster decomposition rates. Agricultural waste can have diverse physical characteristics, influencing its handling, storage, and processing (Xue et al., 2016). Contaminants such as pesticides, herbicides, and heavy metals can be present in agricultural waste due to the use of agricultural chemicals or environmental pollution. Proper management and treatment of agricultural waste are crucial to minimize the potential negative impacts of these contaminants. It is important to consider regional factors, agricultural practices, and specific waste management strategies when assessing and managing agricultural waste in a particular context (Naeem et al., 2020).

13.2.3 Challenges and Opportunities in Utilizing Agricultural Waste

The utilization of agricultural waste in water and wastewater treatment faces several challenges. One of the main challenges is the variability of agricultural waste (Table 13.1), which makes it difficult to develop standardized treatment processes (Chilakamarry et al., 2022; Koul et al., 2022). Additionally, limited infrastructure and technology in regions that generate significant amounts of agricultural waste hinder efficient waste management and treatment. Contamination and safety concerns also pose risks to human health and the environment if agricultural waste is not properly managed and treated (Chilakamarry et al., 2022; Koul et al., 2022). Moreover, the cost-effectiveness of developing and implementing treatment processes is a significant challenge, especially in regions with limited financial resources.

Table 13.1 Characterization of agricultural waste

Agricultural waste	Chemical composition (% w/w)			Ash (%)	References
	Hemicellulose	Cellulose	Lignin		
Soy stalks	34.5	24.8	19.8	10.4	El-Tayeb et al. (2012)
Wheat straw	32.8	38.0	8.9	1.4	Bjerre et al. (1996)
Corn stalks	42.7	23.3	17.5	9.8	Qu et al. (2011)
Rice straw	41.9	25.6	0.8	16.5	Wang et al. (2021)
Palm oil trunk	39.9	21.2	22.6	1.9	Yuliansyah et al. (2010)
Rice husk	32.7	31.7	18.8	16.3	Dizaji et al. (2019)
Sugarcane bagasse	30.2	56.7	13.4	1.9	El-Tayeb et al. (2012)
Palm oil frond	31.5	19.2	14.0	12.3	Hong et al. (2012)

However, there are several opportunities associated with utilizing agricultural waste in water and wastewater treatment. Agricultural waste is a renewable and abundant resource that can reduce reliance on traditional fossil fuel-based treatment processes (Chilakamarry et al., 2022; Koul et al., 2022; Puglia et al., 2021). Nutrient recovery from agricultural waste can be utilized as fertilizers in agricultural practices, closing the nutrient loop and reducing the need for synthetic fertilizers. Certain agricultural waste can be used to generate renewable energy, reducing reliance on fossil fuels. The utilization of agricultural waste in water and wastewater treatment can also contribute to a circular economy approach, where waste is viewed as a resource (Chilakamarry et al., 2022; Koul et al., 2022; Puglia et al., 2021). Lastly, it can create opportunities for local economic development, including job creation and the development of value-added products from agricultural waste.

13.3 Water Treatment Processes

These include physical, chemical, and biological processes.

13.3.1 Physical Treatment Processes

Sedimentation and Filtration

These processes can be implemented in both small-scale and large-scale wastewater treatment systems. Sedimentation is the process of allowing suspended solids to settle down under gravity. Settling basins are large tanks where water is allowed to sit, allowing solids to settle at the bottom. Agricultural waste, such as straw or hay, can be added to enhance the sedimentation process by providing surfaces for solids to adhere to (Harshwardhan & Upadhyay, 2017; Khunt et al., 2019). Inclined plate settlers use inclined plates to increase the surface area available for solids to settle.

Agricultural waste can be used as a media on the plates to enhance the sedimentation process (Moayedi et al., 2011). Filtration is the process of passing water through a medium to remove suspended solids and other impurities. Slow sand filtration involves passing water through a bed of sand at a slow rate. Agricultural waste, such as rice husks or coconut shells, can be added to the sand bed to improve filtration efficiency (Karić et al., 2022). Rapid sand filtration is similar but with a higher flow rate and the addition of coagulants or flocculants. Agricultural waste can be used as a pretreatment media to remove larger particles. Membrane filtration processes use membranes with different pore sizes to remove suspended solids and contaminants (Hoslett et al., 2018). Agricultural waste, such as activated carbon derived from agricultural byproducts, can be used as a pretreatment step. Biofilters, using agricultural waste materials, support the growth of beneficial bacteria and microorganisms that break down organic pollutants and contaminants in the water (Hoslett et al., 2018).

Adsorption

The use of agricultural waste materials as adsorbents in water treatment processes is a cost-effective and environmentally friendly method that can be easily implemented in both rural and urban areas. Common agricultural waste materials such as rice husk, coconut shell, sugarcane bagasse, corn cob, and sawdust have been successfully used for adsorption purposes (Dai et al., 2018; Harshwardhan & Upadhyay, 2017). These materials possess excellent adsorption properties due to their high silica content, surface area, natural carbon, porous structure, and adsorption capacity. To implement the adsorption process, the waste material is pretreated, dried, and grounded into a suitable particle size before being mixed with contaminated water (Dai et al., 2018).

Membrane Filtration

Membrane filtration processes for water and wastewater treatment can be achieved using agricultural waste through various methods. Examples include microfiltration (MF), ultrafiltration (UF), nanofiltration (NF), and reverse osmosis (RO). MF utilizes membranes with pore sizes ranging from 0.1 to 10 micrometers to remove suspended solids, bacteria, and some viruses (Mateo et al., 2021). Low-cost membrane filters for MF can be developed using agricultural waste such as rice husks or coconut shells. UF uses membranes with smaller pore sizes (0.001–0.1 micrometers) to remove colloidal particles, proteins, and larger organic molecules. Agricultural waste-based membranes can be utilized to enhance filtration efficiency and reduce costs. NF membranes have even smaller pore sizes (0.001–0.01 micrometers) and can remove divalent ions, organic matter, and certain smaller molecules (Mateo et al., 2021). Agricultural waste-based membranes can be modified to achieve the desired separation properties for NF. RO is a high-pressure filtration process that uses a semipermeable membrane to remove dissolved salts, heavy metals, and other contaminants from water (Xiang et al., 2022). Agricultural waste-based membranes can be developed for RO applications, providing a cost-effective and sustainable solution. However, to utilize agricultural waste for membrane filtration, the waste material needs to be processed and converted into a suitable membrane material. Techniques such as carbonization, activation, or modification of the waste material may be required to enhance its filtration properties (OboteyEzugbe & Rathilal, 2020). Proper pretreatment of water or wastewater may also be necessary to prevent fouling and prolong the membrane's lifespan.

13.3.2 Chemical Treatment Processes

Coagulation and Flocculation

Coagulation and flocculation are crucial steps in water and wastewater treatment, involving the addition of chemicals to destabilize and aggregate suspended particles for their removal (Ukiwe et al., 2014). However, these processes can be made more cost-effective and sustainable by utilizing agricultural waste. One such example is the use of Moringa seeds, which contain natural coagulants that effectively remove suspended particles from water (Camacho et al., 2017). The seeds are crushed and mixed with water to extract the coagulant, which is then added to the water or wastewater, followed by gentle mixing. Moringa seeds contain cationic proteins that act as coagulants and have been widely studied and used in water treatment due to their excellent properties. They can remove suspended solids, turbidity, bacteria, and even some heavy metals (Shan et al., 2017). Another agricultural waste product that can be used as a coagulant is chitosan, derived from the chitin found in crustaceans like shrimp and crabs. It is biodegradable and renewable, effectively removing organic matter, heavy metals, and certain dyes from water (Camacho et al., 2017; Shan et al., 2017). Tannins, naturally occurring polyphenolic compounds found in various plant sources, also possess coagulation properties and effectively remove turbidity, color, and organic matter from water (de Oliveira et al., 2022; Ibrahim et al., 2021; Tomasi et al., 2022). Rice husk ash, a byproduct of rice milling, contains silica particles that neutralize the charge on suspended particles, promoting their aggregation and settling. It has been used to remove turbidity, color, and heavy metals from water (Anjitha & George, 2016; Nigusie, 2019). Similarly, banana peel, coconut coir, corn starch, and okra mucilage can all be used as natural coagulants in water treatment, effectively removing suspended solids, turbidity, and even heavy metals (Lee et al., 2014).

Precipitation

Various precipitation processes can be employed for water and wastewater treatment utilizing agricultural waste. These processes utilize agricultural waste materials to facilitate the precipitation and elimination of impurities from water. For instance, agricultural waste materials like plant extracts or byproducts can serve as natural coagulants and flocculants. These materials contain organic compounds that can bind with suspended particles and form larger flocs, which can then be easily removed through sedimentation or filtration (Lee et al., 2014). Additionally, certain agricultural plants possess the ability to absorb and accumulate heavy metals and other contaminants from water. These plants, known as hyperaccumulators, can be cultivated in constructed wetlands or other treatment systems to eliminate pollutants through their root systems. Biochar, a type of charcoal derived from agricultural waste, can also be utilized as an adsorbent material to remove organic contaminants from water (Ghosh et al., 2023). With its high surface area and porosity, biochar effectively adsorbs and retains pollutants (A. Kumar & Bhattacharya, 2021; McKay et al., 2022). Furthermore, agricultural waste materials like crop residues or animal manure can be employed as carbon sources in biological nutrient removal processes, providing a food source for microorganisms that can eliminate excess nitrogen and phosphorus from water through processes such as denitrification and phosphorus uptake (Bicudo & Goyal, 2003). Lastly, agricultural waste materials can be utilized in the construction of wetland systems for water and wastewater treatment, creating a natural environment where various physical, chemical, and biological processes can occur to remove contaminants (Verhoeven & Meuleman, 1999).

Oxidation-Reduction Reactions

Agricultural waste can play a significant role in various oxidative and reductive processes used for water and wastewater treatment. Advanced oxidation processes (AOPs) involve the production of highly reactive oxidizing species like hydroxyl radicals, which can degrade organic pollutants (Khunt et al., 2019). Rice husks or corn cobs can serve as precursors for the generation of these oxidizing species. AOPs include processes such as Fenton's reaction, ozonation, and photocatalysis. Electrochemical oxidation utilizes an electric current to generate oxidizing agents like chlorine or ozone, which can degrade organic pollutants. Agricultural waste can be used as an electrode material or as a source of organic matter, enhancing the degradation efficiency. Biological oxidation utilizes agricultural waste as a carbon source for microorganism growth in biological treatment systems. These microorganisms can degrade organic pollutants through aerobic or anaerobic processes, with examples including activated sludge systems, biofilters, and anaerobic digesters (Kumar et al., 2023).

13.3.3 *Biological Treatment Processes*

Aerobic and Anaerobic Digestion

Aerobic and anaerobic digestion processes are commonly used in water and wastewater treatment to break down organic matter and reduce the concentration of pollutants. These processes are also effective in utilizing agricultural waste as a feedstock.

Aerobic digestion is a biological process that occurs in the presence of oxygen. It involves the use of aerobic microorganisms to decompose organic matter and convert it into carbon dioxide, water, and microbial biomass. Agricultural waste such as crop residues, animal manure, and food processing waste can be used as a substrate for aerobic digestion. The process involves pretreatment to remove any large solids or contaminants, digestion in an aerobic digester where the waste is mixed with air or oxygen, microbial decomposition by aerobic microorganisms, and management of resulting sludge which can be further processed or used as a fertilizer (Gavrilescu & Macoveanu, 1999).

On the other hand, anaerobic digestion is a biological process that occurs in the absence of oxygen. It involves the use of anaerobic microorganisms to convert organic matter into biogas and digestate. Agricultural waste, including crop residues, animal manure, and food waste, can be utilized as a substrate for anaerobic digestion. The process involves pretreatment to remove any large solids, contaminants, or inhibitors; introduction of the waste into an anaerobic digester; biogas production through biochemical reactions by microorganisms; and management of the remaining digestate, which can be used as a fertilizer or further processed for nutrient recovery (Kiyasudeen et al., 2016).

Both aerobic and anaerobic digestion processes have their advantages and limitations. Aerobic digestion is faster and more efficient in terms of organic matter removal, but it requires a constant supply of oxygen and energy for aeration. Anaerobic digestion, on the other hand, produces biogas as a valuable byproduct but requires longer retention times and careful control of operating conditions.

Constructed Wetlands

Constructed wetlands are an innovative and sustainable approach to water and wastewater treatment, harnessing the natural purification abilities of wetland plants and microorganisms. By incorporating agricultural waste, these wetlands offer an effective and environmentally friendly solution to water treatment challenges. The process begins with pretreatment, where large solids and debris are removed from the incoming water or wastewater to prevent clogging and damage to the wetland system (Vymazal, 2011). Next, the subsurface flow wetland comes into play, as the wastewater flows horizontally through a bed of porous media, such as gravel or sand, that is planted with wetland vegetation. This vegetation absorbs nutrients like nitrogen and phosphorus from the water, reducing their concentrations. Additionally, agricultural waste can be used to provide nutrients to plants, promoting their growth and enhancing nutrient uptake. Beneath the wetland vegetation, anaerobic or facultative bacteria break down organic matter present in the wastewater, aiding in the removal of pollutants and organic compounds. The porous media in the wetland bed acts as a filter, trapping suspended solids and particulate matter, while the wetland plants and microbial activity further aid in the removal of pollutants and contaminants (Omondi et al., 2020). Oxygenation is crucial for the survival of aerobic bacteria responsible for organic matter degradation and pollutant removal. To ensure sufficient oxygen supply, some constructed wetlands incorporate mechanisms like surface aeration or alternating wet and dry periods. After passing through the wetland bed, the treated water undergoes a final polishing process, which may include disinfection using UV light or chlorination to further reduce pathogens and meet quality standards (Vymazal, 2011). Finally, the treated water can be discharged into bodies of water or reused for irrigation or industrial processes, depending on the achieved quality. Constructed wetlands with agricultural waste integration provide a natural and sustainable solution for water treatment, benefiting both the environment and communities.

13.4 Methods and Technologies Utilizing Agricultural Waste for Water and Wastewater Treatment

These include processes such as adsorption, coagulation and flocculation, membrane filtration, and biological treatments.

13.4.1 Adsorption

Activated Carbon from Agricultural Waste

Activated carbon is a highly effective material for water and wastewater treatment, known for its strong adsorption capacity. To produce activated carbon, various methods and technologies utilize agricultural waste (Fig. 13.1).

One such method is pyrolysis, which involves heating agricultural waste materials like rice husks, coconut shells, or sugarcane bagasse in the absence of oxygen. This process thermally decomposes the waste, leaving behind a carbon-rich residue that can be activated to create activated carbon (Bumajdad & Hasila, 2023; Farah, 2023). Chemical activation is another method that treats agricultural waste with a chemical agent, like an alkali metal hydroxide or acid, to create activated carbon (Farah, 2023). The waste is impregnated with the chemical agent and then heated, creating a porous structure with a high surface area for enhanced adsorption

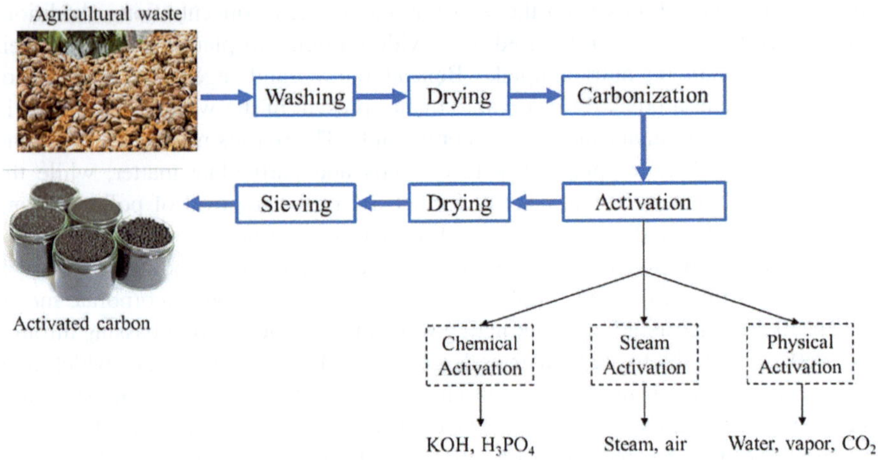

Fig. 13.1 A schematic diagram illustrating the process of producing activated carbon from agricultural waste

capacity. Physical activation, on the other hand, subjects agricultural waste to physical processes like steam activation or carbonization (Farah, 2023). Steam activation exposes the waste to high-temperature steam, while carbonization involves heating the waste in the absence of oxygen (Molina-Sabio et al., 1996). Some methods combine both chemical and physical activation processes to produce activated carbon. Once activated carbon is produced, it can be applied in various adsorption technologies for water and wastewater treatment, such as fixed-bed adsorption, fluidized-bed adsorption, granular activated carbon (GAC) filtration, and powdered activated carbon (PAC) dosing (Jjagwe et al., 2021). These methods offer a sustainable and cost-effective solution for utilizing agricultural waste and improving water and wastewater treatment processes(Farah, 2023).

Biochar

Biochar is a carbon-rich material that is produced through the process of pyrolysis, which involves heating biomass (such as agricultural waste) in the absence of oxygen. It is a highly porous substance with a large surface area, making it effective for various applications, including water and wastewater treatment. There are several methods and technologies that utilize agricultural waste for the production of biochar for water and wastewater treatment(D. Wang et al., 2020).

One method is slow pyrolysis, which involves heating agricultural waste at temperatures ranging from 300 to 700 °C in a slow and controlled manner (Fig. 13.2). This allows for the production of high-quality biochar with enhanced adsorption properties (Wang et al., 2020). Another method is fast pyrolysis, which

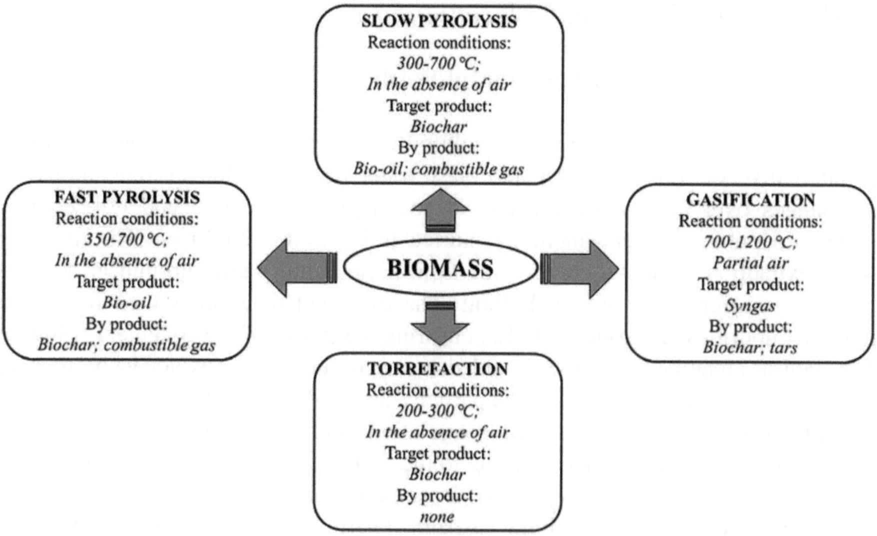

Fig. 13.2 Technologies for the thermochemical conversion of biomass to produce biochar

involves rapid heating of agricultural waste at temperatures above 350 °C (Wang et al., 2020). Fast pyrolysis is often used for large-scale biochar production (Fig. 13.2). Gasification (Fig. 13.2) is another method that involves the partial oxidation of agricultural waste at high temperatures (700–1200 °C), producing a syngas that can be further processed to obtain biochar. Gasification is known for its high energy efficiency and versatility in handling various types of biomass (D. Wang et al., 2020). Hydrothermal carbonization is a relatively fast process that involves the conversion of agricultural waste into biochar under high-pressure, high-temperature conditions in the presence of water. It can be used to produce biochar from wet biomass.

These methods can be combined with various agricultural waste materials to produce biochar. Once produced, biochar can be used in different water and wastewater treatment processes. It has a high adsorption capacity, making it effective in removing organic pollutants, heavy metals, and other contaminants from water (Hama Aziz & Kareem, 2023; Jagadeesh & Sundaram, 2023a). Biochar can also be used as a filtration medium to remove suspended solids and particulate matter. Additionally, it can be used to remove nutrients from agricultural runoff or wastewater and adjust the pH of water or wastewater due to its alkaline nature (Enaime et al., 2020; Jagadeesh & Sundaram, 2023b; Jha et al., 2023).

13.4.2 Coagulation and Flocculation

Natural Coagulants from Agricultural Waste

Natural coagulants derived from agricultural waste are increasingly being explored as alternative options for water and wastewater treatment due to their numerous advantages over conventional chemical coagulants. To extract and utilize these natural coagulants, several methods and technologies are employed. The first step is the extraction of active compounds from agricultural waste through techniques such as grinding, crushing, or boiling. The extracted material is then filtered to remove impurities and solid particles, ensuring the purity of the coagulant solution. The natural coagulant solution is added to the water or wastewater to be treated, causing destabilization of suspended particles and colloids, leading to the formation of larger flocs in a process called coagulation-flocculation. The mixture is then allowed to settle, with the flocs settling to the bottom and forming a layer of sludge, while the clarified water is separated. In some cases, additional filtration may be performed to remove any remaining suspended particles or impurities (Badawi et al., 2023). After the coagulation and filtration processes, the treated water may undergo disinfection to eliminate any remaining pathogens or microorganisms. Common agricultural waste materials used for natural coagulant production include Moringa oleifera seeds, tamarind seeds, banana peels, papaya seeds, and various plant extracts, which contain natural coagulant compounds such as proteins, polysaccharides, and tannins (Abujazar et al., 2022). These sustainable and eco-friendly alternatives have proven to be effective in removing impurities and contaminants from water.

Composite Coagulants

Composite coagulants derived from agricultural waste offer an innovative and sustainable solution for water and wastewater treatment (Balbinoti et al., 2023). These coagulants are created by combining agricultural waste materials with traditional coagulants like alum or ferric chloride, resulting in improved performance and reduced environmental impact. The production and utilization of composite coagulants involve several methods and technologies. First, agricultural waste materials such as rice husks, coconut shells, or corn cobs are collected and processed. These readily available materials are carbonized to eliminate impurities and transform them into carbon-rich substances. The carbonized waste is then activated to enhance its surface area and porosity, thereby increasing its adsorption capacity. The activated carbonized agricultural waste is then mixed with traditional coagulants in specific ratios to form composite coagulants. The combination of agricultural waste and traditional coagulants enhances the coagulation and flocculation properties of the composite coagulant. When added to water or wastewater, the composite coagulant destabilizes suspended particles and colloids, causing them to aggregate and form larger flocs. These flocs can be easily separated from the water through sedimentation or filtration. Following coagulation and flocculation, the water or wastewater undergoes filtration to remove the larger flocs and any remaining particles. Depending on specific requirements, disinfection processes like chlorination or UV treatment may be employed to eliminate harmful microorganisms. Using composite coagulants from agricultural waste offers several benefits, including sustainability, cost-effectiveness, enhanced performance, and versatility in adapting to local agricultural resources and waste streams (Badawi et al., 2023; Balbinoti et al., 2023).

13.4.3 Membrane Filtration

Membranes Modified with Agricultural Waste

Utilizing agricultural waste technology, modified membranes offer a sustainable and cost-effective approach to water and wastewater treatment (Samavati et al., 2023; Shrivastava & Singh, 2022). By leveraging materials like rice husks, coconut shells, and corn cobs, the surface properties of membranes can be enhanced, improving their filtration performance. One method involves incorporating agricultural waste into the membrane matrix during fabrication, where materials like activated carbon or biochar, derived from agricultural waste, are mixed with the membrane material (J. He et al., 2017). This modification enhances the membrane's adsorption capacity, effectively removing organic pollutants and heavy metals from water and wastewater. Another approach entails coating the membrane surface with a thin layer of agricultural waste-derived materials, which enhances hydrophilicity, antifouling properties, and selectivity towards specific contaminants (Ahmad et al., 2022; Shahkaramipour et al., 2017). Chitosan, a natural polymer derived from shrimp and crab shells, can be used for this purpose, improving fouling resistance and antimicrobial properties. Additionally, agricultural waste can serve as a precursor for nanoparticle synthesis, which, when incorporated into the membrane structure, enhances separation efficiency and antimicrobial properties. For instance, silver nanoparticles synthesized from agricultural waste materials inhibit the growth of bacteria and microorganisms within the membrane (Abdelghany et al., 2018; Garg et al., 2020).

Bio-inspired Membranes

Bio-inspired membrane technology involves the creation of membranes that imitate the structure and function of natural biological membranes, aiming to enhance the efficiency and effectiveness of water and wastewater treatment (Abaie et al., 2021). One way to achieve this is by utilizing agricultural waste materials, such as crop residues, animal manure, and food processing byproducts, which can be processed and transformed into membranes suitable for water and wastewater treatment (Abaie et al., 2021). Various methods can be employed to create bio-inspired membranes from agricultural waste, including extracting cellulose from the waste and using it as a base material for membrane fabrication (Ehsani et al., 2022). Cellulose, a biodegradable and renewable polymer, can be modified to enhance its filtration properties (Morales-Jiménez et al., 2023). Another approach involves coating membranes with chitosan, a natural polymer sourced from crustacean shells, which not only improves membrane performance and durability but also possesses antimicrobial properties (Kaushal & Singh, 2023). Additionally, bio-inspired membranes can incorporate natural structures found in biological membranes, such as nanopores created using nanotechnology, enabling more efficient separation and filtration processes. These bio-inspired membranes offer several advantages over traditional ones, including their environmentally friendly nature, reliance on renewable resources, and potential cost-effectiveness due to the availability and affordability of agricultural waste materials (Ehsani et al., 2022; Kaushal & Singh, 2023; Morales-Jiménez et al., 2023).

13.4.4 Biological Treatment

Anaerobic Digestion

Anaerobic digestion is a biological process that breaks down organic materials without oxygen, producing biogas and nutrient-rich effluent (Fig. 13.3).

This technology is effective for water and wastewater treatment, particularly for agricultural waste(Uddin & Wright, 2023). The process involves several steps. First, agricultural waste such as crop residues, animal manure, and food waste is collected and prepared as feedstock. Then, an anaerobic digester is designed and constructed according to the waste characteristics and treatment requirements (Xu & Li, 2017). Different types of digesters are available, including covered lagoons, plug flow digesters, complete mix digesters, and fixed film digesters.

The prepared feedstock is loaded into the digester and mixed to ensure uniform distribution and microbial activity. The anaerobic digestion process occurs in stages, starting with hydrolysis, followed by acidogenesis, and ending with methanogenesis. The produced biogas, mainly composed of methane and carbon dioxide, is collected and stored for energy generation. The effluent still contains nutrients and organic matter, which can be further treated for water quality standards. Benefits of anaerobic digestion include energy generation, waste management, nutrient recovery, and greenhouse gas reduction (Uddin & Wright, 2023; Xu & Li, 2017).

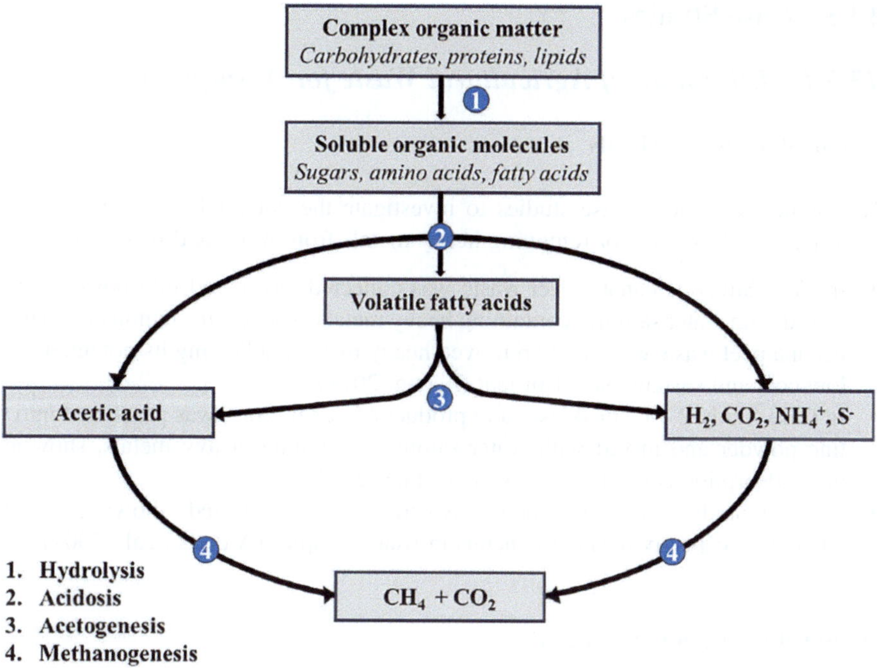

Fig. 13.3 Stages of anaerobic digestion

Constructed Wetlands

Constructed wetlands utilize natural processes to improve water quality by removing pollutants. They consist of shallow basins or channels filled with wetland plants like reeds, cattails, and bulrushes. The wetland cells are lined to prevent seepage. In subsurface flow wetlands, polluted water passes through a gravel or rock bed beneath the plants, where microorganisms break down organic matter and remove pollutants. Surface flow wetlands involve water flowing over the plants' surface, enhancing pollutant removal through plant uptake and microbial processes. Treatment mechanisms include sedimentation, filtration, adsorption, plant uptake, microbial decomposition, and denitrification. Constructed wetlands effectively remove pollutants in agricultural waste, but require regular maintenance and monitoring (Hadidi, 2021). They are cost-effective, energy-efficient, and provide additional benefits like habitat creation and biodiversity conservation. However, they may not remove certain contaminants completely, and efficiency can vary depending on design and environmental conditions (Hadidi, 2021).

13.5 Case Studies

13.5.1 Utilization of Agricultural Waste for Adsorption

Removal of Heavy Metals

Researchers conducted case studies to investigate the potential of different waste materials as adsorbents for removing heavy metals from water and wastewater.

- In Case Study 1, banana peel waste was collected, processed into powder, and mixed with water samples containing heavy metals. The results demonstrated that banana peel waste effectively removed heavy metals, indicating its potential as a low-cost and sustainable adsorbent (Leong, 2018).
- In Case Study 2, rice husk ash, a byproduct of rice milling, was processed into a fine powder and mixed with water samples containing heavy metals, showing high adsorption capacity (Lata & Samadder, 2014).
- In Case Study 3, coconut shell-activated carbon was tested, showing a high adsorption capacity for heavy metals in water samples (Amuda et al., 2007).

Removal of Organic Pollutants

- In Case Study 1, researchers in India utilized rice husk, an agricultural waste material, to remove organic pollutants from wastewater at a treatment plant. By treating the wastewater with varying concentrations of rice husk, the researchers observed its effectiveness in reducing the concentration of organic pollutants, including dyes and heavy metals. The study revealed that rice husk had a significant adsorption capacity for these pollutants (Madhav et al., 2022).
- Similarly, in Case Study 2, coconut shell, another agricultural waste material, was used to remove organic pollutants from water at a treatment plant. The study demonstrated that coconut shells had a high adsorption capacity for organic pollutants like pesticides and pharmaceuticals, effectively reducing their concentration in the water. The researchers also noted that the adsorption capacity increased with a longer contact time between coconut shells and water (Madhav et al., 2022).

13.5.2 Utilization of Agricultural Waste for Coagulation and Flocculation: Removal of Suspended Solids and Turbidity

In two separate case studies, agricultural waste materials were utilized for the removal of suspended solids and turbidity in water and wastewater treatment.

- In the first case study conducted in India, rice husk was collected from local rice mills, washed, dried, and ground into a fine powder. This rice husk powder was then used as a coagulant in the water treatment process, where it was added to contaminated water, stirred, and allowed to settle. The treated water was then filtered to remove any remaining particles, resulting in a significant reduction in suspended solids and turbidity levels (Adams & Mulaba-Bafubiandi, 2014; Ezekoye, 2017).
- Similarly, in the second case study, banana peels were collected from local plantations, prepared, and used as a coagulant in the water treatment process. The banana peel powder was added to contaminated water, mixed, and allowed to settle before filtration. This also led to a notable decrease in suspended solids and turbidity levels. In both cases, the natural properties of the agricultural waste materials facilitated the aggregation and settling of particles, ultimately improving the quality of the water (Azamzam et al., 2022; Mokhtar et al., 2019).

13.5.3 Utilization of Agricultural Waste for Membrane Filtration

Removal of Microorganisms

The researchers examined the application of agricultural waste materials in water and wastewater treatment for microorganism removal in two distinct case studies.

- In the first study, rice straw was used as a natural adsorbent material in a membrane filtration system. The researchers prepared the rice straw by washing, drying, and cutting it into small pieces, which were then packed into a filtration column. The system effectively removed bacteria, viruses, and fungi from the wastewater, with a high removal efficiency for certain microorganisms.
- In the second study, coconut shell-activated carbon was incorporated into a membrane filtration system for water treatment. The activated carbon, prepared by carbonizing and activating coconut shells, demonstrated a high removal efficiency for bacteria, algae, and protozoa (Crites et al., 2014; Smarzewska & Morawska, 2021).

Removal of Organic Matter

- In Case Study 1, rice straw was utilized as agricultural waste material for membrane filtration in water treatment. A composite membrane was developed by incorporating rice straw particles into a polymeric matrix. Its efficiency in removing organic matter from water was tested, resulting in a high removal efficiency for organic matter, with significant reductions in chemical oxygen demand (COD) and total organic carbon (TOC) levels. The membrane also demonstrated good stability and durability during filtration (Hassan et al., 2020).

- Similarly, in Case Study 2, banana peel was used as agricultural waste material for membrane filtration in wastewater treatment. The composite membrane, developed by incorporating banana peel particles into a polymeric matrix, showed excellent performance in organic matter removal, with significant reductions in COD and TOC levels. The membrane also exhibited good mechanical strength and stability during filtration (Ahmad & Danish, 2018; Datta et al., 2012; Farias et al., 2023).

13.5.4 Utilization of Agricultural Waste for Biological Treatment

In this case study, a rural farm successfully implemented an anaerobic digestion system to treat water and wastewater generated from animal manure. The main objective was to utilize agricultural waste and produce biogas as a renewable energy source. With a large number of livestock, the farm faced the challenge of significant daily animal manure production, leading to environmental pollution and foul odors from traditional open lagoon storage. To address these issues and make beneficial use of the waste, the farm installed a series of tanks for anaerobic digestion. This process involved the decomposition of organic matter by bacteria in the absence of oxygen, resulting in the production of biogas and nutrient-rich digestate (Hamilton, 2009; Song et al., 2023). The captured biogas was then used for heating, electricity generation, and as fuel for farm vehicles, reducing the farm's reliance on fossil fuels and lowering their carbon footprint. The nutrient-rich digestate was utilized as a fertilizer for agricultural crops, providing a sustainable source of nutrients and minimizing environmental pollution from synthetic fertilizers. The implementation of the anaerobic digestion system not only improved water and wastewater treatment but also provided economic benefits by reducing energy costs and expenses associated with synthetic fertilizers, while enhancing the overall sustainability of the farm's operations (Hamilton, 2009; Rekleitis et al., 2020; Song et al., 2023).

13.6 Advantages and Limitations

13.6.1 Advantages of Utilizing Agricultural Waste for Water Treatment

Utilizing agricultural waste for water treatment offers several advantages.

- First, it is a cost-effective solution as agricultural waste, such as crop residues or animal manure, is readily available and inexpensive. This reduces the cost of traditional treatment methods.
- Second, repurposing agricultural waste for water treatment promotes sustainability by minimizing the environmental impact of waste disposal and contributing to a circular economy.
- Third, agricultural waste contains valuable nutrients that can be recovered during the treatment process, reducing the need for synthetic alternatives and promoting sustainable agriculture (Green, 2019). Additionally, many agricultural waste materials act as natural filters, promoting biological treatment processes that break down organic pollutants in water and reduce the need for chemical treatment. Moreover, utilizing agricultural waste for water treatment requires less energy compared to conventional methods, especially in biological treatment processes (Crites et al., 2014).
- Lastly, engaging local communities in rural areas where agriculture is a primary economic activity fosters a sense of ownership and promotes environmental stewardship.

13.6.2 Limitations and Challenges

The utilization of agricultural waste for water treatment presents numerous limitations and challenges. First, the composition of agricultural waste varies greatly, including crop residues, animal manure, and food processing byproducts, making it difficult to establish standardized treatment processes. Additionally, agricultural waste may not effectively remove certain contaminants, such as heavy metals or persistent organic pollutants, depending on the specific waste used and the targeted contaminants (Crites et al., 2014; Smarzewska & Morawska, 2021). Furthermore, the availability and scalability of agricultural waste for water treatment can be limited due to factors like seasonal variations, agricultural practices, and geographical location. Some agricultural waste treatment technologies may also require sophisticated equipment or infrastructure, which can be impractical or financially burdensome, particularly in rural areas or developing countries with limited resources. Moreover, the utilization of agricultural waste for water treatment can have unintended environmental consequences, such as the release of excess nutrients leading to eutrophication and ecosystem disruption. Additionally, agricultural waste may contain pathogens, pesticides, or other harmful substances, posing health and safety risks if not properly managed during the treatment process. Adequate precautions must be taken to ensure the safety of workers and the public (Crites et al., 2014; Green, 2019; Smarzewska & Morawska, 2021). Lastly, the utilization of agricultural waste for water treatment may be subject to varying regulatory requirements and policies across jurisdictions, posing compliance challenges, particularly for small-scale or informal treatment systems.

13.6.3 Future Perspectives and Research Directions

The utilization of agricultural waste for water treatment is a promising field with various future perspectives and research directions. With the increasing water scarcity and pollution worldwide, it is crucial to find sustainable and cost-effective methods for water treatment. Agricultural waste, including crop residues, animal manure, and food processing byproducts, can serve as a valuable resource in this regard. One potential future perspective involves the development of innovative technologies that effectively utilize agricultural waste for water treatment. For instance, researchers can explore the use of biochar, a carbon-rich material derived from the pyrolysis of agricultural waste, as a filtration medium to remove contaminants from water. Biochar has shown excellent adsorption properties, effectively eliminating heavy metals, organic pollutants, and nutrients from water. Another research direction involves optimizing treatment processes by enhancing the adsorption capacity and performance of agricultural waste through proper pretreatment methods. Additionally, researchers can investigate the integration of agricultural waste-based treatment systems with other technologies like membrane filtration or electrochemical processes to enhance overall treatment efficiency. Furthermore, it is essential to assess the environmental impacts of utilizing agricultural waste for water treatment, including the potential release of harmful substances or byproducts and their effects on the environment. Understanding the fate and transport of contaminants in agricultural waste-based treatment systems will aid in designing sustainable and environmentally friendly water treatment processes. Moreover, economic feasibility studies are necessary to evaluate the cost-effectiveness of utilizing agricultural waste for water treatment. Researchers can explore ways to optimize the production of agricultural waste-based materials and technologies, making them more affordable and accessible, especially in developing countries where water treatment infrastructure is often inadequate.

13.7 Conclusion

The utilization of agricultural waste for water and wastewater treatment offers advantages and opportunities. It can be a valuable resource for sustainable treatment technologies. Agricultural waste is suitable for various treatment processes. Physical processes like sedimentation, filtration, and adsorption can remove solids and pollutants. Chemical processes can be enhanced using natural and composite coagulants from agricultural waste. Biological processes like digestion and wetlands can degrade organic matter and remove nutrients. Methods and technologies have been developed for water treatment using agricultural waste. Adsorption with activated carbon and biochar removes heavy metals and pollutants. Coagulation and flocculation remove solids and turbidity. Membrane filtration with modified membranes removes microorganisms and matter. Anaerobic digestion and wetlands treat wastewater. Successful case studies include adsorption for heavy metal and pollutant removal, coagulation and flocculation for solids and turbidity removal, membrane filtration for microorganism and matter removal, and anaerobic digestion and wetlands for wastewater treatment. Utilizing agricultural waste has advantages like using renewable and local resources, cost-effectiveness, and environmental sustainability. However, challenges include waste composition variability, pretreatment and modification, and potential contaminant release. Future research should focus on efficient and cost-effective technologies, parameter optimization, exploring new waste types, and evaluating long-term environmental impacts.

References

Abaie, E., Xu, L., & Shen, Y. X. (2021). Bioinspired and biomimetic membranes for water purification and chemical separation: A review. *Frontiers of Environmental Science and Engineering, 15*(6), 1–33. https://doi.org/10.1007/S11783-021-1412-8

Abdelghany, T. M., Al-Rajhi, A. M. H., Al Abboud, M. A., Alawlaqi, M. M., Ganash Magdah, A., Helmy, E. A. M., & Mabrouk, A. S. (2018). Recent advances in green synthesis of silver nanoparticles and their applications: About future directions: A review. *BioNanoScience, 8*(1), 5–16. https://doi.org/10.1007/S12668-017-0413-3

Abujazar, M. S. S., Karaağaç, S. U., Abu Amr, S. S., Alazaiza, M. Y. D., Fatihah, S., & Bashir, M. J. K. (2022). Recent advancements in plant-based natural coagulant application in the water and wastewater coagulation-flocculation process: Challenges and future perspectives. *GlobalNEST International Journal, 24*, 687–705.

Adams, F. V., & Mulaba-Bafubiandi, A. F. (2014). Application of rice hull ash for turbidity removal from water. *Physics and Chemistry of the Earth, Parts A/B/C, 72–75*, 73–76. https://doi.org/10.1016/J.PCE.2014.09.010

Adeyi, O. (2010). Proximate composition of some agricultural wastes in Nigeria and their potential use in activated carbon production. *Journal of Applied Sciences and Environmental Management, 14*(1), 55–58. https://doi.org/10.4314/JASEM.V14I1.56490

Ahmad, T., & Danish, M. (2018). Prospects of banana waste utilization in wastewater treatment: A review. *Journal of Environmental Management, 206*, 330–348. https://doi.org/10.1016/J.JENVMAN.2017.10.061

Ahmad, N. N. R., Mohammad, A. W., Mahmoudi, E., Ang, W. L., Leo, C. P., & Teow, Y. H. (2022). An overview of the modification strategies in developing antifouling nanofiltration membranes. *Membranes, 12*(12), 1276. https://doi.org/10.3390/MEMBRANES12121276

Al Seadi, T., Drosg, B., Fuchs, W., Rutz, D., & Janssen, R. (2013). Biogas digestate quality and utilization. In A. Wellinger, J. Murphy, & D. Baxter (Eds.), *The Biogas handbook: Science, production and applications* (pp. 267–301). Woodhead Publishing. https://doi.org/10.1533/9780857097415.2.267

Amuda, O. S., Giwa, A. A., & Bello, I. A. (2007). Removal of heavy metal from industrial wastewater using modified activated coconut shell carbon. *Biochemical Engineering Journal, 36*(2), 174–181. https://doi.org/10.1016/j.bej.2007.02.013

Anjitha, A., & George, D. (2016). Comparative study using rice husk and its ash as natural coagulants in wastewater treatment. *International Journal of Scientific and Engineering Research, 7*(4), 232–237.

Azamzam, A. A., Rafatullah, M., Yahya, E. B., Ahmad, M. I., Lalung, J., Alam, M., & Siddiqui, M. R. (2022). Enhancing the efficiency of banana peel bio-coagulant in turbid and river water treatment applications. *Water, 14*(16), 2473. https://doi.org/10.3390/W14162473

Badawi, A. K., Salama, R. S., & Mostafa, M. M. M. (2023). Natural-based coagulants/flocculants as sustainable market-valued products for industrial wastewater treatment: A review of recent developments. *RSC Advances, 13*(28), 19335–19355. https://doi.org/10.1039/D3RA01999C

Balbinoti, J. R., dos Santos Junior, R. E., de Sousa, L. B. F., de JesusBassetti, F., Balbinoti, T. C. V., Jorge, R. M. M., & de Matos Jorge, L. M. (2023). Plant-based coagulants for food industry wastewater treatment. *Journal of Water Process Engineering, 52*, 103525. https://doi.org/10.1016/J.JWPE.2023.103525

Bicudo, J. R., & Goyal, S. M. (2003). Pathogens and manure management systems: A review. *Environmental Technology, 24*(1), 115–130. https://doi.org/10.1080/09593330309385542

Bjerre, A. B., Olesen, A. B., Fernqvist, T., Plöger, A., & Schmidt, A. S. (1996). Pretreatment of wheat straw using combined wet oxidation and alkaline hydrolysis resulting in convertible cellulose and hemicellulose. *Biotechnology and Bioengineering, 49*(5), 568–577.

Bumajdad, A., & Hasila, P. (2023). Surface modification of date palm activated carbonaceous materials for heavy metal removal and CO_2 adsorption. *Arabian Journal of Chemistry, 16*(1), 104403. https://doi.org/10.1016/J.ARABJC.2022.104403

Camacho, F. P., Sousa, V. S., Bergamasco, R., & Teixeira, M. R. (2017). The use of *Moringa oleifera* as a natural coagulant in surface water treatment. *Chemical Engineering Journal, 313*, 226–237.

Chang, C. C., & Li, R. (2019). Agricultural waste. *Water Environment Research, 91*(10), 1150–1167. https://doi.org/10.1002/WER.1211

Chilakamarry, C. R., Mimi Sakinah, A. M., Zularisam, A. W., Sirohi, R., Khilji, I. A., Ahmad, N., & Pandey, A. (2022). Advances in solid-state fermentation for bioconversion of agricultural wastes to value-added products: Opportunities and challenges. *Bioresource Technology, 343*, 126065. https://doi.org/10.1016/J.BIORTECH.2021.126065

Crites, R. W., Middlebrooks, E. J., Bastian, R. K., & Reed, S. C. (2014). *Natural wastewater treatment systems* (2nd ed., pp. 1–509). Natural Wastewater Treatment Systems. https://doi.org/10.1201/9781420026443/NATURAL-WASTEWATER-TREATMENT-SYSTEMS-RONALD-CRITES-JOE-MIDDLEBROOKS-SHERWOOD-REED

Dai, Y., Sun, Q., Wang, W., Lu, L., Liu, M., Li, J., Yang, S., Sun, Y., Zhang, K., Xu, J., Zheng, W., Hu, Z., Yang, Y., Gao, Y., Chen, Y., Zhang, X., Gao, F., & Zhang, Y. (2018). Utilizations of agricultural waste as adsorbent for the removal of contaminants: A review. *Chemosphere, 211*, 235–253. https://doi.org/10.1016/J.CHEMOSPHERE.2018.06.179

Datta, S., Karmoker, S., & Sowgath, M. T. (2012). *Membrane development from banana peel fibers for waste water treatment at low cost*. Conference, Environmental Division.

de Oliveira, C., Trevisan, V., & andSkoronski, E. (2022). Application of tannin-based coagulant for high-range turbidity surface water clarification. *Journal of Water, Sanitation and Hygiene for Development, 12*(11), 803–815. https://doi.org/10.2166/WASHDEV.2022.120

Dizaji, B. H., Zeng, T., Hartmann, I., Enke, D., Schliermann, T., Lenz, V., & andBidabadi, M. (2019). Generation of high quality biogenic silica by combustion of rice husk and rice straw combined with pre-and post-treatment strategies—A review. *Applied Sciences, 9*(6), 1083.

Ehsani, M., Kalugin, D., Doan, H., Lohi, A., & Abdelrasoul, A. (2022). Bio-sourced and biodegradable membranes. *Applied Sciences, 12*(24), 12837. https://doi.org/10.3390/APP122412837

Elbeshbishy, E., Nakhla, G., & Hafez, H. (2012). Biochemical methane potential (BMP) of food waste and primary sludge: Influence of inoculum pre-incubation and inoculum source. *Bioresource Technology, 110*, 18–25. https://doi.org/10.1016/J.BIORTECH.2012.01.025

El-Tayeb, T. S., Abdelhafez, A. A., Ali, S. H., & Ramadan, E. M. (2012). Effect of acid hydrolysis and fungal biotreatment on agro-industrial wastes for obtainment of free sugars for bioethanol production. *Brazilian Journal of Microbiology, 43*, 1523–1535.

Enaime, G., Baçaoui, A., Yaacoubi, A., & Lübken, M. (2020). Biochar for wastewater treatment—Conversion technologies and applications. *Applied Sciences, 10*(10), *3492*. https://doi.org/10.3390/APP10103492

Esparza, I., Jiménez-Moreno, N., Bimbela, F., Ancín-Azpilicueta, C., & Gandía, L. M. (2020). Fruit and vegetable waste management: Conventional and emerging approaches. *Journal of Environmental Management, 265*, 110510. https://doi.org/10.1016/J.JENVMAN.2020.110510

Ezekoye, D. (2017). *Coagulation and clarification of wastewater using rice husk ash*. Desalination and Water Treatment https://www.academia.edu/74628553/Coagulation_and_clarification_of_wastewater_using_rice_husk_ash

Farah, A.-A. (2023). The use of agro-waste-based adsorbents as sustainable, renewable, and low-cost alternatives for the removal of ibuprofen and carbamazepine from water. *Heliyon, 9*, e16449.

Farias, K. C. S., Guimarães, R. C. A., Oliveira, K. R. W., Nazário, C. E. D., Ferencz, J. A. P., & Wender, H. (2023). Banana peel powder biosorbent for removal of hazardous organic pollutants from wastewater. *Toxics, 11*(8), 664. https://doi.org/10.3390/TOXICS11080664/S1

Garg, D., Sarkar, A., Chand, P., Bansal, P., Gola, D., Sharma, S., Khantwal, S., Surabhi, Mehrotra, R., Chauhan, N., & Bharti, R. K. (2020). Synthesis of silver nanoparticles utilizing various biological systems: mechanisms and applications—A review. *Progress in Biomaterials, 9*(3), 81. https://doi.org/10.1007/S40204-020-00135-2

Gavrilescu, M., & Macoveanu, M. (1999). Process engineering in biological aerobic waste-water treatment. *Acta Biotechnologica, 19*(2), 111–145. https://doi.org/10.1002/ABIO.370190205

Ghosh, S., Sarkar, B., & Thongmee, S. (2023). Biochar-mediated removal of hazardous dyes. *Current Developments in Biotechnology and Bioengineering, 227–246*. https://doi.org/10.1016/B978-0-323-91902-9.00014-6

Green, A. (2019). Agricultural waste and pollution. In *Waste: A handbook for management* (pp. 531–551). https://doi.org/10.1016/B978-0-12-815060-3.00028-1

Hadidi, L. A. (2021). Constructed wetlands a comprehensive review. *International Journal of Research -GRANTHAALAYAH, 9*(8), 395–417. https://doi.org/10.29121/GRANTHAALAYAH.V9.I8.2021.4176

Hama Aziz, K. H., & Kareem, R. (2023). Recent advances in water remediation from toxic heavy metals using biochar as a green and efficient adsorbent: A review. *Case Studies in Chemical and Environmental Engineering, 8*, 100495. https://doi.org/10.1016/j.cscee.2023.100495

Hamilton, D. W. (2009). *Anaerobic digestion of animal manures: understanding the basic processes*. Oklahoma Cooperative Extension Service.

Harshwardhan, K., & Upadhyay, K. (2017). Effective utilization of agricultural waste: Review. *Journal of Fundamentals of Renewable Energy and Applications, 7*(5), 1–3.

Hassan, M. L., Fadel, S. M., Abouzeid, R. E., Abou Elseoud, W. S., Hassan, E. A., Berglund, L., & Oksman, K. (2020). Water purification ultrafiltration membranes using nanofibers from unbleached and bleached rice straw. *Scientific Reports, 10*(1), 1–9. https://doi.org/10.1038/s41598-020-67909-3

He, J., Song, Y., & Chen, J. P. (2017). Development of a novel biochar/PSF mixed matrix membrane and study of key parameters in treatment of copper and lead contaminated water. *Chemosphere, 186*, 1033–1045. https://doi.org/10.1016/j.chemosphere.2017.07.028

He, C., Liu, Z., Wu, J., Pan, X., Fang, Z., Li, J., & Bryan, B. A. (2021). Future global urban water scarcity and potential solutions. *Nature Communications, 12*(1), 1–11. https://doi.org/10.1038/s41467-021-25026-3

Hong, L. S., Ibrahim, D., & Omar, I. C. (2012). Oil palm frond for the production of bioethanol. *International Journal of Biochemistry and Biotechnology, 1*(1), 7–11.

Hoslett, J., Massara, T. M., Malamis, S., Ahmad, D., van den Boogaert, I., Katsou, E., Ahmad, B., Ghazal, H., Simons, S., Wrobel, L., & Jouhara, H. (2018). Surface water filtration using granular media and membranes: A review. *Science of the Total Environment, 639*, 1268–1282. https://doi.org/10.1016/J.SCITOTENV.2018.05.247

Ibrahim, A., Yaser, A. Z., & Lamaming, J. (2021). Synthesising tannin-based coagulants for water and wastewater application: A review. *Journal of Environmental Chemical Engineering, 9*(1), 105007. https://doi.org/10.1016/J.JECE.2020.105007

Jagadeesh, N., & Sundaram, B. (2023a). Adsorption of pollutants from wastewater by biochar: A review. *Journal of Hazardous Materials Advances, 9*, 100226. https://doi.org/10.1016/j.hazadv.2022.100226

Jagadeesh, N., & Sundaram, B. (2023b). Adsorption of pollutants from wastewater by biochar: A review. *Journal of Hazardous Materials Advances, 9*, 100226. https://doi.org/10.1016/J.HAZADV.2022.100226

Jha, S., Gaur, R., Shahabuddin, S., & Tyagi, I. (2023). Biochar as sustainable alternative and green adsorbent for the remediation of noxious pollutants: A comprehensive review. *Toxics, 11*(2). https://doi.org/10.3390/TOXICS11020117

Jjagwe, J., Olupot, P. W., Menya, E., & Kalibbala, H. M. (2021). Synthesis and application of granular activated carbon from biomass waste materials for water treatment: A review. *Journal of Bioresources and Bioproducts, 6*(4), 292–322. https://doi.org/10.1016/J.JOBAB.2021.03.003

Kamali, M., Appels, L., Kwon, E. E., Aminabhavi, T. M., & Dewil, R. (2021). Biochar in water and wastewater treatment—A sustainability assessment. *Chemical Engineering Journal, 420*, 129946. https://doi.org/10.1016/J.CEJ.2021.129946

Karić, N., Maia, A. S., Teodorović, A., Atanasova, N., Langergraber, G., Crini, G., Ribeiro, A. R. L., & Đolić, M. (2022). Bio-waste valorisation: Agricultural wastes as biosorbents for removal of (in)organic pollutants in wastewater treatment. *Chemical Engineering Journal Advances, 9*, 100239. https://doi.org/10.1016/J.CEJA.2021.100239

Kaushal, N., & Singh, A. K. (2023). Advancement in utilization of bio-based materials including cellulose, lignin, chitosan for bio-inspired surface coatings with special wetting behavior: A review on fabrication and applications. *International Journal of Biological Macromolecules, 246*, 125709. https://doi.org/10.1016/j.ijbiomac.2023.125709

Khunt, H., Dholakiya, H., & Valand, S. (2019). A review on utilization of agriculture waste for wastewater treatment. *Iconic Journal of Engineering Research, 2*, 23–31.

Kiyasudeen, K. S., Ibrahim, M. H., Quaik, S., & Ismail, S. A. (2016). An introduction to anaerobic digestion of organic wastes. *Prospects of Organic Waste Management and the Significance of Earthworms, 23–44*. https://doi.org/10.1007/978-3-319-24708-3_2

Koul, B., Yakoob, M., & Shah, M. P. (2022). Agricultural waste management strategies for environmental sustainability. *Environmental Research, 206*, 112285. https://doi.org/10.1016/J.ENVRES.2021.112285

Kumar, A., & Bhattacharya, T. (2021). Biochar: A sustainable solution. *Environment, Development and Sustainability, 23*(5), 6642–6680. https://doi.org/10.1007/S10668-020-00970-0/METRICS

Kumar, M., Ambika, S., Hassani, A., & Nidheesh, P. V. (2023). Waste to catalyst: Role of agricultural waste in water and wastewater treatment. *Science of the Total Environment, 858*, 159762. https://doi.org/10.1016/J.SCITOTENV.2022.159762

Lata, S., & Samadder, S. R. (2014). Removal of heavy metals using rice husk: A review. *International Journal of Environmental Research and Development, 4*(2), 165–170.

Lee, C. S., Robinson, J., & Chong, M. F. (2014). A review on application of flocculants in wastewater treatment. *Process Safety and Environmental Protection, 92*(6), 489–508.

Leong, K. O. K. (2018). Adsorption of heavy metals using banana peels in wastewater treatment. *The Eurasia Proceedings of Science Technology Engineering and Mathematics, 2*, 312–317.

Madhav, S., Singh, P., Mishra, V., Ahmed, S., & Mishra, P. K. (2022). Recent trends in wastewater treatment. *Recent Trends in Wastewater Treatment, 1–481*. https://doi.org/10.1007/978-3-030-99858-5

Mateo, S., Peinado, S., Morillas-Gutiérrez, F., La Rubia, M. D., & Moya, A. J. (2021). Nanocellulose from agricultural wastes: Products and applications—A review. *Processes, 9*(9), 1594. https://doi.org/10.3390/PR9091594

McKay, G., Parthasarathy, P., Saleem, J., Alherbawi, M., & Sajjad, S. (2022). Dye removal using biochars. *Sustainable Biochar for Water and Wastewater Treatment, 429–471*. https://doi.org/10.1016/B978-0-12-822225-6.00019-1

Mirabella, N., Castellani, V., & Sala, S. (2014). Current options for the valorization of food manufacturing waste: A review. *Journal of Cleaner Production, 65*, 28–41. https://doi.org/10.1016/J.JCLEPRO.2013.10.051

Moayedi, H., Huat, B. B. K., Asadi, A., Salleh, Z. K., & Moghaddas, M. (2011). Surface water treatment process: A review on various methods. *Electronic Journal of Geotechnical Engineering, 16*, 753–761.

Mokhtar, N. M., Priyatharishini, M., & Kristanti, R. A. (2019). Study on the effectiveness of banana peel coagulant in turbidity reduction of synthetic wastewater. *International Journal of Engineering Technology and Sciences, 6*(1), 82–90. https://doi.org/10.15282/IJETS.V6I1.2109

Molina-Sabio, M., González, M. T., Rodriguez-Reinoso, F., & Sepúlveda-Escribano, A. (1996). Effect of steam and carbon dioxide activation in the micropore size distribution of activated carbon. *Carbon, 34*(4), 505–509. https://doi.org/10.1016/0008-6223(96)00006-1

Morales-Jiménez, M., Palacio, D. A., Palencia, M., Meléndrez, M. F., & Rivas, B. L. (2023). Bio-based polymeric membranes: Development and environmental applications. *Membranes, 13*(7), 625. https://doi.org/10.3390/MEMBRANES13070625

Mota, L. S. O., de Oliveira, P. C. O., Peixoto, B. S., & de Moraes, M. C. (2023). Enzyme-coated biochar as a sustainable solution for water and wastewater treatment. *Environmental Science: Water Research and Technology, 9*(11), 2772–2786. https://doi.org/10.1039/D3EW00074E

Mulwa, F., Li, Z., Fangninou, F. F., Mulwa, F., Li, Z., & Fangninou, F. F. (2021). Water scarcity in Kenya: Current status, challenges and future solutions. *Open Access Library Journal, 8*(1), 1–15. https://doi.org/10.4236/OALIB.1107096

Naeem, M., Ansari, A. A., & Gill, S. S. (2020). Contaminants in agriculture: Sources, impacts and management. In M. Naeem, A. A. Abid, & S. G. Sarvajeet (Eds.), *Contaminants in agriculture: Sources, impacts and management* (1st ed., pp. 1–446). Springer. https://doi.org/10.1007/978-3-030-41552-5/COVER

Nigusie, S. (2019). *Physical treatment of drinking water using natural coagulants* (Rice husk, rice husk ash and peanut shell). Master of Science, Jimma Institute of Technology. https://repository.ju.edu.et//handle/123456789/7389

OboteyEzugbe, E., & Rathilal, S. (2020). Membrane technologies in wastewater treatment: A review. *Membranes, 10*(5), 89.

Omondi, D. O., Navalia, A. C., Omondi, D. O., & Navalia, A. C. (2020). Constructed wetlands in wastewater treatment and challenges of emerging resistant genes filtration and reloading. *Inland Waters—Dynamics and Ecology*. https://doi.org/10.5772/INTECHOPEN.93293

Pandey, A., & Soccol, C. R. (2000). Economic utilization of crop residues for value addition: A futuristic approach. *Journal of Scientific and Industrial Research, 59*, 12–22. http://nopr.niscpr.res.in/handle/123456789/26557

Puglia, D., Pezzolla, D., Gigliotti, G., Torre, L., Bartucca, M. L., & Del Buono, D. (2021). The opportunity of valorizing agricultural waste, through its conversion into biostimulants, biofertilizers, and biopolymers. *Sustainability, 13*(5), 2710. https://doi.org/10.3390/SU13052710

Qu, T., Guo, W., Shen, L., Xiao, J., & Zhao, K. (2011). Experimental study of biomass pyrolysis based on three major components: Hemicellulose, cellulose, and lignin. *Industrial and Engineering Chemistry Research, 50*(18), 10424–10433.

Rao, P., & Rathod, V. (2019). Valorization of food and agricultural waste: A step towards greener future. *Chemical Record, 19*(9), 1858–1871. https://doi.org/10.1002/TCR.201800094

Rekleitis, G., Haralambous, K. J., Loizidou, M., & Aravossis, K. (2020). Utilization of agricultural and livestock waste in anaerobic digestion (A.D): Applying the biorefinery concept in a circular economy. *Energies, 13*(17), 4428. https://doi.org/10.3390/EN13174428

Samavati, Z., Goh, P. S., & Ismail, A. F. (2023). Membrane fabrication by solid waste: Opportunities and challenges. *Journal of Resource Recovery, 1*(January–December).

Shahkaramipour, N., Tran, T. N., Ramanan, S., & Lin, H. (2017). Membranes with surface-enhanced antifouling properties for water purification. *Membranes, 7*(1), 13. https://doi.org/10.3390/MEMBRANES7010013

Shan, T. C., Al Matar, M., Makky, E. A., & Ali, E. N. (2017). The use of *Moringa oleifera* seed as a natural coagulant for wastewater treatment and heavy metals removal. *Applied Water Science, 7*, 1369–1376.

Shemer, H., Wald, S., & Semiat, R. (2023). Challenges and solutions for global water scarcity. *Membranes, 13*(6), 1–24. https://doi.org/10.3390/MEMBRANES13060612

Shrivastava, R., & Singh, N. K. (2022). Agro-wastes sustainable materials for wastewater treatment: Review of current scenario and approaches for India. *Materials Today: Proceedings, 60*, 552–558. https://doi.org/10.1016/j.matpr.2022.01.460

Siddiqua, A., Hahladakis, J. N., & Al-Attiya, W. A. K. A. (2022). An overview of the environmental pollution and health effects associated with waste landfilling and open dumping. *Environmental Science and Pollution Research, 29*(39), 58514–58536. https://doi.org/10.1007/S11356-022-21578-Z

Smarzewska, S., & Morawska, K. (2021). Wastewater treatment technologies. *Handbook of Advanced Approaches Towards Pollution Prevention and Control, 1*, 3–32. https://doi.org/10.1016/B978-0-12-822121-1.00001-1

Song, Y., Qiao, W., Westerholm, M., Huang, G., Taherzadeh, M. J., & Dong, R. (2023). Microbiological and technological insights on anaerobic digestion of animal manure: A review. *Fermentation, 9*(5), 436. https://doi.org/10.3390/FERMENTATION9050436

Titiloye, J. O., Abu Bakar, M. S., & Odetoye, T. E. (2013). Thermochemical characterisation of agricultural wastes from West Africa. *Industrial Crops and Products, 47*, 199–203. https://doi.org/10.1016/J.INDCROP.2013.03.011

Tomasi, I. T., Machado, C. A., Boaventura, R. A. R., Botelho, C. M. S., & Santos, S. C. R. (2022). Tannin-based coagulants: Current development and prospects on synthesis and uses. *Science of the Total Environment, 822*, 153454. https://doi.org/10.1016/J.SCITOTENV.2022.153454

Uddin, M. M., & Wright, M. M. (2023). Anaerobic digestion fundamentals, challenges, and technological advances. *Physical Sciences Reviews, 8*(9), 2819–2837. https://doi.org/10.1515/PSR-2021-0068/MACHINEREADABLECITATION/RIS

Ukiwe, L. N., Ibeneme, S. I., Duru, C. E., Okolue, B. N., Onyedika, G. O., & Nweze, C. A. (2014). Chemical and electro-coagulation techniques in coagulation-flocculation in water and wastewater treatment: A review. *Journal of Advances in Chemistry, 9*(3), 2321–2807.

Verhoeven, J. T. A., & Meuleman, A. F. M. (1999). Wetlands for wastewater treatment: Opportunities and limitations. *Ecological Engineering, 12*(1–2), 5–12. https://doi.org/10.1016/S0925-8574(98)00050-0

Vymazal, J. (2011). Constructed wetlands for wastewater treatment: Five decades of experience. *Environmental Science and Technology, 45*(1), 61–69. https://doi.org/10.1021/ES101403Q/SUPPL_FILE/ES101403Q_SI_001.PDF

Wang, D., Jiang, P., Zhang, H., & Yuan, W. (2020). Biochar production and applications in agro and forestry systems: A review. *Science of the Total Environment, 723*, 137775.

Wang, C., Zou, C., Yang, H., Lou, C., Cheng, S., Peng, C., Wang, C., & Zou, H. (2021). Effects of cellulose, hemicellulose, and lignin on the combustion behaviours of biomass under various oxygen concentrations. *Bioresource Technology, 320*, 124375.

Xiang, H., Min, X., Tang, C. J., Sillanpää, M., & Zhao, F. (2022). Recent advances in membrane filtration for heavy metal removal from wastewater: A mini review. *Journal of Water Process Engineering, 49*, 103023. https://doi.org/10.1016/J.JWPE.2022.103023

Xu, F., & Li, Y. (2017). *Biomass digestion* (pp. 197–204). Encyclopedia of Sustainable Technologies. https://doi.org/10.1016/B978-0-12-409548-9.10108-3

Xue, L., Zhang, P., Shu, H., Chang, C.-C., Wang, R., & Zhang, S. (2016). Agricultural waste. *Water Environment Research, 88*(10), 1334–1369. https://doi.org/10.2175/106143016X14696400495019

Yuliansyah, A. T., Hirajima, T., Kumagai, S., & Sasaki, K. (2010). Production of solid biofuel from agricultural wastes of the palm oil industry by hydrothermal treatment. *Waste and BiomassValorization, 1*, 395–405.

Chapter 14
Agriculture Waste for Sustainable Building Materials

Chrysanthos Maraveas

Abbreviations

AI	Artificial intelligence
AKD	Alkyl ketene dimer
ANN	Artificial neural networks
ASCB	Acetylated sugarcane bagasse
BKRHA	Brick kiln rice husk ash
CBR	California bearing ratio
CECS	Cellulose-chitosan
CNFS	Carbon nanofibers
CS	Chitosan
CVD	Chemical vapor deposition
CHP	Coconut husk powder
CSP	Coconut shell powder
CO_2	Carbon dioxide
COP27	Conference of the Parties of the United Nations Framework Convention on Climate Change (UNFCC)
ELM	Extreme learning machine
ESP	Eggshell powder
FA	Fly ash
GEOPAV	Geopolymer paver
GEP	Gene expression programming
GHGS	Greenhouse gases
M5P	M5p model tree

C. Maraveas (✉)
Department of Natural Resources Development and Agricultural Engineering, Agricultural University of Athens, Athens, Greece
e-mail: maraveas@aua.gr

J. Arora et al. (eds.), *Transforming Agriculture Residues for Sustainable Development*, Waste as a Resource, https://doi.org/10.1007/978-3-031-61133-9_14

MTMS	Methyltrimethoxysilane
MS	Mulberry stems
MARS	Multivariate adaptive regression splines
NAS	Natural aggregates
NAC	Natural aggregate concrete
NEP	Nano eggshell powder
N_2O	Nitrous oxide
NRC	Noise reduction coefficient
OTL	Olive tree leaf powder
PAN	Polyacrylonitrile
PGS	Paper grain sand
POFA	Palm oil fuel ash
PP	Polypropylene
POCF	Palm oil clinker fine
POCF-MB	Palm oil clinker fine modified bitumen
QP	Quartz powder
RA	Recycled aggregates
RAC	Recycled aggregate concrete
RFR	Random forest regression
RHA	Rice husk ash
SAA	Sound absorption average
SCB	Sugarcane bagasse
SCBA	Sugarcane bagasse ash
SBW	Sugarcane bagasse waste
SDP	Sawdust powder
SF	Silica fume
SCPS	Sugarcane pulp sand
SLSC	Sustainable lightweight structural concrete
SVM	Support vector machines
UHPC	Ultra-high-performance concrete
WFRB	WashingtoniaFilifera
WG	Waste glass

14.1 Introduction

The rapid increase in the world population and high migration to urban areas leads to pressure on social amenities, including housing (Erkmen & Sari, 2023). As more houses must be constructed to keep up with the growing demand, higher economic costs are further incurred in modern construction projects that rely on traditional materials such as cement, sand, bricks, and stone (Maraveas, 2020a). Significant environmental pollution is attributed to the use of standard traditional materials, including cement and bricks, whose production requires high electrical and thermal energy (Arbelaez Perez et al., 2022). Töbelmann and Wendler (2020) reiterate Arbelaez Perez et al. (2022), who report that the manufacture of 1 tonne of cement consumes electricity at a level of 110 kWh while further generating 800 $kgCO_2$. The insights further establish that the production of cement consumes 14% of energy and also emits 8% of the overall carbon dioxide emissions across the globe (Arbelaez Perez et al., 2022). To address the high economic costs and lower the environmental pollution attributed to the use of traditional construction materials, increased research is being directed to facilitate the identification of suitable alternatives for construction.

Maraveas (2020a) further observed that the disposal of agricultural waste, including rice and coconut husks, sugarcane bagasse, and wheat straw, is a prevalent problem in different countries worldwide. A case in point identified was India, where more than 600 tons of agro-waste were associated with agricultural-based products (Maraveas, 2020a). Bhat et al. (2022) added that strategies adopted for the management of agricultural waste, such as disposal in landfills and incineration, were also contributing to increased land and air pollution, respectively. Elbasiouny et al. (2019) supported Bhat et al. (2022) where they argued that incineration of the agricultural wastes increased the emission of global greenhouse gases (GHGs), including carbon dioxide (CO_2) and nitrous oxide (N_2O). Therefore, an urgency to identify more suitable agricultural waste management strategies is emphasized.

To address both challenges—high costs of modern construction and unsuitable methods for disposal of agricultural wastes—researchers, as early as 1989, revealed the potential application of agricultural wastes as substitutes for traditional construction materials (Erkmen & Sari, 2023). More recent research has shown that agricultural wastes can be used as suitable alternatives in diverse construction projects, for example, sugarcane bagasse ash (SCBA) and coconut shell ash as cement alternatives (Herring et al., 2022). Boudali et al. (2022) shared similar insights as Herring et al. (2022) where they reported that agro-concrete, derived from bio-aggregates, was considered a low-carbon, sustainable, and multifunctional building material that partially replaced cement. As a result, the materials enhance sustainable construction practices while ensuring that agricultural wastes can be disposed of more effectively through reuse in modern construction projects. Furthermore, the widespread and readily accessible agricultural wastes transform them into potent alternatives for the natural aggregates used in different construction materials, including concrete.

Further research also reveals the use of artificial intelligence (AI) methods, such as machine learning to evaluate the properties of the construction materials developed from agro-wastes (Boumaaza et al., 2022). In their study, Boumaaza et al. (2022) demonstrated the utilization of artificial neural networks (ANNs) to predict the flexural strength of biochar cementitious mortars developed from the waste of the *Washingtoniafilifera* (WFRB), where substitutes ranged from 1 to 5%. The results showed that the methods of machine learning were able to make accurate predictions of the optimal substitutes of the WFRB materials. Nnochiri et al. (2023) also demonstrated that ANN models were effective in predicting the California bearing ratio (CBR) of lateritic soil, which was mixed with cement and rice husk ash (RHA). Studies such as Boumaaza et al. (2022) and Nnochiri et al. (2023) reveal that the construction sector is gradually adopting AI to accurately predict the properties of sustainable construction materials developed using agro-wastes. This chapter will expound further on the area and discuss how the use of AI enhances the adoption of agro-waste for construction processes.

This chapter focuses on the production of sustainable construction materials from agricultural waste. This chapter is arranged into six main sections, where the first section reviews the GHG emissions from the construction sector. The second section evaluates strategies for agricultural waste management. The third section elaborates on the utilization of agricultural waste for the construction of sustainable buildings. In the fourth section, the use of AI techniques to examine the properties of the building materials developed from agricultural waste is also evaluated. The fifth section discusses the findings, while the final section synthesizes the key insights from this chapter and also details future directions and research avenues.

14.2 High GHG Emissions in the Construction Industry

The modern construction industry generates numerous adverse consequences on the environment, where the main issue regards the generation of high levels of greenhouse gas emissions (World Green Building Council, 2019). The report by the World Green Building Council (2019) revealed that buildings were responsible for 39% of the carbon emissions related to global energy, where 28% arose from emissions associated with operations in the energy requirements for power and heat for the buildings and 11% was attributed to the materials and construction. El Sheikh (2022) further offered that from the Conference of the Parties of the United Nations Framework Convention on Climate Change (UNFCC) COP27 climate talks in Egypt held in 2022, the construction sector accounted for the generation of high levels of CO_2 emissions at 37% while also accounting for 34% of the energy demand levels. The findings also indicated that the CO_2 emissions had reached 10 gigatonnes of CO_2 equivalent in 2021, a 5% increase over the 2020 levels (El Sheikh, 2022). Sizirici et al. (2021) supported El Sheikh (2022) revealing that in 2004, the GHG building emissions had reached 8.6 billion t-CO_2-e and were expected to increase to 15.6 billion t-CO_2-e by 2030. Close inspection shows that the release of GHGs is a key problem in the construction sector since the emissions not only arise from the production of materials such as cement but also in the operation and use of the final developed buildings. Table 14.1 summarizes the CO_2 emission sources in the construction sector.

From Table 14.1, the emission of GHGs in the construction sector was attributed to two main sources—the production of the raw materials and their transportation. The findings also revealed that the highest percentage of the emissions arose from the production of raw materials.

Miller et al. (2021) observe that the manufacture of cement contributes more than 7% of anthropogenic GHGs, which arise from two main sources; the use of fossil fuels to power the manufacturing activities and the chemical reactions that occur

Table 14.1 Sources of CO_2 emissions in the construction sector (Sizirici et al., 2021; Open access)

Activity	Source of emissions	Percentage of emissions	Total emissions
Production of raw materials	Production of cement, iron, steel, non-metallic minerals	44%	80–93%
	Concrete emissions	78%	
	Production of rebars	10%	
	Production of cement mortar	6%	
Transportation of raw materials	Combustion of fossil fuels	2–3%	2–3%

from limestone decarbonization production processes. Labaran et al. (2022) supported Miller et al. (2021), who argued that similar to the manufacturing activities of raw materials such as cement, high levels of GHGs arose from the clearing of land, demolitions, burning, and the use of dangerous chemicals. In the operation of the buildings, high carbon emissions are generated during the cooling, lighting, heating, and maintenance of the equipment in the buildings (Zhang et al., 2023). Furthermore, Zhang et al. (2023) argued that the sector associated with the construction of buildings was responsible for 15% of CO_2 emissions and 33% of the total global final energy consumption, where energy-consuming appliances, including air conditioners, led to high demands.

To address the high GHG emissions in the construction sector, multiple strategies are recommended, including indirect pricing, adherence to strict pollution regulations, and direct pricing, which encompass emission trading schemes and carbon taxes (Wang et al., 2014). However, Sizirici et al. (2021) argued that the utilization of other building additives for materials and techniques could lead to CO_2 emission minimization by up to 90% at different levels of operations in the construction of the buildings. Ahmed et al. (2020) supported Sizirici et al. (2021) in a study that showed that the utilization of heavyweight cast concrete and alternatives, such as autoclaved aerated concrete, reduced the carbon emissions that were embodied in buildings by 23 and 50%, respectively. In contrast, traditional buildings that used reinforced concrete represented 78% of the overall embodied carbon emissions (Ahmed et al., 2020). Chan et al. (2022) reiterated Ahmed et al. (2020) where they emphasized that the utilization of construction materials that were low in carbon and ranging from precast hollow core slabs to precast half slabs and steel framework systems led to a reduction in GHG emissions in the buildings construction sector. As such, there is a shift to using low-carbon construction materials and aggregates in cement in a bid to lower GHG emissions associated with building construction. The generated insights are important in this chapter, where the replacement of additives in building materials and the use of agro-wastes as alternatives for the sector are identified to significantly reduce overall GHG emissions in the building construction sector.

14.3 Agricultural Waste Management Strategies

The insights from the previous section emphasized that to reduce GHG emissions in the construction sector, it was essential to adopt alternative aggregates as replacements for traditional construction materials. To advance the discussion, this section evaluates the challenges associated with waste management strategies for agricultural wastes.

According to Maraveas (2020a), the disposal of agricultural waste from crops, including wheat straw, rice husks, and sugarcane bagasse, was a significant problem in developing countries. An explanation advanced for the difficulties of agricultural waste disposal regards the high quantities of waste that are generated, responsible for 30% of the total levels of output in agriculture (Luhar et al., 2019). Wang et al. (2021) added that in China, aquaculture waste from bivalve mollusks reached 10.35 million tons in 2010. Nunes et al. (2020) also revealed that countries such as Brazil produced 550,000 tonnes of coconut fruit annually, while the total amount of shells that were thrown away was still unknown. Chen et al. (2021a) supported Nunes et al. (2020), who observed that up to 150 million tons of rice husks were generated per year across the world since rice is considered a staple food. Therefore, high quantities of diverse agricultural waste are generated annually ranging from rice husks and coconut shells to aquaculture. The disposal of these agricultural wastes leads to a further environmental problem.

In support of the argument, Obeng et al. (2020) posit that the disposal of green coconut leads to environmental problems, especially in public spaces where the shells take up to 10 years to degrade. Furthermore, the disposal of coconut shells in landfills and the environment causes unsightliness within the urban environment as they are a focus point for the breeding of mosquitoes and the transmission of diseases (Nunes et al., 2020). Additionally, disposing of agricultural wastes through incineration leads to further adverse environmental problems through the emission of carbon dioxide (Obeng et al., 2020). However, Obeng et al. (2020) observe that in some communities such as Ghana, where firewood is the main source of fuel, burning the shells is viewed as a strategy to dispose of them. In turn, this helps reduce the total waste amounts generated in the environment.

The debate in this section reveals that while there is a large number of agricultural wastes generated annually from farming activities, the methods used for their disposal are still largely ineffective. Disposal of agro-wastes in landfills leads to unsightliness and poses health hazards, especially with coconut shells that may become breeding grounds for mosquitoes. Additionally, incineration methods cause air pollution and contribute to releasing GHG emissions. Therefore, adopting agro-waste to facilitate the development of sustainable construction materials is an effective strategy for disposing of them. A further insight is that by integrating agro-wastes as alternative aggregates for cement in construction projects, this strategy helps address the challenge of greenhouse gas emissions within the sector.

14.4 Agricultural Waste in Sustainable Building Construction

Agricultural wastes have been increasingly adopted in the development of sustainable materials for the construction of buildings based on their effectiveness in addressing climate change problems linked to traditional construction processes and their numerous eco-friendly benefits (Sangmesh et al., 2023). Figure 14.1 summarizes the diverse benefits of using eco-friendly agro-waste as alternative construction materials.

To examine the applications of agro-wastes as alternative construction materials, this section is structured into five subsections based on the application areas. The first section examines the use of agro-wastes in cementitious and binder materials, while the second evaluates thermal insulation applications. In the third section, the applications of agro-wastes in masonry and bricks are elaborated. The fourth section examines the use of agro-wastes in reinforcement materials, while the fifth reviews agro-wastes in particle boards and ceiling tiles.

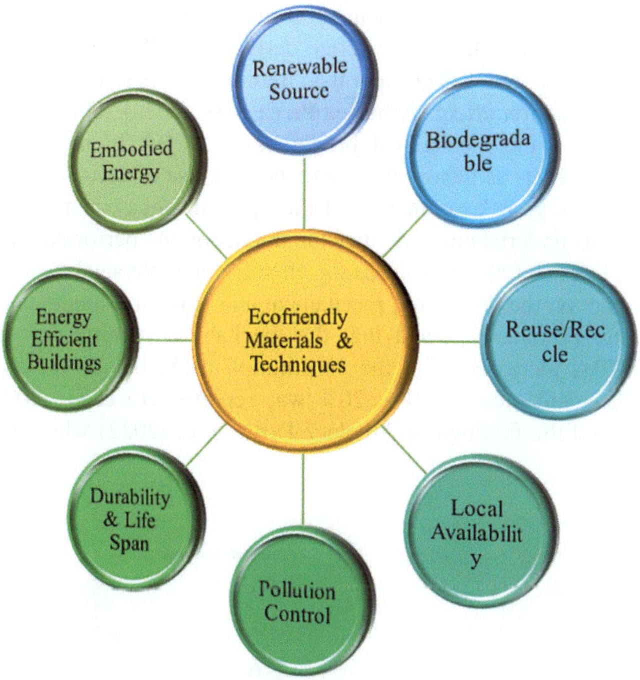

Fig. 14.1 Benefits of eco-friendly agro-wastes in construction (Sangmesh et al., 2023, permission obtained)

14.4.1 Cementitious and Binder Materials

Agro-wastes such as sugarcane bagasse ash (SCBA) have been widely adopted as suitable alternatives for cement in construction projects (Agwa et al., 2022). In their study, Agwa et al. (2022) reported that SCBA was a by-product of the combustion of sugarcane bagasse during sugar processing. However, high volumes of the material were released into the environment through disposal in landfills. To produce SCBA, the sugarcane bagasse has its impurities removed by washing with water. This process also improves reactivity. Thereafter, it is oven-dried at 110 °C to eliminate excess moisture before being heated at 600 °C for 2 hours to produce the SCBA (Amin et al., 2022). Figure 14.2 shows the process of producing SCBA.

Figure 14.2 details the process used in the production of SCBA, which leads to optimal pozzolanic activity. The generated SCBA is also filtered using sieves and mesh to remove the cold particles. From a review of different studies, the use of SCBA in place of cement was identified as an effective construction material due to its high flexural, tensile, and compressive strength qualities (Agwa et al., 2022). Arbelaez Perez et al. (2022) reiterated Agwa et al. (2022) in a study that examined the impact of using SCBA and waste glass (WG) as cement substitutes on concrete features. After 28 days, the SCBA mixed with WG-produced concrete achieved better properties both mechanically and physically than concrete developed using either SCBA or WG alone as a cement substitute (Arbelaez Perez et al., 2022). The evaluation also revealed that the concrete derived from SCBA and WG had high compressive strength, was 8% cheaper than traditional concrete, and further reduced the alkali silica reaction effect (Arbelaez Perez et al., 2022). The findings from the studies highlighted the mechanical properties of SCBA in the development of concrete and low costs as factors that motivated the use of agro-waste.

In a further study, an examination of the optimal replacement ratio for cement using SCBA was undertaken to identify the impact on the performance of concrete. For example, Li et al. (2022) explored the chemical and physical qualities of SCBA and their impact on the durability, mechanical and structural qualities of the developed concrete. Findings showed that the mechanical qualities of concrete that contained SCBA increased with the substitution of SCBA in cement, where an optimal replacement ratio of 10 to 20% was reported (Li et al., 2022). Li et al. (2022) supported the findings of Arbelaez Perez et al. (2022) where they revealed

Fig. 14.2 Production of SCBA (Amin et al., 2022, Open access)

that the integration of SCBA enhanced the resistance of the concrete against sulfate attacks and alkali-silica reactions and also reduced the economic costs in the construction projects. As a result, both economic (low costs) and structural advantages, such as high compressive strength and mechanical strength, arose from the adoption of SCBA.

Zaheer and Tabish (2023) also showed that the optimal levels of replacement of cement by SCBA for qualities such as mechanical and durability were 20%, where the impervious microstructure of the SCBA-concrete to degrading agents enhanced the life span of the material. Bheel et al. (2022) identified similar findings as Zaheer and Tabish (2023), where they demonstrated that replacing fine aggregate using SCBA and cement using coal bottom ash (CBA) enhanced the concrete mixtures' splitting tensile and compressive strength. The findings from the studies are emphatic on the importance of utilizing SCBA-based concrete due to the good structural and mechanical qualities in addition to the high durability of the materials. However, Zaheer and Tabish (2023) argue against the use of SCBA-based concrete especially in the development of slabs and beams. The arguments indicate that SCBA-based concrete is still limited in usage in construction applications. The contradictory arguments also highlight that further research is still needed to ascertain the qualities of SCBA-based concrete before application in high-risk construction projects.

In other studies, utilization of SCBA to replace cement in developing ultra-high-performance concrete (UHPC) was also examined (Wu et al., 2022). The results obtained showed that applying SCBA in UHPC as a cement replacement enhanced the compressive strength of the concrete, the workability of the cementitious materials, and decreased the autogenous shrinkage of the UHPC. Amin et al. (2022) supported Wu et al. (2022) where it was shown that using SCBA and nano eggshell powder (NEP) to replace cementitious materials generated high mechanical qualities in high-strength concrete. The findings showed that the integration of NEP accelerated the setting time of the high-strength concrete, and the best mixture comprised 5% NEP and 15% SCBA. Huang et al. (2023) reiterated Wu et al. (2022) and Amin et al. (2022), who examined the impact of SCBA as a replacement for quartz powder (QP) during the manufacture of green concrete. The study replaced SCBA as an active filler of QP at 20,40,60 and 80% and observed that at a replacement ratio of 60% and under standard curing conditions, the green concrete experienced an increase in flexural strength by 18% and compressive strength by 12% (Huang et al., 2023). Such studies indicated that SCBA was an effective filler and QP replacement in the manufacture of UHPC.

The debate from the synthesized studies highlights the need for using SCBA to replace cement during the manufacture of green concrete. Using the material addresses the current problem regarding the excess generation of waste from agricultural processes whereby the sugarcane bagasse is diverted from landfills and used as a filler material and cement replacement. Based on the reported findings, SCBA was identified as a material with good pozzolanic qualities and high mechanical and structural properties. Therefore, replacing traditional cement with SCBA would lead to improved compressive and flexural qualities both when used individually or in combination with waste materials such as nano eggshells and waste glass. Additionally, the discussion showed that the SCBA was cost-effective and led to lower economic costs during construction projects.

A second material that has been also adopted to replace cement is rice husk ash (RHA) (Channa et al., 2021). In elaboration, Channa et al. (2021) showed that the combination of 5% SCBA and 5% RHA generated high-strength performance, which facilitated the structural applications of the concrete. Jittin and Bahurudeen (2022) also examined the impact of integrating SCBA and RHA as a cement replacement in the production of concrete, where they reported that the optimal compressive strength occurred at 10% SCBA and 5% RHA. The insights from such studies indicated that green concrete developed by mixing SCBA and RHA demonstrated high compressive strength qualities.

In other research, such as Jayaraman et al. (2023), the effect of combining RHA and silica fume (SF) to develop pozzolanic material during composite cement development was investigated. The findings showed that the combination of 5% RHA and 2.5% SF improved the strength qualities of the composite concrete, permeability of chloride ions, and corrosion resistance. Dharmaraj et al. (2023) reiterated Jayaraman et al. (2023) and showed that replacing cement with RHA at 10% improved the compressive strength and bulk resistivity of the materials. Additionally, optimal performance was observed when the RHA proportion was set at 7.5%, leading to a compressive strength of 66 MPa for high-performance concrete materials. Ma et al. (2023) also examined the impact of the replacement ratios of natural aggregates (NAs) by recycled aggregates (RA), cement by RHA, and their interactions on the compressive qualities of the concrete made from recycled aggregate. Findings reported showed that 10% was the optimized ratio of RA and RHA which led to increased compressive strength of 0.4% for the natural aggregate concrete (NAC) and 4.9% for the recycled aggregate concrete (RAC) (Ma et al., 2023). The conclusions reached showed that low RHA proportions were used in the composite mix to generate high qualities of pozzolanic materials. As such, findings emphasized that the parent cementitious material was still highly important in the construction projects since the RHA and substitutes such as SF and SCBA were used in small quantities.

Other studies considered the combination of RHA with glass sand (Su & Xu, 2023a, b). In their study, Su and Xu (2023a) investigated the impact of using 20% glass sand and RHA at 15, 30, and 45% in the replacement of sand from rivers and cement from concrete. The findings obtained revealed that glass sand reduced the slump and concrete's triaxial and uniaxial compressive strength while improving the dynamic sensitivity of the strain rate and compressive strength (Su & Xu, 2023a). The results further indicated that the glass sand concrete was less permeable than ordinary concrete, while subjecting the concrete to a high external also showed that the composite comprising rice husk ash exhibited features of high toughness and plastic deformation which improved with the increase in the replacement rate of rice husk ash. In the second study, Su and Xu (2023b) tested the mechanical features of glass sand concrete and RHA-glass sand concrete and revealed that adding the glass sand reduced the water absorption and compressive strength while enhancing shear and split tensile strength. Such results were indicative that the incorporation of RHA and glass sand in the manufacture of green concrete improved the mechanical properties, including split tensile and shear strength, although degradation of the compressive strength was identified.

The synthesis so far indicates that RHA is a suitable replacement for traditional Portland cement in building construction. The findings from different studies (Channa et al., 2021; Dharmaraj et al., 2023; Jayaraman et al., 2023; Ma et al., 2023; Su & Xu, 2023a, b) showed that small proportions of the RHA were used as partial replacements for the cement material at 5 and 10% leading to high compressive strength qualities. The insights are an indicator that with the RHA agro-wastes, the parent cement material is still relied on during the manufacture of the green concrete. The discussions also showed that RHA could be combined with other materials, including SCBA and glass sand, to improve green concrete's mechanical features.

Research has also established the use of various agro-wastes apart from RHA and SCBA as replacements for cement, such as palm oil fuel ash (Chinnu et al., 2022; Santhosh et al., 2022). In their study, Chinnu et al. (2022) showed that the integration of palm oil fuel ash (POFA) and fly ash improved concrete's workability while also reducing the rate of water absorption up to their optimal levels of replacement. Santhosh et al. (2022) reiterated Chinnu et al. (2022) where they observed that the integration of POFA improved the workability of the blended concrete where a replacement level of 10–20% was reported for the RHA and POFA. Such findings are an indicator that using blended concrete developed from RHA and POFA composites improved the overall mechanical qualities of construction materials. Figure 14.3 details the morphology of the ground POFA and RHA materials.

From Fig. 14.3, the mechanical qualities of the POFA and RHA materials are identified at different time intervals. The results showed that the allocation of more time for the grinding led to the improved fineness of the materials; for instance, the morphology at 120 minutes was smoother as compared to that at 15 minutes.

In summary, the discussion in this section revealed the use of agro-wastes as effective cementitious materials replacements. The agro-wastes included SCBA and RHA as the main composites, although other alternatives, such as POFA and fly ash, were still used. The discussions emphasized the good mechanical and structural features of the green concrete developed using agro-waste materials. Further evaluation also showed that the agro-wastes could be combined with other materials, including glass sand, fly ash, and SF, to improve material properties.

14.4.2 Thermal Insulation

Agro-wastes are further adopted in thermal insulation applications to minimize the transfer of heat among surfaces with dissimilar temperatures and, subsequently, aid in the reduction of energy consumption (Sangmesh et al., 2023). An application of agro-wastes regards the development of aerogels which are porous materials

a) Ground at 150 RPM for 8 h (Karim et al., 2016)
b) Ground POFA till d50 = 2.45μm (Zeyad et al., 2016)
c) Ground POFA (Ranjabir et al., 2015)
d) Ground for 15 min (zain et al., 2011)
e) Ground for 60 min (zain et al., 2011)
f) Ground for 120 min (zain et al., 2011)

Fig. 14.3 Morphology of ground POFA and RHA (Santhosh et al., 2022, permission obtained)

characterized by low density, low thermal conductivity, and larger specific surface area and are subsequently applied in heat insulation areas (Li et al., 2021). In elaboration, Li et al. (2021) demonstrated the manufacture of hydrophobic bagasse aerogels by milling the bagasse and, thereafter, mixing it with an emulsion from alkyl ketene dimer (AKD). The bagasse was further dry-frozen and thermos-crosslinked. The aerogels were generated at an AKD of 10 wt% and slurry consistency of 3 wt%, and the material demonstrated a high oil absorption capacity of (31.65 g g^{-1}) and low thermal conductivity of (36.28 mW m^{-1} K^{-1}), thereby being suitable for thermal insulation applications. Figure 14.4 illustrates the process of the production of aerogels using bagasse agro-waste.

In Fig. 14.4, the production of aerogels through milling of bagasse and mixing it with AKD emulsion was illustrated. The production process involved freezing, sublimating the material, thermo-crosslinking, adsorption, and separation processes. The developed aerogels demonstrated low thermal conductivity and were suitable for thermal insulation applications where they ensured minimal heat loss from the building envelopes.

Chen et al. (2021b) reiterated Li et al. (2021) in a different study where biomass-based carbon nanofibers (CNFs) were prepared by blending acetylated sugarcane bagasse (ASCB) and polyacrylonitrile (PAN) and thereafter, carbonization processes. The generated results showed that the CNFs were suitable for the manufacture of supercapacitors due to their high power density (1.26 kW/ kg), energy density (56.0 W h/ kg), high capacitance retention, and great cycling stability. The similarity between Chen et al. (2021b) and Li et al. (2021) stemmed from the fact that the two

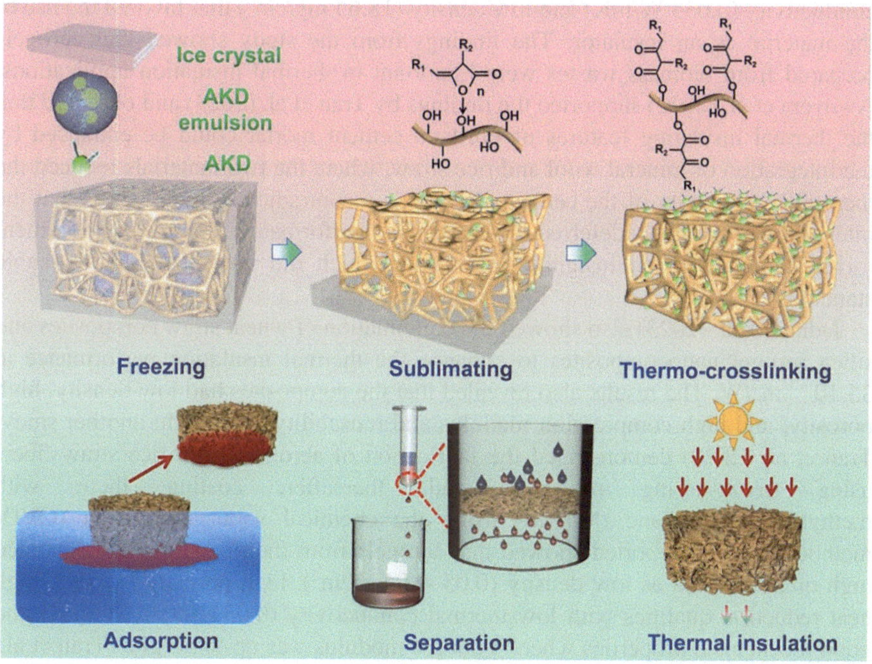

Fig. 14.4 Aerogel production using bagasse agro-wastes (Li et al., 2021, permission obtained)

studies demonstrated the development of thermal insulators using lignocellulosic biomass materials. Mehrzad et al. (2022) also demonstrated the use of sugarcane bagasse waste (SBW) in applications of sound absorption and thermal insulation using guarded hot plate techniques and impedance tube methods. From the results, it was observed that thermal conductivity ranged between 0.034 and 0.042 W/mK, while the sound absorption average (SAA) ranged from 0.26 to 0.64 and noise reduction coefficient (NRC) ranged between 0.27 and 0.62. Zhao et al. (2023) reiterated Mehrzad et al. (2022) where they also showed that bagasse cellulose was suitable for the production of low-cost, flexible film that demonstrated high thermal insulation properties (0.0158 W/m.K). The findings were further indicative that the use of the bagasse cellulosic aerogel films led to reduced loss of energy from windows in modern buildings. Thus, the importance of SCB in thermal insulator production is underscored. Discussing the findings showed that the use of bagasse introduced numerous advantages, including sustainability in terms of thermal and acoustical insulation. The analysis also showed that the bagasse materials were more economical in insulating different buildings due to their cheaper cost.

Further study also revealed that aerogels could be manufactured from other types of agro-wastes such as rice straw. For example, Tran et al. (2023) demonstrated the manufacture of cellulose-chitosan (CECS) aerogel from rice straw and shrimp waste-based chitosan (CS), which was associated with good qualities such as low thermal conductivity (0.035 W/mK) and low density (18.65 mg/cm^3) that favored the use of the material as an insulator. The findings from the study showed that aerogels prepared from biomass wastes were important in thermal insulation applications. Awoyera et al. (2022) supported the findings by Tran et al. (2023) and observed that the thermal insulating features of Portland cement mortar could be enhanced by the integration of mineral wool and rice straw, where the two materials reduced the thermal conductivity of the cement. However, a contradictory finding was that the integration of the fibers deferred the mechanical performance of some of the cement mixes, although in an insignificant manner which did not affect the workable standards.

Jadhav et al. (2023) also showed the combination of wheat straw composites and silica aerogel nanocomposites to enhance the thermal insulation performance at 35 10^{-3} m^2/W. The results also revealed that the composites had low density, high porosity, and high compression modulus and reusability aspects. In another study, Tran et al. (2020) demonstrated the fabrication of aerogels from rice straw fibers using freeze-drying processes and, thereafter, coating them with methyltrimethoxysilane (MTMS) using the chemical vapor deposition (CVD) method. Findings reported showed that aerogels from the rice straw demonstrated high qualities such as low density (0.05–0.06 g/cm^3), high porosity (~97%), high heat reduction qualities with low thermal conductivity (0.034–0.036 W/m.K) and good mechanical properties where Young's modulus was up-to 47 kPa (Tran et al., 2020). Such findings emphasized the importance of agro-wastes such as rice straws in the production of sustainable thermal insulation composites that also depicted capabilities such as oil adsorption up to a capacity of 13 g/g. Additionally, the composites were considered suitable to be adopted in sound insulation applications due to their insulation properties.

Other studies highlighted the use of agro-wastes such as coconut shells in thermal insulation applications where the coconut shell powder (CSP) was used as a pore-forming agent in the manufacture of fired clay bricks (Moujoud et al., 2023). In the study, Moujoud et al. (2023) showed that although the integration of the CSP in the clay bricks did not impact the mineralogy of the bricks, the combination of the CSP and clay reduced the thermal conductivity (0.37 Wm^{-1} K^{-1}) and maintained the tensile strength at 9.88 MPa. As such, the use of the CSP as an agent employed in pore-formation led to elaborate bricks that had high thermal insulating features. Jannat et al. (2022) supported the findings by Moujoud et al. (2023), reporting that the integration of diverse agro-wastes such as Eggshell Powder (ESP), Sawdust Powder (SDP) and Coconut Husk Powder (CHP) in the production of unfired clay blocks improved the strength qualities and thermal insulation capacities. Figure 14.5 illustrates the use of the different agro-waste powders—ESP, SDP, and CHP—in the manufacture of fired clay bricks with high thermal insulation capabilities.

In Fig. 14.5, the integration of agro-wastes, including coconut husk powders, sawdust, and eggshell powders, in the production of unfired clay blocks was demonstrated. The developed composites were observed to demonstrate high thermal insulation capacities while causing minimal effect on the mechanical features of the construction materials.

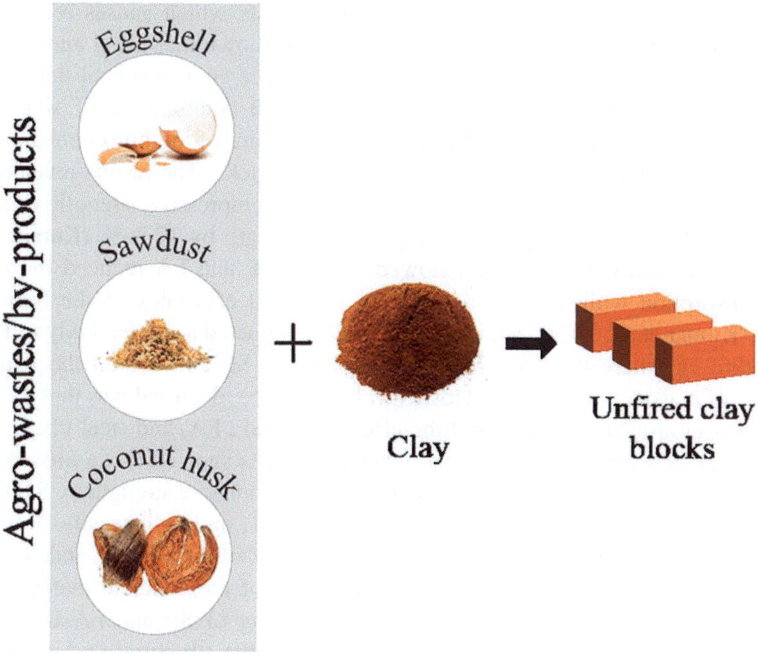

Fig. 14.5 Production of unfired clay bricks from agro-wastes (Jannat et al., 2022, permission obtained)

In summary, the discussion in this section showed that agro-wastes, including sugarcane bagasse, rice straws, coconut husk powder, sawdust, and eggshell powder, were important pore-forming agents that improved the thermal qualities of different types of bricks. The examination of the different types of wastes in this section showed that the integration of the agro-wastes with cement or bricks did not lead to any significant change in structural properties. However, the thermal conductivity was reduced in the materials, which boosted overall thermal performance qualities. The insights also showed that the materials demonstrated good adsorption qualities for oil agents and sound insulation capabilities. The insights highlight the diverse applications of agro-waste materials in the production of sustainable building materials that can be employed in thermal insulation, thereby, leading to a reduction in the energy costs in buildings.

14.4.3 Masonry Composites and Bricks

Agro-wastes have further been adopted in the manufacture of masonry composite materials, for example, burnt clay bricks, where they facilitate the reduction in environmental pollution from the disposal of industrial waste from coal and extraction of topsoil in the sourcing of clay deposits which causes reduced soil fertility (Kumar et al., 2021). Diverse studies have shown that using agro-waste to prepare bricks is a feasible solution in order to reduce environmental degradation. For example, Kumar et al. (2021) showed that the combination of silica fume (SF) and fly ash (FA) with clay during the manufacture of the bricks enhanced the structural and durability features of the burnt clay bricks. The results revealed that adding SF at a percentage range of 4% increased compressive strength values by 27.55%, and FA at 4% improved compressive strength by 17.36% (Kumar et al., 2021). Additionally, the findings showed that the SF and FA reduced the rate of water absorption while also increasing the structural efficiency of the burnt clay bricks (Kumar et al., 2021). Kumar et al. (2022) conducted a similar study to Kumar et al. (2021), where they considered the impact of SF and FA on the structural features of masonry composites. However, a difference identified was that the study by Kumar et al. (2022) considered the effect of the SF, FA, and steel fibers on the mechanical strength of concrete. The generated results showed that a mix of FA at 15%, SF at 12%, and steel fiber at 1% led to high compressive strength by 5.89% and further enhanced the workability, flexural at 20.40%, and split tensile strength at 28.02%. The indication was that a triple blending mix of the SF, FA, and steel fiber enhanced the mechanical strength and workability of the concrete materials. The synthesis of the studies revealed that the embedding of SF and FA in masonry materials, including bricks and concrete, improved the structural and mechanical qualities in terms of compressive and flexural strength.

The review of the literature further showed that more studies were conducted that examined the integration of agro-wastes in masonry materials such as concrete as compared to bricks. In some studies, such as Shupta et al. (2022), an investigation on the impact of embedding rice husk ash (RHA) and fly ash (FA) as partial replacements for cement was undertaken. The generated results showed that the use of RHA at 7.5% and FA at 22.5% as a cement replacement improved compressive strength by 15%, split tensile strength by 0.67%, and flexural strength by 24% (Shupta et al., 2022). Mahdi et al. (2022) supported the findings from Shupta et al. (2022) in a study that investigated the use of brick kiln rice husk ash (BKRHA) in the manufacture of geopolymer paver (GEOPAV) blocks. The findings generated showed that the BKRHA led to insignificant improvement of the compressive strength of the GEOPAV blocks, although they enhanced the split tensile strength and flexural strength of the blocks. Further results indicated that the addition of BKRHA enhanced the resistance to acid and attack by frosts, thereby enhancing the durability of the materials. Figure 14.6 illustrates the split tensile strength qualities of the GEOPAV blocks.

The results in Fig. 14.6 showed that there was an increase in split tensile strength at 3.3 for the mix type 12GFR9505 due to the incorporation of BKRHA agro-waste. The sample (12GFR9505) was identified as the optimal composite mixture based on density and split tensile strength levels. Therefore, the BKRHA enhanced the mechanical properties of the GEOPAV blocks.

Manubothula and Gorre (2022) also investigated the impact of adding RHA in the production of aerated concrete blocks, which were developed using sand, lime, fly

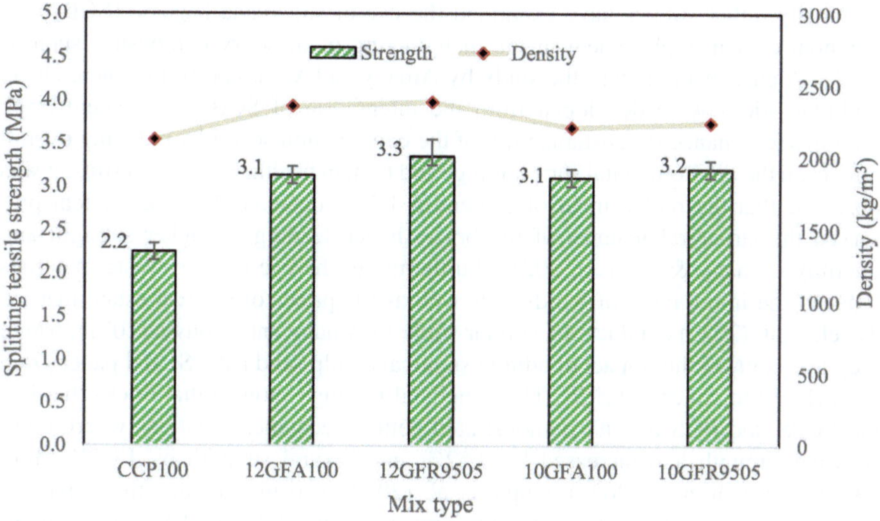

Fig. 14.6 Tensile strength of the GEOPAV blocks after 28 days of curing (Mahdi et al., 2022, open access)

ash, cement, aluminum powder, and superplasticizer. The compressive strength was further tested after 28 days of curing, where findings showed that with the aluminum powder at 0.05% and rice husk ash at 5%, concrete blocks demonstrated higher compressive strength levels. Vivek and Mangai (2023) supported the findings from Manubothula and Gorre (2022) in a study that examined the impact of using rice husk ash and fly ash as a red soil replacement using the sodium hydroxide solution. From the results, it was observed that mixtures using 40% fly ash and 5% RHA demonstrated the highest compressive strength levels and resistance to water absorption. Somna et al. (2022) also utilized RHA and FA in the production of geopolymer hollow blocks based on sodium hydroxide, where the investigation of the mechanical and structural qualities of the materials was done. The generated findings showed that the increase in RHA quantity and fineness led to a subsequent improvement in the compressive strength of the geopolymers. As a result, using RHA during the manufacture of masonry components including geopolymer hollow blocks improves the structural properties by enhancing compressive strength. Additionally, the insights from the studies showed that the materials also reduced the carbon footprint since the hollow geopolymer blocks were smaller in size than those from the Portland cement systems. The findings were an indicator that integrating agro-wastes such as RHA as a replacement for cement in the manufacture of concrete had a positive impact on the mechanical and structural qualities of the masonry composite materials. Furthermore, the continued use of agro-waste as a cement replacement led to reduced pollution of the environment while lowering the overall costs of building construction.

Finally, other studies have examined the use of sugarcane bagasse (SCB) as a potential cement replacement in the manufacture of masonry composites such as bricks. A case example was the study by Amany and Moussa (2021), where cheap building bricks were developed from the integration of SCB with cement brick mixtures to enhance the sustainability of the construction sector by reducing overall effects on the environmental surroundings and human health. From the results, it was observed that the replacement of cement at 1.5% sugarcane bagasse content preserved the structural features of the bricks hence leading to higher strength and ductility (Amany & Moussa, 2021). Furthermore, the use of agro-waste materials reduced the initial costs of building as well as the prices of the real estate market. Tayeh et al. (2023) conducted a similar study to Amany and Moussa (2021), where they investigated the impact of adding sugarcane pulp sand (SCPS) and paper grain sand (PGS) to concrete bricks. The generated findings showed that bricks that had the SCPS and PGS at 10% demonstrated better mechanical qualities where compressive strength had increased by 8.52% and flexural strength by 18.43% after 28 days (Tayeh et al., 2023). Gupta et al. (2022) also investigated the impact of sugarcane bagasse ash as a partial replacement of cement during the manufacture of concrete, where results showed that the material led to improvement in mechanical and structural strength qualities of the mortar and concrete. The synthesis of the different studies showed that sugarcane bagasse was a suitable material in the manufacture of masonry composites such as lightweight bricks due to their positive impact on the reduction of carbon emissions and improvement of flexural strength and compressive strength features.

In summary, the discussion in this section revealed that diverse agro-wastes were utilized in the manufacture of the masonry components, including bricks, and mortar, and as a cement replacement. The use of agro-wastes such as sugarcane bagasse, sugarcane bagasse ash, RHA, FA, and SF led to improvement in the structural qualities of the construction materials. In particular, flexural strength, compressive strength, and split tensile strength were improved. The discussion further showed that the agro-wastes had improved economic benefits where they led to lower costs in the construction of buildings. Furthermore, replacing cement while retaining the structural qualities led to the enhancement of the construction materials.

14.4.4 Reinforcement

A fourth application area of agro-wastes regards reinforcements in construction projects, for example, bamboo rebars within concrete beams (Maraveas, 2020a). In the study, Maraveas (2020a) argued that various agro-wastes, including bamboo, coconut, sisal and hemp, were utilized as cement reinforcement materials. An earlier study by Pacheco-Torgal and Jalali (2011) supported the findings from Maraveas (2020a), where it was reported that vegetable fibers were used as reinforcement in cement-based materials, where they improved the durability performance aspects.

The analysis of these studies (Maraveas, 2020a; Pacheco-Torgal & Jalali, 2011) highlights the use of agro-waste materials that were fiber-reinforced in combination with cement to improve structural properties such as compressive, tensile, and durability. The insight showed that the different fibers were used as supporting beams and served as steel replacements.

However, other studies showed that the cementitious materials could be reinforced by adding agro-waste to the cement (Abdalla et al., 2022). In their study, Abdalla et al. (2022) observed that adding agro-wastes, including hemp, bamboo, and kenaf, in concrete was effective in improving the compressive strength, toughness, and flexural tensile strength of the concrete. Additionally, findings showed that bamboo agro-wastes could be used in varying concentrations when mixed with concrete to attain a targeting ductility. The insight from Abdalla et al. (2022) was the emphasis on natural fibers as an additional component to cement to enhance the structural and mechanical properties, including compressive and tensile strength. The finding was a deviation from Pacheco-Torgal and Jalali (2011) and Maraveas (2020a), who had shown that the bamboo agro-wastes were used as supporting beams and replaced steel rebars. The discussion introduces the novel ways that agro-wastes may be adopted as reinforcement for concrete—as rebars and mixed with cement.

Ganesan et al. (2023) resonated with Abdalla et al. (2022) in a study that investigated the compatibility of bamboo fiber and olive tree leaf powder (OTL) as fillers in epoxy composites. From the results, it was observed that 30-minute cryogenic treatments increased the tensile strength of polymeric structures by 11.34%, impact strength by 26.52%, and flexural strength by 21.76% (Ganesan et al., 2023). Ahmed and Ali (2020) also conducted a study to compare the impact resistance of jute fiber-reinforced concrete walls against those of steel rebars and normal concrete and GFRP rebars and normal concrete. The generated results showed that GFRP rebars and jute fiber-reinforced concrete walls outperformed normal concrete hence being recommended to avoid the spread of concrete fragments against blast impact. Such findings reiterate the arguments that reinforcement of cementitious materials could be achieved by adding fibers into concrete to enhance its strengthening properties.

14.4.5 Particle Boards, Ceiling Tiles, and Bio-based Plastics

Finally, agro-wastes have been employed in diverse applications, including ceiling tiles, particle boards, and bio-based plastics (Sangmesh et al., 2023). In elaboration, Sangmesh et al. (2023) revealed that agro-wastes such as rice straw were increasingly adopted for the manufacture of particle boards where the material led to advantages such as good thermal insulation properties within the buildings. In another study, particle boards made from agro-waste were identified to be more cost-effective (Adediran et al., 2019). A further advantage identified by Adediran et al. (2019) emphasized that particle boards from agro-waste including rice husks were associated with a low rate of water absorption and low density, thereby, being utilized in indoor applications including components of furniture. Martins et al. (2021) examined the effect of integrating lignocellulosic waste on the mechanical and density profiles of particle boards made using eucalypt wood where agro-wastes such as macadamia nut carpel, papaya stem, and coffee husk were used. From the results, it was observed that although the particle boards made from agro-industrial waste demonstrated lower performance, 10%, however, met the requirements for the particle boards for general use (Martins et al., 2021). The insights from the results by Adediran et al. (2019) and Martins et al. (2021) showed that using particle boards from agro-waste was beneficial in not only reducing the disposal of the waste but also in replacing more expensive wood. Gonçalves et al. (2021) reiterated Martins et al. (2021) where they demonstrated that particle boards made from papaya stalk, macadamia nut carpel, and eucalypt sawdust demonstrated resistance to fungi and termites, which indicated that they were suitable in furniture applications. Figure 14.7 illustrates the process of manufacturing the particle boards from the different agro-wastes.

Figure 14.7 shows that the particle boards were manufactured by incorporating different types of agro-wastes, including papaya stalk, macadamia nut carpels, coffee husks, and eucalypt sawdust. By using these agro-wastes, the resultant particle boards were identified to be termite- and fungi-resistant.

Ceiling tiles have also been developed from agro-waste. A study by Guna et al. (2021b) demonstrated that agro-waste including wool and coir could be integrated with gypsum to develop green building materials that had high strength and resistance to moisture. The improved moisture resistance was attributed to the wool fiber, while the coir improved the strength of the gypsum tiles. Additionally, the use of the two agro-wastes boosted the thermal and acoustic properties of the tiles as well as their flame resistance, thereby making them biodegradable and decreasing the environmental burden. Dirisu et al. (2022) reiterated Guna et al. (2021b), who reported that the development of building ceilings using oil bean stalk agro-wastes led to reduced flame propagation based on the low thermal conductivity levels of the composites. The materials were observed to be eco-friendly and flame retardant hence improving the production of sustainable construction.

Finally, agro-wastes, including sugarcane bagasse, rice husks, and bamboo leaf ash, have been utilized in the manufacture of bio-based plastics, where they offer numerous advantages, including low costs and low environmental impact (Maraveas, 2020b). In another study, Mellinas et al. (2020) revealed that pectin derived from the cell walls of many plants was extracted as a biopolymer and isolated as a bioplastic material utilized in diverse applications, including food packaging. Figure 14.8 summarizes the diverse biopolymers extracted from animal and vegetable waste.

From Fig. 14.8, diverse biopolymers identified from animal and vegetable wastes were highlighted, including cellulose, alginate, glycogen, and chitosan, among others. Guna et al. (2021a) identified similar results to Mellinas et al. (2020), where they demonstrated that Mulberry Stems (MS) could be used as substitutes for non-biodegradable polypropylene (PP) composites using fabrication techniques.

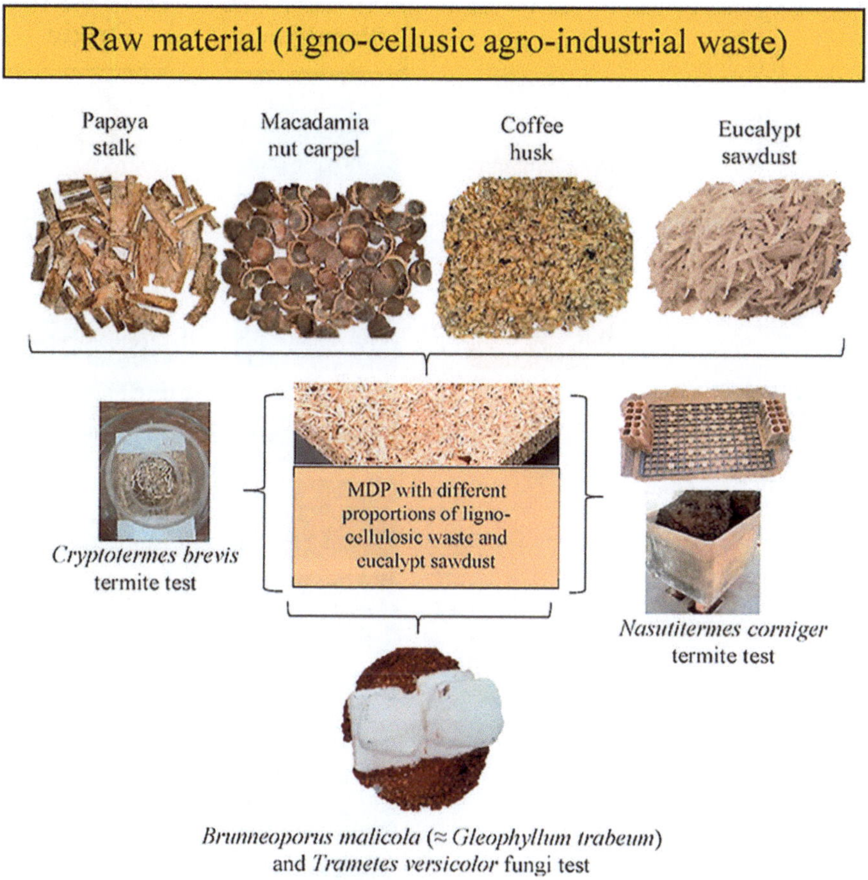

Fig. 14.7 Manufacture of particle boards from agro-wastes (Gonçalves et al., 2021, permission obtained)

From the results, it was observed that the 90/10 MS-PP ratio was the most optimal, which resulted in the highest mechanical strength qualities, while the increase in density improved the sound transmission loss, flame retardancy, thermal insulation, and water stability.

14.5 Artificial Intelligence for Predicting Building Material Properties

Based on the elaborate evaluation of literature in the previous section, insights showed that agro-wastes were suitable for manufacturing various types of sustainable construction materials. The materials ranged from cementitious and binding

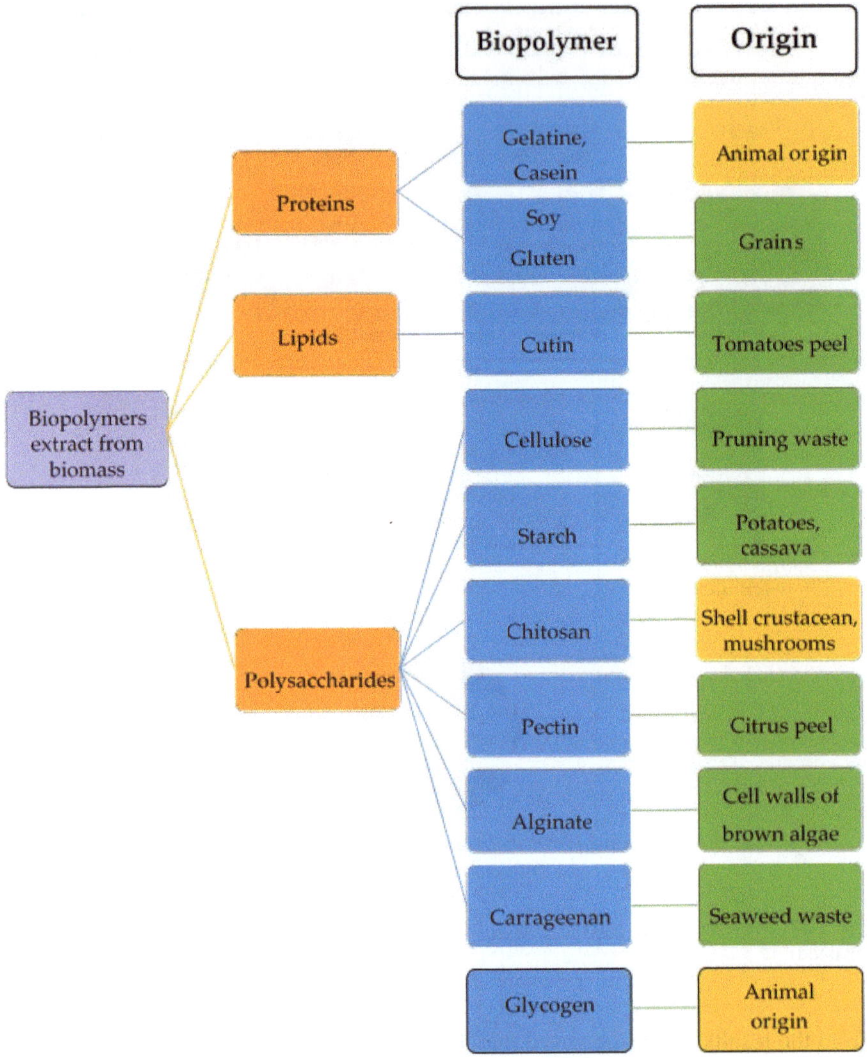

Fig. 14.8 Biopolymers extracted from animal and vegetable wastes (Mellinas et al., 2020, open access)

agents (Channa et al., 2021; Dharmaraj et al., 2023; Jayaraman et al., 2023; Ma et al., 2023; Su & Xu, 2023a, b) to thermal insulators (Chen et al., 2021a; Li et al., 2021; Mehrzad et al., 2022; Zhao et al., 2023), masonry and bricks (Manubothula & Gorre, 2022; Somna et al., 2022; Vivek & Mangai, 2023), reinforcement materials (Abdalla et al., 2022; Ganesan et al., 2023), particle boards (Gonçalves et al., 2021), ceiling tiles (Guna et al., 2021b, b), and bio-based plastics (Mellinas et al., 2020). The findings revealed that the materials offered similar quality and performance to traditional materials, for example, SCBA as an alternative to cement and agro-waste-based particle boards.

However, to further examine the properties of agro-waste-based construction materials, novel technologies such as artificial intelligence (AI) are being utilized (Shah et al., 2022). In their study, Shah et al. (2022) employed ML methods, including support vector machines (SVM), random forest regression (RFR), and gene expression programming (GEP) to estimate the compressive strength of sugarcane bagasse ash concrete. The obtained results showed that the GEP was the most accurate model in providing an estimate of the compressive strength of the SCBA concrete. However, the RFR was effective in providing a strong correlation between estimated and actual outcomes. The conclusions from the research indicated that the ML algorithms outperformed prediction using laboratory tests and literature data.

Shehu et al. (2022) reiterated Shah et al. (2022) where they conducted a study that utilized artificial neural networks (ANN) and response surface methodology (RSM) to optimize and predict the stiffness modulus and rutting features of asphalt mixtures that were prepared using palm oil clinker fine (POCF) modified bitumen (POCF-MB). The ANN results for the R2 stiffness and rutting were 99.972 and 99.880 versus the RSM 99.700 and 99.668, respectively. Therefore, the generated results showed that the ANN outperformed the RSM model in predicting the rutting properties and stiffness modulus of the asphalt mixtures. Ashrafian et al. (2023) conducted a similar study where diverse machine learning algorithms, including Multivariate Adaptive Regression Splines (MARS), Gene Expression Programming (GEP), M5P Model Tree (M5P), and Extreme Learning Machine (ELM) were used to assess the strength of sustainable lightweight structural concrete (SLSC). The generated results showed that the most influential features in predicting the strength of SLSC included gravel (11.84%), oil palm waste aggregate (11.38%), and water-to-binder ratio (11.15%). The discussions from the different studies reveal the growing trend of using AI techniques in predicting the qualities and properties of sustainable construction materials developed from agro-waste.

14.6 Production Issues and Quality Control

Savitha (2012) argues that quality assurance of building materials for construction projects is important to facilitate the development of strong, durable, and cost-effective structures. To ensure quality assurance, diverse approaches are recommended, including selecting the materials in compliance withset construction standards, testing the materials, and verifying whether they conform to set standards. Concerning the agro-waste-based construction materials, discussing the different applications in the previous sections showed that quality control was guaranteed mainly through testing to ensure compliance with traditional materials. A case in point was the use of sugarcane bagasse ash (SCBA) as a cement replacement (Agwa et al., 2022; Arbelaez Perez et al., 2022; Bheel et al., 2022; Li et al., 2022). The findings showed that through testing the developed concrete using SCBA, high compressive and durability qualities were observed that were similar to concrete developed using traditional cement. A similar observation was made with the use of sugarcane bagasse (SCB) in the production of thermal insulators, where the materials demonstrated standard qualities of low thermal conductivity and sound absorption (Chen et al., 2021a; Li et al., 2021; Mehrzad et al., 2022; Zhao et al., 2023). With the use of agro-waste in the development of masonry components such as bricks, different studies showed that quality assurance was achieved mainly through testing the final products and comparing the properties to products developed using standard materials (Manubothula & Gorre, 2022; Somna et al., 2022; Vivek & Mangai, 2023). The direct implication is that quality control for construction materials developed from agro-waste is achieved through testing and comparing the results to traditional construction materials.

14.7 Discussion

The focus of this chapter was to expound on the use of agro-waste in the manufacture of sustainable construction materials. The synthesis of the reviewed literature showed that the use of agro-waste in manufacturing construction materials was justified from two unique viewpoints. First, as emphasized by the World Green Building Council (2019) and El Sheikh (2022), replacing traditional construction materials helped address the GHG emission problem. The construction sector was identified to contribute high levels of carbon emissions both in the production phase (80–93%) and in the transportation processes (2–3%). Furthermore, arguments by Sizirici et al. (2021) indicated that the operation of buildings generated high carbon dioxide emissions through the combustion of fossil fuels. Close inspection showed that where alternatives to traditional materials were used, for instance, SCBA as a replacement for cement, the construction sector would shift away from excessive mining, thereby reducing GHG emissions.

A second factor that contributed to the use of agro-waste in the manufacture of construction materials is the waste management problem prevalent in the construction sector. The synthesis of studies such as Nunes et al. (2020) reported that Brazil produced 550,000 tons of coconut fruit annually, while the total amount of shells that were thrown away was still unknown. Wang et al. (2021) showed that aquaculture production in China reached 10.35 million tons. The inspection of these studies is emphatic on the growing problem of disposing of agricultural waste. The standard processes employed in waste disposal further contribute to high GHG emissions, for example, incineration. Therefore, in hindsight, the findings show that the two problems are complementary—waste disposal and traditional construction materials. High disposal of waste through incineration generates high carbon emissions, while the production and transportation of construction materials, including cement, exacerbate the problem. In this view, replacing the standard construction materials using agro-wastes helps solve the two problems intelligently.

The review of the results further showed that agro-wastes could be adopted in the manufacture of diverse construction materials that demonstrated comparable performance to traditional construction materials. From the findings, diverse studies (Agwa et al., 2022; Arbelaez Perez et al., 2022; Bheel et al., 2022; Li et al., 2022) showed that agro-wastes such as SCBA replaced cement and improved the properties of concrete, for example, compressive strength and durability. In other cases, RHA could be mixed with glass sand or used as a cement replacement to enhance the quality of the construction materials (Channa et al., 2021; Dharmaraj et al., 2023; Jayaraman et al., 2023; Ma et al., 2023; Su & Xu, 2023a, b). Similarly, agro-wastes such as SCB were used for the production of thermal insulators (Chen et al., 2021b; Li et al., 2021; Mehrzad et al., 2022; Zhao et al., 2023). Masonry composites and bricks were also developed from agro-wastes where they demonstrated high compressive, flexural, and split tensile strength (Manubothula & Gorre, 2022; Somna et al., 2022; Vivek & Mangai, 2023). Finally, the applications were identified in areas such as reinforcement (Abdalla et al., 2022; Ganesan et al., 2023), particle boards (Gonçalves et al., 2021; Martins et al., 2021; Sangmesh et al., 2023), ceiling tiles (Guna et al., 2021b, b), and bio-based plastics (Mellinas et al., 2020). The holistic review of the different studies reveals that there are numerous application areas where agro-waste wastes are being employed to replace traditional construction materials. The implication is that more countries are viewing agro-wastes as effective construction materials that are cost-effective, eco-friendly, and possess good structural and mechanical properties.

14.8 Conclusions and Future Perspectives

Several key conclusions have been identified within this chapter. First, environmental problems regarding GHG emissions are still prevalent in the modern world, where the construction sector is a major contributor to the high emission levels. From the findings, the construction industry generates high carbon emissions through production phases as the materials are extracted and during transportation to the construction sites. Furthermore, the operation of buildings also contributes to high carbon emissions through the consumption of fossil fuels to generate electricity for lighting and heating.

The second conclusion is that the current approaches adopted for waste management from agricultural production are still ineffective in handling the total waste generated. Furthermore, since the agro-waste reaches high volumetric levels, such as gigatons, methods such as incineration generate high GHG emissions. Disposal of agro-wastes such as SCB and rice husks in landfills also destroys the environment. Therefore, identifying alternative ways to dispose of the waste materials is essential to address the problems.

Finally, this chapter concludes that shifting from traditional construction materials to agro-waste-based alternatives is an effective strategy for reducing waste disposal and minimizing GHG emissions from construction activities. The findings showed that the agro-waste-based materials could be adopted in diverse construction areas, including bricks, cementitious materials and mortars, thermal insulation, reinforcement, particle boards, ceiling tiles, and bio-based plastics.

In future work, there is a need for scholars to undertake comprehensive studies to investigate how AI can be integrated to predict the properties of agro-waste-based construction materials. The findings from this chapter showed that few studies have explored this area despite its significance in the construction sector. Similarly, future work should examine other methods that are useful in quality control of the agro-waste materials besides testing the final materials and relying on AI to predict the material properties.

Acknowledgments The authors are thankful to the United Greek Council(UGC) for providing the funds to conduct the present investigation.

Conflict of Interest The authors declare that there is no conflict of interest regarding the publication of this article.

References

Abdalla, J. A., Thomas, B. S., & Hawileh, R. A. (2022). Use of hemp, kenaf and bamboo natural fiber in cement-based concrete. *Materials Today: Proceedings, 65*, 2070–2072. https://doi.org/10.1016/j.matpr.2022.06.428

Adediran, A. A., Olawale, O., Ojediran, J., Aladegboye, S., Atoyebi, O. D., Akinlabi, E. T., & Olayanju, T. M. A. (2019). Properties of agro-based hybrid particleboards. *Procedia Manufacturing, 35*, 442–446. https://doi.org/10.1016/j.promfg.2019.05.064

Agwa, I. S., Zeyad, A. M., Tayeh, B. A., Adesina, A., de Azevedo, A. R. G., Amin, M., & Hadzima-Nyarko, M. (2022). *A comprehensive review on the use of sugarcane bagasse ash as a supplementary cementitious material to produce eco-friendly concretes*. Proceedings. https://doi.org/10.1016/j.matpr.2022.03.264

Ahmed, S., & Ali, M. (2020). Use of agriculture waste as short discrete fibers and glass-fiber-reinforced-polymer rebars in concrete walls for enhancing impact resistance. *Journal of Cleaner Production, 268*, 122211. https://doi.org/10.1016/j.jclepro.2020.122211

Ahmed, N., Abdel-Hamid, M., Abd El-Razik, M. M., & El-Dash, K. M. (2020). Impact of sustainable design in the construction sector on climate change. *Ain Shams Engineering Journal.* https://doi.org/10.1016/j.asej.2020.11.002

Amany, M., & Moussa, R. R. (2021). Investigating the economic and environmental effect of integrating sugarcane bagasse (SCB) fibers in cement bricks. *Ain Shams Engineering Journal.* https://doi.org/10.1016/j.asej.2020.12.012

Amin, M., Attia, M. M., Agwa, I. S., Elsakhawy, Y., El-hassan, K. A., & Abdelsalam, B. A. (2022). Effects of sugarcane bagasse ash and nano eggshell powder on high-strength concrete properties. *Case Studies in Construction Materials, 17*, e01528. https://doi.org/10.1016/j.cscm.2022.e01528

Arbelaez Perez, O. F., Florez, D. R., Zapata Vergara, L. M., & Hernández Benavides, K. V. (2022). Innovative use of agro waste cane bagasse ash and waste glass as cement replacement for green concrete. Cost analysis and carbon dioxide emissions. *Journal of Cleaner Production, 379*, 134822. https://doi.org/10.1016/j.jclepro.2022.134822

Ashrafian, A., Panahi, E., Salehi, S., Karoglou, M., & Asteris, P. G. (2023). Mapping the strength of agro-ecological lightweight concrete containing oil palm by-product using artificial intelligence techniques. *Structure, 48*, 1209–1229. https://doi.org/10.1016/j.istruc.2022.12.108

Awoyera, P. O., Akinrinade, A. D., de Sousa Galdino, A. G., Althoey, F., Kirgiz, M. S., & Tayeh, B. A. (2022). Thermal insulation and mechanical characteristics of cement mortar reinforced with mineral wool and rice straw fibers. *Journal of Building Engineering, 53*, 104568. https://doi.org/10.1016/j.jobe.2022.104568

Bhat, M. A., Adil, A. W., Sikander, B., Lone, Y., & Malik, J. (2022). Waste management technology for sustainable agriculture. *IGI Global, 379–398*. https://doi.org/10.4018/978-1-6684-5352-0.ch021

Bheel, N., Khoso, S., Baloch, M. H., Benjeddou, O., & Alwetaishi, M. (2022). Use of waste recycling coal bottom ash and sugarcane bagasse ash as cement and sand replacement material to produce sustainable concrete. *Environmental Science and Pollution Research*. https://doi.org/10.1007/s11356-022-19478-3

Boudali, S., Abdulsalam, B., Soliman, A., Poncet, S., Godbout, S., Palacios, J., & ElSafty, A. (2022). *15—Implementation of agricultural crop wastes toward green construction materials*. ScienceDirect [online]. Available at: https://www.sciencedirect.com/science/article/abs/pii/B978012821730600022X. Accessed 10 July 2023.

Boumaaza, M., Belaadi, A., Bourchak, M., Jawaid, M., & Hamid, S. (2022). Comparative study of flexural properties prediction of *Washingtoniafilifera* rachis biochar bio-mortar by ANN and RSM models. *Construction and Building Materials, 318*, 125985. https://doi.org/10.1016/j.conbuildmat.2021.125985

Chan, M., Masrom, M. A. N., & Yasin, S. S. (2022). Selection of low-carbon building materials in construction projects: Construction professionals' perspectives. *Buildings, 12*(4), 486. https://doi.org/10.3390/buildings12040486

Channa, S. H., Mangi, S. A., Bheel, N., Soomro, F. A., & Khahro, S. H. (2021). Short-term analysis on the combined use of sugarcane bagasse ash and rice husk ash as supplementary cementitious material in concrete production. *Environmental Science and Pollution Research*. https://doi.org/10.1007/s11356-021-15877-0

Chen, R., Congress, S. S. C., Cai, G., Duan, W., & Liu, S. (2021a). Sustainable utilization of biomass waste-rice husk ash as a new solidified material of soil in geotechnical engineering: A review. *Construction and Building Materials, 292*, 123219. https://doi.org/10.1016/j.conbuildmat.2021.123219

Chen, W., Wang, H., Lan, W., Li, D., Zhang, A., & Liu, C. (2021b). Construction of sugarcane bagasse-derived porous and flexible carbon nanofibers by electrospinning for supercapacitors. *Industrial Crops and Products, 170*, 113700. https://doi.org/10.1016/j.indcrop.2021.113700

Chinnu, S. N., Minnu, S. N., Bahurudeen, A., & Senthilkumar, R. (2022). Influence of palm oil fuel ash in concrete and a systematic comparison with widely accepted fly ash and slag: A step towards sustainable reuse of agro waste ashes. *Cleaner Materials, 5*, 100122. https://doi.org/10.1016/j.clema.2022.100122

Dharmaraj, R., Dinesh, M., Sampathkumar, S., Haripprasath, M., & Chandraprakash, V. (2023). High performance concrete using rice husk ash. *Materials Today*. https://doi.org/10.1016/j.matpr.2023.04.104

Dirisu, J. O., Oyedepo, S. O., Fayomi, O. S. I., Joseph, O. O., Akinlabi, E. T., Babalola, P. O., Udoye, N. E., Ajayi, O. O., Aworinde, A. K., Banjo, S. O., & Oluwasegun, K. M. (2022). Thermal-emission assessment of building ceilings from agro-industrial wastes. *Fuel Communications, 10*, 100042. https://doi.org/10.1016/j.jfueco.2021.100042

El Sheikh, S. (2022). *CO_2 emissions from buildings and construction hit new high, leaving sector off track to decarbonize by 2050: UN*. UN Environment [Online]. Available at: https://www.unep.org/news-and-stories/press-release/co2-emissions-buildings-and-construction-hit-new-high-leaving-sector

Elbasiouny, H., Elbanna, B. A., Al-Najoli, E., Alsherief, A., Negm, S., Abou El-Nour, E., Nofal, A., & Sharabash, S. (2019). Agricultural waste management for climate change mitigation: Some implications to Egypt. *Waste Management in MENA Regions, 149–169*. https://doi.org/10.1007/978-3-030-18350-9_8

Erkmen, J., & Sari, M. (2023). Hydrophobic thermal insulation material designed from hazelnut shells, pinecone, paper and sheep wool. *Construction and Building Materials, 365*, 130131. https://doi.org/10.1016/j.conbuildmat.2022.130131

Ganesan, V., Natrayan, L., Chohan, J. S., Vasanthi, P., Angalaeswari, S., Pravin, P., Kaliappan, S., & Arunkumar, D. (2023). Investigation of mechanical and dynamic mechanical analysis of bamboo/olive tree leaves powder-based hybrid composites under cryogenic conditions. *Biomass Conversion and Biorefinery*. https://doi.org/10.1007/s13399-023-04591-1

Gonçalves, F. G., Paes, J. B., Lopez, Y. M., de Alcântara Segundinho, P. G., de Oliveira, R. G. E., Fassarella, M. V., Brito, A. S., Chaves, I. L. S., & Martins, R. S. F. (2021). Resistance of particleboards produced with lignocellulosic agro-industrial wastes to fungi and termites. *International Biodeterioration & Biodegradation, 157*, 105159. https://doi.org/10.1016/j.ibiod.2020.105159

Guna, V., Ilangovan, M., Reddy, N., Radhakrishna, P., Maharaddi, V., Jambunath, A., & Rao, A. P. (2021a). Biobased insulating panels from mulberry stems. *Journal of Thermoplastic Composite Materials, 453–472*. https://doi.org/10.1177/08927057211010884

Guna, V., Yadav, C., Maithri, B. R., Ilangovan, M., Touchaleaume, F., Saulnier, B., Grohens, Y., & Reddy, N. (2021b). Wool and coir fiber reinforced gypsum ceiling tiles with enhanced stability and acoustic and thermal resistance. *Journal of Building Engineering, 41*, 102433. https://doi.org/10.1016/j.jobe.2021.102433

Gupta, C. K., Sachan, A. K., & Kumar, R. (2022). Utilization of sugarcane bagasse ash in mortar and concrete: A review. *Materials Today: Proceedings*. https://doi.org/10.1016/j.matpr.2022.03.304

Herring, T. C., Nyomboi, T., & Thuo, J. N. (2022). Ductility and cracking behavior of reinforced coconut shell concrete beams incorporated with coconut shell ash. *Results in Engineering, 14*, 100401. https://doi.org/10.1016/j.rineng.2022.100401

Huang, P., Huang, B., Li, J., Wu, N., & Xu, Q. (2023). Application of sugarcane bagasse ash as filler in ultra-high performance concrete. *Journal of Building Engineering, 71*, 106447–106447. https://doi.org/10.1016/j.jobe.2023.106447

Jadhav, P. S., Sarkar, A., Pasupathy, S., & Ren, S. (2023). Biogenic straw aerogel thermal insulation materials. *Advanced Engineering Materials, 25*(13). https://doi.org/10.1002/adem.202300037

Jannat, N., Al-Mufti, R. L., Hussien, A., Abdullah, B., & Cotgrave, A. (2022). Influences of agro wastes on the physico-mechanical and durability properties of unfired clay blocks. *Construction and Building Materials, 318*, 126011. https://doi.org/10.1016/j.conbuildmat.2021.126011

Jayaraman, A., Vasudevan, M., Aravind, R., Keerthika, M., & Prakash, J. H. (2023). Optimization of partial replacement of cement by silica fume and rice husk ash for sustainable concrete. *Materials Today*. https://doi.org/10.1016/j.matpr.2023.05.598

Jittin, V., & Bahurudeen, A. (2022). Evaluation of rheological and durability characteristics of sugarcane bagasse ash and rice husk ash based binary and ternary cementitious system. *Construction and Building Materials, 317*, 125965. https://doi.org/10.1016/j.conbuildmat.2021.125965

Kumar, A., Kumar, R., Das, V., Jhatial, A. A., & Ali, T. H. (2021). Assessing the structural efficiency and durability of burnt clay bricks incorporating fly ash and silica fume as additives. *Construction and Building Materials, 310*, 125233. https://doi.org/10.1016/j.conbuildmat.2021.125233

Kumar, M. H., Saikrishnamacharyulu, I., Mohanta, N. R., Ashutosh, A., Mishra, P., & Samantaray, S. (2022). Mechanical behaviour of high strength concrete modified with triple blend of fly ash, silica fume and steel fibers. *Materials Today: Proceedings, 65*, 933–942. https://doi.org/10.1016/j.matpr.2022.03.528

Labaran, Y. H., Mathur, V. S., Muhammad, S. U., & Musa, A. A. (2022). Carbon footprint management: A review of construction industry. *Cleaner Engineering and Technology, 9*, 100531. https://doi.org/10.1016/j.clet.2022.100531

Li, W., Li, Z., Wang, W., Li, Z., Li, Q., Qin, C., & Cao, F. (2021). Green approach to facilely design hydrophobic aerogel directly from bagasse. *Industrial Crops and Products, 172*, 113957. https://doi.org/10.1016/j.indcrop.2021.113957

Li, Y., Chai, J., Wang, R., Zhang, X., & Si, Z. (2022). Utilization of sugarcane bagasse ash (SCBA) in construction technology: A state-of-the-art review. *Journal of Building Engineering, 56*, 104774. https://doi.org/10.1016/j.jobe.2022.104774

Luhar, S., Cheng, T.-W., & Luhar, I. (2019). Incorporation of natural waste from agricultural and aquacultural farming as supplementary materials with green concrete: A review. *Composites Part B: Engineering, 175*, 107076. https://doi.org/10.1016/j.compositesb.2019.107076

Ma, W., Wang, Y., Huang, L., Yan, L., & Kasal, B. (2023). Natural and recycled aggregate concrete containing rice husk ash as replacement of cement: Mechanical properties, microstructure, strength model and statistical analysis. *Journal of Building Engineering, 66*, 105917. https://doi.org/10.1016/j.jobe.2023.105917

Mahdi, S. N., Babu R, D. V., Hossiney, N., & Abdullah, M. M. A. B. (2022). Strength and durability properties of geopolymer paver blocks made with fly ash and brick kiln rice husk ash. *Case Studies in Construction Materials, 16*, 00800. https://doi.org/10.1016/j.cscm.2021.e00800

Manubothula, S., & Gorre, M. (2022). *Influence of rice husk ash on compressive strength of an aerated concrete*. Proceedings. https://doi.org/10.1016/j.matpr.2022.05.320

Maraveas, C. (2020a). Production of sustainable construction materials using agrowastes. *Materials, 13*(2), 262. https://doi.org/10.3390/ma13020262

Maraveas, C. (2020b). Production of sustainable and biodegradable polymers from agricultural waste. *Polymers, 12*(5), 1127. https://doi.org/10.3390/polym12051127

Martins, R. S. F., Gonçalves, F. G., de Alcântara Segundinho, P. G., Lelis, R. C. C., Paes, J. B., Lopez, Y. M., Chaves, I. L. S., & de Oliveira, R. G. E. (2021). Investigation of agro-industrial lignocellulosic wastes in fabrication of particleboard for construction use. *Journal of Building Engineering, 43*, 102903. https://doi.org/10.1016/j.jobe.2021.102903

Mehrzad, S., Taban, E., Soltani, P., Samaei, S. E., & Khavanin, A. (2022). Sugarcane bagasse waste fibers as novel thermal insulation and sound-absorbing materials for application in sustainable buildings. *Building and Environment, 211*, 108753. https://doi.org/10.1016/j.buildenv.2022.108753

Mellinas, C., Ramos, M., Jiménez, A., & Garrigós, M. C. (2020). Recent trends in the use of pectin from agro-waste residues as a natural-based biopolymer for food packaging applications. *Materials, 13*(3), 673. https://doi.org/10.3390/ma13030673

Miller, S. A., Habert, G., Myers, R. J., & Harvey, J. T. (2021). Achieving net zero greenhouse gas emissions in the cement industry via value chain mitigation strategies. *One Earth, 4*(10), 1398–1411. https://doi.org/10.1016/j.oneear.2021.09.011

Moujoud, Z., Harrati, A., Manni, A., Naim, A., El Bouari, A., & Tanane, O. (2023). Study of fired clay bricks with coconut shell waste as a renewable pore-forming agent: Technological, mechanical, and thermal properties. *Journal of Building Engineering, 68*, 106107. https://doi.org/10.1016/j.jobe.2023.106107

Nnochiri, E. S., Okokpujie, I. P., & Tartibu, L. K. (2023). Artificial neural network models for predicting California bearing ratio of lateritic soil admixed with reinforce and rice husk ash. *International Information and Engineering Technology Association, 37*(2), 305–313. https://doi.org/10.18280/ria.370208

Nunes, L. A., Silva, M. L. S., Gerber, J. Z., & de Kalid, R. A. (2020). Waste green coconut shells: Diagnosis of the disposal and applications for use in other products. *Journal of Cleaner Production, 255*, 120169. https://doi.org/10.1016/j.jclepro.2020.120169

Obeng, G. Y., Amoah, D. Y., Opoku, R., Sekyere, C. K. K., Adjei, E. A., & Mensah, E. (2020). Coconut wastes as bioresource for sustainable energy: Quantifying wastes, calorific values and emissions in Ghana. *Energies, 13*(9), 2178. https://doi.org/10.3390/en13092178

Pacheco-Torgal, F., & Jalali, S. (2011). Cementitious building materials reinforced with vegetable fibres: A review. *Construction and Building Materials, 25*(2), 575–581. https://doi.org/10.1016/j.conbuildmat.2010.07.024

Sangmesh, B., Patil, N., Jaiswal, K. K., Gowrishankar, T. P., Selvakumar, K. K., Jyothi, M. S., Jyothilakshmi, R., & Kumar, S. (2023). Development of sustainable alternative materials for the construction of green buildings using agricultural residues: A review. *Construction and Building Materials, 368*, 130457. https://doi.org/10.1016/j.conbuildmat.2023.130457

Santhosh, K. G., Subhani, S. M., & Bahurudeen, A. (2022). Recycling of palm oil fuel ash and rice husk ash in the cleaner production of concrete. *Journal of Cleaner Production, 354*, 131736. https://doi.org/10.1016/j.jclepro.2022.131736

Savitha, R. (2012). *Importance of quality assurance of materials for construction work* [Online] Available at: https://nbro.gov.lk/images/content_image/publications/symposia/2010/importance_of_quality_assurance.pdf

Shah, M. I., Javed, M. F., Aslam, F., & Alabduljabbar, H. (2022). Machine learning modeling integrating experimental analysis for predicting the properties of sugarcane bagasse ash concrete. *Construction and Building Materials, 314*, 125634. https://doi.org/10.1016/j.conbuildmat.2021.125634

Shehu, N., Sutanto, M. H., Habib, N. Z., Napiah, M., Usman, A. D., & Muhammad, A. H. (2022). Comparison of response surface methodology and artificial neural network approach in predicting the performance and properties of palm oil clinker fine modified asphalt mixtures. *Construction and Building Materials, 324*, 126618–126618. https://doi.org/10.1016/j.conbuildmat.2022.126618

Shupta, G., Goyal, A., Shetty, A., & Kanoungo, A. (2022). Effect of agro waste as a partial replacement in cement for sustainable concrete production. In *Proceedings of International Conference on Innovative Technologies for Clean and Sustainable Development (ICITCSD—2021)* (pp. 447–458). https://doi.org/10.1007/978-3-030-93936-6_37

Sizirici, B., Fseha, Y., Cho, C.-S., Yildiz, I., & Byon, Y.-J. (2021). A review of carbon footprint reduction in construction industry, from design to operation. *Materials, 14*(20), 6094. https://doi.org/10.3390/ma14206094

Somna, R., Saowapun, T., Somna, K., & Chindaprasirt, P. (2022). Rice husk ash and fly ash geopolymer hollow block based on NaOH activated. *Case Studies in Construction Materials, 16*, e01092. https://doi.org/10.1016/j.cscm.2022.e01092

Su, Q., & Xu, J. (2023a). Compression behavior and permeability of concrete composed of glass sand and rice husk ash. *Journal of Building Engineering, 76*, 107095–107095. https://doi.org/10.1016/j.jobe.2023.107095

Su, Q., & Xu, J. (2023b). Mechanical properties of concrete containing glass sand and rice husk ash. *Construction and Building Materials, 393*, 132053–132053. https://doi.org/10.1016/j.conbuildmat.2023.132053

Tayeh, B. A., Ahmed, S. M., & Hafez, R. D. A. (2023). Sugarcane pulp sand and paper grain sand as partial fine aggregate replacement in environment-friendly concrete bricks. *Case Studies in Construction Materials, 18*, e01612. https://doi.org/10.1016/j.cscm.2022.e01612

Töbelmann, D., & Wendler, T. (2020). The impact of environmental innovation on carbon dioxide emissions. *Journal of Cleaner Production, 244*, 118787. https://doi.org/10.1016/j.jclepro.2019.118787

Tran, D. T., Nguyen, S. T., Do, N. D., Thai, N. N. T., Thai, Q. B., Huynh, H. K. P., Nguyen, V. T. T., & Phan, A. N. (2020). Green aerogels from rice straw for thermal, acoustic insulation and oil spill cleaning applications. *Materials Chemistry and Physics, 253*, 123363. https://doi.org/10.1016/j.matchemphys.2020.123363

Tran, V. T., Le, T. M., Trang, T., & C.L., Yen, Huynh, V.Q., Le, D.T. and Le, P.K. (2023). Development of facile and green fabrication of cellulose–chitosan composite aerogel and lignin/silica hybrid from agro wastes. *Fibers and Polymers, 24*(2), 403–411. https://doi.org/10.1007/s12221-023-00097-9

Vivek, S. S., & Mangai, B. A. (2023). Study on interlocking geo-polymer interlocking earth blocks made with residual rice husk ash and fly ash. *Materials Today: Proceedings*. https://doi.org/10.1016/j.matpr.2023.03.068

Wang, T., Foliente, G., Song, X., Xue, J., & Fang, D. (2014). Implications and future direction of greenhouse gas emission mitigation policies in the building sector of China. *Renewable and Sustainable Energy Reviews, 31*, 520–530. https://doi.org/10.1016/j.rser.2013.12.023

Wang, W., Wei, W., Gao, S., Chen, G., Yuan, J., & Li, Y. (2021). Agricultural and aquaculture wastes as concrete components: A review. *Frontiers in Materials, 8*. https://doi.org/10.3389/fmats.2021.762568

World Green Building Council. (2019). *Embodied carbon*. World Green Building Council [Online] Available at: https://worldgbc.org/advancing-net-zero/embodied-carbon/#:~:text=Buildings%20are%20currently%20responsible%20for

Wu, N., Ji, T., Huang, P., Fu, T., Zheng, X., & Xu, Q. (2022). Use of sugar cane bagasse ash in ultra-high performance concrete (UHPC) as cement replacement. *Construction and Building Materials, 317*, 125881. https://doi.org/10.1016/j.conbuildmat.2021.125881

Zaheer, M. M., & Tabish, M. (2023). The durability of concrete made up of sugar cane bagasse ash (SCBA) as a partial replacement of cement: A review. *Arabian Journal for Science and Engineering, 48*(4), 4195–4225. https://doi.org/10.1007/s13369-023-07698-9

Zhang, Y., Teoh, B. K., Wu, M., Chen, J., & Zhang, L. (2023). Data-driven estimation of building energy consumption and GHG emissions using explainable artificial intelligence. *Energy, 262*, 125468. https://doi.org/10.1016/j.energy.2022.125468

Zhao, P., Xie, Y., Hu, S., Li, Y., Xue, X., Sun, N., Hu, T., Zhang, S., & Wang, L. (2023). Flexible and transparent bagasse aerogels for thermal regulation glazing. *ACS Sustainable Chemistry & Engineering, 11*(26), 9711–9720. https://doi.org/10.1021/acssuschemeng.3c01600

Chapter 15
Transformation of Cassava Wastes into Useful Products

Omojola Awogbemi, Daramy Vandi Von Kallon, and Ramesh C. Ray

Abbreviations

AD	Anaerobic digestion
ARC	Agricultural Research Council
BOD	Biological oxygen demand
COD	Chemical oxygen demand
CP	Cassava peels
FAO	Food and Agriculture Organization
FFA	Free fatty acid
GHG	Greenhouse gases
HRT	Hydraulic retention time
MMT	Metric million tons
NIP	National industrial policy
TDS	Total dissolved solids
TS	Total solids
WCO	Waste cooking oil

O. Awogbemi (✉) · D. V. Von Kallon
Department of Mechanical and Industrial Engineering Technology, University of Johannesburg, Johannesburg, South Africa

R. C. Ray
Center for Industrial Biotechnology Research, Siksha 'O' Anusandhan (SOA) deemed to be University, Khandagiri, Bhubaneswar, India

Formerly, ICAR-Central Tuber Crops Research Institute (Regional Center), Bhubaneswar, Odisha, India

J. Arora et al. (eds.), *Transforming Agriculture Residues for Sustainable Development*, Waste as a Resource, https://doi.org/10.1007/978-3-031-61133-9_15

15.1 Introduction

The persistent rise in global population has directly led to industrialization, demand for food, water, and energy, and waste generation. Of all the direct impacts of population explosion, the demand for timely, quality, and adequate food appears primary and sacrosanct. To meet the increased demand for food to feed the global population and raw materials for the ever-growing industrial sector, there have been increased activities in the agricultural sector. Cassava (*Manihot esculenta* Krantz) also known as manioc or yuca, is the world's most essential food item after rice, wheat, and maize, in that order (Adebayo, 2023). Cassava is believed to have originated in present-day Brazil in South America and is being cultivated and consumed in Africa, Asia, South America, and other parts of the world. The global cassava production has continued to increase in recent decades. Available data showed that global cassava production rose from 175.84 metric million tons (MMT) in 2000 to 230.05 MMT in 2010,287.53 MMT in 2015, and further to 314.81 in 2021 and has been projected to rise to 357.26 MMT by 2028 (Imarc Group, 2023, FAO, 2023). During the same period, about 70% of the global cassava production was derived from Nigeria, Brazil, Thailand, Indonesia, and the Democratic Republic of Congo. Nigeria is the greatest producer of cassava globally with about 32.01 MMT, 44.58 MMT, 57.41 MMT, and 63.03 MMT in 2000, 2010, 2015, and 2021, respectively. Nigeria is predicted to continue to be a major producer of cassava due to the favorable climatic conditions, soil conditions, and abundant government support to small-holder farmers who are the major cultivators of cassava (Oghenejoboh et al., 2021). Figure 15.1 shows the top ten producers of cassava globally in 2000, 2010, 2015, and 2021.

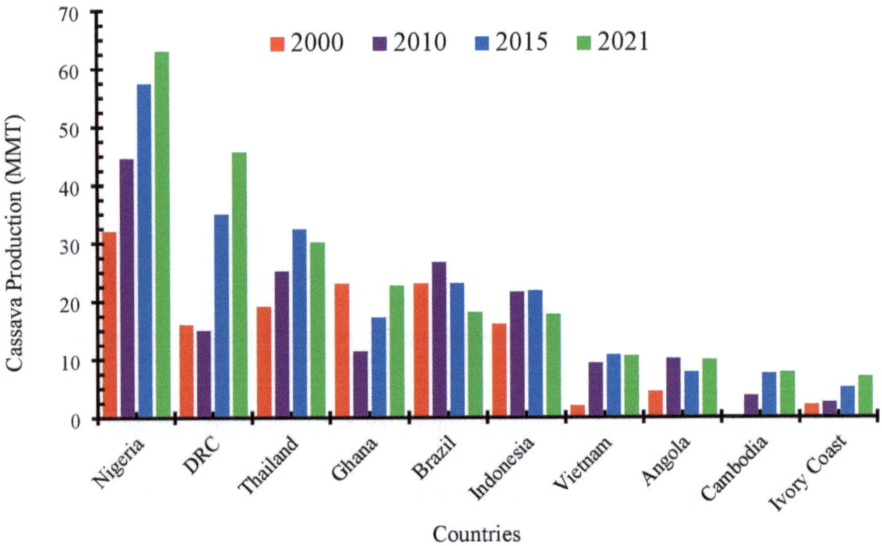

Fig. 15.1 Ten best producers of cassava in 2017 and 2021. (Data compiled from FAO, 2023)

The root of the cassava plant is the consumed part of the plant. Cassava is classified either as sweet or bitter and neither of these species should be consumed raw due to the high content of cyanide, a toxic chemical. Soaking and boiling cassava tuber significantly reduces the cyanide contents and makes it safe for consumption (Ray & Shivkumar, 2009; Oghenejoboh et al., 2021). Cassava can also be grated or ground into flour and processed into flour, gari, tapioca, fufu, bread, crackers, starch, or other food items, depending on the locality (Ray et al., 2010). Though goats and other ruminant animals can consume cassava, the high economic value of cassava and its contribution to food security makes this practice unpopular. When consumed, cassava domestically is a veritable source of energy, calcium, fiber, vitamin C, carbs, thiamine, and copper. Regular consumption of cassava can boost immunity, serve as an antioxidant, enhance collagen production, reduce the risk of cancer and diabetes, contribute to blood sugar management, relieve headaches, and help improve skin and gut health (Dey & Ray, 2020). Cassava can be converted to fuel ethanol, sorbitol, pastries, alcohol, chemicals, and animal feed. In recent years, cassava has been mostly used as raw materials for producing medicines, cosmetics, biodegradable materials, adhesives and glues, and other non-food products (Ray & Swain, 2011). Notwithstanding its multifaceted applications, cassava is low in protein content, has a low shelf life, contains cyanide, and generates a high volume of wastewater during processing (Sigüenza-Andrés et al., 2021, Ikuemonisan et al., 2020).

Processing of cassava for domestic and industrial use involves a series of activities including peeling, grating, washing of cassava tubers, fermentation, dewatering, and sieving. Leading to the generation of wastes, in both solid and liquid forms (Saranraj et al., 2019). Wastes such as leaves, stems, peels, and wastewater are generated during domestic or industrial processing of cassava. These wastes have continued to constitute a challenge to cassava processing and pose a significant nuisance to the environment and humanity in general. This has provoked investigations into the possible utilization of these wastes. Over the years, various researchers have investigated potential avenues for transforming crop residues such as eggshells, corn cobs, sugarcane bagasse, and rice husk into useful products. Recently, the benefits of the pretreatment and conversion of agricultural wastes and crop residues to biochar, biofuels, catalysts, and other useful products were reported to include low-cost waste management, environmental sanitation, circular economy, and resource recovery (Awogbemi & Kallon, 2023a, b). More specifically, Veiga et al. (2016), Ekop et al. (2019), Oghenejoboh et al. (2021), and Kumar et al. (2021) demonstrated how wastes generated from cassava processing can be transformed into useful products. The authors enumerated biochar, bioethanol, activated carbon, livestock feeds, and other bio-products as some of the outcomes of cassava waste conversion. Cassava waste has also been used for microalgae cultivation, manures, feedstock for biogas generation, and biorefinery. The authors were not in doubt that effective conversion of cassava waste will contribute to a cleaner environment, ensure a circular economy, provide additional income to cassava growers, and boost efforts at attaining environmental sustainability (Padi & Chimphango, 2021; Bancal & Ray, 2022).

However, bearing in mind the enormous impact of inappropriate management and utilization of cassava waste, it is believed that not enough studies have been done to expose and escalate the benefits of conversion of cassava waste into useful products. The pertinent question to ask is whether enough research has been conducted in the research space and whether there is adequate awareness in the public domain of the various products that can be derived from cassava wastes. This forms the motivation for the current intervention.

This chapter aims to escalate applications of cassava waste to useful products and the benefits derivable therefrom. The objective of the study is to bring to the fore the various useful products that can be derived from cassava waste. Specifically, this chapter highlights recent outcomes of the conversion of cassava peels, cassava wastewater, cassava pulp, cassava bagasse, cassava straw, cassava stalks, etc. into useful products. As a prelude, this chapter also discusses the generation of waste from cassava processing, cassava waste characterization, and major conversion technologies for cassava waste. The implications, challenges, and prospects of transformation of cassava wastes into valuable products. The outcome of the current study will enrich the body of knowledge by providing updated information on the diverse products that can be obtained from cassava wastes. The current intervention is, however, limited to a desktop review of recent information on the conversion of cassava peels, wastewater, cassava pulp, and cassava bagasse into useful products as published in the literature. The actual conversion of cassava waste into useful products is beyond the scope of this chapter.

15.2 Cassava Processing, Waste Generation, and Conversion Technologies

Cassava is processed into various food products such as starch, gari, and fufu and in this process, a lot of waste is generated in the form of peels, bagasse, and wastewater,

15.2.1 Cassava Processing and Wastes Generation

The processing and conversion of cassava tuber into useful products generates wastes in solid and liquid forms. Improper management of these wastes greatly contaminates and impacts the environment with grave health implications. Generally, cassava tuber is processed into gari, cassava flour, cassava starch, cassava chips, fufu, etc. The Food and Agricultural Organization (FAO) has established processing procedures for the conversion of cassava tuber into starch and gari to include harvesting of cassava tuber, peeling, washing, grating, dewatering, fermentation, sieving, sifting, frying, etc. (FAO, 2001).

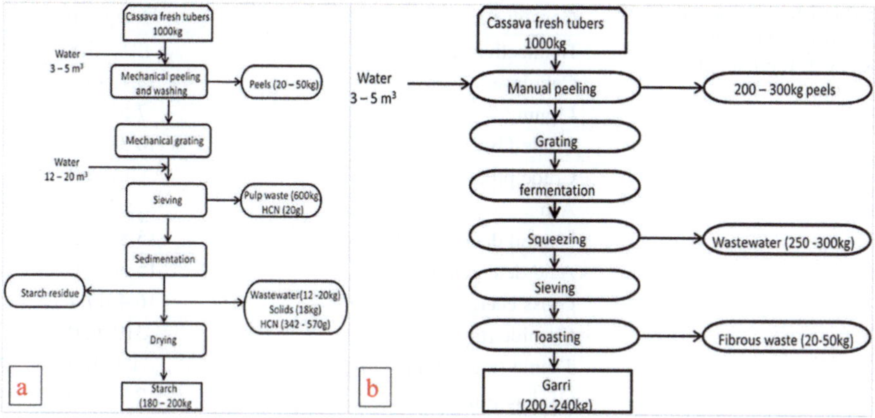

Fig. 15.2 Flowchart for the production of starch and gari from cassava tuber. (Sourced from FAO, 2001)

As shown in Fig. 15.2, cassava peels, wastewater, cyanide, and cassava pulp are the major wastes generated during the processes. Cassava peel is generated during cassava processing and if left untreated oozes an offensive odor, attracts flies, and contaminates surface and underground water. Similarly, about 20,000–30,000 L of highly organic and acidic wastewater is generated during the processing of one ton of cassava tuber. Other forms of residues including straw, stalk, bran, fiber, and bagasse are also generated from cassava processing (Nizzy & Kannan, 2022). In the past, the processing of cassava was mainly done manually but with a significant increment in the volume of cassava being processed and advancements in agricultural mechanization and industrialization, many of the processes have been mechanized. For example, harvesters are being deployed for cassava harvesting while peeling, washing, grating, dewatering, etc. are also mechanized. In recent years, the processing of cassava tuber to gari and starch has been automated to reduce the severity of the labor required, ensure increased production, and prevent product contamination. The wastes generated during the processing must be quickly removed from the production line to prevent product contamination and environmental pollution.

15.2.2 Cassava Wastes Composition

Cassava peel is generated when the outer covering of the cassava tuber is removed. Cassava peel is a major solid waste generated from the consumption and processing of cassava. Based on the huge volume of cassava produced and processed yearly, about 15 million tonnes of solid cassava wastes are generated in Nigeria and about 50 million tonnes in Africa annually. Cassava peel energy content is put at between

Table 15.1 Composition of cassava peel (Okoli, 2020; Nizzy & Kannan, 2022)

Constituent	Composition (%)
Hemicellulose	21–37.8
Cellulose	13.7–40.5
Lignin	7.5–21.0
Crude fiber	10.3–31.8
Crude protein	3.7–5.9
Ash	3.4–8.0
Neutral detergent fiber	15.8–35.5
Acid detergent fiber	15.5–18.6
Gross energy(MJ/kg)	16.4–17.7
Cyanide (mg/kg)	310–650
Tannin (mg/100 mg)	1.1–120
Oxalate (mg/kg)	200–330

Table 15.2 Composition of cassava wastewater

Constituents	Composition
pH	1.43–2.33
COD (mg/L)	1570–1571
BOD (mg/L)	546.34–577.48
TDS (mg/L)	510–890
Conductivity (μs/cm)	1090–1970
Total cyanide (g/L)	0.002–0.09
Potassium (g/L)	0.35–5.9
Calcium (g/L)	0.02–2.03
Phosphorus (g/L)	0.03–0.74

16.4 and 17.7 MJ/kg while its hemicellulose, cellulose, and lignin contents are 21–37.8%, 13.7–40.5%, and 7.5–21%, respectively (Okoli, 2020, Nizzy & Kannan, 2022). Table 15.1 compiles the major constituents of cassava peel.

Cassava wastewater is generated from the water used for the washing of the cassava tuber during processing. The liquid effluent generated during the dewatering process is also categorized as wastewater. The wastewater generated from cassava processing contains high concentrations of biological oxygen demand (BOD), chemical oxygen demand (COD), starch content, total dissolved solids (TDS), and total suspended solids. The BOD of cassava in most samples of wastewater is between 546.34 and 577.48 mg/L, the COD is about 1570 mg/L, while the pH ranges between 1.43 and 2.33 (Okpozu et al., 2019). The composition of nitrogen, potassium, magnesium, phosphorus, calcium, iron, and sodium in cassava wastewater from starch and flour production is reported to be 3.3, 5.9, 0.62, 0.74, 0.38, 0.11, and 0.74 g/L, respectively. The concentration of copper, manganese, and zinc are, however, about 0.02 g/L (de Oliveira Schmidt et al., 2023). Table 15.2 shows the outcome of the characterization of cassava wastewater.

15.2.3 Conversion Technologies for Cassava Wastes

Technologies for the conversion of cassava wastes can be broadly classified into two integrated processes, namely thermochemical and biochemical. The thermochemical processes consist of the use of thermal and chemical tools to degrade and convert cassava wastes into useful products and they include combustion, pyrolysis, and gasification or their combinations (Odeyemi et al., 2023). Direct combustion helps in converting cassava peels into heat, electricity, and other forms of energy. The combustion process usually takes place in stoves, boilers, ovens, turbo generators, and steam turbines. However, direct combustion of cassava peels has been discouraged due to its crudeness and the generation of carbon monoxide, smoke, and other toxic gases. Pyrolysis is considered one of the most commonly used thermochemical technologies for converting cassava peels and other lignocellulosic biomass into useful products. Pyrolysis whether slow, intermediate, fast, or flash can be achieved at temperatures between 300 and 1200 °C, a heating rate of 1–2 °C/s, and a residence time of 2 s to 300 s, depending on the type of pyrolysis adopted. When cassava peel is pyrolyzed, it yields biochar, bio-oil, and syngas (consisting mostly of hydrogen, methane, and some higher hydrocarbons) (Awogbemi & Kallon, 2023b). Gasification involves the application of heat and chemicals, usually called gasifying agents for the conversion of cassava peel and other carbonaceous materials into useful products. In gasification, the four major processes of drying, pyrolysis, combustion, and reduction occur almost simultaneously. During the process, the reactor, usually called the gasifier, is maintained at an elevated temperature above 700 °C in the presence of a selected gasifying agent. The gasification of cassava peels yields methane, hydrogen, acetylene, ethylene, and unwanted by-products. The high moisture content of cassava peels, the emission of toxic gasses, and the requirement of high process temperature are some of the major drawbacks of the thermochemical conversion of cassava peels (Awogbemi et al., 2023).

The biochemical conversion technologies involve the application of heat and biological means such as enzymes, biocatalysts, and chemicals to convert cassava peels into useful products. When compared with the thermochemical conversion technologies, the biochemical conversion technique is more eco-friendly, easily achievable, and requires a lower temperature (Akinyele et al., 2017). The two major biochemical conversion techniques for cassava peels are anaerobic digestion (AD) and fermentation. During the AD process, microorganisms decompose cassava peels in the absence of oxygen and convert the feedstock to a methane-rich gas called biogas. The AD process occurs in an airtight reactor called a digester and involves four different stages hydrolysis, acidogenesis, acetogenesis, and methanation. The production of biogas in an AD process can be boosted when two or more feedstock are mixed in the digester in a process called co-digestion (Ismail et al., 2022). Fermentation involves the microbial breaking down of complex molecules of cassava peels into simple intermediate molecules such as sugar and liquid biofuels and other gases as final products. The fermentation of cassava peels is activated by fungi, yeasts, and enzymes and is achieved at a moderate temperature of about 60 °C and a pH of 7 (Anigboro et al., 2023;r El Sheikha & Ray, 2023).

However, to ease the degradation process and enhance the digestibility of the solid cassava wastes, there is a need for some form of pretreatment procedures. Various physical, chemical, biological, and physicochemical techniques have been adopted to improve biodegradability and digestibility and enhance their conversion into the desired products (Awogbemi & Kallon, 2022).

15.3 Cassava Wastes and Their Products

One of the eco-friendly, cost-effective, and sustainable measures for the management of cassava peel, cassava wastewater, and other wastes generated during the consumption and processing of cassava.

15.3.1 Cassava Peel

The conversion of cassava peel into useful products has been demonstrated by researchers from diverse disciplines in recent years. Biochar is one of the easily achievable products from cassava peels. Biochar is black charcoal-like solid fuel produced from the thermochemical conversion of cassava peels in an oxygen-regulated environment. Cassava peels collected from farms, households, and cassava processing plants are converted to biochar through various thermochemical technologies including pyrolysis, and carbonization, and used for carbon sequestration, greenhouse gas (GHG) reduction, and enhancement of soil fertility (Hamissou et al., 2023, Egbosiuba, 2022, Odeyemi et al., 2023). Cassava peels are also converted to biogas by digesting the peels in an airtight biodigester through the AD process. The biogas produced through this process is used for cooking, heating, and electricity generation, while the spent substrates are used for horticultural purposes and to grow vegetables (Ismail et al., 2022, Aduba et al., 2023, Fagbenle & Olukanni, 2022). Cassava peels collected from various sources are also cleaned, dried, pulverized, sieved, and sometimes subjected to other thermal modification processes such as calcination and used as adsorbents for the removal of heavy metals, dyes, organic, and inorganic contaminants in wastewater. The biochar derived from cassava peels is effective in the treatment of wastewater from households, industries, the agricultural sector, and stormwater, and the water is used for irrigation, cleaning, and other purposes (Gajendiran et al., 2023, Phetrungnapha et al., 2023, Owamah, 2014). The conversion of cassava peels to bioethanol is one of the avenues of conversion of waste to fuel. During the process, cassava peels are cleaned and hydrolyzed to bioethanol by the action of microorganisms such as *Saccharomyces cerevisiae* (Aruwajoye et al., 2020, Baki et al., 2020, Mustafa et al., 2019). The conversion of cassava peels to biochar, biogas, and bioethanol demonstrates feasible strategies for effective waste disposal and management, supports waste to energy, and contributes to efforts at achieving increased penetration of renewable energy. Table 15.3 highlights the outcomes of recent research on the conversion of cassava peel into useful products.

Table 15.3 Conversion of cassava peels to useful products

Conversion technology/process parameters	Products	Remark/application	References
Slow pyrolysis at 400 °C for 3 h	Biochar	The product contains nutrients that can improve soil fertility and other agronomic applications	Hamissou et al. (2023)
Pyrolysis at 300–600 °C for 6 h	Biochar	Product demonstrates improved fuel properties	Egbosiuba (2022)
Carbonization at 338 °C for 160 min	Biochar	Conversion of CP for carbon sequestration, GHG reduction, and enhancement of soil fertility	Odeyemi et al. (2023)
AD in a 2 L reaction at 35 °C and HRT of 25 days	Biogas	High methane yield for heat, fuel, and electricity generation	Ismail et al. (2022)
AD in 5 L bioreactor for HRT of 15 days	Biogas	Production of bioelectricity, biogas, and bio-fertilizer	Aduba et al. (2023)
200 L biodigester for AD of CP for HRT 30 days	Biogas	Clean biogas for cooking and spent substrate as fertilizer	Fagbenle and Olukanni (2022)
AD in a fixed bed bioreactor	Biogas	High conversion rate of CP into quality biogas	Sarah and Husin (2023)
Grinding, and adsorption	Adsorbent	Removal of heavy metals, organic, and inorganic pollutants from wastewater	Gajendiran et al. (2023)
Grinding and calcination at 200 °C for 2 h	Adsorbent	Removal of FFA from WCO	Phetrungnapha et al. (2023)
Drying, pulverization, and carbonization/activation	Adsorbent	Removal of Pb(II) and Cu(II) from hospital wastewater	Owamah (2014)
Hydrolyzed with hydrochloric acid at 69.68 °C for 2.57 h	Bioethanol	Conversion of cassava peel to bioethanol contributes to increased penetration of bioenergy	Aruwajoye et al. (2020)
Hydrolyzed with Aspergillus niger at 69.68 °C for 2.57 h	Bioethanol	Effective conversion of CP to bioethanol	Baki et al. (2020)
Hydrolyzed with sulfuric acid	Bioethanol	Reduction in GHG emission and contribution to circular economy	Mustafa et al. (2019)

CP cassava peels, *AD* anaerobic digestion, *GHG* greenhouse gases, *HRT* hydraulic retention time, *FFA* free fatty acid, *WCO* waste cooking oil

15.3.2 Cassava Wastewater

The huge volume of wastewater generated during cassava processing needs to be properly managed to avert untoward environmental and human health crises. One of the effective management strategies often recommended is the conversion of such effluent to useful products. Investigations in this important research domain have progressed considerably in recent decades. Anaerobic digestion is one of the popular and most employed technologies for converting cassava wastewater to biogas and biohydrogen. The reported works of Jiraprasertwong et al. (2019), Watthier et al. (2019), and Andrade Cruz et al. (2022) on the production of biogas from wastewater generated from cassava processing is a testament to the feasibility of this process. The conversion of cassava wastewater supports the concept of a circular economy, offers techno-economic benefits, reduces environmental degradation, and contributes to renewable energy production from wastewater (Wattanasilp et al., 2021). Similarly, cassava wastewater has been successfully converted to biohydrogen for various applications. Previous studies have shown how cassava wastewater is effectively converted to biohydrogen, bioethanol, biomethane, biodiesel, etc. through fermentation and AD processes. The products generated from these processes are of high quality, meet international specifications, and represent feasible biological strategies for biofuel production (Mari et al., 2020; de Gois Araújo Tavares et al., 2023). Other products such as organic acids (da Silva et al., 2021), volatile fatty acids (Hasan et al., 2015), and carotenoids (Ribeiro et al., 2019) have been successfully produced from wastewater generated from cassava processing. These processes contribute to effective wastewater management, help in resource recovery, provide raw materials for industries, and have overall positive effects on socioeconomic and environmental well-being. Table 15.4 shows the summary of some of the outcomes of the conversion of cassava wastewater to useful products.

15.3.3 Other Cassava Wastes

Cassava pulp is the solid waste generated when producing starch from cassava. It contains about 50–60% starch content and impacts the environment if not properly disposed and managed. Cassava pulp has been effectively converted to biobutanol (Johnravindar et al., 2019), biohydrogen (Chantawan et al., 2022), bioethanol (Khanpanuek et al., 2022), poultry feed (Sugiharto, 2019), cattle feed (Dagaew et al., 2022), etc. by various researchers. The conversion of cassava pulp into these useful products demonstrates the contribution of cassava pulp to renewable energy, animal husbandry, and circular economy. Cassava bagasse is a fibrous material and one of the residues of cassava processing. It largely contains starch (40–60%), fiber (15–50%), and a small fraction of proteins and lipids. It is rich in organic nature and can therefore be easily digested, fermented, and converted to biofuel and other value-added products. As part of the contribution to waste management and circular economy strategies, researchers have successfully produced biomethane (Dupont et al., 2023), bioethanol and carotene (Pham et al., 2022), bioadsorbent (Diehl et al., 2023), animal feed (Shinya et al., 2022), and biodegradable food packaging composite (Weligama Thuppahige et al., 2023). More studies are ongoing on other avenues for cassava pulp and bagasse utilization. Table 15.5 summarizes products derived from cassava pulp and cassava bagasse.

Table 15.4 Conversion of cassava wastewater to useful products

Conversion technologies/process parameters	Product	Remark/application	References
Three-stage upflow anaerobic digester	Biogas	Effective methane production	Jiraprasertwong et al. (2019)
AD in a fixed bed reactor for 4 days	Biogas	COD removal of 99% and TS removal efficiencies of 86.2%	Watthier et al. (2019)
AD in a bioreactor for 21 days	Biogas	Effective conversion process with digestate used as manure	Andrade Cruz et al. (2022)
AD process in a digester	Biogas	The process presents technical, cost, and environmental benefits	Wattanasilp et al. (2021)
AD in an up-flow fixed-bed reactor	Biohydrogen	Ecofriendly and effective production process	Corbari et al. (2019)
Dark fermentation in a 2 L bioreactor at 105 °C for 3 h	Biohydrogen	Applicability of wastewater conversion to biohydrogen	Chantawan et al. (2022)
2-stage AD process in a 4.3 L batch biofilm reactor	Biohydrogen and biomethane	High-quality biohydrogen and biomethane production	Mari et al. (2020)
Anaerobic fluidized bed reactor at 65 °C for 2 h	Biohydrogen and bioethanol	High product yield at moderate temperature	de Gois Araújo Tavares et al. (2023)
2 L bubble column bioreactor	Biodiesel	Biodiesel complies with international specifications	Neves et al. (2016)
Fermentation in a batch reactor	Lactic acid, acetic acid, and propionic acid	High productivity and quality of the acids	da Silva et al. (2021)
AD and fermentation in a batch reactor at 30 °C for 2 days	Volatile fatty acids	Effective generation of acetic acid, butyric acid, and propionic acid	Hasan et al. (2015)
Fermentation	Carotenoids and fatty acids	Low-cost production of carotenoids and fatty acids	Ribeiro et al. (2019)

AD anaerobic digestion, *COD* chemical oxygen demand, *TS* total solids

15.4 Cassava Wastes Generation and Utilization in South Africa

South Africa is not among the cassava producers in the world. In the African continent where cassava is extremely popular and enjoys a wide following, South Africa is not mentioned among the producers of the important root tuber. Cassava production occurs in a few commercial farms of less than 5000 hectares of farmland in scattered and fragmented locations in South Africa. Though cassava is produced in parts of KwaZulu-Natal, Limpopo, and Mpumalanga where the total

Table 15.5 Conversion of other cassava wastes to useful products

Cassava waste	Conversion technologies	Product	Remark/application	References
Cassava pulp	AD and fermentation in a 5 L bioreactor	Biobutanol	8.01 g/L of butanol generated	Johnravindar et al. (2019)
	Dark fermentation	Biohydrogen	High yield of biohydrogen from cassava pulp	Chantawan et al. (2022)
	Simultaneous saccharification and fermentation	Bioethanol	High conversion efficiency and product yield	Khanpanuek et al. (2022)
	Solid-state fermentation	Poultry ration	Improvements in chicken growth and health	Sugiharto (2019)
	Fermentation and drying	Cattle feed	Increased digestibility and better cattle health	Dagaew et al. (2022)
Cassava bagasse	AD	Biomethane	High product yield	Dupont et al. (2023)
	Solid-state fermentation	Bioethanol and carotone	Biofuel and animal feed production	Pham et al. (2022)
	Fermentation	Bioabsorbent	Removal of methylene blue from contaminated aqueous medium	Diehl et al. (2023)
	Hot-pressing and film-forming	Food packaging films	Effective development of biodegradable food packaging materials	Weligama Thuppahige et al. (2023)
	Fermentation	Animal feed	Enrichment of animal feed	Shinya et al. (2022)

AD anaerobic digestion

arable land suitable for cassava cultivation is less than two million hectares (Adebayo, 2023; Manganyi et al., 2023), the total annual rainfall of less than 500 mm in these Provinces can hardly support effective large-scale cassava production. This has resulted in cassava cultivation being restricted to a few smallholder farmers. Other constraints of commercial cassava production in South Africa include inadequate arable land, low input farming system, the absence of high-yield varieties, and inclement weather conditions for cassava cultivation in most parts of the country (Marx & Nquma, 2013; Amelework et al., 2021; Otun et al., 2023). Due to these drawbacks, cassava and its products are not considered a popular staple food in South Africa and therefore not commercially cultivated. This is not to say that cassava is not being consumed or utilized for various applications in South Africa. Available records show that cassava is only cultivated on a small scale and traded informally and in small quantities in South Africa as a minor crop.

Currently, there is a dearth of reliable data on the production of cassava and cassava waste in South Africa. While the potential of cassava to contribute to food security, reduce hunger and poverty, and use it as feedstock for industrial processes, biofuel, and bioenergy production, etc., is recognized, little attention is still given to the development of the sector. In order to develop and promote cassava production and utilization in South Africa, The National Industrial Policy, in conjunction with the Agricultural Research Council, in its bid to promote exportation and reduce the importation of traditional crop commodities in the country, has intensified research and development, awareness campaigns, and other strategies to popularize the cultivation, consumption, and other utilization avenues of cassava in South Africa (Amelework et al., 2021; Manganyi et al., 2023).

Cassava waste such as peels, bagasse, leaves, stems, pulp, and rejected tubers has been used as animal feeds, manures, adhesive, paper, and textile grade starches, and feedstock for biofuels production. Keeping in mind that South Africa derived about 90% of its primary energy from fossil-based sources and fossil-based energy sources is a major contributor to the global greenhouse gases particularly from transportation engines. The desire to reduce the reliance on coal for electricity to about 31% by 2050, as specified by the Integrated Resource Plan, can only be realized by the diversification of the energy sources to renewable. Production of biogas, bioethanol, biomethane, biohydrogen, biodiesel, and bioenergy from cassava wastes remains a feasible strategy and a step toward energy security and environmental sustainability (Mkruqulwa et al., 2017; Amelework et al., 2021). The development of cassava waste biorefineries has led to the conversion of cassava waste into bioplastics and other biopolymer products, organic acid, adsorbents, and other value-added products. This has paved the way for sustainable economic development, strategic waste management, job creation, poverty reduction, and improved industrial activity in South Africa using the circular economy principles (Hierro-Iglesias et al., 2022).

There is an urgent need to evolve sustainable agricultural practices and innovative research that will develop the cassava production sector in South Africa. Such a system led to the development of high-yielding, disease-resistant, and drought-resistant cassava varieties. Sustainable agricultural practices such as irrigation, hydroponics and aquaponics, crop rotation, biodynamic agriculture, and integrated pest management should be adopted and used in cassava production. Cassava growers must be assisted with tools, knowledge, finance, farm inputs, machineries, and other incentives to ease their work. Public investment in cassava production, processing, and utilization will play a vital role in product development, food security, and job creation, and reduce malnutrition, insecurity, and urban migration. Policymakers, researchers, and other stakeholders must collaborate to enact, implement, and execute policies and interventions that address all concerns in the cassava production and utilization value chain.

15.5 Implications, Challenges, and Future Prospects

The huge volume of solid resides and wastewater generated during cassava processing needs to be appropriately managed to preserve humanity and the ecosystem. If well managed, waste generated during cassava processing offers feasible opportunities for effective sanitation, resource recovery, pollution prevention, biofuel production, and generation of manure for growing vegetables. Farmers can also earn extra income by selling the wastes generated from cassava farms thereby boosting their socioeconomic status. Effective management of residues from cassava processing contributes to achieving a pollution-free environment, reduces the amount of waste at dumpsites, and provides eco-friendly raw materials for adsorbents, nanomaterials, and fertilizers.

However, the acidic nature of the cassava waste particularly the high cyanide content of cassava peels and wastewater remains a potent drawback in their handling and conversion. The cyanide released during cassava processing ends up in the wastewater, while the dust and bad odor pollute the atmospheric air. Currently, there is no adequate infrastructure such as appropriate protective wear, waste collection structures, and cassava waste recycling to make the collection, sorting, transportation, and recycling of wastes generated during cassava processing hazardless.

There is a need to intensify research and development efforts to improve the conversion of residues and wastewater from cassava processing. Novel, cost-effective, and low-energy consumption pretreatment techniques should be developed to aid the digestibility and conversion of solid cassava residues. The use of nanocatalysts and other additives that can significantly promote the conversion of cassava peels and other cassava processing wastes into biofuels and other chemicals should be intensified. Molecular techniques should be developed to breed disease-resistant, high-yielding, and low-cyanide-content cassava variants to reduce the environmental impact of cassava processing.

15.6 Conclusion

Cassava is a multipurpose root crop that is processed and consumed by huge global population, despite its highly acidic and toxic nature. The enormous cassava peels generated can be treated and converted to biochar, biogas, bioethanol, and bioadsorbents, while the wastewater is a low-cost raw material for the production of bioethanol, biobutanol, biohydrogen, organic acids, carotenoid, and fatty acids. Other cassava residues such as cassava pulp and cassava bagasse are converted to biobutanol, bioethanol, biohydrogen, poultry and animal feed, bioadsorbent, and biodegradable packaging materials. The current study has brought to the fore the enormous potential inherent in cassava wastes and wastewater if properly handled. Going forward, more multidisciplinary studies are needed to come up with low-cost and eco-friendly conversion technologies for cassava waste. The scientific community should intensify efforts at developing cassava variants with less acidity and low cyanide content.

References

Adebayo, W. G. (2023). Cassava production in Africa: A panel analysis of the drivers and trends. *Heliyon, 9*, e19939.

Aduba, C. C., Ndukwe, J. K., Onyejiaka, C. K., Onyeiwu, S. C., & Moneke, A. N. (2023). Integrated valorization of cassava wastes for production of bioelectricity, biogas and biofertilizer. *Waste and Biomass Valorization, 14*, 4003–4019.

Akinyele, B., Olaniyi, O., & Jeff-Agboola, Y. (2017). Effect of fermentation on chemical composition of cassava peels. *Asian Journal of Plant Science and Research, 7*, 31–38.

Amelework, A. B., Bairu, M. W., Maema, O., Venter, S. L., & Laing, M. (2021). Adoption and promotion of resilient crops for climate risk mitigation and import substitution: A case analysis of cassava for South African agriculture. *Frontiers in Sustainable Food Systems, 5*, 617783.

Andrade Cruz, I., Andrade, L. R. S., Jesus, A. A. D., Vasconcelos, B. R. D., Bharagava, R. N., Bilal, M., Figueiredo, R. T., Souza, R. L. D., & Romanholo Ferreira, L. F. (2022). Potential of eggshell waste derived calcium for sustainable production of biogas from cassava wastewater. *Journal of Environmental Management, 321*, 116000.

Anigboro, A. A., Ajoh, A. I., Avwioroko, O. J., Ehwarieme, D. A., & Tonukari, N. J. (2023). Solid-state fermentation of cassava (*Manihot esculenta*) peels using *Rhizopus oligosporus:* Application of the fermented peels in yeast production and characterization of α-amylase enzyme produced in the process. *Chemistry Africa, 6*, 1669–1678.

Aruwajoye, G. S., Sewsynker-Sukai, Y., & Kana, E. B. G. (2020). Valorisation of cassava peels through simultaneous saccharification and ethanol production: Effect of prehydrolysis time, kinetic assessment and preliminary scale up. *Fuel, 278*, 118351.

Awogbemi, O., & Kallon, D. V. V. (2022). Pretreatment techniques for agricultural waste. *Case Studies in Chemical and Environmental Engineering, 6*, 100229.

Awogbemi, O., & Kallon, D. V. V. (2023a). Application of biochar derived from crop residues for biofuel production. *Fuel Communications, 15*, 100088.

Awogbemi, O., & Kallon, D. V. V. (2023b). Progress in agricultural waste derived biochar as adsorbents for wastewater treatment. *Applied Surface Science Advances, 18*, 100518.

Awogbemi, O., Ojo, A. A., & Adeleye, S. A. (2023). Advanced thermochemical conversion approaches for green hydrogen production from crop residues. *Journal of Renewable Materials*, 1–25.

Baki, A., Bande, Y., & Bello, A. (2020). Comparative studies on bioethanol production from cassava peels using *Saccharomyces cerevisiae* and *Zymomonas mobilis*. *Open Access Journal of Biomedical Engineering and Biosciences, 3*, 373–378.

Bancal, V., & Ray, R. C. (2022). Overview of food loss and waste in fruits and vegetables: From issue to resources. In R. C. Ray (Ed.), *Fruits and vegetable wastes: Valorization to bioproducts and platform chemicals* (pp. 3–29). Springer.

Chantawan, N., Moungprayoon, A., Lunprom, S., Reungsang, A., & Salakkam, A. (2022). High-solid dark fermentation of cassava pulp and cassava processing wastewater for hydrogen production. *International Journal of Hydrogen Energy, 47*, 40672–40682.

Corbari, S. D. M. L., Andreani, C. L., Torres, D. G. B., Eng, F., & Gomes, S. D. (2019). Strategies to improve the biohydrogen production from cassava wastewater in fixed-bed reactors. *International Journal of Hydrogen Energy, 44*, 17214–17223.

Da Silva, D. B., Fernandes, B. S., & Da Silva, A. J. (2021). Effect of initial pH and substrate concentration on the lactic acid production from cassava wastewater fermentation by an enriched culture of acidogenic microorganisms. *Water Environment Research, 93*, 1925–1933.

Dagaew, G., Wongtangtintharn, S., Suntara, C., Prachumchai, R., Wanapat, M., & Cherdthong, A. (2022). Feed utilization efficiency and ruminal metabolites in beef cattle fed with cassava pulp fermented yeast waste replacement soybean meal. *Scientific Reports, 12*, 16090.

de Gois Araújo Tavares, T., Peiter, F. S., Chaves, T. C., Gois, G. N. S. B., & de Amorim, E. L. C. (2023). Effect of thermophilic temperatures on hydrogen and ethanol production in anaerobic fluidized bed reactor from cassava wastewater. *Brazilian Journal of Chemical Engineering, 40*, 115–127.

De Oliveira Schmidt, V. K., De Vasconscelos, G. M. D., Vicente, R., De Souza Carvalho, J., Della-Flora, I. K., Degang, L., De Oliveira, D., & De Andrade, C. J. (2023). Cassava wastewater valorization for the production of biosurfactants: Surfactin, rhamnolipids, and mannosileritritol lipids. *World Journal of Microbiology and Biotechnology, 39*, 65.

Dey, G., & Ray, R. C. (2020). Biotransformations of tuber crops ingredients with lactic acid bacteria: Focus on functional foods and disease prevention. In K. Shetty & D. Sarkar (Eds.), *Functional foods and biotechnology: Biotransformations and analysis of functional foods and ingredients* (pp. 41–56). CRC Press.

Diehl, M., Silva, L. F., Schnorr, C., Netto, M. S., Bruckmann, F. S., & Dotto, G. L. (2023). Cassava bagasse as an alternative biosorbent to uptake methylene blue environmental pollutant from water. *Environmental Science and Pollution Research, 30*, 51920–51931.

Dupont, G. K., Oliveira, M. M., Clerici, N. J., Serafini, C. G., Daroit, D. J., & Wenzel, B. M. (2023). Kinetic modelling and improvement of methane production from the anaerobic co-digestion of swine manure and cassava bagasse. *Biomass and Bioenergy, 176*, 106900.

Egbosiuba, T. C. (2022). Biochar and bio-oil fuel properties from nickel nanoparticles assisted pyrolysis of cassava peel. *Heliyon, 8*, e10114.

Ekop, I., Simonyan, K., & Evwierhoma, E. (2019). Utilization of cassava wastes for value added products: An overview. *International Journal of Scientific Engineering and Science, 3*, 31–39.

El Sheikha, A. F., & Ray, R. C. (2023). Bioprocessing of horticultural wastes by solid-state fermentation into value-added/innovative bioproducts: A review. *Food Reviews International, 39*, 3009–3065.

Fagbenle, E., & Olukanni, D. (2022). Production and purification of biogas from cassava peel using cow dung as inoculum. In *IOP conference series: Earth and environmental science, 2022 center for research, innovation and discovery* (p. 012012). Covenant University, Nigeria. IOP Publishing.

FAO. (2001). *Strategic environmental assessment. an assessment of the impact of cassava production and processing on the environment and biodiversity*. Available online https://www.fao.org/3/y2413e/y2413e0d.htm.

FAO. (2023). *Cassava production, 1997 to 2021*. Available online https://ourworldindata.org/grapher/cassava-production?tab=chart&time=1997..latest&country

Gajendiran, V., Deivasigamani, P., Sivamani, S., & Sivakumar, P. M. (2023). A review on cassava residues as adsorbents for removal of organic and inorganic contaminants in water and wastewater. *Journal of Chemistry, 2023*, 7891518.

Hamissou, I. G. M., Appiah, K. E. K., Sylvie, K. A. T., Ousmaila, S. M., Casimir, B. Y., & Benjamin, Y. K. (2023). Valorization of cassava peelings into biochar: Physical and chemical characterizations of biochar prepared for agricultural purposes. *Scientific African, 20*, e01737.

Hasan, S. D. M., Giongo, C., Fiorese, M. L., Gomes, S. D., Ferrari, T. C., & Savoldi, T. E. (2015). Volatile fatty acids production from anaerobic treatment of cassava waste water: Effect of temperature and alkalinity. *Environmental Technology, 36*, 2637–2646.

Hierro-Iglesias, C., Chimphango, A., Thornley, P., & Fernández-Castané, A. (2022). Opportunities for the development of cassava waste biorefineries for the production of polyhydroxyalkanoates in Sub-Saharan Africa. *Biomass and Bioenergy, 166*, 106600.

Ikuemonisan, E. S., Mafimisebi, T. E., Ajibefun, I., & Adenegan, K. (2020). Cassava production in Nigeria: Trends, instability and decomposition analysis (1970–2018). *Heliyon, 6*, e05089.

IMARC Group. (2023). *Cassava processing market report*. Available online https://www.imarcgroup.com/cassava-processing-plant

Ismail, N., Fauzi, N., Salehabadi, A., Latif, S., Awiszus, S., & Müller, J. (2022). A study on biogas production from cassava peel and stem in anaerobic digestion reactor. *International journal of Environmental Science and Technology, 19*, 1695–1704.

Jiraprasertwong, A., Maitriwong, K., & Chavadej, S. (2019). Production of biogas from cassava wastewater using a three-stage upflow anaerobic sludge blanket (UASB) reactor. *Renewable Energy, 130*, 191–205.

Johnravindar, D., Elangovan, N., Gopal, N. O., Muthaiyan, A., & Fei, Q. (2019). Biobutanol production from cassava waste residue using *Clostridium* sp. AS3 in batch culture fermentation. *Biofuels*, 1259–1266.

Khanpanuek, S., Lunprom, S., Reungsang, A., & Salakkam, A. (2022). Repeated-batch simultaneous saccharification and fermentation of cassava pulp for ethanol production using amylases and *Saccharomyces cerevisiae* immobilized on bacterial cellulose. *Biochemical Engineering Journal, 177*, 108258.

Kumar, V., Al-Gheethi, A., Asharuddin, S. M., & Othman, N. (2021). Potential of cassava peels as a sustainable coagulant aid for institutional wastewater treatment: Characterisation, optimisation and techno-economic analysis. *Chemical Engineering Journal, 420*, 127642.

Manganyi, B., Lubinga, M. H., Zondo, B., & Tempia, N. (2023). Factors influencing cassava sales and income generation among cassava producers in South Africa. *Sustainability, 15*(19), 14366.

Mari, A. G., Andreani, C. L., Tonello, T. U., Leite, L. C. C., Fernandes, J. R., Lopes, D. D., Rodrigues, J. A. D., & Gomes, S. D. (2020). Biohydrogen and biomethane production from cassava wastewater in a two-stage anaerobic sequencing batch biofilm reactor. *International Journal of Hydrogen Energy, 45*, 5165–5174.

Marx, S., & Nquma, T. Y. (2013). Cassava as feedstock for ethanol production in South Africa. *African Journal of Biotechnology, 12*(31), 4975–4983.

Mkruqulwa, U. L., Okudoh, V. I., & Oyekola, O. O. (2017). Biomethane potential from co-digestion of cassava and winery waste in South Africa. In *9th international conference on advances in science, engineering, technology and waste management (ASETWM-17)*, November 27–28, 2017 Parys, South Africa.

Mustafa, H. M., Salihu, D., Bashir, A., & Ibrahim, A. (2019). Bio-ethanol production from cassava (*Manihot esculenta*) waste peels using acid hydrolysis and fermentation process. *Science World Journal, 14*, 45–50.

Neves, C., Maroneze, M. M., Santos, A. M. D., Francisco, E. C., Wagner, R., Zepka, L. Q., & Jacob-Lopes, E. (2016). Cassava processing wastewater as a platform for third generation biodiesel production. *Scientia Agricola, 73*, 412–416.

Nizzy, A. M., & Kannan, S. (2022). A review on the conversion of cassava wastes into value-added products towards a sustainable environment. *Environmental Science and Pollution Research, 29*, 69223–69240.

Odeyemi, S. O., Iwuozor, K. O., Emenike, E. C., Odeyemi, O. T., & Adeniyi, A. G. (2023). Valorization of waste cassava peel into biochar: An alternative to electrically-powered process. *Total Environment Research Themes, 6*, 100029.

Oghenejoboh, K. M., Orugba, H. O., Oghenejoboh, U. M., & Agarry, S. E. (2021). Value added cassava waste management and environmental sustainability in Nigeria: A review. *Environmental Challenges, 4*, 100127.

Okoli, I. C. (2020). *Cassava: Cassava peel production, processing, and nutrient composition* (pp. 1–7). Research Tropica.

Okpozu, O. O., Ogbonna, I. O., Ikwebe, J., & Ogbonna, J. C. (2019). Phycoremediation of cassava wastewater by *Desmodesmus armatus* and the concomitant accumulation of lipids for biodiesel production. *Bioresource Technology Reports, 7*, 100255.

Otun, S., Escrich, A., Achilonu, I., Rauwane, M., Lerma-Escalera, J. A., Morones-Ramírez, J. R., & Rios-Solis, L. (2023). The future of cassava in the era of biotechnology in Southern Africa. *Critical Reviews in Biotechnology, 43*(4), 594–612.

Owamah, H. (2014). Biosorptive removal of Pb (II) and Cu (II) from wastewater using activated carbon from cassava peels. *Journal of Material Cycles and Waste Management, 16*, 347–358.

Padi, R. K., & Chimphango, A. (2021). Comparative sustainability assessments for integrated cassava starch wastes biorefineries. *Journal of Cleaner Production, 290*, 125171.

Pham, T. A., Tran, L. N., Dam, T. H., & To, K. A. (2022). Valorization of cassava bagasse using co-culture of *Aspergillus oryzae* VS1 and *Sporidiobolus pararoseus* O1 in solid-state fermentation. *Waste and Biomass Valorization, 13*, 3003–3012.

Phetrungnapha, A., Wiengnak, N., & Maikrang, K. (2023). Removal of free fatty acid from waste cooking oil using an adsorbent derived from cassava peels. *Korean Journal of Chemical Engineering, 40*, 2253–2262.

Ray, R. C., & Shivkumar, P. S. (2009). Traditional and novel fermented foods and beverages from tropical root and tuber crops: Review. *International Journal of Food Science and Technology, 44*, 1073–1087.

Ray, R. C., & Swain, M. R. (2011). Bio-ethanol, bioplastics and other fermented industrial products from cassava starch and flour. In C. M. Pace (Ed.), *Cassava: Farming, uses and economic impacts* (pp. 1–32). Nova Science Publishers.

Ray, R. C., Bari, M. L., & Isobe, S. (2010). Agro–industrial bio-processing of tropical root and tuber crops: Current research and future prospects. In D. K. Maheshwari, R. C. Dubey, & R. S. Saravanamuthu (Eds.), *Industrial exploitation of microorganisms* (pp. 352–374). I.K. International Publishers House (P) Ltd.

Ribeiro, J. E. S., Sant'ana, A. M. D. S., Martini, M., Sorce, C., Andreucci, A., Melo, D. J. N. D., & Silva, F. L. H. D. (2019). Rhodotorula glutinis cultivation on cassava wastewater for carotenoids and fatty acids generation. *Biocatalysis and Agricultural Biotechnology, 22*, 101419.

Sarah, M., & Husin, A. (2023). Biogas production from cassava peels utilizing fixed bed bioreactor with silica gel media. In *Journal of physics: Conference series, 2023 2nd TALENTA-International conference on science and technology* 02/12/2020–04/12/2020 Sumatera Utara, Indonesia. IOP Publishing, 012020.

Saranraj, P., Behera, S. S., & Ray, R. C. (2019). Traditional foods from tropical root and tuber crops: Innovations and challenges. In C. M. Galanakis (Ed.), *Innovations in traditional foods* (pp. 159–191). Woodhead Publishing.

Shinya, T. Y., Pereira Elsner, V. H., De Lima Júnior, D. S., De Barros Ranke, F. F., Escaramboni, B., Da Paixão, G., Melo, W., Fernández Núñez, E. G., & De Oliva Neto, P. (2022). Bioprocess development with special yeasts for cassava bagasse enrichment nutritional to use in animal feed. *Animal Feed Science and Technology, 290*, 115338.

Sigüenza-Andrés, T., Gallego, C., & Gómez, M. (2021). Can cassava improve the quality of gluten free breads? *LWT, 149*, 111923.

Sugiharto, S. (2019). A review on fungal fermented cassava pulp as a cheap alternative feedstuff in poultry ration. *Journal of World's Poultry Research, 9*, 1–6.

Veiga, J. P. S., Valle, T. L., Feltran, J. C., & Bizzo, W. A. (2016). Characterization and productivity of cassava waste and its use as an energy source. *Renewable Energy, 93*, 691–699.

Wattanasilp, C., Songprakorp, R., Nopharatana, A., & Khompatraporn, C. (2021). Techno-cost-benefit analysis of biogas production from industrial cassava starch wastewater in Thailand for optimal utilization with energy storage. *Energies, 14*, 416.

Watthier, E., Andreani, C. L., Torres, D. G., Kuczman, O., Tavares, M. H., Lopes, D. D., & Gomes, S. D. (2019). Cassava wastewater treatment in fixed-bed reactors: Organic matter removal and biogas production. *Frontiers in Sustainable Food Systems, 3*, 1–6.

Weligama Thuppahige, V. T., Moghaddam, L., Welsh, Z. G., & Karim, A. (2023). Investigation of morphological, chemical, and thermal properties of biodegradable food packaging films synthesised by direct utilisation of cassava (*Manihot esculanta*) bagasse. *Polymers, 15*, 767.

Part IV
Public Policy and Circular Economy

Chapter 16
Challenges and Future Perspectives in Agriculture Waste Valorization: Sugarcane Trash as a Case Study

Noé Aguilar-Rivera

Abbreviations

CE	Circular economy
GHG	Greenhouse gases
SMS	Spent mushroom substrate

16.1 Introduction

Agriculture is the most important anthropogenic activity for food, livestock, and industrialization of a growing world population. However, agriculture is a major generator of greenhouse gas (GHG) emissions, and emits large amounts of so-called "non-CO_2 gases" including N_2O and CH_4 with a higher warming power, respectively, compared to CO_2. Otherwise, agroecosystems also have great GHG mitigation potential when agroecological management practices are implemented, nature-based solutions are applied, crop residues are conserved as cover crops, tillage is reduced and cover, rotation or diversification crops can be introduced to improve sustainability or are used as raw materials for conversion into bioproducts, closing the cycle of the circular economy (CE) and reaching zero-wastes (Yang et al., 2024).

Nevertheless, the sustainability of agriculture and the socioeconomic activities related to this activity, in a country or region depends fundamentally on the adequate, rational, and productive management of the agricultural, livestock, forestry lands and waste on which these activities are being carried out, with particular

N. Aguilar-Rivera (✉)
Facultad de Ciencias Biológicas y Agropecuarias, Universidad Veracruzana, Córdoba, Veracruz, Mexico
e-mail: naguilar@uv.mx

J. Arora et al. (eds.), *Transforming Agriculture Residues for Sustainable Development*, Waste as a Resource, https://doi.org/10.1007/978-3-031-61133-9_16

importance to the formulation and application of public policies on a scientific basis, for decision-makers and stakeholders be leaders in achieving sustainability, profitability and the welfare of the population (Steponavičienė et al., 2023).

Kannan et al. (2024), Bandh et al. (2023), and Phiri et al. (2023) concluded that collaterally agricultural activities are generators of by-products and residues of organic, perishable nature and in large quantities (agro-waste), but with a wide potential to be industrialized due to their availability, heterogeneity, and the growing need to reduce the consumption of non-renewable sources of energy and raw materials and products derived from mining, oil, gas, coal, etc., which are continuously impacting ecosystems and the world's population. These materials, considered waste, derived from agriculture, potentially through sustainable technologies based on CE, Industry 4.0 and Smart waste management 4.0, and others can be valorized and converted into the main option for the survival of future generations, avoiding their confinement in landfills or deposition in water sources (rivers, lakes, ocean) or their burning in agricultural fields generating higher GHG emissions.

Concerning the aforementioned, agricultural residues (plants, branches, chips, leaves and others of forest origin, lignocellulosic biomass, fruits, and vegetables) generated from harvesting in agricultural fields, transportation, industrialization, or wholesale or retail sale and even in the homes of the population, it is possible through strategic management plans first to reduce their generation or later for treatment, with available efficient and sustainable technologies, to convert these wastes into high-value goods for the production of food, animal feed, solid and liquid biofuels, chemicals, bioproducts, bioplastics, biomaterials, pharmaceuticals with numerous and varied applications that will contribute significantly to regional sustainable development (Kumar Sarangi et al., 2023; Wang & Tester, 2023; Roy, 2023; Ginni et al., 2021).

In other words, the purposes of public agricultural policies for agricultural waste should be:

1. Reducing waste production by improving technological and management processes.
2. Ensuring that wastes are properly disposed of when it is technologically not possible to convert them into other goods.
3. Recycle, reutilize, and reuse to give them added value as the establishment of biorefineries based on a CE (Tanveer et al., 2022).

For instance, biorefineries are multiproduct systems that utilize green biomass and waste from different sources in agricultural and agro-industrial activities and fruits and vegetable waste as surplus renewable resources and with broad potential such as feedstock for industrial production of bioproducts as biohydrogen, biofertilizers, bricks, biochar, bio-plastics, paper, biofuels, industrial enzymes, organic acids, etc., to decouple environmental pressures from economic growth currently obtained from non-renewable sources as oil, gas, carbon, and minerals (Karp et al., 2022; Ray, 2022).

16.2 Sugarcane Trash as an Agricultural Waste: Scope for Valorization

The world's production of cane sugar is derived from one of the main crops for geographical extension and is also one of the largest generators of agricultural residues. Its great socioeconomic importance derives from the fact that it meets one of the basic dietary needs of the population as a simple carbohydrate (sucrose), as well as sugarcane by-products and their processing as inputs for industries that manufacture food, soft drinks, juices and confectionery in general, paper and cardboard pulp, chipboard, animal feed, bioethanol, bioelectricity by cogeneration, etc., contributing to the industrial development of cane sugar-producing countries.

Nonetheless, sugar cane has been cultivated with monoculture-based green revolution technology packages, and unsustainable management practices such as intensive use of agrochemicals, harvesting by burning, and the low level of utilisation of processing by-products (trash, bagasse, filter sludge, molasses and vinasse) have caused particularly severe impacts on the environment and the population (Friedl et al., 2023).

This is derived from the fact that sugarcane has been conventionally harvested manually, and burning the leaves and tops before harvesting has been a common practice to improve crop productivity after cutting at harvest; it is a widespread practice for the elimination of vegetative cover and weed control that facilitates soil preparation and replanting with reduced harvesting costs, facilitates transportation, and protects field workers from poisonous animals. However, soil humidity is reduced and at the same time neighboring ecosystems are affected, organic matter is destroyed, and soil structure is lost with increased drying and erosion. These effects, together with soil compaction caused by the passage of machinery, are the main causes of the decrease in productivity and profitability in sugarcane cultivation. (Srivastava, 2020).

Therefore, improvements in sugarcane crop production, as part of environmental conservation programs and a new approach to industrial production based on the CE, aimed at the sustainability of each subsystem in the medium and long term, are the most important challenges and bottlenecks for the world sugar industry. The challenge arises of identifying diversified, integrated, and sustainable industrial systems that generate higher productivity, based on the use of own resources and production residues that, without causing the transformation of ecosystems, allow to meet the transformation needs of the traditional sugar industry in crisis (Singh et al., 2021; Meghana & Shastri, 2020).

Concerning the above, Powar et al. (2022) commented that the sugarcane harvest residue (trash) is the remainder of the sugarcane stalks, leaves, tops, or flowers after the sugarcane has been harvested for subsequent transport to the mill or biorefinery. It is a fibrous by-product formed mainly by leaves, abundant and available since it generates approximately 25–30% of the total weight of the productivity of a field of cane stalk harvest in green cut (approximately 8–10 tons of trash per ha), so the higher the productivity or field yield (t ha-1), the generation of this waste will be greater. Trash has a high content of structural carbohydrates, lignin, and minerals, which makes it a raw material with a chemical composition very similar to wood and bagasse and constitutes a lignocellulosic material that could be used to manufacture various value-added derivatives with broad application possibilities in the areas of soil protection, biofertilizers, food production, animal feed, biofuels, fiber, construction materials, textiles, nanomaterials, solid fermentation products, etc.

However, this lignocellulosic residue with abundant availability and viable technological routes for its conversion into bioproducts is burned in its entirety during the harvesting of sugarcane stalks, emitting greenhouse gases, particulate matter and dioxins that are highly harmful to the environment, agricultural workers and the populations near the sugarcane fields (Martins et al., 2023).

The amount of trash depends on factors such as type of varieties, crop management, environmental and agroecological characteristics of the region or production areas, and age of the cane, among many others, and these residues consist of leaves, straw, buds, tops, etc. In green harvest or raw cane, about 20–25% is residue. In burned cane, about 16% is residue. The most sustainable potential uses for this biomass are:

1. Incorporate them in sugarcane soils to improve their organic matter content considering the amount of trash, available surface, and the technological options to accelerate their mineralization in the soil.
2. Packaging them for their use as livestock feed (preparation of rations and silage).
3. Extract them from the field for energy generation (use of biomass) in cogeneration systems.

Purchase et al. (2008) concluded that there are some factors to consider for this purpose

1. Relative costs of manual harvesting of burnt and green cane.
2. Extended delays and increased rates of cane deterioration due to burning.
3. Long-term yield decline.
4. Nutrient recycling.
5. Herbicide cost savings due to weed suppression by trash blankets.
6. Water savings and associated yield increases, or irrigation cost savings, due to trash blankets.
7. Loading and transport costs as affected by additional trash on green cane.
8. Milling costs and recoveries as affected by extraneous matter and cane deterioration.

The poor management of this waste and a combination of various factors, including political factors, has been responsible for the physical, chemical, and biological degradation of soil including the growth of the cane crop, reduction in productivity and profits as a result of continuous monoculture, the extensive use of inorganic fertilizers, insecticides, and herbicides and several constraints (Freitas et al., 2021; Ortiz Laurel et al., 2012). Furthermore, due to its low density, it is necessary to handle large volumes of trash with complex logistics, constituting an obstacle to the generalization of its use.

Nevertheless, it is an attractive feedstock for application in many fields, such as bioenergy, biofuels, and composites due to its composition, abundant availability, energetic capacity, and relatively low cost. Its utilization as feedstock could increase the economic value of products and process sustainability, preventing the accumulation of this residue uncontrolled in the field (Aguiar et al., 2021).

16.3 Cane Crop Field Burning

Currently, according to Lima and Beacorn (2022), there are constraints that are affecting the transition to a green harvesting system in most sugar cane producing countries:

1. The culture and tradition of pre-harvest and post-harvest burning of the sugarcane field.
2. The cost of production of green cane & burned cane without considering the deterioration of the quality of the cane stalks during transportation to the sugar mill.
3. The low level of Research+Development+Industry+Innovation+ Technology Transfer (R + D + I + I + TT), related to beneficial agroecological practices for the use of trash in the management of soil cultivated with sugarcane.
4. The lack of coordination between the livestock sector and sugarcane growers for the management of trash (green and brown leaves and tops), as survival livestock feed in times of drought, in addition to the cost of machinery for the packaging of this waste and its transportation.
5. The scarce use of trash in sugar mills to produce bioelectricity and steam by cogeneration.
6. The low diffusion of simple technologies for the use of trash by the sugarcane grower.
7. The grower's perception of low level of profitability of trash management due to the cane market, technological impedance, harvesting difficulties, and inadequate extension activities by sugar mill technicians, university researchers, unions, or the government.
8. The absence of public policies or their ambiguity without a prospective plan for the reduction of harvesting with burning and the sustainable management of trash to obtain higher value-added products.

Therefore, future-oriented strategies are required to promote the development of various technological levels for the management of trash and its use as a raw material in bioprocesses (Aquino et al., 2019).

In relation to productive activities that the sugarcane growers could carry out on their farms to contribute to regional rural development and their well-being, it is possible to mention mulching with trash, which brings direct benefits such as the reduction in soil moisture loss by evaporation and minimizes drought and soil

Fig. 16.1 Use of trash as mulch in sugarcane-producing soils. (Photograph taken in Veracruz, Mexico)

degradation, reduction in pests and weeds and, in the medium term, the incorporation of organic carbon by the degradation of the trash. However, this agroecological practice is not yet accepted by most of the growers (Fig. 16.1).

Besides, Shanthy et al. (2020) and Dietrich et al. (2019) mentioned the advantages of mulch:

1. Soil protection, water dynamics, and nutrient fluxes.
2. Conservation of soil moisture.
3. Overcome drought situation to a greater extent.
4. Restoration of soil fertility.
5. Improvement in humus content of the soil.
6. Increased cane yield.
7. Smothering of weeds.
8. Improvement in soil pH.
9. Reduces the soil EC.
10. Improves soil aeration.
11. Better soil structure.
12. Helps to maintain microclimate in the fields.
13. Enhances soil microflora and fauna.
14. Carbon pool.

However, the amount of trash that can be removed without prejudicing the sustainability of the sugarcane production system has still been little studied. The removal of part of the trash adversely affected some of the soil quality attributes. This suggests the need for monitoring over the years in sugarcane areas where part of the trash is removed for industrial purposes, to understand the cumulative impact on soil quality better and to ensure the sustainability of bioenergy production (Vieira et al., 2021; Silva et al., 2019).

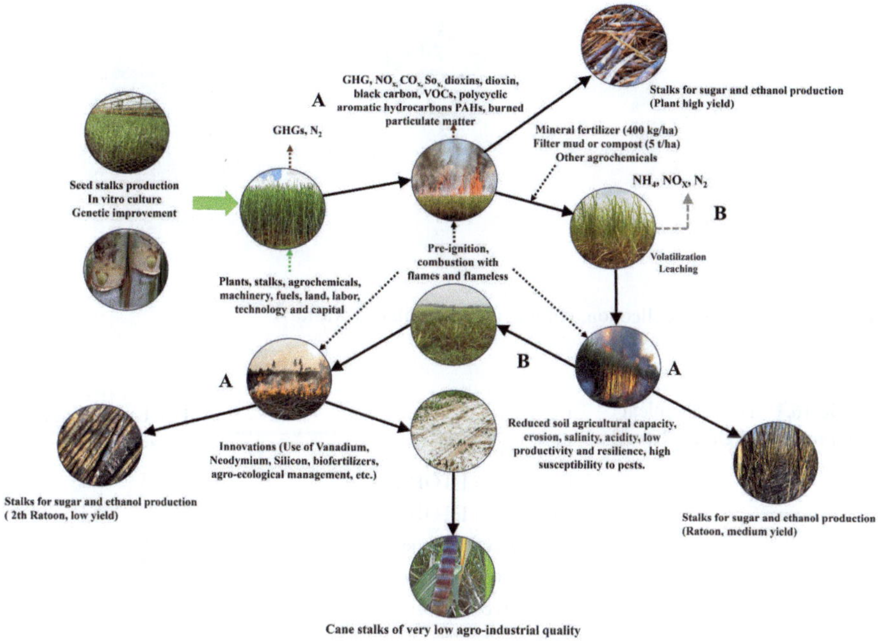

Fig. 16.2 Cycle of cane cultivation and burning harvest (**a**) generation of dioxins and (**b**) N volatilization

Therefore, the options available for the use of trash in the industrial sector are varied; however, the actual options are scarce and only the use in the field as mulch and energy in sugar mills or biorefineries in cogeneration systems is recognized. Nevertheless, despite these potentially profitable uses, the sugarcane producer still has enough trash available for other immediate uses other than sugarcane cultivation. However, the practice of burning continues due to various factors, mainly economic, technological, inefficient public policies, and grower culture causing a cycle of sugarcane cultivation and harvesting with burning (Fig. 16.2).

Concerning the previous analysis of the physicochemical characteristics, technological, socioeconomic, and environmental factors related to the generation and potential uses of trash. The objective of this work was to evaluate the sustainable use of trash with simple technologies to develop on cane small farms the production of biodegradable pots, handmade paper, organic biofertilizer and edible mushrooms highly valued in the market of products based on nature.

The trash was collected in the Cordoba-Golfocane-producing region located in Veracruz Mexico in plantations of the Mex 69–290 variety corresponding to the harvest season 2022–2023.

The straw and leaves of the trash were reduced from size to 4 cm with a manual cutter (machete) to be dried in the sun to an average humidity of 10%; later half of the trash was paragraph in a manual mill of grooved discs (defibrillator) to obtain material fibrous (Fig. 16.3).

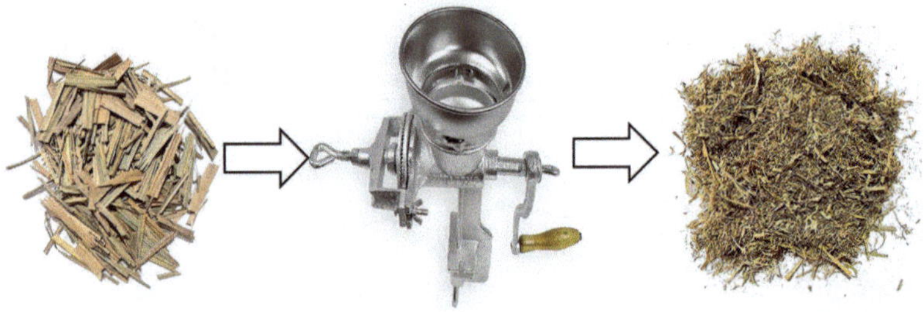

Fig. 16.3 Trash after collection, cutting, and defibrillated waste

Table 16.1 Trash chemical composition (Aguilar-Rivera, 2010)

Variable	Dry base weight (%)
Moisture	12.75
Extractives ETOH-toluene	4.55
Extractives ETOH	7.46
Extractives toluene	5.46
Extractives benzene	2.89
Extractives hot water	6.56
Ashes	5.66
Lignin	20.29
Holocellulose	74.67
α-Cellulose	52.3
Pentosans	22.4
Proteins (N × 6.25)	7.43
Free N	1.64
Fat	0.98
Total reducer sugars	0.042
Crude fiber	71.47
N-free extract	33.40

The analysis of the chemical composition of the trash, from the sugar mills cane fields of the Cordoba-Golfo Veracruz Mexico region of the Mex 69–290 variety, was reported by Aguilar-Rivera (2010) and is structured by the determination of its chemical composition, specifically polysaccharides (holocellulose), lignin, ashes, moisture among other elements of the bromatological analysis (Table 16.1).

16.4 Bioproducts from Sugarcane Trash: Case Studies

Sugarcane crop trash (green and brown leaves and tops) had a chemical composition similar to cane bagasse or hardwoods and therefore is a biomaterial with the potential for several bioproducts based on lignocellulosic materials. After obtaining the chopped and defibrated trash, the following bioproducts were obtained:

- Biodegradable pots.
- Artisanal handmade papers.
- Cultivation of edible mushrooms.
- Composting of spent mushroom waste.

16.4.1 Case Study 1: Biodegradable Pots

The defibrated trash was weighed on a digital scale. The trash was the 75% of the total weight of the fibrous material calculated for the final weight of the pot, the rest corresponds to the glue, which was prepared with wheat flour, tap water, and salt according to the formula used by the artisans to make Mexican handicrafts. Once the glue was obtained, it was mixed with trash to obtain a homogeneous mixture which was placed in plastic pots of different shapes and sizes previously lubricated with edible vegetable oil to facilitate the separation of the mould. After moulding, the pots were dried under sunlight and demoulded as a biodegradable pot (Fig. 16.4). This technique was developed based on the findings of Jarupan et al. (2023); Paudel et al. (2022) and Fuentes et al. (2021).

Fig. 16.4 Use of trash as a substrate for the production of biodegradable pots. (Photography taken in Veracruz, Mexico)

Ecological pots present homogeneous matrix, pot formability, stability permanence, and resistance to environmental factors and water penetration as well as other pots made of other residues such as rice or coffee husks (Fig. 16.4).

As a bioproduct, the container of various sizes and shapes its use is directed to the production of cane seedlings, germination, and sale of ornamental, aromatic, and vegetables which constitutes a business opportunity and income for cane producers with a product of biological origin in addition to reducing the disadvantageous properties of plastic and plastic wastes.

16.4.2 Case Study 2: Artisanal Handmade Paper

Artisanal paper has been mainly used to communicate artistic, cultural, and spiritual expressions through painting, engraving, molding, or printing with digital means. This type of paper has characteristics such as rusticity, and the visual presence of fibers of diverse size, thickness, and texture which makes it a product of great added value.

The origins of this paper are located in China with the production of bamboo-based paper, in Egypt with papyrus, and in Mesoamerica, it was called amatl paper with tree bark. (Jain & Gupta, 2021).

The use of trash fiber has been evaluated by researchers (Madhukar Thakker & Sun, 2023; Velázquez et al., 2020; García-Alcocer et al., 2019) to obtain industrial use products such as cellulose for paper and chemical derivatives.

To obtain handmade trash paper, the dry defibrated material was weighed on a digital scale and washed with tap water to eliminate dust, grease, or any contaminant. It was then deposited in a metal container and an alkaline solution was added at a 5 to 1 ratio (water/fiber) to completely cover the fibrous material. The alkaline solution was previously prepared with a concentration of calcium hydroxide [$Ca(OH)_2$] at 10% by weight of fibers. Once the trash fiber was homogenized with the alkaline solution inside the metal container, a butane gas burner was used to boil the mixture for 90 min at 85 °C for the cooking process to eliminate most of the lignin from the trash fiber and expose the cellulose. After this time the container was removed from the fire, cooled to room temperature, and the fibers were washed with tap water until total elimination of the solution where the trash fibers were delignified and deposited in a container with water. Subsequently, the fibers were defibrated in a manual mill with grooved discs to obtain a material suitable for the formation of sheets of different textures.

This defibrated material was molded on a smooth timber frame supported by wooden rollers to manually form a homogeneous sheet favoring the union of fibers and eliminating as much water as possible. Once the sheet was formed, a felt was placed on top of it to absorb the water with the help of the pressure of a wooden roller. Once most of the moisture had been removed, the felt was removed and the sheet was dried in the sun. The dried sheet was then removed from the wooden molding frame (Fig. 16.5).

Fig. 16.5 Use of trash for the production of handmade paper. (Photography taken in Veracruz, Mexico)

16.4.3 Case Study 3: Sugarcane Trash for Edible Mushroom Production

Worldwide, there is a marked change in land use based on deforestation and an increase in the agricultural border for production of several crops such as conventional grains (wheat, barley, rice, oat, sorghum, corn), oil crops (safflower, soy, sesame, sunflower, peanut, canola, etc.), sugar crops (sugarcane and beet), horticultural and ornamental crops. However, all crops generate a large number of by-products and wastes. In this context, edible mushrooms, as an option for processing lignocellulosic biomass, have been produced for many years as survival food or agribusiness due to multiple properties and potential applications besides increasing the circularity (Ekun et al., 2024; Östbring et al., 2023).

Balan et al. (2022) carried out an extensive review of the properties, methods of obtaining, pasteurization techniques, substrates, and uses of the main commercial edible mushroom worldwide. Oyster mushroom (*Pleurotus ostreatus*) is one of the most economically viable and cost-effective organic recycling processes. The success of edible *Pleurotus* mushroom production requires a fundamental understanding of the physical, chemical, biological, and enzymatic chemical properties of substrates and a strategy for utilization of wastes prior production of edible mushroom and after strategies for waste management (Kumar et al., 2021).

Raman et al. (2021) carried out an overview of the recent literature on the use of different agri-food residues as growth substrates for *Pleurotus* spp. production with attention to their effects on the growth and chemical composition of the cultivated mushrooms. In this farm-level research, the production of edible mushrooms on sugarcane trash, as substrate, was evaluated using the conventional hot water immersion pasteurisation method, in addition to innovative treatments such as steam, alkaline immersion at room temperature and washing, and alkaline peroxide [$Ca(OH)_2$] according to the findings of Kalita, (2015); González et al. (2022) and De

Mastro et al. (2023). The trash (leaves and tops) was collected in a crop field in the region Cordoba-Golfo in Veracruz, Mexico of the variety Mex 69–290. The trash was completely dried under the sun, chopped, washed, soaked in cold water for 12 h, subsequently it was washed with potable water to eliminate organics acids, and sterilized by four methods: 1) immersion in hot water (85 °C) for 30 min, 2) Vigorous washing with tap water and hydration for 24 h with 10 % of Ca $(OH)_2$ and posterior washing with drinking water to avoid contamination; 3) steam treatment for 30 min in a commercial steamer 100 °C and 4) immersion in cold water by 60 min with addition of 10% of $Ca(OH)_2$ and 1% of H_2O_2 and washing with potable water. The four treatments were drained and later, when the substrate was at room temperature, it was inoculated in metallic container with approximately 3 kg of substrate (wet weight) and 0,150 kg of spawn (5%). Incubation took place in darkness at room temperature (25 °C) and relative humidity of 65%. Later, when fruiting bodies or primordia were observed, the containers were transferred to a production room until the harvest. Subsequently for the evaluation of the trash as a substrate, a joint experiment was carried out using a regional control (wheat straw), widely used for its competitive performance, determining biological efficiency, yield, and productivity considering results only from the first harvest of three obtained (Figs. 16.6 and 16.7 and table 16.2).

Biological efficiency (BE) (%) varied significantly from 40 to 80% for trash as a substrate, however, trash is considered a competitive substrate the production parameters of edible mushroom production of a substrate is perceived as acceptable from 50% biological efficiencies. In comparison with the control treatment since BE (%) for trash was greater for steam, alkaline hydration and washing treatments that require less energy, time and water reducing costs and the generation of contaminants. The same trend was observed for productivity (%) and yield (%). In relation to the size of the fruiting body of the mushroom they vary from 4 to 13 cm (Fig. 16.8).

$$\text{BE (\%)} = \frac{\text{Fresh edible mushroom (g)}}{\text{dry weight of the substrate}} \text{ x } 100$$

$$\text{Productivity} = \frac{\text{BE (\%)}}{\text{Days of evaluation}} \text{ x } 100$$

$$\text{Yield (\%)} = \frac{\text{Fresh edible mushroom (g)}}{\text{Wet weight of the substrate}} \text{ x } 100$$

Fig. 16.6 Equations for the determination of productive parameters of edible mushrooms. *BE* biological efficiency

Fig. 16.7 Use of trash as a substrate for the production of edible mushroom

Table 16.2 Yield of oyster mushroom (fruiting body) of trash and control

Substrate	Biological efficiency (BE) (%)	Productivity (%)	Yield (%)
Trash immersion in hot water	50	2.27	7.69
Trash in steam	80	2.05	20.51
Trash alkaline hydration and washing	65	2.57	6.96
Trash alkaline peroxide	40	2.57	6.29
Wheat straw immersion in hot water	80	3.64	12.90
Wheat straw in steam	40	1.25	10.67
Wheat straw alkaline hydration and washing	47	2.00	8.41
Wheat straw alkaline peroxide	42	2.49	8.01

Trash, according to the results obtained with the modification of the conventional hot water immersion method, constitutes a medium-to-high capacity substrate for food production via solid-state fermentation by *Pleurotus ostreatus*; however, it is necessary to evaluate the addition of protein supplements or mixtures with other substrates such as cereal straws, legumes, or regional agricultural residues to increase productivity without the need to increase treatment times that consume inputs such as water and energy.

Fig. 16.8 Production of edible mushrooms in trash as a substrate

16.4.4 Case Study 4: Composting of the Spent Substrate of Edible Mushroom Produced from Sugarcane Trash

The production of edible mushrooms generates a large amount of waste called spent mushroom substrate (SMS). Even if bioplastic or reusable containers are used, it is necessary to develop or use techniques for the conversion of this residue into added-value products. Due to its biological nature, degraded lignocellulosic biomass content, and availability, this waste can be used in various applications such as biofertilizers, animal feeds, bioenergy, bio-remediators, bioethanol, production of enzymes (e.g., manganese peroxidase, laccase, and lignin peroxidase), the green synthesis of metal nanoparticles, such as nano-silver, nano-TiO_2, and nano-ZnO with smart green technologies (Mahari et al., 2020; El-Ramady et al., 2022; Sabaratnam et al., 2024).

However, at the farm level, the most promising application for growers is the production of solid and liquid composts and biofertilizers using the individual spent substrate or in combination with other by-products such as filter mud, manures, fruits and vegetable waste, forest, and agriculture residues or coffee pulp.

The composite of agricultural, agroindustrial, domestic, or urban waste is the tendency to sensitize all involved not to ignore the main factor of the sustainability of food production (soil) (Martín et al., 2023).

Organic fertilizers are a crucial element for the regulation of many processes related to agricultural productivity; their main functions are well known, such as a substrate or growing medium, cover or mulch, maintenance of the original levels of soil organic matter, and complement or replacement of synthetic fertilizers; this last aspect is of great importance, due to the increase in their implementation in clean and ecological production systems. Organic fertilizer is the material resulting from the natural decomposition of organic matter by the action of microorganisms present in the environment, which digest the materials, transforming them into other beneficial materials that provide nutrients to the soil and, as a result, to the plants that grow in it. It is a controlled and accelerated process of waste decomposition, which can be aerobic or anaerobic, resulting in a stable product of high value as a soil improver (Chojnacka et al., 2020).

Organic fertilizers have high contents of mineral nitrogen and significant amounts of other macro and micronutrients for plants. Depending on the level applied, they cause an increase in soil organic matter content, moisture retention capacity, and pH, as well as an increase in available potassium, calcium, and magnesium. In terms of physical properties, they improve water infiltration, soil structure, and hydraulic conductivity; they decrease bulk density and evaporation rate, thus promoting in general a better phytosanitary condition of growing plants (Diacono et al., 2019; Du et al., 2020).

Mushroom-spent substrate composting is an alternative for food production, improving soils affected by monocultures, having the by-product effectively, and closing economic cycles avoiding new pathways of pollution. However, if the spent waste of *Pleurotus* is not degraded, they are not effectively incorporated into the soil. Using composting techniques, it is possible to obtain organic manure to improve the soil structure, nutrient availability, fertility, and pH, thereby promoting good crop development and productivity, which makes it suitable for various agricultural applications such as vegetable, fruit, and plantation crops; it also improves various types of acid and saline soils and can be an alternative to generate income on farms and in supply areas, because composting is not a method of waste disposal, but it is a method of waste recycling and is also used for agricultural purposes.

The remaining substrate of mushroom production was chopped with a cutter by hand to have a homogeneous size and favor degradation in the composting process. The waste was deposited in a plastic container in the facilities of a greenhouse. The waste was homogenized and water was added until it has a humidity close to 60% starting the composting process. The homogenization process was carried out on a weekly basis, adding water and turning with an agricultural shovel besides introduce air for the microorganisms coming from the spent substrate itself degrade the organic material for 90 days. The average greenhouse temperature during composting time was 28 °C. The average greenhouse temperature during the composting time was 28 °C. Likewise, this process can be carried out abroad in an installation with proper management of moisture and weekly homogenization of the compost. The compost obtained was evaluated for its nutritional content according to NMX-FF-109-SCFI-2007 test methods, and Official Mexican Standard NOM-021-RECNAT-2000, which establishes fertility specifications. Samples were analyzed in triplicate to determine moisture, pH, electrical conductivity, ash, organic matter, total carbon, total nitrogen, C/N ratio, calcium (CaO), magnesium (MgO), sodium (Na_2O), potassium (K_2O), phosphorus (P_2O_5), iron (Fe), copper (Cu), zinc (Zn) and manganese (Mn). Results are shown in Fig. 16.9 and Table 16.3.

The process of converting spent substrate into compost and the analytical determinations carried out it, following the NMX-FF-109-SCFI-2007 and NOM-021-RECNAT-2000 standards, demonstrated that this organic biofertilizer can be applied to various crops and soil types, favoring the growth and development of crops,

Fig. 16.9 Use of spent substrate for the production of compost. (Photography taken in Veracruz, Mexico)

Table 16.3 Compost from the spent substrate of production of *Pleurotus ostreatus* on sugarcane trash

Variable	Unit	Spent substrate compost
Moisture	%	59.405
pH	–	7945
Soil electrical conductivity	dsm^{-1}	5711
Ash	(%)	23.185
Organic matter	(%)	76.815
Total carbon C	(%)	44.558
Total nitrogen N	(%)	1.31
C/N	–	34.013
Ca	(%)	4022
Mg	(%)	0.549
Na	(%)	0.127
K_2O	(%)	2669
P_2O_5	(%)	0.908
Fe	(%)	0.0458
Cu	(%)	0.0013
Zn	(%)	0.0109
Mn	(%)	0.0388

especially in acidic soils, characteristic of the production of monocultures. The organic biofertilizer can be applied to various crops and soil types, favoring the growth and development of plants especially in acidic soils, characteristic of the production of monocultures. In addition, for agricultural growers and edible mushroom producers it constitutes a productive option to diversify their income, enhance their well-being and improve their quality of life by selling it to other farmers or applying it in their crops decreasing or eliminating the use of chemical fertilizer, improving soil quality over time, obtaining food, and significantly reducing the accumulation of waste.

16.5 Conclusions and Future Prospects

Currently with the expectative to reach 9 billion people in 2050 the problems related to organic solid waste disposal will increase critically and technological solutions are required for their reduction and transformation into useful and non-polluting products. Therefore, the sustainable production of food, fodder, fiber, biofuels and bioproducts from waste for local and industrial use is a priority in terms of improving food and energy security, environmental protection, and the well-being of the population. We present an analysis of importance of waste agricultural management and four case studies. The bioproducts have suitable properties for use without depending at production stage on new raw materials, fossil fuels, inputs, or foreign technologies. They are competitive and sustainable because close the life cycle. The four bioproducts, namely handmade paper, edible mushroom, compost, and ecological pots, showed properties like commercial products. Their success and profitability will depend on a forward-looking approach, based on knowledge of bioproducts, preferences, and innovations. This work showed that even with conventional, proven, and low-cost technologies, agro-industrial waste can be successfully transformed into value-added products and new value chains can be starting point to reduce poverty.

References

Aguiar, A., Milessi, T. S., Mulinari, R., Lopes, M. S., da Costa, S. M., & Candido, R. G. (2021). Sugarcane straw as a potential second generation feedstock for biorefinery and white biotechnology applications. *Biomass and Bioenergy, 144*, 105896. https://doi.org/10.1016/j.biombioe.2020.105896

Aguilar-Rivera, N. (2010). Modelo cinético de la hidrólisis del residuo de cosecha cañero. *Ciencia e Ingeniería Neogranadina, 20*(2), 5–18.

Aquino, G. S., Medin, C. C., Cunha, A. C. B., Kussab, D. A. O., & Moreira, A. (2019). Vegetative waste management on sugarcane yield and industrial quality. *Communications in Soil Science and Plant Analysis, 50*(2), 132–140. https://doi.org/10.1080/00103624.2018.1556279

Balan, V., Zhu, W., Krishnamoorthy, H., Benhaddou, D., Mowrer, J., Husain, H., & Eskandari, A. (2022). Challenges and opportunities in producing high-quality edible mushrooms from lignocellulosic biomass in a small scale. *Applied Microbiology and Biotechnology, 106*(4), 1355–1374. https://doi.org/10.1007/s00253-021-11749-2

Bandh, S. A., Mall, F. A., Wan, S. A., & Hoang, A. T. (2023). Waste management and circular economy. In S. A. Bandh & F. A. Malla (Eds.), *Waste management in the circular economy*. Springer. https://doi.org/10.1007/978-3-031-42426-7_1

Chojnacka, K., Moustakas, K., & Witek-Krowiak, A. (2020). Bio-based fertilizers: A practical approach towards circular economy. *Bioresource Technology, 295*, 122223. https://doi.org/10.1016/j.biortech.2019.122223

De Mastro, F., Traversa, A., Matarrese, F., Cocozza, C., & Brunetti, G. (2023). Influence of growing substrate preparation on the biological efficiency of *Pleurotus ostreatus*. *Horticulturae, 9*(4), 439. https://doi.org/10.3390/horticulturae9040439

Diacono, M., Persiani, A., Testani, E., Montemurro, F., & Ciaccia, C. (2019). Recycling agricultural wastes and by-products in organic farming: Biofertilizer production, yield performance and carbon footprint analysis. *Sustainability, 11*(14), 3824. https://doi.org/10.3390/su11143824

Dietrich, G., Recous, S., Pinheiro, P. L., Weiler, D. A., Schu, A. L., Rambo, M. R. L., & Giacomini, S. J. (2019). Gradient of decomposition in sugarcane mulches of various thicknesses. *Soil and Tillage Research, 192*, 66–75. https://doi.org/10.1016/j.still.2019.04.022

Du, C., Munir, S., Abad, R., & Lu, D. (2020). Valorization of organic waste into biofertilizer and its field application. *Waste Biorefinery*, 179–198. https://doi.org/10.1016/B978-0-12-818228-4.00007-1

Ekun, V. S., Adenipekun, C. O., Idowu, O., & Etaware, P. M. (2024). Mushroom husbandry: A tool for pollution control and waste management with job opportunities and revenue for rural communities and farm settlements. *Waste Management Bulletin, 1*(4), 15–22. https://doi.org/10.1016/j.wmb.2023.08.001

El-Ramady, H., Abdalla, N., Fawzy, Z., Badgar, K., Llanaj, X., Törős, G., & Prokisch, J. (2022). Green biotechnology of oyster mushroom (*Pleurotus ostreatus* L.): A sustainable strategy for myco-remediation and bio-fermentation. *Sustainability, 14*(6), 3667. https://doi.org/10.3390/su14063667

Freitas, J. V., Bilatto, S., Squinca, P., Pinto, A. S., Brondi, M. G., Bondancia, T. J., & Farinas, C. S. (2021). Sugarcane biorefineries: Potential opportunities towards shifting from wastes to products. *Industrial Crops and Products, 172*, 114057. https://doi.org/10.1016/j.indcrop.2021.114057

Friedl, J., Warner, D., Wang, W., Rowlings, D. W., Grace, P. R., & Scheer, C. (2023). Strategies for mitigating N2O and N2 emissions from an intensive sugarcane cropping system. *Nutrient Cycling in Agroecosystems, 125*(2), 295–308. https://doi.org/10.1007/s10705-023-10262-4

Fuentes, R. A., Berthe, J. A., Barbosa, S. E., & Castillo, L. A. (2021). Development of biodegradable pots from different agroindustrial wastes and byproducts. *Sustainable Materials and Technologies, 30*, e00338. https://doi.org/10.1016/j.susmat.2021.e00338

García-Alcoce, S. K., Salgado-García, S., Córdova-Sánchez, S., Rincón-Ramírez, J. A., Bolio-López, G. I., Ceja, R. C., & Valerio-Cárdenas, C. (2019). Blanqueo de la fibra de celulosa de paja de caña de azúcar (Saccharumspp.) con peróxido de hidrógeno. *Agro Productividad, 12*(7). https://doi.org/10.32854/agrop.v0i0.1434

Ginni, G., Kavitha, S., Kannah, Y., Bhatia, S. K., Kumar, A., Rajkumar, M., & Chi, N. T. L. (2021). Valorization of agricultural residues: Different biorefinery routes. *Journal of Environmental Chemical Engineering, 9*(4), 105435. https://doi.org/10.1016/j.jece.2021.105435

González, A. A. M., Zafra, L. M. C., Bordons, A., & Rodriguez-Porrata, B. (2022). Pasteurization of agricultural substrates for edible mushroom production. *Journal of Microbiology, Biotechnology and Food Sciences, 12*(1), e5729–e5729. https://doi.org/10.55251/jmbfs.5729

Jain, P., & Gupta, C. (2021). A sustainable journey of handmade paper from past to present: a review. *Problemy Ekorozwoju, 16*(2), 234–244. https://doi.org/10.35784/pe.2021.2.25

Jarupan, L., Kruakam, C., & Bumbudsanpharoke, N. (2023). Valorization of brewer's spent grains to produce nutrient biodegradable plant pot. *J Mater Cycles WasteManag, 26*, 501. https://doi.org/10.1007/s10163-023-01845-y

Kalita, M. K. (2015). Impact of various sterilization methods on growth and yield of oyster mushroom (Pleurotus Florida). *International Journal of Agricultural Sciences, 11*(1), 104–107.

Kanna, D., Khademolqorani, S., Janatyan, N., & Alavi, S. (2024). Smart waste management 4.0: The transition from a systematic review to an integrated framework. *Waste Management, 174*, 1–14. https://doi.org/10.1016/j.wasman.2023.08.041

Karp, S. G., Schmitt, C. C., Moreira, R., de Oliveira, P. R., de Mello, A. F. M., Herrmann, L. W., & Soccol, C. R. (2022). Sugarcane biorefineries: Status and perspectives in bioeconomy. *Bioenergy Research, 15*(4), 1842–1853. https://doi.org/10.1007/s12155-022-10406-4

Kumar Sarangi, P., Subudhi, S., Bhatia, L., Sah, K., Mudgil, D., Prasad Shadangi, K., & Arya, R. K. (2023). Utilization of agricultural waste biomass and recycling toward circular bioeconomy. *Environmental Science and Pollution Research, 30*(4), 8526–8539. https://doi.org/10.1007/s11356-022-20669-1

Kumar, H., Bhardwaj, K., Sharma, R., Nepovimova, E., Cruz-Martins, N., Dhanjal, D. S., & Kuča, K. (2021). Potential usage of edible mushrooms and their residues to retrieve valuable supplies for industrial applications. *Journal of Fungi, 7*(6), 427. https://doi.org/10.3390/jof7060427

Lima, I. M., & Beacorn, J. A. (2022). Targeting a sustainable sugar crops processing industry: A review (part I)—By-product applications. *Sugar Tech, 24*, 970–991. https://doi.org/10.1007/s12355-022-01169-5

Madhukar Thakker, A., & Sun, D. (2023). Handmade papers: Innovation, technology, and design. *Journal of Natural Fibers, 20*(1), 2187505. https://doi.org/10.1080/15440478.2023.2187505

Mahari, W. A. W., Peng, W., Nam, W. L., Yang, H., Lee, X. Y., Lee, Y., & Lam, S. S. (2020). A review on valorization of oyster mushroom and waste generated in the mushroom cultivation industry. *Journal of Hazardous Materials, 400*, 123156. https://doi.org/10.1016/j.jhazmat.2020.123156

Martín, C., Zervakis, G. I., Xiong, S., Koutrotsios, G., & Strætkvern, K. O. (2023). Spent substrate from mushroom cultivation: Exploitation potential toward various applications and value-added products. *Bioengineered, 14*(1), 2252138. https://doi.org/10.1080/21655979.2023.2252138

Martins, M. B., Filho, A. C. M., Santana, L. S., de Almeida Prado Bortlheiro, F. P., & da Silva, K. G. P. (2023). Sugarcane harvester: A bibliometric review. *Sugar Tech, 25*(6), 1316–1327.

Meghana, M., & Shastri, Y. (2020). Sustainable valorization of sugar industry waste: Status, opportunities, and challenges. *Bioresource Technology, 303*, 122929. https://doi.org/10.1016/j.biortech.2020.122929

Ortiz Laurel, H., Salgado García, S., Castelán Estrada, M., & Córdova, S. S. (2012). Perspectivas de la cosecha de la caña de azúcar cruda en México. *Revista Mexicana de Ciencias Agrícolas, 3*(SPE4), 767–773.

Östbring, K., Lager, I., Chagas, J. C. C., Ramin, M., Ahlström, C., & Hultberg, M. (2023). Use of oyster mushrooms (*Pleurotus ostreatus*) for increased circularity and valorization of rapeseed residues. *Journal of Environmental Management, 344*, 118742. https://doi.org/10.1016/j.jenvman.2023.118742

Paudel, B., Basak, J. K., Kaushalya Madhavi, B. G., Kim, N. E., Lee, G. H., Choi, G. M., & Kim, H. T. (2022). Properties of paper-based biodegradable pots for growing seedlings. *Horticulture, Environment, and Biotechnology, 63*(6), 793–807. https://doi.org/10.1007/s13580-022-00457-z

Phiri, R., Rangappa, S. M., & Siengchin, S. (2023). Agro-waste for renewable and sustainable green production: A review. *Journal of Cleaner Production*. https://doi.org/10.1016/j.jclepro.2023.139989

Powar, R. V., Mehetre, S. A., Powar, T. R., & Patil, S. B. (2022). End-use applications of sugarcane trash: A comprehensive review. *Sugar Tech, 24*(3), 699–714. https://doi.org/10.1007/s12355-022-01107-5

Purchase, B. S., Wynne, A. T., Meyer, E., & Van Antwerpen, R. (2008). Is there profit in cane trash? Another dimension to the assessment of trashing versus burning. *Proceedings of the South African Sugar Technology Association, 81*, 86–99.

Raman, J., Jang, K. Y., Hyun-Jae, S., Kong, W. S., & Lakshmanan, H. (2021). Industrial and biotechnological applications of *Pleurotus*. In *Advances in macrofungi* (pp. 89–107). CRC Press.

Ray, R. C. (2022). *Fruits and vegetable wastes: Valorization into bioproducts and platform chemicals* (p. 448). Springer.

Roy, P. (2023). Nothing is 'waste ' in agriculture: From nanotechnology and bioprocesses perspectives. In *Agriculture waste management and bioresource: The circular economy perspective* (pp. 131–148). https://doi.org/10.1002/9781119808428.ch6

Sabaratnam, V., Phan, C. W., Lakshmanan, H., & Raman, J. (2024). Spent mushroom substrate as an alternative source for the production of chemical substitutes. In *Chemical substitutes from agricultural and industrial by-products: Bioconversion, bioprocessing, and biorefining* (pp. 87–101). https://doi.org/10.1002/9783527841141.ch5

Shanthy, T. R., Gurumurthy, D., & Saravanan, L. (2020). Trash management practices in sugarcane cultivation: A socio-economic analysis. *Indian Research Journal of Extension Education, 20*(4), 1–8.

Silva, A. G., Lisboa, I. P., Cherubin, M. R., & Cerri, C. E. (2019). How much sugarcane straw is needed for covering the soil? *Bioenergy Research, 12*, 858–864. https://doi.org/10.1007/s12155-019-10008-7

Singh, S. P., Jawaid, M., Chandrasekar, M., Senthilkumar, K., Yadav, B., Saba, N., & Siengchin, S. (2021). Sugarcane wastes into commercial products: Processing methods, production optimization and challenges. *Journal of Cleaner Production, 328*, 129453. https://doi.org/10.1016/j.jclepro.2021.129453

Srivastava, S. (2020). Diversification of sugar and sugarcane industry: Agro-industrial alternatives. In N. Mohan & P. Singh (Eds.), *Sugar and sugar derivatives: Changing consumer preferences*. Springer. https://doi.org/10.1007/978-981-15-6663-9_10

Steponavičienė, V., Rudinskienė, A., Žiūraitis, G., & Bogužas, V. (2023). The impact of tillage and crop residue incorporation systems on agrophysical soil properties. *Plants, 12*(19), 3386. https://doi.org/10.3390/plants12193386

Tanveer, M., Khan, S. A. R., Umar, M., Yu, Z., Sajid, M. J., & Haq, I. U. (2022). Waste management and green technology: Future trends in circular economy leading towards environmental sustainability. *Environmental Science and Pollution Research, 29*(53), 80161–80178. https://doi.org/10.1007/s11356-022-23238-8

Velázquez, L. Y. L., Salgado-García, S., Turrado-Saucedo, J., Hidalgo-Moreno, C. I., Ortiz-García, C. F., Córdova-Sánchez, S., & Canché-Escamilla, G. (2020). Celulosa y microcelulosa de residuos del cultivo de cana de azucar (Saccharumspp.). *AGRO Productividad, 13*(4), 11–18. https://doi.org/10.32854/agrop.vi.1591

Vieira, R. F., Ramos, N. P., & Pazianotto, R. A. A. (2021). Different amounts of sugarcane trash left on the soil: Effects on microbial and enzymatic indicators in a short-term experiment. *Soil Use and Management, 37*(3), 658–666. https://doi.org/10.1111/sum.12584

Wang, K., & Tester, J. W. (2023). Sustainable management of unavoidable biomass wastes. *Green Energy and Resources*. https://doi.org/10.1016/j.gerr.2023.100005

Yang, X., Xiong, J., Du, T., Ju, X., Gan, Y., Li, S., & Butterbach-Bahl, K. (2024). Diversifying crop rotation increases food production, reduces net greenhouse gas emissions and improves soil health. *Nature Communications, 15*(1), 198. https://doi.org/10.1038/s41467-023-44464-9

Chapter 17
Economic Viability and Policy Implications of Agriculture Waste Valorization: Social and Economic Impacts, Policy Frameworks, and Regulations

Noé Aguilar-Rivera

Abbreviations

AUC	Area under curve
AWB	Agro-food waste biomass
CE	Circular economy
FVW	Fruit vegetable waste
FW	Fruit wastes
GHG	Greenhouse gases
SDG	Sustainable development goals
TEA	Techno-economic analysis

17.1 Introduction

There are a lot of economic, social, environmental, technological, and political factors that are a potential risk for the development of sustainable societies at least in the medium term and for future generations. Among these factors, one of the most important is the unstable, fragile, and low resilience food and biological raw materials production from crops and agroindustry besides the large number of by-products and waste generated in the value chain.

The above is founded on the 20th century linear economic model of production-consumption-waste process due to the wide availability of abundant natural

N. Aguilar-Rivera (✉)
Facultad de Ciencias Biológicas y Agropecuarias, Universidad Veracruzana, Córdoba, Veracruz, México
e-mail: naguilar@uv.mx

J. Arora et al. (eds.), *Transforming Agriculture Residues for Sustainable Development*, Waste as a Resource, https://doi.org/10.1007/978-3-031-61133-9_17

resources and fossil energy, rapid population growth and technology. Therefore, this model results in the increased consumption of goods and products derived from the unsustainable use of natural resources and the application of the factors of production land, labour, capital and technology. At the same time, our day-to-day disposable objects, which were not designed for the end of their useful life, have been deposited as waste in landfills and in the environment (soil, rivers, lakes, sea, etc.).

Thereby, the successful transition from a linear to a circular economy model, broadly compatible with the environment and with the production of sustainable bio-based products from waste (agricultural, food, municipal solid waste, sewage sludge and non-food biomass) must be driven by all stakeholders (academia, producers, industry, society and policy makers) to set the goals and objectives of the decision-making processes that will impact society in the transition process (Facchini et al. 2023).

Nevertheless, anthropogenic productive systems are based on extraction of natural resources mainly petroleum, gas and minerals, and plantation crops, besides the long-term resilience of ecosystems is considerably affected by the contamination of industrial processes waste. Therefore, the efficient waste management through valorization, as raw material and reuse is critical in circular economy and bioeconomy due to its renewability, complex structure, and immediate availability throughout the year. Waste biomass can be a beneficial substrate to produces several high value bioproducts such as biofuels, biofertilizers, composts, biochar, pharmaceuticals, bioplastics, and food additives (Wijerathna et al. 2023).

Abd-Aziz et al. (2024) have defined agricultural wastes as unwanted products as a result of agricultural activities, while agro-industrial by-products are leftover from industrial activities with abundant organic and inorganic nutrient composition.

17.2 Bioresources as Raw Material for Bioproducts

Clark, (2019) concluded the production of bio-based goods, the stakeholders can help with green management within the circular bioeconomy because lignocellulosic wastes are a potential source of components such as fibre, biobased material, phytochemicals, biomolecules, etc. besides are available in large quantities in all regions of the world and can be transformed into bioproducts with different technological levels (Fig. 17.1).

Koul et al. (2022) discusses the factors influencing the recycling and utilization of agricultural wastes besides pros and cons/merits and demerits of the current management strategies. It also emphasizes the need of effective public policies towards circular economy. Araújo et al (2023) concluded the agri-food systems are the anthropogenic activities of greater impact on environmental, economic, and social challenges, due the degradation of ecosystems, biodiversity loss, reduction in soil fertility, water scarcity, land degradation, lack of sustainability and social welfare.

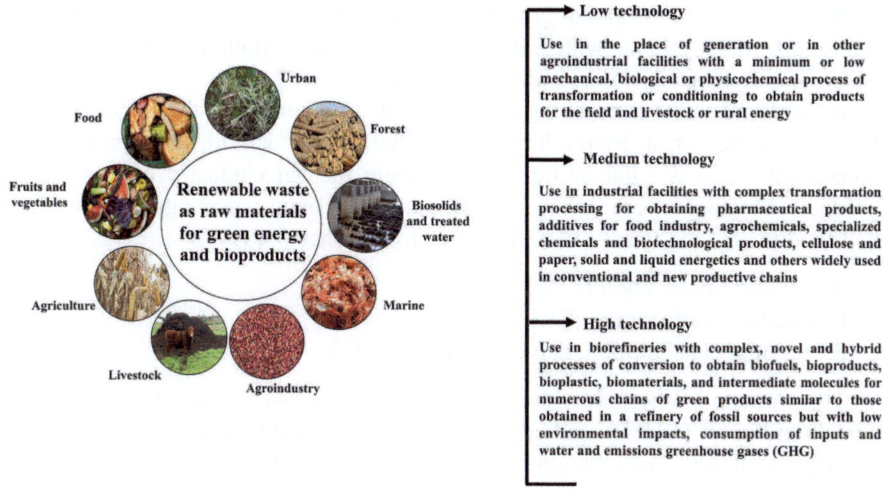

Fig. 17.1 Potential sources of biological waste and technological levels of transformation into bioproducts

Moreover, four perspectives are necessary to reduce unsustainability due to their unacceptable environmental impacts and are causing the depletion of non-renewable resources and reduce food insecurity:

1. The increase of efficiency by changes in production
2. Demand restraint by changes in consumption
3. Food system transformation by changes in food system functioning and governance
4. In the case of fruits and vegetables, reduction of loss and waste due to consumer behavior (wholesale and retail) (Araújo et al., 2023)

The global amount of solid waste has dramatically increased as a result of rapid population growth, accelerated urbanization, agricultural demand, and industrial development.

According to Romero-Perdomo and González-Curbelo (2023), the agri-food waste biomass (AWB) is a highly available bio-resource in the circular bio-economy for the creation of new productive chain and diversification (multi-output production chains). Nevertheless, its collection and processing still have numerous constraints that must be resolved because there are few crop residues with robust research and the applications are only organic fertilizers, biogas, bioethanol, biodiesel, bioelectricity, etc. Despite the few commercial applications, for some biomass sources, other options that could have a high added value are not yet considered by the stakeholders. However, social impacts and environmental improvement are weakened.

Besides, 25–30% of global greenhouse gas (GHG) emissions are attributable to food production systems, which also result in significant pollution of air, soil, and water, and biodiversity losses (Kiran et al., 2023).

Agricultural waste is cropping field waste that remains after the target crop has been harvested as a by-product of post-harvesting activities, whereas food-processing waste is solid or liquid waste from the agro-industry of agricultural products. Metabolites are AWB is divided into two main types: agricultural waste and food-processing waste. There are three major stages of AWB:

- Agricultural production (33%)
- Post-harvest and storage (54%)
- Processing, consumption, and distribution (46%)

The primary AWB sources are beverages (26%), fruits and vegetables (14.8%), cereals and seeds (12.9%), and edible oils (3.9%), among others.

Gupta et al. (2022) carried out a systematic review of the use and valorization of agricultural residues to obtain bioproducts such as construction material, biofertilizers, paper and packaging products, heating applications (pellets and briquettes), composites, nano cellulosic materials, soil stabilizers, biofuel, and dye removal.

Waste plants, fruits, and vegetables contain metabolites of great economic interest (Ray, 2022). The primary biomolecules are lignin, and carbohydrates (cellulose, hemicellulose, starch, sucrose, etc.), which are feedstock for liquid, gaseous, and solid biofuels, and the secondary metabolites are sugars, gums, resins, rubber, terpenoids, steroids, triglycerides, etc. that can be used to produce value-added chemicals for the food, pharmaceuticals and biotechnological industries (Banerjee, 2023; Alan & Köker, 2023).

Riseh et al. (2024) concluded the smart management of agricultural waste, as a sustainable source of cellulose and upgrading the conversion by innovative products such as cellulose nanomaterials and nanocomposites can have high environmental and economic benefits for agriculture, industry, medicine, and energy.

Therefore, the conversion of underutilized agricultural, agro-industrial, and lignocellulosic biomass from a perspective of productive integration and diversification has the potential to direct environmental goals of whole process with different configurations, looking to be more attractive to investors and agritech businesses (Cuadrado-Osorio et al. 2022).

Currently, the use of agricultural waste for bioproducts and bioenergy is the major alternative, however, detailed analysis of feedstock, characterization, bioconversion, and contemporary pre-treatment procedures, appear to be vital for successful utilization (Kumar et al. 2023).

17.3 Agriculture Waste Management

Agricultural waste has great expectations for its use as a partial and long-term substitute for non-renewable resources. Despite these advantages, there are no laws, or specific mandates that direct the management and treatment of agricultural waste, since each type of residue, by-product, or waste of agriculture activities (organic, plastics, phytosanitary, etc.) must be recycled and managed concretely.

Nonetheless, selective collection, transport, storage, recycling, and processing of agricultural waste can be hosted by current specific management systems while new public policies are established. However, according to the review of Peng et al. (2023), the approach, design, and implementation of industrial processing facilities such as biorefinery as value-added use of biomass and waste and applications are complicated because the variability of the source of the feedstock, vegetation type in biomass, quality, composition, and properties, collection, segregation, classification, standardization, transportation and logistics, pretreatment, technologies, cost of production and economic feasibility, and limitations and future perspectives is complex.

First, waste classification has numerous aspects to consider, according to origin, physical state (liquid or solid), chemical composition, availability, or source. Secondly (i) Promote the complete application of value-added reliable waste technologies to relieve pressure on waste disposal; (ii) Encourage recycling and reuse mainly agricultural waste; (iii) Mitigate the adverse environmental impacts of waste; (iv) Conserve natural resources and expedite the achievement at least the 3R strategy — Reduce, Recycle, and Reuse or the 10R of circular economy.

Biomass of all types, sources, and compositions as agricultural residues is an attractive feedstock to produce products of high profit as biofuels, biomaterials, and biochemicals due to its versatility, renewable nature, and low environmental impacts in the value chain.

According to Awasthi et al. (2023), the lignocellulosic biomass mainly from agro-industrial waste such as sugarcane, corn, soy, and cereals accounts for 181 billion tons/year remains underutilized, accounting for 50% of worldwide biomass production. However, the total biomass worldwide is utilized (10%) as feedstock for biofuel production relatively very less without contributing to carbon neutrality.

Therefore, the application of Research, Technology Development, Innovation (R + D + I), and Technology Transfer (TT) principles are necessary to convert lignocellulosic biomass from agricultural waste and by-products into an efficient raw material for implementation at large scale at integrated biorefineries for biofuels, biocompounds, and biochemicals. Therefore, researchers and policymakers must work together to solve three challenges to the sustainable use of biomass:

1. Techno-economic analysis (TEA) of biomass biorefineries installed across the globe
2. Bottlenecks haunting wider dissemination of biorefineries
3. Need for an integrated approach to make biomass-based circular bio-economy feasible

For the case of green agricultural crop residues such as leaves, grasses, stems or roots, immature cereals from extensive land cultivation or attacked by pests or various types of field weeds due to its the ability to produce higher yields of biomass and flexibility to fit into various crop cycles as feedstocks for the green biorefinery technological strategy can be a profitable option for obtaining bioproducts because the valuable biomolecules of fresh biomass are proteins, soluble sugars, and the fiber fractions (cellulose, hemicelluloses, and parts of lignin degradation). Therefore, it is necessary to make some considerations for the success of green biorefinery (Ray, 2020; Kataki & Kataki, 2022; Abd-Aziz et al., 2024).

Gaffey, J., Rajauria, (2023) and Xiu and Shahbazi, (2015) concluded the green juice of fresh biomass is rich in proteins, free amino acids, organic acids, dyes, enzymes, hormones, other organic substances, and minerals. Therefore it is a potential source for high value chain bioproducts. The press cake of green biomass can be used as solid fuels, green feed pellets/fodder pellets, as a raw material for the production of chemicals, and biofuels.

There, the options for green biorefineries should consider the following aspects to achieve greater sustainability:

- Feedstock selection and evaluation
- The development of innovative technologies for biofuels and bioproducts from the press cake of green biomass
- Life cycle analysis and sustainability of processes derived from green press cake
- Increase local rural resilience, and create diversification opportunities for farmers

The goal for sustainability is processing the whole green and dry biomass such as grass, harvest residues, food and vegetable, forest and garden waste, and various other sources to obtain a variety of bioproducts that are either valuable products themselves or form the basis for further production lines for specific biorefineries.

However, agro-waste or agricultural residues are a great economic problem for the farmers due to the lack of dumping lands, proper transportation, removal facilities, lack of planned utilization, or training for conversion into higher value-added products (Srivastava et al., 2023). The physicochemical composition of waste biomass and analysis of its properties, are key factors in making decisions regarding the selection or design of technologies to add value to the waste which will then be used as a raw material (Matumoto-Pintro & Saraiva, 2023).

17.4 Circular Economy (CE) and Bio-Economy for Management of Agricultural Waste

The CE as a framework to achieve the sustainability of processes and products is aimed at reducing the dependency on finite resources extraction (oil, gas, coal, minerals, etc.) and increasing the life cycle of the final products through the 10Rs for economic, environmental, and social benefits.

Donner et al. (2020) mentioned that the most complex challenge of implementing a circular economy for the management of agricultural waste from agri-food systems, as a potential source for bioproducts production, is a change of current production paradigms and the participation of all actors within the value chains (suppliers, manufacturers, retailers, consumers) and an enterprise level, incorporating innovative business models for new products. Besides, they explained that the implementation of circular economy needs several steps: (1) repair and maintenance, (2) reuse and redistribution, (3) refurbishment and remanufacturing, (4) recycling, (5) cascading and repurposing, and (6) organic feedstock. These patterns correspond to the technical (1–4) and biological (5–6) reverse cycles:

Other aspects of CE are:

1. Reducing input of resources, especially scarce ones (minerals and fossils)
2. Reducing emission levels (soil, water, air, and impact on populations)
3. Reducing material losses/waste (with the application of AGRICULTURE 4.0 and INDUSTRY 4.0)
4. Increasing input of renewable, waste, and recycled resources (development of public policies and participation of stakeholders)
5. Maximizing the utility and durability of products (reduce the use of inputs, water, and energy)
6. Creating local jobs at all skill levels (corporate social responsibility)
7. Value-added creation and distribution (Maximize productive diversification)
8. Increase social well-being (creation of new jobs in the region)

Besides the CE models in the agri-food systems can reduce the amount of food loss and waste generated in the agri-food supply chain by using them as a new resource taking advantage of the competitive advantages of this biomaterial as wide availability throughout the year, heterogeneity, low cost, chemical composition, and existing technologies for conversion into numerous bioproducts.

In this regard, the waste and by-products collection in the agri-food systems is the major challenge because, as first activity of the new bioproduct, implies new costs and accounting-efficiency, creation of long-term partnerships with local farmers, quality monitoring and handling perishability, seasonality, and variability of agricultural waste. The next step, the transformation of low-value agricultural waste and by-products into marketable end-products (valorization pathways) requires specific value creation strategies such as biomass value cascading and complex bioprocess.

In relation to the above, circular bioeconomy, as a inter and transdisciplinary framework, is a set of strategies developed to foster societal and economic transformation. The bioeconomy itself relies on the management of renewable biomass resources to produce biomaterials and bioenergy besides the traditional supply of food, feed, and fiber to support the Sustainable Development Goals (SDGs) and the Paris Agreement.

Therefore, bioeconomy is a contemporary approach that proposes develop visions, strategies, and roadmaps of stakeholders to find best solutions to the major global challenges such as food security, natural resource scarcity, fossil resource dependence, and climate change while achieving sustainable economic development with the increased use of biological resources, biomass, organic waste, and energy crops.

Nevertheless, for the success of bioeconomy circular are required innovative public policies, as direct tools for regulation and economic instruments, through the promotion of incentives, taxes, and subsidies, environmental regulations, voluntary approaches, information exchange, advisory systems, and market-based signaling approaches in the areas of climate and energy, water, biodiversity and minimization of pollution, and the management of waste and residues from food, feed, forestry and fisheries production (Szarka et al., 2023; Duque-Acevedo et al., 2020).

The bio-economy, as the renewable partnership of the CE, promotes the use and sustainable recovery of AWB and food waste (FW), and fruits and vegetable wastes (FVW) as an essential supply. This bio-based economic model has become one of the main tools for drawing up new development policies based on the SDGs (Fig. 17.2).

The CE and the circular bio-economy are frameworks and paradigms that are making the transition to new economic activities, sustainable production, and consumption as well as their associated services related to the invention, development, production, and use of biological products and processes for bioenergy, biomaterials, and biochemicals. Both concepts aim to create economic, natural, and social capital based on four key principles:

1. Design waste and pollution-free processes and products
2. Keep products and materials closing loops
3. Regenerate natural an industrial system that is restorative or regenerative through the production of renewable energy and non-toxic materials, compost, biogas, and cascading of bioproducts and
4. Increase environment quality, and the social aspect of sustainability as human well-being to support the decarbonization of the economy (net-zero emission targets)

In contrast to the current economic linear model, the CE looks for economic growth in the function of the consumption of finite resources and minimizing waste to landfills or burned.

The sustainable bio-economy is not just about substituting fossil resources with renewable resources; it will require sustainable biomass feedstock production, conversion technology, processes, and final products. The goal is to use fewer resources to achieve the same purpose (Tan & Lamers, 2021).

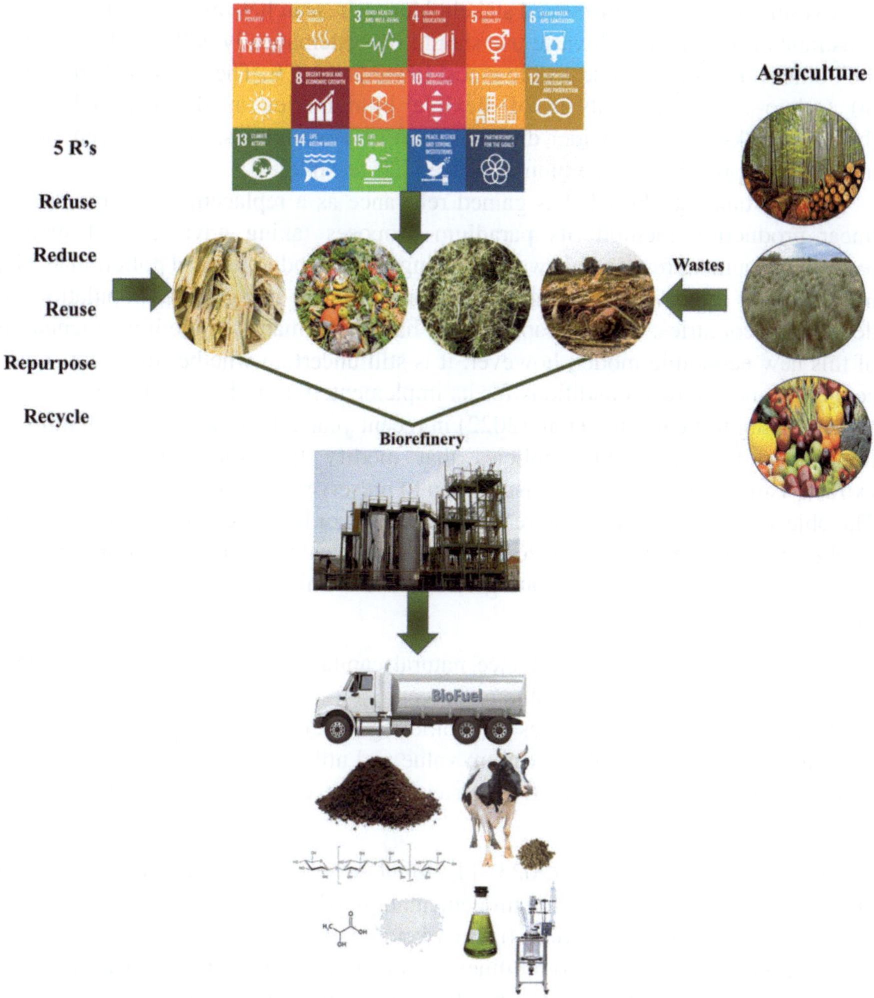

Fig. 17.2 5R circular economy strategies and circularity of agricultural waste, biorefineries, bioproducts, and sustainable development

17.5 Public Policies and Challenges for Sustainable Management of Agricultural Waste

The environmental policies of policymakers should be effective for the sustainable management of resources and nature and prevent human activities from crossing the biophysical boundaries at the local, regional, national, and international levels (Arias et al., 2022).

Awasthi et al. (2023) discussed stakeholders as researchers and policymakers the constraints for agricultural waste use should be systematically addressed to resolve the bottlenecks as high yield and low cost towards the CE. The bottleneck areas are (a) techno-economic analysis of flexible biorefineries at the regional level, (b) bottlenecks haunting wider dissemination of biorefineries, and (c) need for an integrated approach to make biomass-based CE feasible.

Notwithstanding, the CE has gained relevance as a replacement for the classic linear production method; its paradigm proposes taking advantage of natural resources in a more responsible way, reducing waste production and pollution levels, in addition to improving the living conditions and well-being of the population. In developing countries, different approaches have been made for the implementation of this new economic model; however, it is still uncertain whether there are socio-economic and political conditions for its implementation in these countries.

According to Leal Filho et al. (2022) in recent years, there have been attempts to implement different public policies that modify the conventional model of extractivism, production, and consumption to preserve and protect the environment. The objective of the CE is not to reduce production or limit the generation of wealth; on the contrary, the objective is to optimize resources, respect the relationship with nature, generate wealth, and create greater value within the productive cycles with the following steps:

- Principle 1: preserve and enhance natural capital by controlling the stock and balancing the flow of renewable resources
- Principle 2: optimize resources by promoting the circularity of products, components, and materials with maximum value and utility
- Principle 3: adopt an effective system that reveals and eliminates negative externalities

Nevertheless, Jesus et al. (2023) identified several barriers to the adoption of environmental strategies in the implementation of the CE: (i) limited financial capacity, (ii) lack of information, (iii) market and environmental results and uncertainty benefits, (iv) lack of commitment from top management, (v) lack of clarity regarding strategic and business objectives, (vi) little awareness and qualification among employees, and (vii) lack or scarcity of technology and equipment.

Notwithstanding, Medina-Abad and Freire-Pesántez (2023) concluded that there are several constraints to the successful implementation of CE principles mainly in developing countries:

17.5.1 Technological

These limiting factors consider the lack of technological innovation that allows the implementation of a circular production model. Technological infrastructure includes equipment capable of efficiently taking advantage of the resources used in production, and after use can be reused as part of a new production process. The constraints are lack of technological infrastructure, supply chain, and technical knowledge and low availability of technology.

17.5.2 Structural/Organizational

The ability of companies to support this paradigm shift is very complex. Public policies play a fundamental role in the transition to a circular model mainly in developing countries with economies that depend strongly on extractive industries such as oil and mining and there are no policies that support the practice of sustainable strategies. The main constraints are:

1. High costs for the acquisition of equipment, materials, and implementation of processes compatible with the environment.
2. In the short term, the transition to sustainable processes can translate into low economic yields.
3. Difficulties in control and monitoring.
4. Poor management capacity for transition to a circular model (change resistance).
5. Difficult application of innovative productive systems and processes.
6. There is no organizational culture that supports productive change and promotes cooperation between companies.
7. Resistance to change to a circular model.
8. Few companies have the economic resources to change the production manner.
9. The benefits of the implementation of a circular model at the business level are not known or underestimated.
10. There are no materials available for reuse or are difficult to access.
11. Shareholders and investors consider the change to a circular model.
12. Companies prefer to maximize sales above other objectives.
13. Greenwashing.

17.5.3 Institutional/Governmental

The following are the institutional and governmental constraints.

1. There is no planning to overcome the barriers to the effective implementation of a circular model.
2. Waste collection and treatment is largely handled by the informal sector, without techniques that guarantee the quality of the processes.
3. The collection of materials for use does not have the necessary quality due to failures in collection systems and the heterogeneity of materials.
4. Cities with higher population density do not have the space for waste management plants.

5. Information on materials and waste is incomplete and unreliable.
6. There is no support from governmental entities for financing the change to a circular economy model.
7. The change to a sustainable production model is not a priority for governments and public policies in developing countries.
8. Environmentally unfriendly materials continue to be an economically viable option due to subsidies to extractive industries.
9. Policies and laws that do not promote the implementation of the Circular model and the change of production mode.
10. Lack of awareness programs, education, and promotion of a circular culture.
11. Economies are based on the extraction of raw materials.
12. Developing policies that take into account the environment is precarious due to the low academic skills of politicians.

17.5.4 Cultural

The cultural barriers are as follows.

1. Lack of coordination between government, consumers, suppliers, and society.
2. Consumers are not interested in production processes, raw materials, and environmental impacts of products.
3. The traditional culture of consumerism.
4. The benefits of implementing a circular model at a social level are not known or are underestimated.
5. No culture supports the 3Rs or 5Rs and the use of previously used materials.
6. Uncertainty about the benefits of CE.
7. There is insufficient demand for environmental technology or low development of such technology financed by governments.

17.5.5 Integration of Regions with Agro-Ecological Potential for the Production of Biofuels and Bioproducts Supported by Public Policies

Public policies are the instruments available to a government to direct objectives for the welfare of its population through decrees, laws, agreements, mandates, regulations, and norms, which should structure the legal regime in this case of agriculture, agro-industry, and biorefineries from the management of natural resources (water, energy, land use, etc.), planting, cultivation, harvesting, industrialization and marketing of products and by-products, environmental impacts and relations between growers, unions, entrepreneurs and government seeking to increase productivity and profitability through the industrialization and marketing of products and by-products (Surya et al., 2021).

Nowadays, the lack of coordination between growers, government, and stakeholders to create and implement efficient public policies towards the sustainable crop production and the management of agricultural waste, has incalculable social, economic, and environmental consequences, such as soil degradation, water, and air pollution in the areas close to crop fields. Nonetheless, there are other factors that constrained a competitive environment such as the non-technological-based use of agrochemicals such as herbicides, pesticides, and fertilizers; the uncertainty in the national and international price of commodities due to different geopolitical and regional causes such as the absence of planning of potential harvests based on inventories or the opacity of agricultural policies.

Notwithstanding the evidence, this lack of institutionalization, knowledge management, and governance of public policies in the agricultural sector has direct effects on the impoverishment of families due to the increase in production costs and the reduction in productivity that affects the income of producers, discouraging re-investment. In addition, environmental deterioration is evident and, consequently, a decrease in the quality of life and standard of living of the inhabitants of agro-industrial regions, where assistance policies do not reach the beneficiaries for whom they were created and are used as an electoral tool without revitalizing productive areas (Barrera-Perales & Burgos, 2023; Liszbinski et al., 2023).

Consequently, one of the main failures in agricultural public policies, in producing regions of developing countries, has been land management and the location of agricultural activities with objectives of creation of jobs, food security and sovereignty, production of raw materials for diversified agro-industries, and the reduced use of agricultural by-products, residues and wastes. For these reasons, it is necessary to develop policies based on management tools for the optimal regionalization of agricultural and agro-industrial activities and the sustainable management of wastes (Short Jr. et al., 2023; Jiménez et al., 2023).

Consequently, productive regional planning in developing countries such as Mexico and its regions has occupied a privileged place in governmental discourse and plans, based on potentialities or comparative or competitive advantages (social, natural, political, and financial). However, the analysis of the location of economic activities in Mexico, at the geographic and sectoral level, poses numerous problems to be solved to achieve regional sustainable development, making it necessary to reactivate integrally the weak and poorly articulated agricultural activities determined by historical processes, with their dynamics and worldview, which correspond to societies with socioeconomic and cultural characteristics. In its methodological approach, the experience of studies carried out in developed countries is useful but cannot be applied mechanically to developing countries and their regions, since they have different degrees of maturity and progress, but are of a different and complex nature.

17.5.6 Case Study in Veracruz, Mexico

This work was carried out in a region of southeastern Mexico, specifically in Veracruz.

The state of Veracruz is located in the coastal strip of the Gulf of Mexico, with an area of 72,815 km2, occupying 3.7% of the national territory. Veracruz is considered one of the states with the greatest biodiversity in Mexico with 2000 plant species, with 8 types of vegetation among the most important, the mountain mesophyll forest, formed by 2279 communities distributed in 57 municipalities, with an area of 6035 km^2, used for agricultural activity with high cultivated extension of several agro-industrial crops but with low productivity and competitiveness. Therefore, the application of the maximum entropy method (Maxent) (Fitzgibbon et al., 2022), as a tool for decision-makers, for the formulation of public policies for the determination of the optimal areas for crops with the potential to produce biofuels and bioproducts based on agricultural residues generated as biomass (https://www.biodiversidad.gob.mx/region/EEB/estudios/ee_veracruz).

According to the Agri-Food and Fisheries Information Service (SIAP, 2023) of the Mexican government, those crops rich in sugars or starches, such as corn, sorghum, barley, sugar cane, cassava, wheat, and beet are considered strategic crops to produce biofuels and bioproducts. The collected data (geographic coordinates) of the cultivated areas in Veracruz; those data that presented location inconsistencies were not considered for the analysis.

Therefore, to develop bioclimatic modeling, 19 layers of temperature and precipitation were used, one topographic, at a resolution of 30 seconds of arc ($\approx$1 Km^2 at the equatorial level) (Hijmans et al., 2005, version 1.4) and nine variables of soil characteristics at a scale of 1: 1000000 (Cruz-Cárdenas et al., 2014) (Table 17.1). The information corresponding to each cell of the layer represents the average values interpolated from environmental data collected in meteorological stations between 1950 and 2000. These layers were cut to the extreme coordinates of the state of Veracruz, guaranteeing the inclusion of the territory (Cruz-Cárdenas et al., 2014; Ramirez-Villegas et al., 2013).

Using the MaxEnt software (Merow et al., 2013; Phillips & Dudík, 2008; Fitzgibbon et al., 2022), a map of potential distribution for each potential crop for biofuel and bioproducts was developed, using the geolocations of the zones of greater production and the bioclimatic variables that intersect with those corresponding to the areas of the crop, to generate graphically a probability map (from 0 to 1, where 0 is low probability and 1 maximum probability) of the agro-ecological conditions of each crop (Phillips et al., 2004, 2006). Once the model of the agro-ecological zones for each crop was obtained, it was evaluated with the values of the area under the curve (AUC), having as criterion of an AUC $>$ 0.9 as a good fit of the model, which characterizes the performance of the distribution model (Phillips et al., 2006). The model itself corresponds to a graphic output (map) that shows the ability to discriminate a presence (sensitivity) versus the ability to discriminate an absence (specificity, Phillips et al., 2004). The color scale of the

Table 17.1 Important values of the layers of soil, climatic and topographic variables used in agroecological zoning models of potential crops for biofuels and bioproducts from biomass

Variables	Sugar cane	Maize	Beet	Sorghum	Wheat	Cassava	Barley
Bio1 Annual Mean Temperature(ºC)	0	0	0	0,4	0	0,9	1,9
Bio2 Mean Diurnal Range	0	0	0	15,5	5,4	0,3	4,3
Bio3 Isothermality	7,3	0	0,3	1,4	25,8	0,1	2,8
Bio4 Temperature Seasonality	0	50	0	8,8	3,7	5	19,1
Bio5 Max Temperature of Warmest Month (ºC)	0	5,9	11,1	0,4	0	0	0
Bio6 Min Temperature of Coldest Month (ºC)	7,2	4,9	0	4,3	9,9	0,9	17,2
Bio7 Temperature Annual Range (ºC)	0	0	5,3	0,9	0	0,1	0
Bio8 Mean Temperature of Wettest Quarter (ºC)	0	0	0	0,2	0	0,1	0
Bio9 Mean Temperature of Driest Quarter (ºC)	6,1	4,9	5,2	1,8	0,1	1,8	0
Bio10 Mean Temperature of Warmest Quarter (ºC)	0	0	5,2	0,7	0	0,7	0
Bio11 Mean Temperature of Coldest Quarter (ºC)	0	0	1	0	0	1,4	0
Bio12 Annual Precipitation	25	17,8	0	0,2	3,3	0,5	0
Bio13 Precipitation of Wettest Month	0	0	3,7	0,1	0	0,8	0
Bio14 Precipitation of Driest Month	23,5	0	0,7	1,6	14,1	20,6	0,1
Bio15 Precipitation Seasonality	0	0	1,8	5	0	3	15,1
Bio16 Precipitation of Wettest Quarter	15,6	0	0	0,3	2,6	0,1	0,6
Bio17 Precipitation of Driest Quarter	7,5	1,7	0	0,1	0,2	1,4	0,3
Bio18 Precipitation of Warmest Quarter	0	0	21,2	1,1	0	0,3	0
Bio19 Precipitation of Coldest Quarter	0	3,1	2,3	1,6	2,1	0,9	0
DEM Digital Elevation Model	6,2	2,8	9,9	4,6	20,3	4,1	12,1
K	0	2,6	1,8	3	4,4	0,8	1,7
Mg	0	0	20,9	2,2	1,8	0,3	1
Organic matter	0	0	0	3,2	0,1	1,3	0
Na	0	1,9	2	0,2	1,4	0,9	3,1
pH	1,4	4,3	0,6	0,1	3,6	1,3	0,5
CE (electrical conductivity) mS/cm	0	0	1,8	0	0	0,2	0
OC (Organic carbon) (%)	0	0	0,7	2,8	0	0,8	0
Evaporación an nual	0	0	0,6	1	0	46,5	0
Evaporation of wettest months	0	0	0	2,5	0	0,9	0
Evaporation of driest months	0	0	0	0	0,7	1	0,1
Annual precipitation	0	0	0	0	0	1	0
Precipitation of wettest months	0	0	0	35,5	0	0,2	0
Precipitation of driest months	0	0	0,5	0,3	0,7	0,1	0
Temperature of wettest months	0	0	1,7	0,1	0	1,1	20,3
Annual mean temperature	0	0	0	0	0	0	0
Temperature of driest months	0	0	0	0,1	0	0,5	0

logistic output map indicates the probability of coincidence of suitable environmental conditions for the development of crops. That is, Maxent's logistic output was used since it approximates the true likelihood of crop geolocations (Fithian & Hastie, 2013) and can be interpreted as the suitability of the agroecological conditions for the crop. The maps were georeferenced, and the cartographic design was applied with ArcMap v. Software. 10.3.

17.5.7 Results

Table 1 shows the agroecological modeling of each crop, besides the impact of the variables used in the model. The crops in the study region, showed different responses to climatic conditions, for example, to the diurnal temperature range, which is important for photosynthesis, the precipitation of the driest month that is interpreted than in the dry season, crops can adapt to these terms. In the geographical aspect, the altitude played a preponderant role the crops can only be established at altitudes below 1600 msnm until 100 msnm, where maximum production can be guaranteed. The edaphological variables showed that Mg, Na, K and pH are important in most crops, therefore, before establishing plantations which are intended to the production of biofuels and bioproducts, the basic elements must be considered. In this regard, the projection made in the Veracruz territory, shows the high, medium and low potential (Table 17.1 and 17.2) was differential for each of the crops, as well as the geographical regions with suitable characteristics (Figs. 17.3, 17.4, 17.5, 17.6, 17.7, 17.8, and 17.9).

By applying the Maxent Method, as a tool for decision-makers, to existing crops in the state of Veracruz Mexico with potential for bioproducts and bioenergy, broad possibilities were identified for the use of by-products and agricultural residues from these crops in each region, besides their current use in the food industry. However, there is a lack of a strategic project based on the circular bio-economy to move towards a development policy based on renewable resources and local environmental, technological, and necessary organizational changes, as well as improvements in process technologies and products and social capacities. Studies by academic, industrial, and political decision-makers require the design of regulatory and political mechanisms that guarantee diversified land use, social profitability for growers, and sufficient environmental measures to reverse the intense degradation and pollution processes linked to the production of traditional crops and bioenergy potentials in the region. However, without adequate political regulations, the intensive production of bioenergy and bioproducts can intensify social conflicts over the control of natural resources, land dispossession, water control and contamination, higher food prices, and intensified migration. A prospective plan is needed to determine the susceptible supply areas of raw materials (agroecological zoning) and consider the total use of their residues, avoiding competition with the already fragile food production.

Table 17.2 Characteristics of the edaphoclimatic zoning model of different potential crops to produce biofuels and bioproducts

			Potential of agroecological conditions			Area by agroecological conditions			
Culture	N	AUC	Low (%)	Medium (%)	High (%)	Low	Medium	High	Total area
Sugarcane	116	0.942	40	25	35	31.53 km^2	19.7 km^2	27.58 km^2	78.81 km^2
Maize	721	0.910	60	16	24	47.29 km^2	12.61 km^2	18.91 km^2	
Beet	43	0.977	30	33	37	23.64 km^2	26.01 km^2	29.16 km^2	
Sorghum	270	0.865	10	20	70	7.88 km^2	15.76 km^2	55.17 km^2	
Wheat	17	0.971	75	15	10	59.11 km^2	11.8 km^2	7.88 km^2	
Cassava	186	0.942	15	30	55	11.8 km^2	23.64 km^2	43.35 km^2	
Barley	20	0.980	80	7	13	63.06 km^2	5.5 km^2	10.24 km^2	

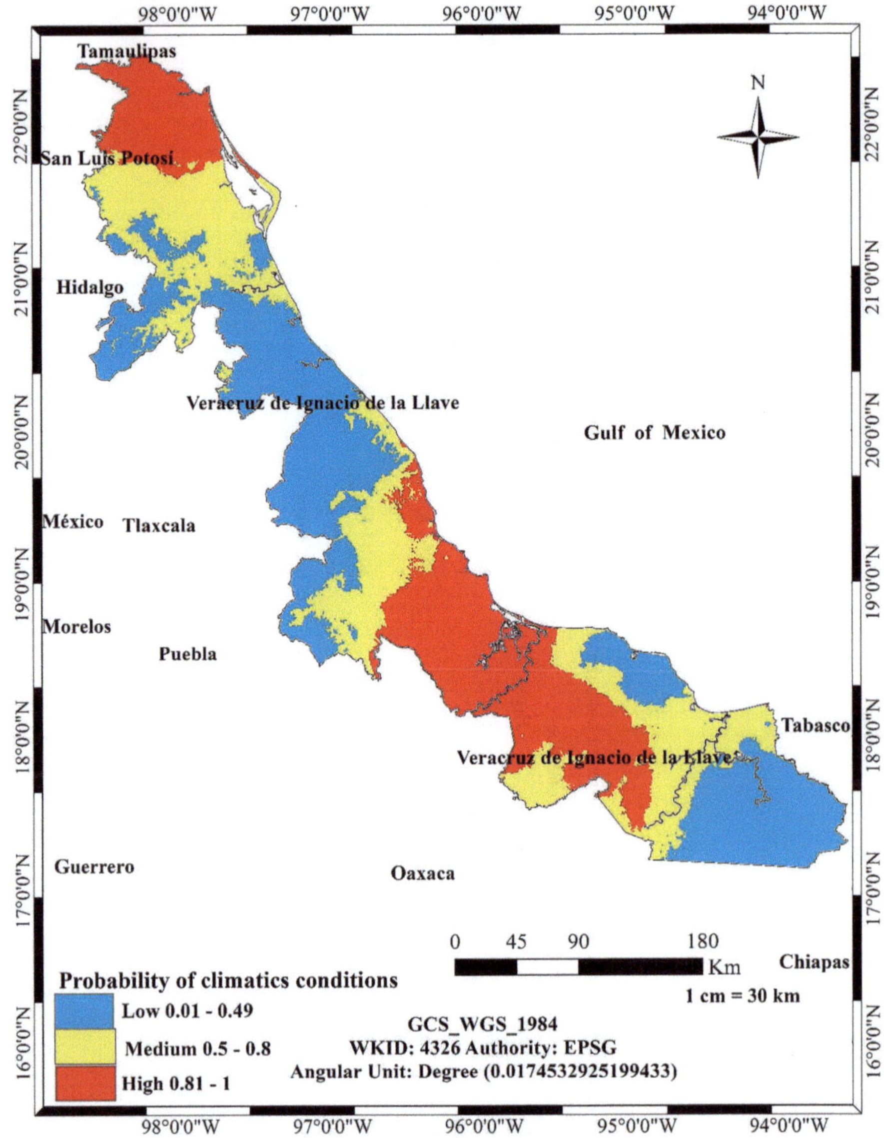

Fig. 17.3 Area with edaphoclimatic potential for the production of *Beta vulgaris* (beet)

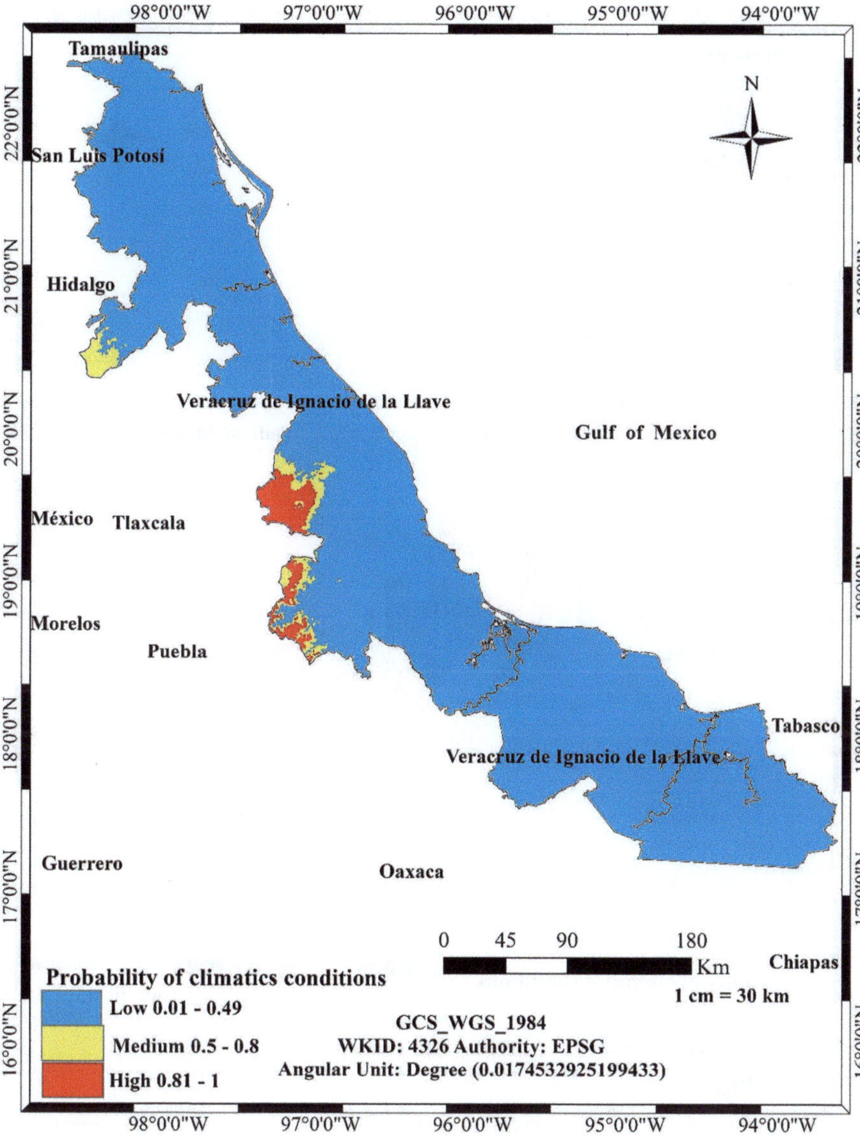

Fig. 17.4 Area with edaphoclimatic potential for the production of *Hordeum vulgare* (barley)

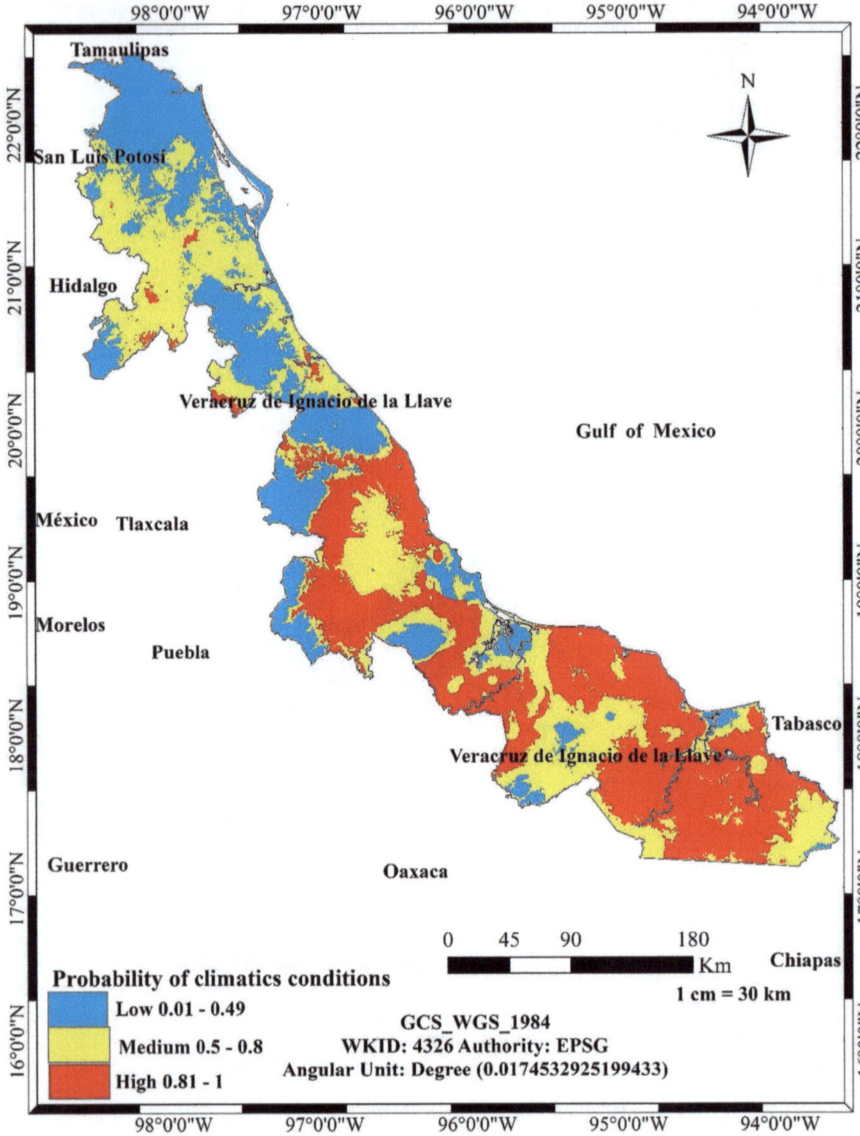

Fig. 17.5 Area with edaphoclimatic potential for the production of *Manihot esculenta* (Cassava)

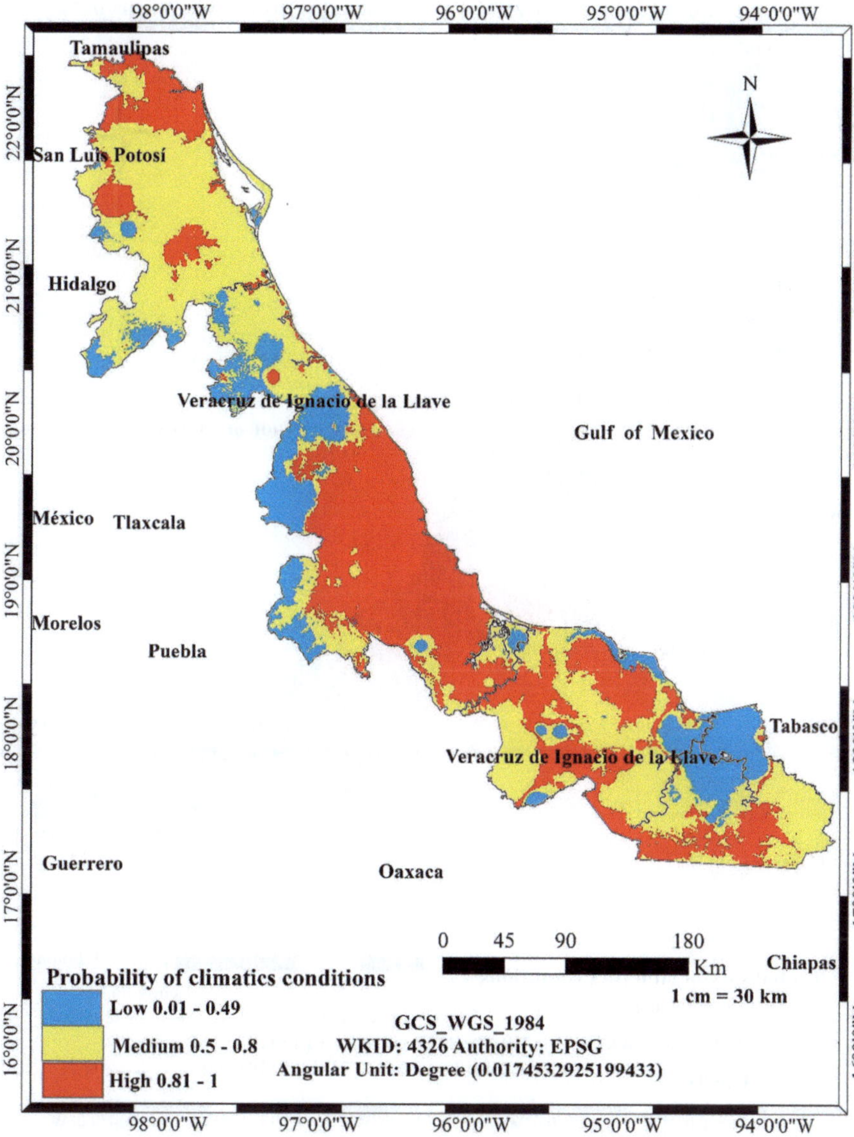

Fig. 17.6 Area with edaphoclimatic potential for the production of *Sorghum bicolor* (sorghum)

17.6 Conclusions

Biomass from agricultural activities as an alternate raw material is potentially very cost-effective if it is developed commercially in novel industrial facilities and is a viable solution to several challenges and waste management. Developing a CE with innovative technologies and profitable business practices addresses agricultural

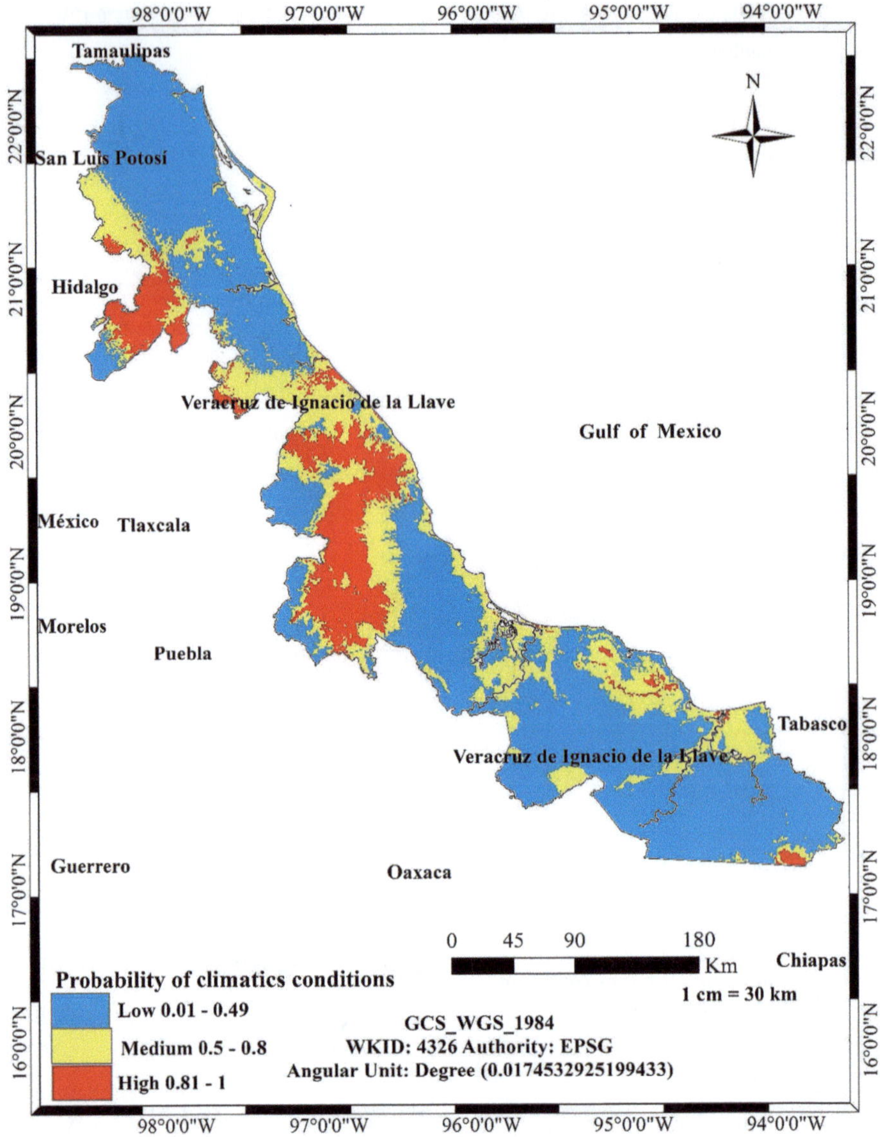

Fig. 17.7 Area with edaphoclimatic potential for the production of *Saccharum officinarum* (sugarcane)

waste recycling for bioproducts. However, new sustainable business models are necessary for the transition to economic growth while reducing negative effects on the natural environment and society. In this regard, agricultural biomass is available throughout all regions worldwide which allows green biorefineries as a multiproduct system of any kind (energy, pulp, and paper, pharmaceutical, sucrochemical,

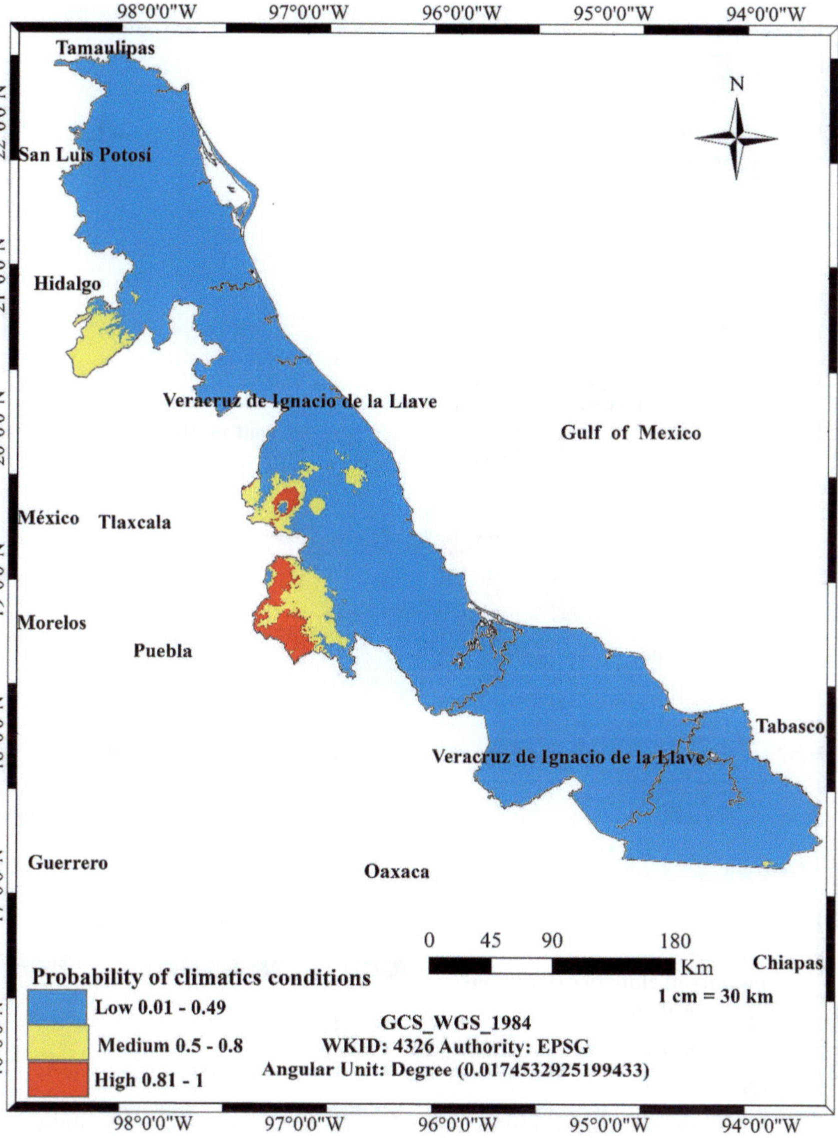

Fig. 17.8 Area with edaphoclimatic potential for the production of *Triticum aestivum* (wheat)

lignochemistry, ethanolchemistry, etc.) to be established at any geographical location concerning available agricultural biomass throughout the year. Biomass as a source of basic chemicals (cellulose, lignin, sugars, phytochemicals, etc.) can fulfill the biofuels, materials, bioplastics, and biochemicals needs.

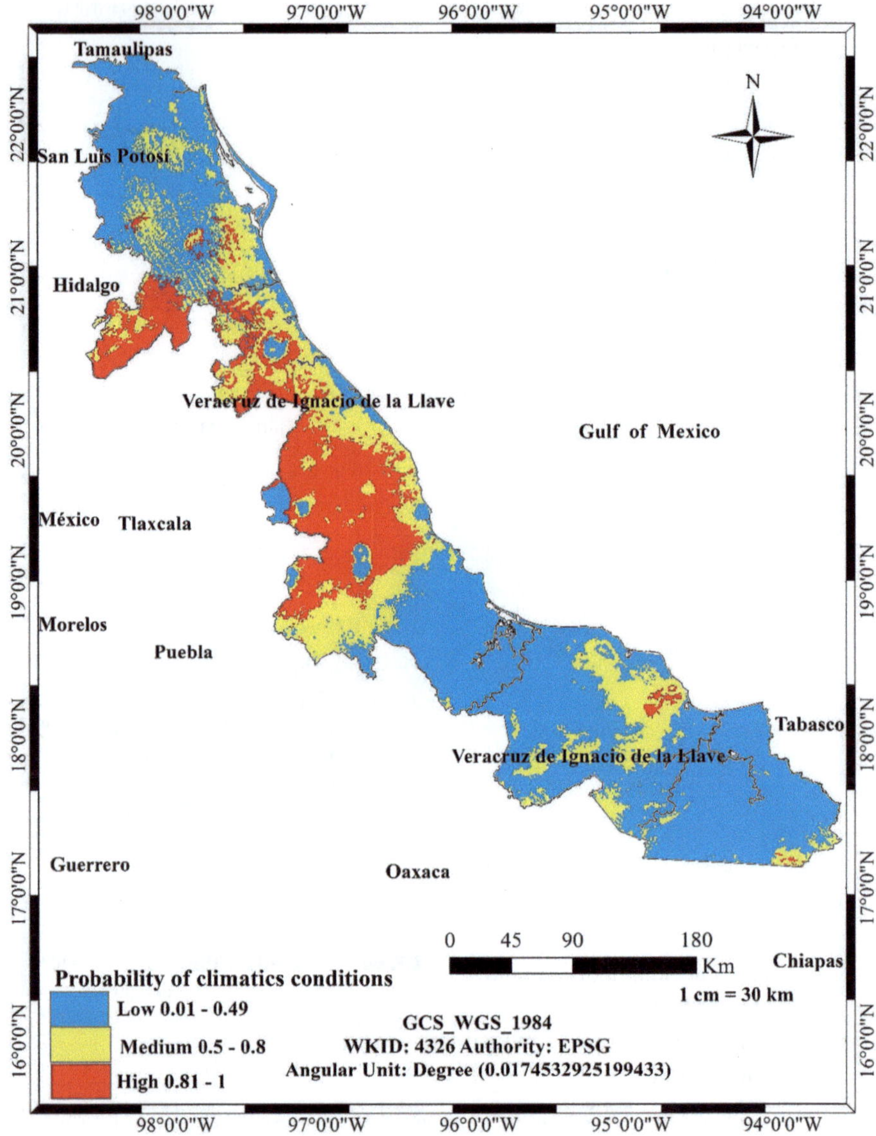

Fig. 17.9 Area with edaphoclimatic potential for the production of *Zea mays* (maize or corn)

However, public policies, decision-makers, and stakeholders are required to standardize tools for analysis of current and future land use, potential production of crops and waste, the sources of biomass feedstock in the steps of emissions involved with input production, feedstock harvest or collection, distribution, transportation, pretreatment, storage, processing and conversion technologies (biochemical and thermochemical pathways), modeling, and optimization, product transportation, and consumption on markets, cost of production, environmental impact, and sustainability. More emphasis is needed to improve corporate social responsibility and generate welfare and well-being for the population.

References

Abd-Aziz, S., Gozan, M., Ibrahim, M. F., Phang, L. Y., & Jenol, M. A. (2024). A glance on biorefinery of chemical substitutes from agriculture and industrial by-products. In *Chemical substitutes from agricultural and industrial by-products: Bioconversion, bioprocessing, and biorefining* (pp. 1–17). https://doi.org/10.1002/9783527841141.ch1

Alan, H., & Köker, A. R. (2023). Analyzing and mapping agricultural waste recycling research: An integrative review for conceptual framework and future directions. *Resources Policy, 85*, 103987. https://doi.org/10.1016/j.resourpol.2023.103987

Araújo, R. G., Chavez-Santoscoy, R. A., Parra-Saldívar, R., Melchor-Martínez, E. M., & Iqbal, H. M. (2023). Agro-food systems and environment: Sustaining the unsustainable. *Current Opinion in Environmental Science & Health, 31*, 100413. https://doi.org/10.1016/j.coesh.2022.100413

Arias, O., Pulgar, J. A., & Soto, M. (2022). Application of organic wastes to soils and legislative intricacies in a circular economy context. *Clean Technologies and Environmental Policy, 24*(6), 1871–1888. https://doi.org/10.1007/s10098-022-02293-2

Awasthi, M. K., Sar, T., Gowd, S. C., Rajendran, K., Kumar, V., Sarsaiya, S., & Taherzadeh, M. J. (2023). A comprehensive review on thermochemical, and biochemical conversion methods of lignocellulosic biomass into valuable end product. *Fuel, 342*, 127790. https://doi.org/10.1016/j.fuel.2023.127790

Banerjee, N. (2023). Biomass to energy—An analysis of current technologies, prospects, and challenges. *Bioenergy Research, 16*(2), 683–716. https://doi.org/10.1007/s12155-022-10500-7

Barrera-Perales, O. T., & Burgos, A. L. (2023). Conceptual distinction between agricultural innovation and rural innovation: Implications for scientific research and public policy. *Innovation and Development, 13*(3), 471–492. https://doi.org/10.1080/2157930X.2022.2049534

Clark, J. H. (2019). Green biorefinery technologies based on waste biomass. *Green Chemistry, 21*(6), 1168–1170. https://doi.org/10.1039/C9GC90021G

Cruz-Cárdenas, G., López-Mata, L., Ortiz-Solorio, C. A., Villaseñor, J. L., Ortiz, E., Silva, J. T., & Estrada-Godoy, F. (2014). Interpolation of Mexican soil properties at a scale of 1: 1,000,000. *Geoderma, 213*, 29–35.

Cuadrado-Osorio, P. D., Ramírez-Mejía, J. M., Mejía-Avellaneda, L. F., Mesa, L., & Bautista, E. J. (2022). Agro-industrial residues for microbial bioproducts: A key booster for bio-economy. *Bioresource Technology Reports, 20*, 101232. https://doi.org/10.1016/j.biteb.2022.101232

Donner, M., Gohier, R., & de Vries, H. (2020). A new circular business model typology for creating value from agro-waste. *Science of the Total Environment, 716*, 137065. https://doi.org/10.1016/j.scitotenv.2020.137065

Duque-Acevedo, M., Belmonte-Ureña, L. J., Cortés-García, F. J., & Camacho-Ferre, F. (2020). Recovery of agricultural waste biomass: A sustainability strategy for moving towards a circular bio-economy. In *Handbook of solid waste management: Sustainability through circular economy* (pp. 1–30). https://doi.org/10.1007/978-981-15-7525-9_25-1

Facchini, F., Silvestri, B., Digiesi, S., & Lucchese, A. (2023). Agri-food loss and waste management: Win-win strategies for edible discarded fruits and vegetables sustainable reuse. *Innovative Food Science & Emerging Technologies, 83*, 103235. https://doi.org/10.1016/j.ifset.2022.103235

Fithian, W., & Hastie, T. (2013). Finite-sample equivalence in statistical models for presence-only data. *The Annals of Applied Statistics, 7*, 1917–1939. https://doi.org/10.1214/13-AOAS667

Fitzgibbon, A., Pisut, D., & Fleisher, D. (2022). Evaluation of maximum entropy (Maxent) machine learning model to assess relationships between climate and corn suitability. *Land, 11*(9), 1382. https://doi.org/10.3390/land11091382

Gaffey, J., RajauriaG, M. M. H., Ravindran, R., Dominguez, C., Ambye-Jensen, M., & Sanders, J. P. (2023). Green biorefinery systems for the production of climate-smart sustainable products from grasses, legumes and green crop residues. *Biotechnology Advances, 66*, 108168. https://doi.org/10.1016/j.biotechadv.2023.108168

Gupta, N., Mahur, B. K., Izrayeel, A. M. D., Ahuja, A., & Rastogi, V. K. (2022). Biomass conversion of agricultural waste residues for different applications: A comprehensive review. *Environmental Science and Pollution Research, 29*(49), 73622–73647. https://doi.org/10.1007/s11356-022-22802-6

Hijmans, R. J., Cameron, S. E., Parra, J. L., Jones, P. G., & Jarvis, A. (2005). Very high resolution interpolated climate surfaces for global land areas. *International Journal of Climatology, 25*, 1965–1978. https://doi.org/10.1002/joc.1276

Jesus, G. M. K., Jugend, D., Paes, L. A. B., Siqueira, R. M., & Leandrin, M. A. (2023). Barriers to the adoption of the circular economy in the Brazilian sugarcane ethanol sector. *Clean Technologies and Environmental Policy, 25*(2), 381–395. https://doi.org/10.1007/s10098-021-02129-5

Jiménez, A. A., Marceleño Flores, S. M. L., González, O. N., & Vilchez, F. F. (2023). Potential coffee distribution in a Central-Western region of Mexico. *Ecologies, 4*(2), 269–287. https://doi.org/10.3390/ecologies4020018

Kataki, R., & Kataki, M. D. (2022). Weeds as a renewable bioresource: Prospects for bioconversion to biofuels and biomaterials through a cascade of approaches. In *Biofuels and bioenergy* (pp. 437–461). Elsevier. https://doi.org/10.1016/B978-0-323-85269-2.00021-6

Kiran, B. R., Prasad, M. N. V., & Mohan, S. V. (2023). Farm to fork: Sustainable agrifood systems. In *Sustainable and circular management of resources and waste towards a green deal* (pp. 25–38). Elsevier. https://doi.org/10.1016/B978-0-323-95278-1.00012-7

Koul, B., Yakoob, M., & Shah, M. P. (2022). Agricultural waste management strategies for environmental sustainability. *Environmental Research, 206*, 112285. https://doi.org/10.1016/j.envres.2021.112285

Kumar, J. A., Sathish, S., Prabu, D., Renita, A. A., Saravanan, A., Deivayanai, V. C., & Hosseini-Bandegharaei, A. (2023). Agricultural waste biomass for sustainable bioenergy production: Feedstock, characterization and pre-treatment methodologies. *Chemosphere, 331*, 138680. https://doi.org/10.1016/j.chemosphere.2023.138680

Leal Filho, W., Vidal, D. G., Chen, C., Petrova, M., Dinis, M. A. P., Yang, P., & Neiva, S. (2022). An assessment of requirements in investments, new technologies, and infrastructures to achieve the SDGs. *Environmental Sciences Europe, 34*(1), 1–17. https://doi.org/10.1186/s12302-022-00629-9

Liszbinski, B. B., Brizolla, M. M. B., & Patias, T. Z. (2023). Driving factors for the involvement of agroindustries in the sustainable development goals. *Journal of Cleaner Production, 410*, 137279. https://doi.org/10.1016/j.jclepro.2023.137279

Matumoto-Pintro, P. T., & Saraiva, B. R. (2023). High value-added products from food waste. *Food, 12*(21), 4038. https://doi.org/10.3390/foods12214038

Medina-Abad, J., & Freire-Pesántez, A. (2023). Barreras para la implementación de la economía circular en países en vías de desarrollo. *Estudios de la Gestión, 14*, 101–123. https://doi.org/10.32719/25506641.2023.14.6

Merow, C., Smith, M. J., & Slander, J. A. (2013). A practical guide to MaxEnt for modeling species' distributions: What it does, and why inputs and settings matter. *Ecography, 36*(10), 1058–1069. https://doi.org/10.1111/j.1600-0587.2013.07872.x

Peng, X., Jiang, Y., Chen, Z., Osman, A. I., Farghali, M., Rooney, D. W., & Yap, P. S. (2023). Recycling municipal, agricultural and industrial waste into energy, fertilizers, food and construction materials, and economic feasibility: A review. *Environmental Chemistry Letters, 21*(2), 765–801. https://doi.org/10.1007/s10311-022-01551-5

Phillips, S. J., & Dudík, M. (2008). Modeling of species distributions with Maxent: New extensions and a comprehensive evaluation. *Ecography, 31*(2), 161–175. https://doi.org/10.1111/j.0906-7590.2008.5203.x

Phillips, S. J., Dudík, M., & Schapire, R. E. (2004). A maximum entropy approach to species distribution modeling. In *Proceedings of the twenty-first international conference on machine learning* (p. 83). ACM. https://doi.org/10.1145/1015330.1015412

Phillips, S. J., Anderson, R. P., & Schapire, R. E. (2006). Maximum entropy modeling of species geographic distributions. *Ecological Modelling, 190*(3), 231–259. https://doi.org/10.1016/j.ecolmodel.2005.03.026

Ramirez-Villegas, J., Jarvisa, A., & Läderachd, P. (2013). Empirical approaches for assessing impacts of climate change on agriculture: The EcoCrop model and a case study with grain sorghum. *Agricultural and Forest Meteorology, 170*, 67–78. https://doi.org/10.1016/j.agrformet.2011.09.005

Ray, R.C.(2020). *Sustainable biofuels: Challenges and opportunities* (p. 502). Academic Press (Elsevier).

Ray, R. C. (2022). *Fruits and vegetable wastes: Valorization into bioproducts and platform chemicals* (p. 448). Springer.

Riseh, R. S., Vazvani, M. G., Hassanisaadi, M., & Thakur, V. K. (2024). Agricultural wastes: A practical and potential source for the isolation and preparation of cellulose and application in agriculture and different industries. *Industrial Crops and Products, 208*, 117904. https://doi.org/10.1016/j.indcrop.2023.117904

Romero-Perdomo, F., & González-Curbelo, M. Á. (2023). Integrating multi-criteria techniques in life-cycle tools for the circular bio-economy transition of Agri-food waste biomass: A systematic review. *Sustainability, 15*(6), 5026. https://doi.org/10.3390/su15065026

Short, N. M., Jr., Woodward-Greene, M. J., Buser, M. D., & Roberts, D. P. (2023). Scalable knowledge management to meet global 21st century challenges in agriculture. *Land, 12*(3), 588. https://doi.org/10.3390/land12030588

SIAP. (2023). https://nube.siap.gob.mx/avance_agricola/

Srivastava, R. K., Sarangi, P. K., Shadangi, K. P., Sasmal, S., Gupta, V. K., Govarthanan, M., & Subudhi, S. (2023). Biorefineries development from agricultural byproducts: Value addition and circular bio-economy. *Sustainable Chemistry and Pharmacy, 32*, 100970. https://doi.org/10.1016/j.scp.2023.100970

Surya, B., Saleh, H., Idris, M., & Ahmad, D. N. A. (2021). Rural agribusiness-based agropolitan area development and environmental management sustainability: Regional economic growth perspectives. *International Journal of Energy Economics and Policy, 11*(1), 142–157.

Szarka, N., García Laverde, L., Thrän, D., Kiyko, O., Ilkiv, M., Moravčíková, D., & Martín, J. I. (2023). Stakeholder engagement in the co-design of regional bio-economy strategies. *Sustainability, 15*(8), 6967. https://doi.org/10.3390/su15086967

Tan, E. C., & Lamers, P. (2021). Circular bio-economy concepts—A perspective. *Frontiers in Sustainability, 2*, 701509. https://doi.org/10.3389/frsus.2021.701509

Wijerathna, P. A. K. C., Udayagee, K. P. P., Idroos, F. S., & Manage, P. M. (2023). Waste biomass valorization and its application in the environment. In D. B. Pal & A. K. Tiwari (Eds.), *Sustainable valorization of agriculture & food waste biomass. Clean energy production technologies*. Springer. https://doi.org/10.1007/978-981-99-0526-3_1

Xiu, S., & Shahbazi, A. (2015). Development of green biorefinery for biomass utilization: A review. *Trends in Renewable Energy, 1*(1), 4–15. https://doi.org/10.17737/tre.2015.1.1.008

Correction to: Biotechnological Approaches for Agriculture Waste Transformation

Cristina Soares, Olena Dorosh, Filipe Fernandes, Clara Grosso, Manuela Moreira, and Cristina Delerue-Matos

Correction to:
Chapter 4 in: J. Arora et al. (eds.), ***Transforming Agriculture Residues for Sustainable Development*****, Waste as a Resource,**
https://doi.org/10.1007/978-3-031-61133-9_4

The original version of the chapter "Biotechnological Approaches for Agriculture Waste Transformation" was inadvertently published with incorrect spelling of the author's name. The spelling of the author, Olena Dorosh, has been updated in the chapter.

The updated version of this chapter can be found at
https://doi.org/10.1007/978-3-031-61133-9_4

J. Arora et al. (eds.), *Transforming Agriculture Residues for Sustainable Development*, Waste as a Resource, https://doi.org/10.1007/978-3-031-61133-9_18

Index

J. Arora et al. (eds.), *Transforming Agriculture Residues for Sustainable Development*, Waste as a Resource, https://doi.org/10.1007/978-3-031-61133-9

MIX
Papier aus verantwortungsvollen Quellen
Paper from responsible sources
FSC® C105338
FSC
www.fsc.org

If you have any concerns about our products,
you can contact us on
ProductSafety@springernature.com

In case Publisher is established outside the EU,
the EU authorized representative is:
Springer Nature Customer Service Center GmbH
Europaplatz 3, 69115 Heidelberg, Germany

Printed by Libri Plureos GmbH
in Hamburg, Germany